Measure and Integration

Heinz König

Measure and Integration

An Advanced Course
in Basic Procedures and Applications

 Springer

Heinz König
Universität des Saarlandes
Fakultät für Mathematik und Informatik
D-66123 Saarbrücken
Germany
hkoenig@math.uni-sb.de

The cover picture shows a section of the skeleton of a marine
diatomee arachnoidiscus. (SEM micrograph reproduced by
courtesy of Manfred P. Kage).
This is a beautiful example of structure in nature,
which reoccurs in man-made buildings and is a model
for structure in mathematics.

ISBN 978-3-642-08277-1   e-ISBN 978-3-540-89502-2

DOI 10.1007/978-3-540-89502-2

Library of Congress Cataloging-in-Publication Data available.

Die Deutsche Bibliothek – CIP-Einheitsaufnahme
König, Heinz: Measure and integration: an advanced course in basic procedures and applications/Heinz
König. – Berlin; Heidelberg; New York; Barcelona; Budapest; Hong Kong; London; Milan; Paris; Santa
Clara; Singapore; Tokyo: Springer, 1997

Mathematics Subject Classification (2000): 28-02

Corrected, 2nd printing 2009

© 2010 Springer-Verlag Berlin Heidelberg

Cover design: WMX Design GmbH, Heidelberg

Printed on acid-free paper

9 8 7 6 5 4 3 2 1

springer.com

Meiner Frau Karin
in Liebe und Dankbarkeit gewidmet

Preface

Ich schaffe, was ihr wollt, und schaffe mehr;
Zwar ist es leicht, doch ist das Leichte schwer.
Es liegt schon da, doch um es zu erlangen,
Das ist die Kunst! Wer weiss es anzufangen?
GOETHE, Faust II

The present text centers around a fundamental task of measure and integration theory, which has not found an adequate solution so far. It is the task to produce, with unified and universal means, true contents and above all measures from more primitive data, in order to extend elementary contents and to represent so-called elementary integrals. The traditional main tools are the Carathéodory extension theorem and the Daniell-Stone representation theorem. These theorems are much too restrictive in order to fulfil the needs.

Around 1970 a new development started in the work of Topsøe and others. It was based on the notion of regularity, which for a set function means to determine its values from a particular set system by approximation from above or below. In traditional measure theory this notion is linked to topology.

The present text wants to be a systematic treatment of the context in the new spirit. It is based to some extent on personal work of the author. The main results are equivalence theorems for the existence and uniqueness of extensions and representations, which are not more complicated than the traditional ones but much more powerful. With these results the text clarifies and unifies the entire context. The main instruments are certain new envelope formations which resemble the traditional Carathéodory outer measure.

The systematic theory has numerous applications. The most important application is the full extension of the classical Riesz representation theorem in terms of Radon measures, from locally compact to arbitrary Hausdorff topological spaces. As another application we note an extension and at the same time simplification of the Choquet capacitability theorem, which shows that the new formations can be useful for so-called non-additive set functions as well. Some of the applications are treated without pronounced technical sophistication. We rather want to demonstrate that certain basic ideas and results are natural outflows from the new theory.

The central parts of the text are chapters II and V. Their main substance as well as their history and motivation are outlined in the introduction below. It is an elaboration of a lecture which the author delivered at several places, in the present form for the first time at the symposium in honour of Adriaan C.Zaanen in Leiden in September 1993.

Chapters I and IV are filled with preparations. We need certain standard material in unconventional versions which have to be developed. We also need several new notations.

The application to the Riesz representation theorem is in chapter V section 16. The other applications are in chapters III, VI and VII. We emphasize that chapter VII develops an abstract product formation which comprises the Radon product measure of Radon measures. The final chapter VIII is an appendix which is independent of the central chapters II and V. It wants to demonstrate that the unconventional notions of content and measure introduced in chapter I can be useful in other areas of measure theory as well.

All this says that the central themes of the present text are the fundamentals of measure and integration theory. The author hopes that its readers will find it less technical than it looks at first sight. He thinks that the text can be read with appreciation by anyone who has struggled through the traditional abstract and topological theories. However, it is different from a textbook in the usual sense. The presentation is ab ovo, though more like in a book of research. The author hopes that the text will be used in future courses. An ideal prerequisite would be the recent small book of Stroock [1994], because on the one hand it provides the concrete material which should precede this one, and on the other hand it does not take the reader onto the traditional paths of abstract measure and integration theory which the present work wants to restructure.

The author wants to express his warmest thanks to Gustave Choquet, Jean-Paul Pier, Reinhold Remmert, Klaus D.Schmidt, Maurice Sion, and Flemming Topsøe for insightful comments, encouragement, and good advice. Likewise he thanks Robert Berger and Gerd Wittstock for constant help with the resistful machine into which he typed the final version of the text. He extends his thanks to the former and present directors of the Mathematical Research Institute Oberwolfach, Martin Barner and Matthias Kreck, for several periods of quiet work in the unique atmosphere of the Institute.

August 1996   Heinz König

Contents

Introduction

The textbooks on measure and integration theory can often be subdivided into two parts of almost equal size: One part describes what can be done when one is in possession of measures of one or another type. The other part describes how to obtain these measures from more primitive data, which as a rule are elementary contents or elementary integrals. The former part is based on some famous theories. But the latter part is in less favourable state, because its main theorems do not fit the actual needs in certain central points. We shall explain this statement, and then describe how the situation can be repaired. To do this we sketch the main ideas and results of our chapters II and V, which form the central parts of the present text.

Construction of Measures from Elementary Contents

The classical theorem on the existence of measure extensions reads as follows. Our technical terms are either familiar or obvious.

THEOREM. *Let $\varphi : \mathfrak{S} \to [0, \infty]$ be a content on a ring $\mathfrak{S}$ of subsets in a nonvoid set X. Then φ can be extended to a measure $\alpha : \mathfrak{A} \to [0, \infty]$ on a σ algebra $\mathfrak{A}$ iff φ is upward σ continuous.*

There are few situations where this theorem can be applied without complications. The reason is that the natural set systems which carry elementary contents are almost never rings, but at most lattices. This is in particular true for the basic set systems in topological spaces. Even to construct the Lebesgue measure via rings forces us to work with the unnatural half-open intervals, which might be adequate in order to produce sophisticated counterexamples, but not for the foundations of one of the most basic theories in analysis.

Like the theorem itself, also its usual proof due to Carathéodory [1914] does not fit the actual needs as it stands. Let us recall that it is based on two formations. On the one hand one defines for a set function $\varphi : \mathfrak{S} \to [0, \infty]$ on a set system $\mathfrak{S}$ with $\varnothing \in \mathfrak{S}$ and $\varphi(\varnothing) = 0$ the so-called **outer measure** $\varphi^\circ : \mathfrak{P}(X) \to [0, \infty]$ to be

$$\varphi^\circ(A) = \inf \left\{ \sum_{l=1}^{\infty} \varphi(S_l) : (S_l)_l \text{ in } \mathfrak{S} \text{ with } A \subset \bigcup_{l=1}^{\infty} S_l \right\}.$$

On the other hand one defines for a set function $\phi : \mathfrak{P}(X) \to [0, \infty]$ with $\phi(\varnothing) = 0$ the so-called **Carathéodory class**

$$\mathfrak{C}(\phi) := \{A \subset X : \phi(S) = \phi(S \cap A) + \phi(S \cap A') \; \forall S \subset X\} \subset \mathfrak{P}(X).$$

Then for the nontrivial direction of the theorem one verifies that $\varphi^\circ | \mathfrak{C}(\varphi^\circ)$ is a measure on a σ algebra and an extension of φ.

We shall see that the formation $\mathfrak{C}(\cdot)$ is so felicitous that it will survive the upheaval to come, at least within the present step of abstraction. In contrast, we shall see that the specific form of the outer measure must be blamed for the deficiencies around the extension theorem which will now be described in more detail.

1) The outer measure is a beautiful tool in the frame of rings, but it ceases to work beyond this frame. It does not even allow to extend the theorem to the particular lattices $\mathfrak{S}$ which fulfil $B \setminus A \in \mathfrak{S}^\sigma$ for all $A \subset B$ in $\mathfrak{S}$, where the assertion will be seen to persist. The class of these lattices is much more realistic than the class of rings. For example, it includes the lattices of the closed subsets and of the compact subsets of a metric space.

2) The outer measure is an outer regular formation: The definition shows that

$$\varphi^\circ(A) = \inf\{\varphi^\circ(S) : S \in \mathfrak{S}^\sigma \text{ with } S \supset A\} \quad \text{for all } A \subset X,$$

that is φ° is **outer regular** $\mathfrak{S}^\sigma$. Now present-day analysis requires inner regular formations perhaps even more than outer regular ones. However, the definition of the outer measure is such that no inner regular counterpart is visible.

The need for inner regular formations comes from the predominant role of compactness in topological measure theory. It became clear that the most important class of measures on an arbitrary Hausdorff topological space X are the **Radon measures**, defined to be the Borel measures $\alpha : \text{Bor}(X) \to [0, \infty]$ which are finite on the system $\text{Comp}(X)$ of the compact subsets of X and inner regular $\text{Comp}(X)$. It is then an immediate problem to characterize those set functions $\varphi : \text{Comp}(X) \to [0, \infty[$ which can be extended to (of course unique) Radon measures, the so-called **Radon premeasures**. We see that the classical extension theorem does not help in this problem for at least two reasons.

3) The outer measure is a formation of *sequential* type. But present-day analysis also requires non-sequential formations, once more for topological reasons. However, the definition of the outer measure is such that no non-sequential counterpart is visible.

4) It is a sad fact that the methods employed for contents and measures have not much in common with those for so-called *non-additive* set functions like capacities. Now the outer measure has a certain built-in *additive* character. One can be suspicious that this fact is responsible for the imperfections which we speak about.

There were of course attempts to improve the situation. The main results of Pettis [1951] were complicated and hard to use because, as it seems now, regularity had not yet attained its true position. Srinivasan [1955] was restricted to the extension from rings, but was able to develop a symmetric outer/inner extension procedure and anticipated the later expressions in this frame. Around 1970 deliberate efforts started in order to develop improved extension methods in terms of lattices, outer and inner regularity, and sequential and non-sequential procedures. A decisive prelude was the characterization of the Radon premeasures due to Kisyński [1968]. The main achievements came from Topsøe [1970ab], albeit restricted to the inner situation, from Kelley-Srinivasan [1971] and Kelley-Nayak-Srinivasan [1973], Ridder [1971][1973], and later from Sapounakis-Sion [1983][1987] and others. But the new methods were less simple and coherent than the traditional ones and therefore did not find access to the textbooks. The reason was that there were no universal substitutes for the outer measure. It is a surprise that one did not resume the expressions of Srinivasan [1955] (as a result the author himself did not look at that paper earlier than while he wrote the present text). Also there was no adequate symmetric treatment of the outer and inner cases. The basic symmetric formations were the crude outer and inner envelopes $\varphi^\star, \varphi_\star : \mathfrak{P}(X) \to [0, \infty]$, defined for an isotone set function $\varphi : \mathfrak{S} \to [0, \infty]$ with $\varnothing \in \mathfrak{S}$ and $\varphi(\varnothing) = 0$ to be

$$\varphi^\star(A) = \inf\{\varphi(S) : S \in \mathfrak{S} \text{ with } S \supset A\},$$
$$\varphi_\star(A) = \sup\{\varphi(S) : S \in \mathfrak{S} \text{ with } S \subset A\},$$

which are adequate for contents but not for measures (otherwise the outer measure would not have come into existence).

At this point we postpone further historical comments and turn to the vita of the present author on which the plan for this text is based. In an analysis course [1969/70] I wanted to construct the Lebesgue measure without use of half-open intervals. I observed that the old proof extends without further efforts from rings to the particular lattices described in 1), provided that instead of the outer measure one uses the formation $\varphi^\sigma :$ $\mathfrak{P}(X) \to [0, \infty]$, defined for an isotone set function $\varphi : \mathfrak{S} \to [0, \infty]$ to be

$$\varphi^\sigma(A) = \inf \left\{ \lim_{l\to\infty} \varphi(S_l) : (S_l)_l \text{ in } \mathfrak{S} \text{ with } S_l \uparrow \text{ some subset } \supset A \right\}.$$

The formations φ° and φ^σ are of course close relatives, and are in fact identical for contents on rings (as in elementary analysis infinite series are equivalent to infinite sequences), but need not be identical beyond. We see that φ^σ continues to work where φ° does not.

At that time I was content with this. But fifteen years later I returned to the context and observed that besides 1) the new formation also removes the deficiencies described in 2) and 3). In fact, the formation φ^σ has an obvious inner regular counterpart $\varphi_\sigma : \mathfrak{P}(X) \to [0, \infty]$, defined via decreasing sequences in $\mathfrak{S}$. Furthermore φ^σ and φ_σ have obvious non-sequential counterparts $\varphi^\tau, \varphi_\tau : \mathfrak{P}(X) \to [0, \infty]$, defined via upward/downward directed set systems instead of sequences in $\mathfrak{S}$. Then another five years later I observed that the new formations permit to improve certain concepts and results related to capacities, and thus contribute to 4) as well.

After this it is no surprise that the envelope formations $\varphi^\star \geqq \varphi^\sigma \geqq \varphi^\tau$ and $\varphi_\star \leqq \varphi_\sigma \leqq \varphi_\tau$ permit to develop comprehensive extension theories which fulfil the requirements described above. The theories are of uniform structure in $\bullet = \star\sigma\tau$, and the outer and inner developments are parallel in all essentials. For historical reasons the outer version looks more familiar, but the inner version is perhaps more important. The Carathéodory class $\mathfrak{C}(\cdot)$ is a basic notion in all cases.

There remains one more step. I observed that the outer and inner theories are not only parallel, with their typical little peculiarities, but are in fact identical. However, this presupposes a drastic step of extension and abstraction: One has to admit lattices which avoid the empty set like the entire space, and isotone set functions with values in $\mathbb{R}$ or $\overline{\mathbb{R}}$ instead of $[0, \infty[$ or $[0, \infty]$ (not to be confused with the familiar signed measures which of course need not be isotone). The previous envelope formations retain their basic structure, but the Carathéodory class $\mathfrak{C}(\cdot)$ requires an essential reformulation. I consider this extension to be quite essential for theoretical reasons, but it is too technical for an introduction. Thus we return to the previous step. We choose the inner situation for a short description of the basic concepts and results.

Let $\varphi : \mathfrak{S} \to [0, \infty[$ be an isotone set function on a lattice $\mathfrak{S}$ with $\varnothing \in \mathfrak{S}$ and $\varphi(\varnothing) = 0$. The basic idea is to concentrate on a particular class of extensions of φ. For each choice of $\bullet = \star\sigma\tau$ we define an **inner $\bullet$ extension** of φ to be an extension of φ which is a content $\alpha : \mathfrak{A} \to [0, \infty]$ on a ring $\mathfrak{A}$, with the properties that $\mathfrak{A}$ also contains $\mathfrak{S}_\bullet$ (:=the system of the respective intersections), and that

α is inner regular $\mathfrak{S}_\bullet$,
$\alpha|\mathfrak{S}_\bullet$ is downward $\bullet$ continuous (this is void when $\bullet = \star$).

Thus we impose a characteristic combination of inner regularity and downward continuity. We define φ to be an **inner $\bullet$ premeasure** iff it admits inner $\bullet$ extensions. Our aim is to characterize those φ which are inner $\bullet$ premeasures, and then to describe all inner $\bullet$ extensions of φ. We shall obtain a natural and beautiful solution.

The solution will be in terms of the inner envelopes $\varphi_\bullet : \mathfrak{P}(X) \to [0, \infty]$. First note that $\varphi_\star|\mathfrak{S} = \varphi$, while for $\bullet = \sigma\tau$ we have $\varphi_\bullet|\mathfrak{S} = \varphi$ iff φ is downward $\bullet$ continuous. This is of course a necessary condition in order that φ be an inner $\bullet$ premeasure. Likewise $\varphi_\bullet(\varnothing) = 0$ iff φ is (of course downward) $\bullet$ continuous at $\varnothing$. This weaker condition is much easier and sometimes even trivial, for example when $\varphi : \mathrm{Comp}(X) \to [0, \infty[$ on a Hausdorff topological space X. Also $\varphi_\bullet(\varnothing) = 0$ ensures that the traditional $\mathfrak{C}(\varphi_\bullet)$ is defined. We turn to the main results.

PROPOSITION. *If φ has inner $\bullet$ extensions then all these $\alpha : \mathfrak{A} \to [0, \infty]$ are restrictions of $\varphi_\bullet | \mathfrak{C}(\varphi_\bullet)$.*

THEOREM. *Assume that φ is supermodular. Then the following are equivalent.*

1) *φ has inner $\bullet$ extensions, that is φ is an inner $\bullet$ premeasure.*

2) *$\varphi_\bullet | \mathfrak{C}(\varphi_\bullet)$ is (defined and) an inner $\bullet$ extension of φ. Furthermore*

$$\text{if } \bullet = \star \quad : \quad \varphi_\bullet | \mathfrak{C}(\varphi_\bullet) \text{ is a content on the algebra } \mathfrak{C}(\varphi_\bullet),$$
$$\text{if } \bullet = \sigma\tau \quad : \quad \varphi_\bullet | \mathfrak{C}(\varphi_\bullet) \text{ is a measure on the } \sigma \text{ algebra } \mathfrak{C}(\varphi_\bullet).$$

3) *$\varphi_\bullet | \mathfrak{C}(\varphi_\bullet)$ is (defined and) an extension of φ in the crude sense, that is $\varphi_\bullet | \mathfrak{S} = \varphi$ and $\mathfrak{S} \subset \mathfrak{C}(\varphi_\bullet)$.*

4) *$\varphi(B) = \varphi(A) + \varphi_\bullet(B \setminus A)$ for all $A \subset B$ in $\mathfrak{S}$.*

5) *$\varphi_\bullet | \mathfrak{S} = \varphi$; and $\varphi(B) \leqq \varphi(A) + \varphi_\bullet(B \setminus A)$ for all $A \subset B$ in $\mathfrak{S}$.*

5') *$\varphi_\bullet(\varnothing) = 0$; and $\varphi(B) \leqq \varphi(A) + \varphi_\bullet^B(B \setminus A)$ for all $A \subset B$ in $\mathfrak{S}$. Here $\varphi_\bullet^B := \big(\varphi | \{S \in \mathfrak{S} : S \subset B\}\big)_\bullet$ for $B \in \mathfrak{S}$.*

We define φ to be **inner $\bullet$ tight** iff it fulfils the second partial condition in 5').

It follows that an inner $\bullet$ premeasure φ has a unique maximal inner $\bullet$ extension, which is $\varphi_\bullet | \mathfrak{C}(\varphi_\bullet)$. The above theorem and its outer counterpart are our substitutes for the classical extension theorem. It is obvious that the present characterizations and explicit representations stand and fall with the new envelopes.

Construction of Measures from Elementary Integrals

This time we start with the traditional Daniell-Stone representation theorem. It is the counterpart and also an extension of the classical measure extension theorem.

THEOREM. *Let $I : H \to \mathbb{R}$ be a positive (:=isotone) linear functional on a Stonean lattice subspace $H \subset \mathbb{R}^X$ of real-valued functions on a nonvoid set X. Then the following are equivalent.*

i) *There exists a measure $\alpha : \mathfrak{A} \to [0, \infty]$ on a σ algebra $\mathfrak{A}$ which represents I, that is all $f \in H$ are integrable α with $I(f) = \int f d\alpha$.*

ii) *I is σ continuous at 0, that is for each sequence $(f_l)_l$ in H with pointwise $f_l \downarrow 0$ one has $I(f_l) \downarrow 0$.*

More famous than this is perhaps the traditional Riesz representation theorem from topological measure theory.

THEOREM. *Let X be a locally compact Hausdorff topological space, and*

$$\mathrm{CK}(X, \mathbb{R}) := \{f \in \mathrm{C}(X, \mathbb{R}) : f = 0 \text{ outside of some } K \in \mathrm{Comp}(X)\}.$$

Then there is a one-to-one correspondence between the positive linear functionals $I : \mathrm{CK}(X, \mathbb{R}) \to \mathbb{R}$ and the Radon measures $\alpha : \mathrm{Bor}(X) \to [0, \infty]$.

The correspondence is

$$I(f) = \int f \, d\alpha \quad \text{for all } f \in \mathrm{CK}(X, \mathbb{R}).$$

The drawbacks of the traditional Daniell-Stone theorem are like those of the classical measure extension theorem. Thus it is of no visible use for the proof of the traditional Riesz theorem. But this latter theorem does not fulfil the needs either, because in present-day analysis one is often forced to exceed the frame of local compactness. Then $\mathrm{CK}(X, \mathbb{R})$ becomes too small, so that the theorem breaks down and has to be filled with new substance. On the measure side one wants to adhere to the Radon measures. As to the functional side, one observes that on each Hausdorff topological space X there is a wealth of *semicontinuous* real-valued functions which vanish outside of compact subsets, for example the multiples of the characteristic functions χ_K of the $K \in \mathrm{Comp}(X)$. But this leads to function classes which are *lattice cones* and as a rule not lattice subspaces. Thus it seems natural to search for an extended Riesz theorem on appropriate lattice cones of upper semicontinuous functions on X with values in $[0, \infty[$.

With this in mind we return to the Daniell-Stone theorem in the abstract theory. We want to develop the context in the spirit and scope of the previous part on measure extensions. The above look at the Riesz theorem confirms our intuitive impression that the former transition from rings to lattices should reappear as a transition from lattice subspaces to lattice cones. In fact, we shall see that the final Riesz theorem will become a direct specialization of the final Daniell-Stone theorem.

We fix a lattice cone $E \subset [0, \infty[^X$ of $[0, \infty[$-valued functions on a nonvoid set X. E is called **primitive** iff $v - u \in E$ for all $u \leq v$ in E; equivalent is $E = H^+ := \{f \in H : f \geq 0\}$ for some (unique) lattice subspace $H \subset \mathbb{R}^X$. It is of utmost importance that E need not be primitive. We assume E to be *Stonean*, defined to mean that $f \in E \Rightarrow f \wedge t, (f - t)^+ \in E$ for all real $t > 0$. In view of $f = f \wedge t + (f - t)^+$ this is the familiar notion when E is primitive. For E we define at once the set system

$$\mathfrak{T}(E) := \{[f \geq t] : f \in E \text{ and } t > 0\} = \{[f \geq 1] : f \in E\},$$

which is a lattice with $\varnothing \in \mathfrak{T}(E)$.

Next we fix an **elementary integral** on E, defined to be an isotone positive-linear functional $I : E \rightarrow [0, \infty[$. We are interested in integral representations of I. We want to define a **representation** of I to be a content $\alpha : \mathfrak{A} \rightarrow [0, \infty]$ on a ring $\mathfrak{A}$ such that

$$\text{for all } f \in E : f \text{ is measurable } \mathfrak{A} \text{ and } I(f) = \int f \, d\alpha.$$

This has to be made precise, except in the special case that $\mathfrak{A}$ is a σ algebra and α is a measure. We do this in that we require

$$\text{for all } f \in E : [f \geq t] \in \mathfrak{A} \; \forall t > 0 \text{ and } I(f) = \int\limits_{0\leftarrow}^{\rightarrow\infty} \alpha([f \geq t])dt.$$

The first part of the condition means that $\mathfrak{T}(E) \subset \mathfrak{A}$. Therefore α produces the restriction $\alpha|\mathfrak{T}(E)$. The set function $\alpha|\mathfrak{T}(E)$ is of obvious importance, because it suffices to reproduce I by the second part of the condition.

It is a Hahn-Banach consequence that I admits representations iff it has the truncation properties

(0) $I(f \wedge t) \downarrow 0$ for $t \downarrow 0$ and $I(f \wedge t) \uparrow I(f)$ for $t \uparrow \infty$ for all $f \in E$.

But the assumption that I is downward σ continuous does not enforce that it admits measure representations, except in case that E is primitive where this follows from the traditional Daniell-Stone theorem. All this shows that the present notion is too superficial in order to be the central one in our enterprise.

We turn to the true central notion. For $\bullet = \star\sigma\tau$ we define a $\bullet$ **representation** of I to be a representation $\alpha : \mathfrak{A} \to [0,\infty]$ of I such that α is an inner $\bullet$ extension of $\alpha|\mathfrak{T}(E)$. This time the word *inner* is redundant, because there will be no outer counterpart. Our aim is to characterize those I which admit $\bullet$ representations, and then to describe all $\bullet$ representations of I.

We start to define the crude outer and inner envelopes $I^\star, I_\star : [0,\infty]^X \to [0,\infty]$ of I to be

$$\begin{aligned} I^\star(f) &= \inf\{I(u) : u \in E \text{ with } u \geq f\}, \\ I_\star(f) &= \sup\{I(u) : u \in E \text{ with } u \leq f\}. \end{aligned}$$

These envelopes induce set functions $\Delta, \nabla : \mathfrak{T}(E) \to [0,\infty[$, defined to be

$$\Delta(A) = I^\star(\chi_A) \text{ and } \nabla(A) = I_\star(\chi_A) \quad \text{for } A \in \mathfrak{T}(E).$$

Of course $I_\star \leq I^\star$ and $\nabla \leq \Delta$. One proves the criterion which follows.

PROPOSITION. *Assume that I fulfils* (0). *A content $\alpha : \mathfrak{A} \to [0,\infty]$ on a ring $\mathfrak{A}$ which contains $\mathfrak{T}(E)$ is a representation of I iff $\nabla \leq \alpha|\mathfrak{T}(E) \leq \Delta$. If furthermore $\alpha|\mathfrak{T}(E)$ is downward σ continuous then $\alpha|\mathfrak{T}(E) = \Delta$.*

This makes clear that the cases $\bullet = \sigma\tau$ and $\bullet = \star$ fall apart. In the present introduction we shall restrict ourselves to the case $\bullet = \sigma\tau$, which is the simpler and the more important one. From the former main theorem we obtain at once what follows.

CONSEQUENCE (for $\bullet = \sigma\tau$). *I admits $\bullet$ representations iff it fulfils* (0) *and Δ is an inner $\bullet$ premeasure. Then the $\bullet$ representations of I are the inner $\bullet$ extensions of Δ. In particular I has the unique maximal $\bullet$ representation $\Delta_\bullet|\mathfrak{C}(\Delta_\bullet)$.*

This is not yet the desired characterization, because it is not in terms of I itself. In order to achieve this we form for $\bullet = \sigma\tau$ the precise counterparts $I_\bullet : [0, \infty]^X \to [0, \infty]$ of the previous inner $\bullet$ envelopes, that is

$$I_\sigma(f) = \sup \left\{ \lim_{l \to \infty} I(u_l) : (u_l)_l \text{ in } E \text{ with } u_l \downarrow \text{ some function } \leqq f \right\},$$

and the respective $I_\tau(f)$. We also form for $v \in E$ the satellites $I_\bullet^v : [0, \infty]^X \to [0, \infty[$ in the same sense as before. In these terms our main theorem then reads as follows.

THEOREM (for $\bullet = \sigma\tau$). *For an elementary integral* $I : E \to [0, \infty[$ *the following are equivalent.*

1) I *admits* $\bullet$ *representations.*

2) $I(v) = I(u) + I_\bullet(v - u)$ *for all* $u \leqq v$ *in* E.

3) $I_\bullet | E = I$*; and* $I(v) \leqq I(u) + I_\bullet(v - u)$ *for all* $u \leqq v$ *in* E.

3') $I_\bullet(0) = 0$*; and* $I(v) \leqq I(u) + I_\bullet^v(v - u)$ *for all* $u \leqq v$ *in* E.

The two last results are the precise counterpart of the main theorem on measure extensions for $\bullet = \sigma\tau$. It is our substitute for the traditional Daniell-Stone theorem. We note that

$$I_\star \leqq I_\sigma \leqq I_\tau, \text{ and } I_\star(f) = I_\star^v(f) \leqq I_\sigma^v(f) \leqq I_\tau^v(f) \quad \text{for } 0 \leqq f \leqq v \in E.$$

Also $I_\star | E = I$, and for $\bullet = \sigma\tau$ the equivalents to $I_\bullet | E = I$ and $I_\bullet(0) = 0$ are as before. We define I to be $\bullet$ **tight** iff it fulfils the second partial condition in 3'). The former crude envelope $I_\star$ allows to define I to be $\star$ **tight** iff

$$I(v) \leqq I(u) + I_\star(v - u) \quad \text{for all } u \leqq v \text{ in } E.$$

An earlier result due to Topsøe [1976] after Pollard-Topsøe [1975] was that 3') with $\star$ tight instead of $\bullet$ tight implies 1). But the converse is not true.

In order to obtain the traditional Daniell-Stone theorem we assume for a moment that E is primitive. Then each I is $\star$ tight and hence $\bullet$ tight. Thus I *admits* $\bullet$ *representations iff* $I_\bullet(0) = 0$. In this case it has the unique maximal $\bullet$ representation $\Delta_\bullet | \mathfrak{C}(\Delta_\bullet)$, which in particular is a measure representation of I. Thus we obtain for $\bullet = \sigma$ much more than the nontrivial direction in the traditional Daniell-Stone theorem.

We next attempt to incorporate the Riesz representation theorem. We assume X to be a Hausdorff topological space. Let $E \subset [0, \infty[^X$ be a Stonean lattice cone. We need certain conditions on E in order to relate E to the compact subsets of X. One assumption is that E be **concentrated on compacts**, defined to mean that $\mathfrak{T}(E) \subset \text{Comp}(X)$. It implies that E is contained in the class $\text{USC}^+(X)$ of $[0, \infty[$-valued upper semicontinuous functions on X, and that its members are bounded. On the other hand, when E is contained in the subclass

$$\text{USCK}^+(X) := \{f \in \text{USC}^+(X) : f = 0 \text{ outside of some } K \in \text{Comp}(X)\},$$

then E is of course concentrated on compacts. The other assumption is that E be **rich**, defined to mean that

$$\chi_K = \inf\{f \in E : f \geqq \chi_K\} \quad \text{for all } K \in \text{Comp}(X).$$

To see the relevance of this condition note that $\mathrm{CK}^+(X, \mathbb{R})$ is rich iff X is locally compact.

Then the $\bullet = \tau$ version of our Daniell-Stone theorem, combined with the classical Dini theorem, has as an almost immediate consequence the Riesz type theorem which follows.

THEOREM. *Assume that the Stonean lattice cone E is concentrated on compacts and rich. For an elementary integral $I : E \to [0, \infty[$ then the following are equivalent.*

0) *I admits a Radon measure representation* (note that $\mathfrak{T}(E) \subset \mathrm{Comp}(X) \subset \mathrm{Bor}(X)$).

1) *I admits τ representations.*

2) *$I(v) = I(u) + I_\tau(v - u)$ for all $u \leqq v$ in E.*

3') *$I(f \wedge t) \downarrow 0$ for $t \downarrow 0$ for all $f \in E$* (this is redundant when $E \subset \mathrm{USCK}^+(X)$); *and I is τ tight.*

In this case I has the unique Radon measure representation $\Delta_\tau | \mathrm{Bor}(X)$ with $\mathrm{Bor}(X) \subset \mathfrak{C}(\Delta_\tau)$, which therefore is a τ representation of I.

Let us look at the particular case $E \subset \mathrm{USCK}^+(X)$. Then each Radon measure $\alpha : \mathrm{Bor}(X) \to [0, \infty]$ defines an elementary integral I on E via $I(f) = \int f d\alpha$ for $f \in E$. This I is τ tight in view of 0)$\Rightarrow$3'). Thus we obtain what follows.

THEOREM. *Assume that the Stonean lattice cone $E \subset \mathrm{USCK}^+(X)$ is rich. Then there is a one-to-one correspondence between the elementary integrals $I : E \to [0, \infty[$ which are τ tight and the Radon measures $\alpha : \mathrm{Bor}(X) \to [0, \infty]$. The correspondence is $I(f) = \int f d\alpha$ for all $f \in E$.*

It seems that this is the first Riesz representation theorem which applies to all Hausdorff topological spaces X and contains the traditional Riesz theorem as a direct specialization. In fact, if E is primitive then each I is $\star$ tight and hence τ tight, as we have seen above. Thus for locally compact X and $E = \mathrm{CK}^+(X, \mathbb{R})$ we obtain the the traditional Riesz theorem.

CHAPTER I

Set Systems and Set Functions

The present text requires the usual concepts for set systems and set functions, but with some nontrivial modifications and extensions. The first chapter serves to introduce these notions and to develop their properties as needed in the sequel. We also include certain classical extension theorems for set functions which are not part of the later systematic development.

1. Set Systems

Basic Notions and Notations

Let X be a nonvoid set, and let as usual $\mathfrak{P}(X)$ denote its power set. For $A \subset X$ the complement will be written A'. Besides the usual operations with subsets we define the new formation

$$U|A|V := (U \cap A') \cup (V \cap A) \quad \text{for } U, V, A \subset X.$$

Thus $U|A|V$ is the unique subset of X which coincides with U on A' and with V on A. We list without proof some simple properties.

1.1. PROPERTIES. 1) $U|A|V$ *is isotone in* U *and in* V. 2) $U \cap V \subset U|A|V \subset U \cup V$. 3) $U|A|V = V|A'|U$. 4) $(U|A|V)' = U'|A|V' = V'|A'|U'$. 5) *For* $U, V, A \subset X$ *form* $P := U|A|V$ *and* $Q := U|A'|V$. *Then* $P|A|Q = U$ *and* $P|A'|Q = V$.

We understand a **set system** in X to be a collection of subsets of X, that is a subset of $\mathfrak{P}(X)$. A nonvoid set system is called a **paving** in X. We emphasize that a paving need not have the member $\varnothing \subset X$. We shall meet different kinds of set systems. Basic properties of a set system $\mathfrak{S}$ are

$\cup : A, B \in \mathfrak{S} \Rightarrow A \cup B \in \mathfrak{S};$
$\cap : A, B \in \mathfrak{S} \Rightarrow A \cap B \in \mathfrak{S};$
$\setminus : A, B \in \mathfrak{S}$ with $A \subset B \Rightarrow B \setminus A := B \cap A' \in \mathfrak{S};$
$\perp : A \in \mathfrak{S} \Rightarrow A' \in \mathfrak{S}.$

We define $\mathfrak{S}^\star$ to consist of the unions $\bigcup\limits_{S \in \mathfrak{M}} S$ for the finite pavings $\mathfrak{M} \subset \mathfrak{S}$, and $\mathfrak{S}_\star$ to consist of the respective intersections. Thus $\mathfrak{S}^\star = \mathfrak{S}$ iff $\mathfrak{S}$ fulfils $\cup$, and $\mathfrak{S}_\star = \mathfrak{S}$ iff $\mathfrak{S}$ fulfils $\cap$. We also form $\mathfrak{S} \perp := \{A' : A \in \mathfrak{S}\}$. We define a paving $\mathfrak{S}$ to be

a **lattice**   iff it has the properties $\cup\cap$;
an **oval**   iff $U,V,A \in \mathfrak{S} \Rightarrow U|A|V \in \mathfrak{S}$;
a **ring**   iff it fulfils $\cup\cap\setminus$;
an **algebra** iff it fulfils $\cup\cap\perp$.

We have thus added the new class of ovals to some familiar ones. We shall see that the ovals are similar to rings, but form a more symmetric class. The remarks and examples below contain some simple assertions which are left as exercises.

1.2. REMARKS AND EXAMPLES. 1) We have $\cup\perp \Leftrightarrow \cap\perp$. Thus an algebra can be defined to fulfil $\cup\perp$, and likewise to fulfil $\cap\perp$. Furthermore we have $\cup\setminus \Rightarrow \cap\setminus$, but not $\Leftarrow$. Thus a ring can be defined to fulfil $\cup\setminus$, but not to fulfil $\cap\setminus$. 2) If $\mathfrak{S}$ is a ring then $A,B \in \mathfrak{S} \Rightarrow B\cap A' \in \mathfrak{S}$. 3) We have $\mathfrak{S}$ algebra $\Rightarrow$ $\mathfrak{S}$ ring $\Rightarrow$ $\mathfrak{S}$ oval $\Rightarrow$ $\mathfrak{S}$ lattice . 4) $\mathfrak{S}$ algebra $\Leftrightarrow$ $\mathfrak{S}$ ring with $X \in \mathfrak{S}$. Likewise $\mathfrak{S}$ ring $\Leftrightarrow$ $\mathfrak{S}$ oval with $\varnothing \in \mathfrak{S}$. 5) If $\mathfrak{S}$ is a lattice then $\mathfrak{S}\perp$ is a lattice as well. The same holds true for ovals and for algebras, but not for rings.

6) $\{\varnothing,X\}$ and $\mathfrak{P}(X)$ are algebras. $\{\varnothing\}$ is a ring, but not an algebra. Furthermore $\{A\}$ is an oval for each $A \subset X$, but not a ring when $A \neq \varnothing$. 7) A set system which is totally ordered under inclusion $\subset$ is a lattice. But it is not an oval when it has at least three members.

8) Let X be a topological space. Then the system $\mathrm{Op}(X)$ of its open subsets and the system $\mathrm{Cl}(X)$ of its closed subsets are lattices which contain $\varnothing$, but as a rule are not ovals.

9) Each paving $\mathfrak{S}$ in X is contained in a unique smallest lattice $\mathrm{L}(\mathfrak{S})$, called the lattice generated by $\mathfrak{S}$. It is as usual the intersection of all lattices which contain $\mathfrak{S}$. We likewise obtain the oval $\mathrm{O}(\mathfrak{S})$, the ring $\mathrm{R}(\mathfrak{S})$, and the algebra $\mathrm{A}(\mathfrak{S})$ generated by $\mathfrak{S}$. Thus $\mathrm{L}(\mathfrak{S}) \subset \mathrm{O}(\mathfrak{S}) \subset \mathrm{R}(\mathfrak{S}) \subset \mathrm{A}(\mathfrak{S})$. 10) If $\mathfrak{S}$ has $\cap$ then $\mathrm{L}(\mathfrak{S}) = \mathfrak{S}^\star$.

11) The most important examples on $\mathbb{R}$ are the systems of the closed bounded intervals $[a,b]$ with real $a \leq b$ and of the open bounded intervals $]a,b[$ with real $a < b$, each time combined with $\varnothing$, and their obvious counterparts in $\mathbb{R}^n$. The generated lattices consist of closed/open sets and hence cannot be rings. 12) Another example on $\mathbb{R}$ is the system of the half-open bounded intervals $[a,b[$ with real $a < b$, as above combined with $\varnothing$, and its obvious counterpart in $\mathbb{R}^n$. This time the generated lattice turns out to be a ring. However, we shall not use this example for systematic purposes.

We turn to the relevant infinite formations. For a set system $\mathfrak{S}$ we define $\mathfrak{S}^\sigma \subset \mathfrak{S}^\tau$ to consist of the unions $\bigcup_{S \in \mathfrak{M}} S$ for the countable/arbitrary pavings $\mathfrak{M} \subset \mathfrak{S}$, and $\mathfrak{S}_\sigma \subset \mathfrak{S}_\tau$ to consist of the respective intersections. In the sequel we shall use a simple and practical shorthand notation: The symbol $\bullet$ means that in a fixed context it can be read as one of the symbols $\star\sigma\tau$, at times with restrictions as noted in the respective context. Thus we define $\mathfrak{S}$ to fulfil $\cup\bullet / \cap\bullet$ iff $\mathfrak{S}^\bullet = \mathfrak{S}/\mathfrak{S}_\bullet = \mathfrak{S}$. In this connection we denote the finite/countable/arbitrary pavings to be of type $\bullet = \star\sigma\tau$.

Next let $(S_l)_l$ be a sequence of subsets of X. We write

$S_l \uparrow$ iff the sequence increases,

$S_l \uparrow S$ iff it increases with union S,

$S_l \uparrow\supset A$ iff it increases with some union $S \supset A$.

A set system $\mathfrak{S}$ in X is defined to fulfil $\uparrow\sigma$ iff for each sequence $(S_l)_l$ in $\mathfrak{S}$ with $S_l \uparrow S$ one has $S \in \mathfrak{S}$. All this also applies to $\downarrow$ alike, that is with decrease to the intersection instead of increase to the union.

The nonsequential counterparts are as follows. A paving $\mathfrak{M}$ in X is called **upward directed** iff for each pair $U, V \in \mathfrak{M}$ there exists $W \in \mathfrak{M}$ with $U, V \subset W$. Note that $\mathfrak{M}$ is upward directed when it has $\cup$. We write

$\mathfrak{M} \uparrow$ iff $\mathfrak{M}$ is upward directed,

$\mathfrak{M} \uparrow S$ iff $\mathfrak{M}$ is upward directed with union S,

$\mathfrak{M} \uparrow\supset A$ iff $\mathfrak{M}$ is upward directed with some union $S \supset A$.

A set system $\mathfrak{S}$ in X is defined to fulfil $\uparrow\tau$ iff for each paving $\mathfrak{M} \subset \mathfrak{S}$ with $\mathfrak{M} \uparrow S$ one has $S \in \mathfrak{S}$. As before all this applies to $\downarrow$ alike, that is with downward directed to the intersection instead of upward directed to the union.

1.3. EXERCISE. 1) Let $\mathfrak{M}$ be a finite paving. Then $\mathfrak{M} \uparrow S$ means that $S \in \mathfrak{M}$ and $M \subset S \; \forall M \in \mathfrak{M}$, that is that $\mathfrak{M}$ has the maximum S under inclusion. 2) Let $\mathfrak{M}$ be a countable paving. Then $\mathfrak{M} \uparrow S$ means that there exists a sequence $(S_l)_l$ in $\mathfrak{M}$ with $S_l \uparrow S$ such that each $M \in \mathfrak{M}$ is contained in some S_l.

1.4. EXERCISE. Assume that $\bullet = \sigma\tau$. 0) If $\mathfrak{S}$ is a lattice then $\mathfrak{S}^\bullet$ and $\mathfrak{S}_\bullet$ are lattices as well. 1) Let $\mathfrak{S}$ be a set system with $\cup$. If $S \in \mathfrak{S}^\bullet$ then there exists a paving $\mathfrak{M} \subset \mathfrak{S}$ of type $\bullet$ such that $\mathfrak{M} \uparrow S$. Thus in case $\bullet = \sigma$ there exists a sequence $(S_l)_l$ in $\mathfrak{S}$ such that $S_l \uparrow S$. 2) For each set system $\mathfrak{S}$ one has $\cup\bullet \Leftrightarrow \cup$ and $\uparrow\bullet$.

1.5. EXERCISE. 1) Prove that $\mathfrak{M} \uparrow S \Leftrightarrow (\mathfrak{M}\perp) \downarrow S'$. 2) For each set system $\mathfrak{S}$ one has $(\mathfrak{S}^\bullet)\perp = (\mathfrak{S}\perp)_\bullet$.

We define a **σ lattice/ σ oval/σ ring/σ algebra** to be a lattice/oval/ ring/ algebra with $\uparrow\sigma$ and $\downarrow\sigma$; in view of 1.4.2 this means that $\mathfrak{S}^\sigma = \mathfrak{S}_\sigma = \mathfrak{S}$.

1.6. REMARKS AND EXAMPLES. 1) For lattices and ovals none of the properties $\uparrow\sigma$ and $\downarrow\sigma$ implies the other. For rings we have $\uparrow\sigma \Rightarrow \downarrow\sigma$, but not $\Leftarrow$. For algebras we have $\uparrow\sigma \Leftrightarrow \downarrow\sigma$. 2) $\{\varnothing, X\}$ and $\mathfrak{P}(X)$ are σ algebras. There are other obvious examples.

3) Each paving $\mathfrak{S}$ in X is contained in a unique smallest σ lattice $L\sigma(\mathfrak{S})$, called the σ lattice generated by $\mathfrak{S}$. It is the intersection of all σ lattices which contain $\mathfrak{S}$. We likewise obtain the σ oval $O\sigma(\mathfrak{S})$, the σ ring $R\sigma(\mathfrak{S})$, and the σ algebra $A\sigma(\mathfrak{S})$ generated by $\mathfrak{S}$. Thus $L\sigma(\mathfrak{S}) \subset O\sigma(\mathfrak{S}) \subset$

$R\sigma(\mathfrak{S}) \subset A\sigma(\mathfrak{S})$. However, the definitions of these set systems are so indirect that their explicit description even for the simplest $\mathfrak{S}$ can produce extreme difficulties. Thus in order to handle these formations delicate methods have been invented. Two of them will be presented later in this section.

4) Let X be a topological space. One defines the **Borel σ algebra** of X to be
$$\mathrm{Bor}(X) := A\sigma(\mathrm{Op}(X)) = A\sigma(\mathrm{Cl}(X));$$
its members are called the **Borel subsets** of X. Furthermore besides $\mathrm{Op}(X)$ and $\mathrm{Cl}(X)$ one considers the smaller pavings
$$\mathrm{COp}(X) := \{[f \neq 0] : f \in C(X, \mathbb{R})\}, \quad \mathrm{CCl}(X) := \{[f = 0] : f \in C(X, \mathbb{R})\},$$
which are lattices as well. The definitions have numerous obvious variants, of which we shall make free use. In case X is semimetrizable the new pavings coincide with the former ones, but as a rule they are different. One defines the **Baire σ algebra** of X to be
$$\mathrm{Baire}(X) := A\sigma(\mathrm{COp}(X)) = A\sigma(\mathrm{CCl}(X));$$
its members are called the **Baire subsets** of X.

1.7. EXERCISE. $\star$) If $\mathfrak{S}$ is a ring then $\mathfrak{S} \cup (\mathfrak{S}\bot)$ is an algebra. σ) If $\mathfrak{S}$ is a σ ring then $\mathfrak{S} \cup (\mathfrak{S}\bot)$ is a σ algebra.

We need one more notation. Let $\mathfrak{T}$ be a set system in X. A subset $A \subset X$ is called

  upward enclosable $\mathfrak{T}$ iff $A \subset T$ for some $T \in \mathfrak{T}$,
  downward enclosable $\mathfrak{T}$ iff $A \supset T$ for some $T \in \mathfrak{T}$.

Let $\sqsubset \mathfrak{T}$ and $\sqsupset \mathfrak{T}$ consist of all these subsets. Also we form for set systems $\mathfrak{S}$ and $\mathfrak{T}$ the set system
$$\mathfrak{S} \sqsubset \mathfrak{T} := (\sqsupset \mathfrak{S}) \cap (\sqsubset \mathfrak{T}) = \{A \subset X : S \subset A \subset T \text{ for some } S \in \mathfrak{S} \text{ and } T \in \mathfrak{T}\}.$$
These set systems can of course be void.

1.8. EXERCISE. Let $\mathfrak{S}$ be a paving. 1$\star$) If $\mathfrak{S} \uparrow$ then $\sqsubset \mathfrak{S}$ is a ring. If $\mathfrak{S} \downarrow$ then $\sqsupset \mathfrak{S}$ is an oval. 2$\star$) We have
$$O(\mathfrak{S}) \subset (\mathfrak{S}_\star \sqsubset \mathfrak{S}^\star);$$
$$R(\mathfrak{S}) \subset (\sqsubset \mathfrak{S}^\star);$$
$$A(\mathfrak{S}) \subset \{A \subset X : A \text{ or } A' \text{ in } \sqsubset \mathfrak{S}^\star\}.$$
1σ) If $\mathfrak{S}^\sigma = \mathfrak{S}$ then $\sqsubset \mathfrak{S}$ is a σ ring. If $\mathfrak{S}_\sigma = \mathfrak{S}$ then $\sqsupset \mathfrak{S}$ is a σ oval. 2σ) We have
$$O\sigma(\mathfrak{S}) \subset (\mathfrak{S}_\sigma \sqsubset \mathfrak{S}^\sigma);$$
$$R\sigma(\mathfrak{S}) \subset (\sqsubset \mathfrak{S}^\sigma);$$
$$A\sigma(\mathfrak{S}) \subset \{A \subset X : A \text{ or } A' \text{ in } \sqsubset \mathfrak{S}^\sigma\}.$$

Inverse Images of Pavings

This is the first of the two methods announced in 1.6.3) in order to handle the formations for set systems introduced above. Let X and Y be nonvoid sets and $\vartheta : X \to Y$ be a map. It is common to form

for $A \subset X$ the image set $\vartheta(A) := \{\vartheta(x) : x \in A\} \subset Y$,

for $B \subset Y$ the inverse image set $\overset{-1}{\vartheta}(B) := \{x \in X : \vartheta(x) \in B\} \subset X$.

The behaviour of the inverse images is much better than that of the direct images. Thus one has

$$\overset{-1}{\vartheta}\left(\bigcup_{B \in \mathfrak{B}} B\right) = \bigcup_{B \in \mathfrak{B}} \overset{-1}{\vartheta}(B) \quad \text{and} \quad \overset{-1}{\vartheta}\left(\bigcap_{B \in \mathfrak{B}} B\right) = \bigcap_{B \in \mathfrak{B}} \overset{-1}{\vartheta}(B)$$

for each paving $\mathfrak{B}$ in Y, and $\overset{-1}{\vartheta}(B') = (\overset{-1}{\vartheta}(B))'$ for each $B \subset Y$, but only

$$\vartheta\left(\bigcup_{A \in \mathfrak{A}} A\right) = \bigcup_{A \in \mathfrak{A}} \vartheta(A) \quad \text{and} \quad \vartheta\left(\bigcap_{A \in \mathfrak{A}} A\right) \subset \bigcap_{A \in \mathfrak{A}} \vartheta(A)$$

for each paving $\mathfrak{A}$ in X, and no relation between $\vartheta(A')$ and $(\vartheta(A))'$ for $A \subset X$. Therefore our prime interest is to form for a paving $\mathfrak{B}$ in Y the inverse image paving $\overset{-1}{\vartheta}(\mathfrak{B})$ in X, defined to consist of the $\overset{-1}{\vartheta}(B)$ for all $B \in \mathfrak{B}$.

1.9. REMARK. *If the paving $\mathfrak{B}$ in Y is a lattice/oval/ring/algebra or a σ lattice/σ oval/σ ring/σ algebra then the same holds true for the paving $\overset{-1}{\vartheta}(\mathfrak{B})$ in X.*

Proof. i) The above rules for inverse image sets show that the properties $\mathfrak{B}^\bullet = \mathfrak{B}$ and $\mathfrak{B}_\bullet = \mathfrak{B}$ carry over to $\overset{-1}{\vartheta}(\mathfrak{B})$. Thus we obtain the assertion for lattices and for σ lattices. ii) For $U, V, B \subset Y$ we have

$$\overset{-1}{\vartheta}(U|B|V) = \overset{-1}{\vartheta}(U \cap B') \cup \overset{-1}{\vartheta}(V \cap B)$$
$$= \left(\overset{-1}{\vartheta}(U) \cap (\overset{-1}{\vartheta}(B))'\right) \cup \left(\overset{-1}{\vartheta}(V) \cap \overset{-1}{\vartheta}(B)\right) = \overset{-1}{\vartheta}(U)|\overset{-1}{\vartheta}(B)|\overset{-1}{\vartheta}(V).$$

Thus we obtain the assertion for ovals and hence for σ ovals. iii) It is obvious that $\varnothing \in \mathfrak{B} \Rightarrow \varnothing \in \overset{-1}{\vartheta}(\mathfrak{B})$ and $Y \in \mathfrak{B} \Rightarrow X \in \overset{-1}{\vartheta}(\mathfrak{B})$. Thus we obtain the assertion for rings and algebras and hence for σ rings and σ algebras.

Now as a substitute for the direct image of a paving $\mathfrak{A}$ in X we form the set system $\vartheta[\mathfrak{A}]$ in Y, defined to consist of the $B \subset Y$ such that $\overset{-1}{\vartheta}(B) \in \mathfrak{A}$. Note that $\vartheta[\mathfrak{A}]$ can be void.

1.10. REMARK. *Let $\mathfrak{A}$ be a paving in X such that $\vartheta[\mathfrak{A}]$ is nonvoid. If $\mathfrak{A}$ is a lattice/oval/ring/algebra or a σ lattice/σ oval/σ ring/σ algebra then the same holds true for the paving $\vartheta[\mathfrak{A}]$ in Y.*

Proof. i) As above one verifies that the rules for inverse image sets imply that the properties $\mathfrak{A}^\bullet = \mathfrak{A}$ and $\mathfrak{A}_\bullet = \mathfrak{A}$ carry over to $\vartheta[\mathfrak{A}]$. Thus we obtain the assertion for lattices and for σ lattices. ii) If $\mathfrak{A}$ is an oval then for $U, V, B \in \vartheta[\mathfrak{A}]$ we have as above

$$\overset{-1}{\vartheta}(U|B|V) = \overset{-1}{\vartheta}(U)|\overset{-1}{\vartheta}(B)|\overset{-1}{\vartheta}(V) \in \mathfrak{A} \text{ and hence } U|B|V \in \vartheta[\mathfrak{A}].$$

Thus we obtain the assertion for ovals and hence for σ ovals. iii) It is obvious that $\varnothing \in \mathfrak{A} \Rightarrow \varnothing \in \vartheta[\mathfrak{A}]$ and $X \in \mathfrak{A} \Rightarrow Y \in \vartheta[\mathfrak{A}]$. Thus we obtain the assertion for rings and algebras and hence for σ rings and σ algebras.

The two remarks combine to furnish the desired result.

1.11. THEOREM. *Let* T *denote one of the operations* LORA *or* LσOσRσ Aσ. *Then* $\mathrm{T}(\overset{-1}{\vartheta}(\mathfrak{B})) = \overset{-1}{\vartheta}(\mathrm{T}(\mathfrak{B}))$ *for each paving* $\mathfrak{B}$ *in* Y.

Proof. $\subset$) By definition $\mathrm{T}(\mathfrak{B})$ is a lattice$/\cdots$ which contains $\mathfrak{B}$. Hence $\overset{-1}{\vartheta}(\mathrm{T}(\mathfrak{B}))$ contains $\overset{-1}{\vartheta}(\mathfrak{B})$, and is a lattice$/\cdots$ by 1.9. It follows that $\mathrm{T}(\overset{-1}{\vartheta}(\mathfrak{B})) \subset \overset{-1}{\vartheta}(\mathrm{T}(\mathfrak{B}))$. $\supset$) By definition the paving $\mathfrak{A} := \mathrm{T}(\overset{-1}{\vartheta}(\mathfrak{B}))$ in X is a lattice$/\cdots$ which contains $\overset{-1}{\vartheta}(\mathfrak{B})$. By its definition thus $\vartheta[\mathfrak{A}]$ contains $\mathfrak{B}$ and hence is nonvoid, and is a lattice$/\cdots$ by 1.10. It follows that $\mathrm{T}(\mathfrak{B}) \subset \vartheta[\mathfrak{A}]$. Since by definition $\overset{-1}{\vartheta}(\vartheta[\mathfrak{A}]) \subset \mathfrak{A}$ we obtain $\overset{-1}{\vartheta}(\mathrm{T}(\mathfrak{B})) \subset \overset{-1}{\vartheta}(\vartheta[\mathfrak{A}]) \subset \mathfrak{A} = \mathrm{T}(\overset{-1}{\vartheta}(\mathfrak{B}))$. The proof is complete.

1.12. EXAMPLE. Let $X \subset Y$ be nonvoid. Then each paving $\mathfrak{B}$ in Y produces a paving $\mathfrak{B} \cap X$ in X, defined to consist of the $B \cap X$ for all $B \in \mathfrak{B}$, and called the **trace** of $\mathfrak{B}$ on X. If $\vartheta : X \to Y$ denotes the injection then $\mathfrak{B} \cap X = \overset{-1}{\vartheta}(\mathfrak{B})$ in the above sense. Thus if $\mathfrak{B}$ is a lattice$/\cdots$ then by 1.9 the trace $\mathfrak{B} \cap X$ is a lattice$/\cdots$ as well. Furthermore we have by 1.11 with the same abbreviation $\mathrm{T}(\mathfrak{B} \cap X) = (\mathrm{T}(\mathfrak{B})) \cap X$ for each paving $\mathfrak{B}$ in Y. Here of course the operation T is understood to be relative to X on the left side, and relative to Y on the right side.

1.13. EXERCISE. Let Y be a topological space, and let the nonvoid subset $X \subset Y$ be equipped with the relative topology. 1) Prove that $\mathrm{Bor}(X) = \mathrm{Bor}(Y) \cap X$. 2) We have

$$\mathrm{Bor}(X) \supset \{A \in \mathrm{Bor}(Y) : A \subset X\}.$$

Here we have equality iff X is in $\mathrm{Bor}(Y)$. 3) Can the counterpart of 1) for $\mathrm{Baire}(\cdot)$ be answered in a similar manner? Deduce from a standard theorem in topology that $\mathrm{Baire}(X) = \mathrm{Baire}(Y) \cap X$ at least if Y is normal and X is closed in Y.

The Transporter

We turn to the second of the two methods announced in 1.6.3). The basic idea seems to be due to Sierpiński; see Hoffmann-Jørgensen [1994] section 1.6. In the more recent literature it appears under the names monotone class theorem and Dynkin systems; see for example Halmos [1950] section 6 and Bauer [1992] section 2. It will here be based on the notion of transporter as introduced in König [1991].

For two pavings $\mathfrak{M}$ and $\mathfrak{N}$ in the nonvoid set X we define the **transporter** $\mathfrak{M}\top\mathfrak{N}$ to consist of all subsets $A \subset X$ such that $M \in \mathfrak{M} \Rightarrow A \cap M \in \mathfrak{N}$. Note that this set system can be void. In particular we write $\mathfrak{M}\top\mathfrak{M} =: \mathfrak{M}\top$. The members of $\mathfrak{M}\top$ are often called the **local $\mathfrak{M}$ sets**.

1.14. PROPERTIES. 1) $X \in \mathfrak{M}\top\mathfrak{N} \Leftrightarrow \mathfrak{M} \subset \mathfrak{N}$. *In particular we have* $X \in \mathfrak{M}\top$. 2) $A \in \mathfrak{M}\top$ *and* $B \in \mathfrak{M}\top\mathfrak{N} \Rightarrow A \cap B \in \mathfrak{M}\top\mathfrak{N}$. *In other words* $\mathfrak{M}\top \subset (\mathfrak{M}\top\mathfrak{N})\top$. *In particular the paving* $\mathfrak{M}\top$ *fulfils* $\cap$. 3) $\mathfrak{M}\top \subset \mathfrak{M} \Leftrightarrow X \in \mathfrak{M}$. *Furthermore* $\mathfrak{M}\top \supset \mathfrak{M} \Leftrightarrow \mathfrak{M}$ *fulfils* $\cap$. *In particular* $\mathfrak{M}\top\top = \mathfrak{M}\top$. 4) $\mathfrak{N}$ *fulfils* $\setminus \Rightarrow \mathfrak{M}\top\mathfrak{N}$ *fulfils* $\setminus$. 5) *Let* $\bullet = \sigma\tau$. *Then* $\uparrow\bullet$ *carries over from* $\mathfrak{N}$ *to* $\mathfrak{M}\top\mathfrak{N}$. *The same holds true for* $\downarrow\bullet$.

Proof. All properties except 4) are obvious. To see 4) note that for $A, B, M \subset X$ with $A \subset B$ one has $(B \setminus A) \cap M = B \cap A' \cap M = (B \cap M) \cap (A' \cup M') = (B \cap M) \setminus (A \cap M)$.

We prepare the main theorem with a useful lemma.

1.15. LEMMA. *Let* $\mathfrak{M}$ *be a paving.* $\star$) *If* $\mathfrak{M}$ *fulfils* $\setminus$ *then* $\mathfrak{M}\top$ *is an algebra.* σ) *If* $\mathfrak{M}$ *fulfils* $\setminus$ *and one of the properties* $\uparrow\sigma \downarrow\sigma$ *then* $\mathfrak{M}\top$ *is a σ algebra.*

Proof. Assume that $\mathfrak{M}$ has $\setminus$. $\star$) We have $X \in \mathfrak{M}\top$ by the above property 1), and $\mathfrak{M}\top$ has $\setminus$ by 4). Thus $\mathfrak{M}\top$ has $\bot$. Furthermore $\mathfrak{M}\top$ has $\cap$ by 2). Thus $\mathfrak{M}\top$ is an algebra. σ) If $\mathfrak{M}$ has one of the properties $\uparrow\sigma \downarrow\sigma$ then $\mathfrak{M}\top$ has the same by 5). Thus $\mathfrak{M}\top$ is a σ algebra.

1.16. THEOREM (The Transporter Theorem). *Let* $\mathfrak{N}$ *be a paving.* $\star$) *If* $\mathfrak{N}$ *fulfils* $\setminus$ *then*

$$A(\mathfrak{M}\top) \subset \mathfrak{M}\top\mathfrak{N} \quad \text{for all pavings } \mathfrak{M} \subset \mathfrak{N}.$$

σ) *If* $\mathfrak{N}$ *fulfils* $\setminus$ *and one of the properties* $\uparrow\sigma \downarrow\sigma$ *then*

$$A\sigma(\mathfrak{M}\top) \subset \mathfrak{M}\top\mathfrak{N} \quad \text{for all pavings } \mathfrak{M} \subset \mathfrak{N}.$$

Proof. Assume that $\mathfrak{N}$ has $\setminus$, and fix a paving $\mathfrak{M} \subset \mathfrak{N}$. $\star$) $\mathfrak{M}\top\mathfrak{N}$ has $\setminus$ by the above property 4). Hence by 1.15.$\star$) $(\mathfrak{M}\top\mathfrak{N})\top$ is an algebra. Now $\mathfrak{M}\top \subset (\mathfrak{M}\top\mathfrak{N})\top$ by 2) and hence $A(\mathfrak{M}\top) \subset (\mathfrak{M}\top\mathfrak{N})\top$. Furthermore $X \in \mathfrak{M}\top\mathfrak{N}$ by 1) and hence $(\mathfrak{M}\top\mathfrak{N})\top \subset \mathfrak{M}\top\mathfrak{N}$ by 3). This combines to produce the assertion. σ) If $\mathfrak{N}$ has one of the properties $\uparrow\sigma \downarrow\sigma$ then $\mathfrak{M}\top\mathfrak{N}$ has the same by 5). Hence by 1.15.σ) $(\mathfrak{M}\top\mathfrak{N})\top$ is a σ algebra. It follows as above that $A\sigma(\mathfrak{M}\top) \subset (\mathfrak{M}\top\mathfrak{N})\top \subset \mathfrak{M}\top\mathfrak{N}$.

The transporter theorem will find substantial applications in later parts. For the moment we continue with some simple but typical consequences. The reader is warned that the transporter theorem becomes false in both parts $\star$) and σ), and even for $\mathfrak{M} = \mathfrak{N}$, if on the left side one writes $\mathfrak{M}$ instead of $\mathfrak{M}\top$.

1.17. REMARK. *Let $\mathfrak{S}$ be a paving. Then*

$\star$) $R(\mathfrak{S}) = A(\mathfrak{S}) \cap (\sqsubset \mathfrak{S}^\star)$ *and* $A(\mathfrak{S}) = R(\mathfrak{S}) \cup (R(\mathfrak{S})\bot)$;

σ) $R\sigma(\mathfrak{S}) = A\sigma(\mathfrak{S}) \cap (\sqsubset \mathfrak{S}^\sigma)$ *and* $A\sigma(\mathfrak{S}) = R\sigma(\mathfrak{S}) \cup (R\sigma(\mathfrak{S})\bot)$.

In particular if $\mathfrak{S}$ is a σ ring then $A(\mathfrak{S})$ is a σ algebra.

Proof of $\star$). First assertion: The inclusion $\subset$ is contained in 1.8.2$\star$); thus we have to prove $\supset$. Since $R(\mathfrak{S})$ has $\setminus$ we conclude from 1.15.$\star$) that $(R(\mathfrak{S}))\top$ is an algebra; therefore $A(\mathfrak{S}) \subset (R(\mathfrak{S}))\top$. Now $(R(\mathfrak{S}))\top = R(\mathfrak{S})\top R(\mathfrak{S}) \subset \mathfrak{S}^\star \top R(\mathfrak{S})$. Thus we have $A(\mathfrak{S}) \subset \mathfrak{S}^\star \top R(\mathfrak{S})$. It follows that $A \in A(\mathfrak{S}) \cap (\sqsubset \mathfrak{S}^\star) \Rightarrow A \in R(\mathfrak{S})$. Second assertion: To see $\subset$ combine the last assertion in 1.8.2$\star$) with the first assertion above. The implication $\supset$ is obvious. The proof of σ) is left as an exercise.

1.18. PROPOSITION. *Let $\mathfrak{S}$ be a paving. Then*

$\star$) $\{A \in A(\mathfrak{S}\top) : A \text{ or } A' \text{ in } \sqsubset \mathfrak{S}^\star\} \subset A(\mathfrak{S})$,

σ) $\{A \in A\sigma(\mathfrak{S}\top) : A \text{ or } A' \text{ in } \sqsubset \mathfrak{S}^\sigma\} \subset A\sigma(\mathfrak{S})$.

Proof of σ). We can assume that $A \in A\sigma(\mathfrak{S}\top)$ is in $\sqsubset \mathfrak{S}^\sigma$. Thus $A \subset \bigcup_{l=1}^\infty S_l$ or $A = \bigcup_{l=1}^\infty A \cap S_l$ for a sequence $(S_l)_l$ in $\mathfrak{S}$. From 1.16.σ) applied to $\mathfrak{N} := A\sigma(\mathfrak{S})$ and $\mathfrak{M} := \mathfrak{S}$ we obtain $A \in \mathfrak{S}\top A\sigma(\mathfrak{S})$. Therefore $A \cap S_l \in A\sigma(\mathfrak{S}) \; \forall l \in \mathbb{N}$ and hence $A \in A\sigma(\mathfrak{S})$.

1.19. CONSEQUENCE. *Let $\mathfrak{S}$ and $\mathfrak{K}$ be pavings such that $\mathfrak{K} \subset \mathfrak{S} \subset \mathfrak{K}\top$. Then*

$\star$) $A(\mathfrak{K}) = \{A \in A(\mathfrak{S}) : A \text{ or } A' \text{ in } \sqsubset \mathfrak{K}^\star\}$;

σ) $A\sigma(\mathfrak{K}) = \{A \in A\sigma(\mathfrak{S}) : A \text{ or } A' \text{ in } \sqsubset \mathfrak{K}^\sigma\}$.

Proof of σ). The inclusion $\subset$ follows from 1.8.2σ). The inclusion $\supset$ follows from 1.18.σ) since $\mathfrak{S} \subset \mathfrak{K}\top$.

1.20. EXAMPLE. Let X be a topological space, and let $\mathrm{Comp}(X)$ consist of its compact subsets. We assume that X is Hausdorff, so that $\mathrm{Comp}(X) \subset \mathrm{Cl}(X)$. Then 1.19.$\sigma$) applied to $\mathfrak{S} := \mathrm{Cl}(X)$ and $\mathfrak{K} := \mathrm{Comp}(X)$ furnishes the important relation

$$A\sigma(\mathrm{Comp}(X)) = \{A \in \mathrm{Bor}(X) : A \text{ or } A' \text{ in } \sqsubset (\mathrm{Comp}(X))^\sigma\}.$$

Another application of the transporter theorem 1.16.$\star$) is a useful description of the ring generated by a lattice which contains $\varnothing$.

1.21. EXERCISE. Let $\mathfrak{S}$ be a lattice with $\varnothing \in \mathfrak{S}$. Then for each $M \in R(\mathfrak{S})$ there exists a finite sequence $S_1 \subset T_1 \subset \cdots \subset S_r \subset T_r$ in $\mathfrak{S}$ such that

$$M = \bigcup_{l=1}^r (T_l \setminus S_l) \quad \text{(the converse is obvious)}.$$

Hint: 1) For subsets $A \subset B$ and $S \subset T$ one has

$$(B \setminus A) \cap (T \setminus S)' = \Big((A \cup (B \cap S)) \setminus A\Big) \cup \Big(B \setminus (A \cup (B \cap T))\Big).$$

Note that $A \subset A \cup (B \cap S) \subset A \cup (B \cap T) \subset B$. 2) Define $\mathfrak{H}$ to consist of all subset $H \subset X$ which can be written in the above form. Deduce from 1) that $H \in \mathfrak{H}$ implies $H \cap (T \setminus S)' \in \mathfrak{H}$ for all $S \subset T$ in $\mathfrak{S}$. Conclude that $\mathfrak{H}$ fulfils $\setminus$. 3) Use 1.16.$\star$) for $\mathfrak{S} \subset \mathfrak{H}$ and note 1.8.2$\star$).

Complements for Ovals and σ Ovals

The final subsection contains further material on these unfamiliar notions, in particular on their relations to the more familiar ones.

1.22. PROPOSITION. $\star$) Let $\mathfrak{S}$ be an oval. Then $U|A|V \in \mathfrak{S}$ for all $U, V \in \mathfrak{S}$ and $A \in A(\mathfrak{S})$. σ) Let $\mathfrak{S}$ be a σ oval. Then $U|A|V \in \mathfrak{S}$ for all $U, V \in \mathfrak{S}$ and $A \in A\sigma(\mathfrak{S})$.

Proof. Let $\mathfrak{S}$ be an oval. Define $\mathfrak{A}$ to consist of all subsets $A \subset X$ such that $U|A|V \in \mathfrak{S} \ \forall U, V \in \mathfrak{S}$. Then $\mathfrak{S} \subset \mathfrak{A}$, and $\mathfrak{A}$ has $\perp$. $\star$) We show that $\mathfrak{A}$ has $\cup$ and hence is an algebra. To see this let $A, B \in \mathfrak{A}$. For $U, V \in \mathfrak{S}$ then $W := U|B|V \in \mathfrak{S}$ since $B \in \mathfrak{A}$. It follows that

$$\begin{aligned}
U|A \cup B|V &= \big(U \cap (A' \cap B')\big) \cup \big(V \cap (A \cup B)\big) \\
&= (U \cap A' \cap B') \cup (V \cap A) \cup (V \cap B) \\
&= (U \cap B' \cap A') \cup (V \cap B \cap A') \cup (V \cap A) \\
&= (W \cap A') \cup (V \cap A) = W|A|V,
\end{aligned}$$

which is in $\mathfrak{S}$ since $A \in \mathfrak{A}$. Thus we have indeed $A \cup B \in \mathfrak{A}$. σ) We show that $\mathfrak{A}$ fulfils $\uparrow \sigma$ and hence is a σ algebra. To see this let $(A_l)_l$ in $\mathfrak{A}$ with $A_l \uparrow A$. For $U, V \in \mathfrak{S}$ then $B_l := U|A_l|V \in \mathfrak{S} \ \forall l \in \mathbb{N}$. Now $B_l \cap U, B_l \cap V \in \mathfrak{S}$ since $\mathfrak{S}$ is a lattice. We have

$$\begin{aligned}
B_l \cap U = (U \cap A_l') \cup (U \cap V \cap A_l) \ &= \ (U \cap V) \cup (U \cap A_l') \\
&\downarrow \ (U \cap V) \cup (U \cap A'),
\end{aligned}$$

$$\begin{aligned}
B_l \cap V = (U \cap V \cap A_l') \cup (V \cap A_l) \ &= \ (U \cap V) \cup (V \cap A_l) \\
&\uparrow \ (U \cap V) \cup (V \cap A),
\end{aligned}$$

and hence by assumption $P := (U \cap V) \cup (U \cap A') \in \mathfrak{S}$ and $Q := (U \cap V) \cup (V \cap A) \in \mathfrak{S}$. Now

$$P \cup Q = (U \cap V) \cup (U \cap A') \cup (V \cap A) = (U \cap A') \cup (V \cap A) = U|A|V,$$

which is in $\mathfrak{S}$ since $\mathfrak{S}$ is a lattice. Thus we have indeed $A \in \mathfrak{A}$.

1.23. CONSEQUENCE. Let $\mathfrak{S}$ be a paving. Then
$\star$) $O(\mathfrak{S}) = A(\mathfrak{S}) \cap (\mathfrak{S}_\star \sqsubset \mathfrak{S}^\star)$;
σ) $O\sigma(\mathfrak{S}) = A\sigma(\mathfrak{S}) \cap (\mathfrak{S}_\sigma \sqsubset \mathfrak{S}^\sigma)$.

Proof. This corresponds to part of 1.17. Let us prove σ). The inclusion $\subset$ is contained in 1.8.2σ). Thus we have to prove $\supset$. Let $A \in A\sigma(\mathfrak{S}) \cap (\mathfrak{S}_\sigma \sqsubset \mathfrak{S}^\sigma)$, that is $A \in A\sigma(\mathfrak{S})$ and $U \subset A \subset V$ with $U \in \mathfrak{S}_\sigma \subset O\sigma(\mathfrak{S})$ and $V \in \mathfrak{S}^\sigma \subset O\sigma(\mathfrak{S})$. From 1.22.$\sigma$) applied to $O\sigma(\mathfrak{S})$ we obtain $U|A|V \in O\sigma(\mathfrak{S})$. But $U|A|V = A$ since $U \subset A \subset V$. Thus $A \in O\sigma(\mathfrak{S})$.

The next assertions deal with the ring $R(\mathfrak{S})$ generated by an oval $\mathfrak{S}$.

1.24. LEMMA. *Let $\mathfrak{S}$ be an oval. Define $a(\mathfrak{S}) \subset A(\mathfrak{S})$ to consist of those $A \in A(\mathfrak{S})$ which are upward enclosable $\mathfrak{S}$ and upward enclosable $\mathfrak{S}\bot$, that is*

$$a(\mathfrak{S}) = A(\mathfrak{S}) \cap (\sqsubset \mathfrak{S}) \cap (\sqsubset (\mathfrak{S}\bot)).$$

1) $a(\mathfrak{S})$ is a ring $\subset R(\mathfrak{S})$. 2) If $\varnothing \in \mathfrak{S}$ then $a(\mathfrak{S}) = \mathfrak{S}$. If $\varnothing \notin \mathfrak{S}$ then $a(\mathfrak{S})$ and $\mathfrak{S}$ are disjoint. 3) For $A \in a(\mathfrak{S})$ and $S \in \mathfrak{S}$ we have $S \cup A, S \cap A' \in \mathfrak{S}$ and $(S \cup A) \setminus (S \cap A') = A$.
4) $a(\mathfrak{S}) = \{V \cap U' : U, V \in \mathfrak{S}\} = \{V \setminus U : U, V \in \mathfrak{S} \text{ with } U \subset V\}$.
5) $a(\mathfrak{S}) \cup \mathfrak{S}$ is a ring and hence $= R(\mathfrak{S})$.

Proof. 1) Follows from 1.8.1$\star$) and 1.17.$\star$). 2) If $\varnothing \in \mathfrak{S}$ then $X \in \mathfrak{S}\bot$ and hence $a(\mathfrak{S}) = A(\mathfrak{S}) \cap (\sqsubset \mathfrak{S})$ by 1.17.$\star$). On the other hand, if $A \in a(\mathfrak{S})$ is in $\mathfrak{S}$, then $A \subset V'$ for some $V \in \mathfrak{S}$ imples that $\varnothing = A \cap V \in \mathfrak{S}$. 3) We have $A \subset U$ and $A' \supset V$ for some $U, V \in \mathfrak{S}$. Thus $S|A|U = (S \cap A') \cup A = S \cup A$ and $S|A|V = (S \cap A') \cup \varnothing = S \cap A'$ are in $\mathfrak{S}$ by 1.22.$\star$). The last relation is an obvious identity. 4) The first inclusion $\supset$ is clear by definition, and the first inclusion $\subset$ follows from 3). 5) Let $P, Q \in a(\mathfrak{S}) \cup \mathfrak{S}$. Then $P \cup Q \in a(\mathfrak{S}) \cup \mathfrak{S}$ in all cases by 1)3). On the other hand, in case $Q \in a(\mathfrak{S})$ we have $Q \cap P' \in a(\mathfrak{S})$ by definition. In case $Q \in \mathfrak{S}$ we have $Q \cap P' \in \mathfrak{S}$ for $P \in a(\mathfrak{S})$ by 3) and $Q \cap P' \in a(\mathfrak{S})$ for $P \in \mathfrak{S}$ by 4).

1.25. REMARK. *If $\mathfrak{S}$ is a σ oval then $a(\mathfrak{S})$ and $R(\mathfrak{S})$ are σ rings. Hence $A(\mathfrak{S})$ is a σ algebra by 1.17.*

Proof. 1) Let $(P_l)_l$ be a sequence in $a(\mathfrak{S})$ with $P_l \uparrow P$. Then for $S \in \mathfrak{S}$ we have by 1.24.3) $S \cup P_l, S \cap P_l' \in \mathfrak{S}$ and hence $S \cup P \in \mathfrak{S}^\sigma = \mathfrak{S}$ and $S \cap P' \in \mathfrak{S}_\sigma = \mathfrak{S}$. It follows that $P \in a(\mathfrak{S})$. 2) Let now $(P_l)_l$ be a sequence in $R(\mathfrak{S}) = a(\mathfrak{S}) \cup \mathfrak{S}$ with $P_l \uparrow P$. Then either $P \in a(\mathfrak{S})$ by 1), or $P \in \mathfrak{S}$ by assumption. Thus $P \in R(\mathfrak{S})$.

2. Set Functions

Basic Properties of Set Functions

We consider set functions $\varphi : \mathfrak{S} \to \overline{\mathbb{R}}$ which are defined on lattices $\mathfrak{S}$ in nonvoid sets X and take values in $\overline{\mathbb{R}} := \mathbb{R} \cup \{-\infty, \infty\}$. We recall that $\overline{\mathbb{R}}$ carries a natural total order, and a natural topology which is metrizable and compact and is the order topology. But it is a problem to extend the addition from $\mathbb{R}$ to $\overline{\mathbb{R}}$. We recall that for $u, v \in \overline{\mathbb{R}}$ the sum $u + v \in \overline{\mathbb{R}}$ is well-defined except when u and v have opposite values $\pm\infty$. Our method to handle this problem is expressed in the next remark.

2.1. REMARK. *There are exactly two binary operations $\overline{\mathbb{R}} \times \overline{\mathbb{R}} \to \overline{\mathbb{R}}$ on $\overline{\mathbb{R}}$ which are associative and commutative and produce the usual addition in all doubtless cases. These operations are $\overset{.}{+} : \infty \overset{.}{+}(-\infty) = \infty$ and $\underset{.}{+} :$ $\infty \underset{.}{+}(-\infty) = -\infty$.*

In the sequel the symbol $\dot{+}$ means that in a fixed context it can be read as of one of these additions.

2.2. PROPERTIES. 1) *For $u, v \in \overline{\mathbb{R}}$ we have $u \dot{+} v \in \mathbb{R} \Leftrightarrow u, v \in \mathbb{R}$. 2) Both $\dot{+}$ and $+$ are isotone in each argument. 3) $-(u \dot{+} v) = (-u) \dot{+} (-v)$ for all $u, v \in \overline{\mathbb{R}}$.*

Proof of 2.1 and 2.2. i) The operations $\dot{+}$ and $+$ defined in 2.1 are commutative and fulfil 2.2.1). We prove the associativity for $+$. To be shown is $(u+v)+w = u+(v+w)$ for $u, v, w \in \overline{\mathbb{R}}$. If at least one of u, v, w is $= -\infty$ then both sides are $= -\infty$ by definition. Thus we can assume that $u, v, w > -\infty$. If then at least one of u, v, w is $= \infty$ then both sides are $= \infty$. It remains the case $u, v, w \in \mathbb{R}$ which is clear. ii) Assume now that $\circ$ is a binary operation on $\overline{\mathbb{R}}$ as described in 2.1. To be shown is that either $\circ = \dot{+}$ or $\circ = +$. If not then $(-\infty) \circ \infty = \infty \circ (-\infty) =: c \in \mathbb{R}$. For each $x \in \mathbb{R}$ then

$$c = (-\infty) \circ \infty = \big(x \circ (-\infty) \big) \circ \infty = x \circ \big((-\infty) \circ \infty \big) = x \circ c = x + c,$$

and hence $x = 0$. This proves the assertion. iii) We prove that $+$ is isotone in the first argument. To be shown is $u \leq v \Rightarrow u + a \leq v + a$ for $u, v, a \in \overline{\mathbb{R}}$. This is clear for $a = -\infty$. Thus we can assume that $a > -\infty$. The assertion is also clear for $u = -\infty$. Thus we can assume that $u > -\infty$ and hence $v > -\infty$ as well. But in this case the assertion is obvious. iv) The proof of 2.2.3) is left as an exercise.

After this we return to the set functions $\varphi : \mathfrak{S} \to \overline{\mathbb{R}}$. The first notion to be defined is that φ be additive in the appropriate sense; the traditional name for this is modular. We define $\varphi : \mathfrak{S} \to \overline{\mathbb{R}}$ to be **modular** $\dot{+}$ iff

$$\varphi(A \cup B) \dot{+} \varphi(A \cap B) = \varphi(A) \dot{+} \varphi(B) \quad \text{for all } A, B \in \mathfrak{S};$$

and furthermore to be **submodular/supermodular** $\dot{+}$ iff

$$\varphi(A \cup B) \dot{+} \varphi(A \cap B) \leq/\geq \varphi(A) \dot{+} \varphi(B) \quad \text{for all } A, B \in \mathfrak{S}.$$

No specification $\dot{+}$ is needed when φ attains at most one of the values $\pm\infty$. But otherwise the notions for $\dot{+}$ and $+$ do not coincide; see exercise 2.8 below. We also recall the older relatives of these notions. A set function $\varphi : \mathfrak{S} \to [0, \infty]$ on a lattice $\mathfrak{S}$ with $\varnothing \in \mathfrak{S}$ is called **additive** iff

$$\varphi(A \cup B) = \varphi(A) + \varphi(B) \quad \text{for all } A, B \in \mathfrak{S} \text{ with } A \cap B = \varnothing,$$

and **subadditive/superadditive** iff

$$\varphi(A \cup B) \leq/\geq \varphi(A) + \varphi(B) \quad \text{for all } A, B \in \mathfrak{S} \text{ with } A \cap B = \varnothing.$$

2.3. EXAMPLES. 1) For fixed $x \in X$ define the Dirac set function $\delta_x : \mathfrak{P}(X) \to [0, \infty[$ to be

$$\delta_x(A) = \left\{ \begin{array}{ll} 1 & \text{if } x \in A \\ 0 & \text{if } x \notin A \end{array} \right\} = \chi_A(x) \text{ for } A \subset X,$$

where as usual χ_A denotes the characteristic function of A. Then δ_x is modular. 2) Assume that $\mathfrak{S}$ is totally ordered under inclusion; see 1.2.7). Then each set function $\varphi : \mathfrak{S} \to \overline{\mathbb{R}}$ is modular $\dot{+}$.

2.4. EXERCISE. If $\varphi : \mathfrak{S} \to \overline{\mathbb{R}}$ is submodular $\dotplus$ and $\neq \infty$ then $[\varphi < \infty] := \{A \in \mathfrak{S} : \varphi(A) < \infty\}$ is a lattice. If φ is supermodular $\dotplus$ and $\neq -\infty$ then $[\varphi > -\infty] := \{A \in \mathfrak{S} : \varphi(A) > -\infty\}$ is a lattice.

2.5. EXERCISE. 1) Let $\varphi : \mathfrak{S} \to \mathbb{R}$ be modular. Prove for $A(1), \cdots A(r) \in \mathfrak{S}$ with $A := \bigcup_{l=1}^{r} A(l)$ that

$$\varphi(A) = \sum_{\varnothing \neq T \subset \{1, \cdots, r\}} (-1)^{\#(T)-1} \varphi\Big(\bigcap_{l \in T} A(l)\Big),$$

where $\#(T)$ denotes the number of elements of T. 2) Combine this with 2.3.1) to obtain

$$\chi_A = \sum_{\varnothing \neq T \subset \{1, \cdots, r\}} (-1)^{\#(T)-1} \prod_{l \in T} \chi_{A(l)}.$$

Next we define the set function $\varphi : \mathfrak{S} \to \overline{\mathbb{R}}$ to be **isotone** iff $\varphi(A) \leqq \varphi(B)$ for all $A \subset B$ in $\mathfrak{S}$. In the present text all set functions in the mainstream will be isotone until chapter VIII. Also the subsequent definitions will be phrased for isotone set functions.

The most important notion is that of a regular set function. Assume that $\mathfrak{M} \subset \mathfrak{S}$ and $\mathfrak{T} \subset \mathfrak{S}$. Then an isotone set function $\varphi : \mathfrak{S} \to \overline{\mathbb{R}}$ is called **outer regular $\mathfrak{M}$ at $\mathfrak{T}$** iff

$$\varphi(S) = \inf\{\varphi(M) : M \in \mathfrak{M} \text{ with } M \supset S\} \quad \text{for all } S \in \mathfrak{T},$$

and is called **inner regular $\mathfrak{M}$ at $\mathfrak{T}$** iff

$$\varphi(S) = \sup\{\varphi(M) : M \in \mathfrak{M} \text{ with } M \subset S\} \quad \text{for all } S \in \mathfrak{T},$$

with the usual conventions $\inf \varnothing := \infty$ and $\sup \varnothing := -\infty$. In the most frequent case $\mathfrak{T} = \mathfrak{S}$ we call φ **outer/inner regular $\mathfrak{M}$**.

Next we define an isotone set function $\varphi : \mathfrak{S} \to \overline{\mathbb{R}}$ to be **bounded above** iff $\varphi \leqq c$ for some $c \in \mathbb{R}$, to be **finite above** iff $\varphi < \infty$, and to be **semifinite above** iff it is inner regular $[\varphi < \infty]$. We likewise define the notions **bounded/finite/semifinite below**.

We turn to the important notion of a continuous set function. An isotone set function $\varphi : \mathfrak{S} \to \overline{\mathbb{R}}$ is called **upward σ continuous** iff

$$\varphi(S_l) \uparrow \varphi(A) \text{ for all sequences } (S_l)_l \text{ in } \mathfrak{S} \text{ with } S_l \uparrow A \in \mathfrak{S};$$

and **almost upward σ continuous** iff this holds true whenever $\varphi(S_l) > -\infty \,\forall l \in \mathbb{N}$. φ is called **upward τ continuous** iff

$$\sup_{S \in \mathfrak{M}} \varphi(S) = \varphi(A) \text{ for all pavings } \mathfrak{M} \subset \mathfrak{S} \text{ with } \mathfrak{M} \uparrow A \in \mathfrak{S};$$

and **almost upward τ continuous** iff this holds true whenever $\varphi(S) > -\infty \forall S \in \mathfrak{M}$. We likewise define the obvious **downward** counterparts. Here the exceptional value is of course ∞.

2.6. EXERCISE. Let $\bullet = \star\sigma\tau$. Let us redefine an isotone set function $\varphi : \mathfrak{S} \to \overline{\mathbb{R}}$ to be **upward $\bullet$ continuous** iff

$$\sup_{S \in \mathfrak{M}} \varphi(S) = \varphi(A) \text{ for all pavings } \mathfrak{M} \subset \mathfrak{S} \text{ of type } \bullet \text{ with } \mathfrak{M} \uparrow A \in \mathfrak{S},$$

and **almost upward $\bullet$ continuous** iff this holds true whenever $\varphi(S) > -\infty \; \forall S \in \mathfrak{M}$. Then these definitions coincide with the former ones when $\bullet = \sigma\tau$. In case $\bullet = \star$ the definitions are void since the required properties are always fulfilled, but are practical for the sake of a uniform treatment of the three cases. The **downward** counterparts are obvious as before.

At last we form for a set function $\varphi : \mathfrak{S} \to \overline{\mathbb{R}}$ the **upside-down transform** $\varphi\bot : \mathfrak{S}\bot \to \overline{\mathbb{R}}$ to be $\varphi\bot(T) = -\varphi(T')$ for $T \in \mathfrak{S}\bot$. It is obvious that $\varphi\bot\bot = \varphi$.

2.7. EXERCISE. Let $\varphi : \mathfrak{S} \to \overline{\mathbb{R}}$ be a set function. Prove the following equivalences. 1) φ modular $\dotplus \Leftrightarrow \varphi\bot$ modular $+$. Furthermore φ submodular/supermodular $\dotplus \Leftrightarrow \varphi\bot$ supermodular/submodular $+$. 2) φ isotone $\Leftrightarrow \varphi\bot$ isotone. For the remainder we assume that φ be isotone. 3) Let $\mathfrak{M} \subset \mathfrak{S}$ and $\mathfrak{T} \subset \mathfrak{S}$. Then φ outer regular $\mathfrak{M}$ at $\mathfrak{T} \Leftrightarrow \varphi\bot$ inner regular $\mathfrak{M}\bot$ at $\mathfrak{T}\bot$. 4) φ bounded/finite/semifinite above $\Leftrightarrow \varphi\bot$ bounded/finite/semifinite below. 5) φ (almost) upward $\bullet$ continuous $\Leftrightarrow \varphi\bot$ (almost) downward $\bullet$ continuous.

2.8. EXERCISE. 1) Define on a set X of two elements an isotone set function $\varphi : \mathfrak{P}(X) \to \overline{\mathbb{R}}$ which is modular $+$ but not submodular $\dotplus$. In particular the notions modular $\dotplus$ and $+$ do not coincide. 2) Prove that each isotone set function $\varphi : \mathfrak{S} \to \overline{\mathbb{R}}$ fulfils

$$\varphi \text{ submodular } \dotplus \Rightarrow \varphi \text{ submodular } +.$$

Note that the converse implication is false by 1). Thus it seems that the combination of submodular with $\dotplus$ is in a sense superior to the combination of submodular with $+$. See also 2.4 and 4.1.5) below.

2.9. EXERCISE. Construct examples of set functions $\varphi : \mathfrak{S} \to \{0, 1\}$ on algebras $\mathfrak{S}$ which are isotone and modular and fulfil

  1) φ is not upward σ continuous;
  2) φ is upward σ continuous but not upward τ continuous.

Hint for 1): Let X be an infinite countable set, and let $\mathfrak{S}$ consist of those subsets which are either finite or cofinite (:=of finite complement). Define $\varphi : \mathfrak{S} \to \{0, 1\}$ to be $\varphi(A) = 0$ if A is finite and $\varphi(A) = 1$ if A is cofinite. Hint for 2): Similar construction on an uncountable set X. Define the cocountable subsets of X!

Contents and Measures

We come to the two central notions of measure theory. In the present text the conventional versions will be specializations of more comprehensive new versions. However, it will be seen that in a sense the two versions are equivalent.

We define a **content** $\dot{+}$ (or $\dot{+}$ content) to be a set function $\varphi : \mathfrak{S} \to \overline{\mathbb{R}}$ on an oval $\mathfrak{S}$ which is modular $\dot{+}$ and isotone and attains at least one finite value. No specification $\dot{+}$ is needed when φ attains at most one of the values $\pm\infty$. In the special case that

$$\varnothing \in \mathfrak{S}, \qquad \text{which means that } \mathfrak{S} \text{ is a ring, and}$$
$$\varphi(\varnothing) = 0, \qquad \text{which implies that } \varphi : \mathfrak{S} \to [0, \infty],$$

we speak of a **conventional content**, in short **ccontent**.

2.10. REMARK. *Let $\mathfrak{S}$ be a ring. A set function $\varphi : \mathfrak{S} \to [0, \infty]$ is a ccontent iff it is additive and attains at least one finite value.*

Proof. We have to show that the two conditions are sufficient. i) Let $A \in \mathfrak{S}$ with $\varphi(A) \in \mathbb{R}$. For $B := \varnothing$ then $\varphi(A) = \varphi(A) + \varphi(\varnothing)$ and hence $\varphi(\varnothing) = 0$. ii) For $A, B \in \mathfrak{S}$ we have the disjoint decompositions $A \cup B = A \cup (B \cap A')$ and $B = (A \cap B) \cup (B \cap A')$, the members of which are in $\mathfrak{S}$ since $\mathfrak{S}$ is a ring. It follows that

$$\varphi(A \cup B) + \varphi(A \cap B) = \big(\varphi(A) + \varphi(B \cap A')\big) + \varphi(A \cap B)$$
$$= \varphi(A) + \big(\varphi(B \cap A') + \varphi(A \cap B)\big) = \varphi(A) + \varphi(B).$$

iii) For $A \subset B$ in $\mathfrak{S}$ we have the disjoint decomposition $B = A \cup (B \setminus A)$ and $\varphi(B) = \varphi(A) + \varphi(B \setminus A) \geq \varphi(A)$. Thus φ is isotone.

We define a **measure** $\dot{+}$ (or $\dot{+}$ measure) to be a set function $\varphi : \mathfrak{S} \to \overline{\mathbb{R}}$ on a σ oval $\mathfrak{S}$ which is a content $\dot{+}$ and both almost upward σ continuous and almost downward σ continuous. In the special case that

$$\varnothing \in \mathfrak{S}, \qquad \text{which means that } \mathfrak{S} \text{ is a } \sigma \text{ ring, and}$$
$$\varphi(\varnothing) = 0, \qquad \text{which implies that } \varphi : \mathfrak{S} \to [0, \infty],$$

we speak of a **conventional measure**, in short **cmeasure**. In this special case it is well-known that the assumption that φ be almost downward σ continuous is redundant. This fact extends to the present context as follows.

2.11. PROPOSITION. *Let $\varphi : \mathfrak{S} \to \overline{\mathbb{R}}$ be a content $\dot{+}$ on an oval. If φ is semifinite below then*

$$\varphi \text{ almost upward } \sigma \text{ continuous} \Rightarrow \varphi \text{ almost downward } \sigma \text{ continuous.}$$

If φ is semifinite above then

$$\varphi \text{ almost upward } \sigma \text{ continuous} \Leftarrow \varphi \text{ almost downward } \sigma \text{ continuous.}$$

Proof. In view of the upside-down transform method it suffices to prove the first assertion. Let us fix $(S_l)_l$ in $\mathfrak{S}$ with $S_l \downarrow S \in \mathfrak{S}$ and $\varphi(S_l) < \infty \; \forall l \in \mathbb{N}$. To be shown is $\varphi(S_l) \downarrow \varphi(S)$. i) We first prove this under the assumption that $\varphi(S) > -\infty$. Then $\varphi(S_l), \varphi(S) \in \mathbb{R}$. We form $D_l := S_1 | S_l | S = (S_1 \cap S_l') \cup S \in \mathfrak{S}$. Then $S \subset D_l \subset S_1$ and hence $\varphi(D_l) \in \mathbb{R}$. Furthermore $S_l \cup D_l = S_1$ and $S_l \cap D_l = S$ and hence

$$\varphi(S_1) + \varphi(S) = \varphi(S_l) + \varphi(D_l).$$

Now $D_l \uparrow (S_1 \cap S') \cup S = S_1$ and hence by assumption $\varphi(D_l) \uparrow \varphi(S_1)$. It follows that $\varphi(S_l) \downarrow \varphi(S)$. ii) Now we assume that $\varphi(S) = -\infty$. We can also assume that $\varphi(S_l) > -\infty$ and hence $\varphi(S_l) \in \mathbb{R} \, \forall l \in \mathbb{N}$, since otherwise the assertion were obvious. We fix a set $T \in \mathfrak{S}$ such that $T \supset S$ and $\varphi(T) \in \mathbb{R}$. Since φ is modular $\dotplus$ we then have $\varphi(S_l \cup T), \varphi(S_l \cap T) \in \mathbb{R}$ and

$$\varphi(S_l) + \varphi(T) = \varphi(S_l \cup T) + \varphi(S_l \cap T).$$

From i) applied to $S_l \cup T \downarrow T$ we obtain $\varphi(S_l \cup T) \downarrow \varphi(T)$. It follows that

$$\lim_{l \to \infty} \varphi(S_l) = \lim_{l \to \infty} \varphi(S_l \cap T) \leqq \varphi(T).$$

Now since φ is downward semifinite the values $\varphi(T)$ have the infimum $-\infty$. Therefore $\varphi(S_l) \downarrow -\infty = \varphi(S)$.

2.12. EXERCISE. The two semifiniteness assumptions in 2.11 cannot be interchanged, even if one fortifies semifinite to bounded. As an example construct a ccontent $\varphi : \mathfrak{P}(\mathbb{N}) \to [0, \infty]$ which is almost downward σ continuous but not (almost) upward σ continuous.

2.13. EXERCISE. 1) Let $\alpha : \mathfrak{A} \to [0, \infty]$ be a ccontent on an algebra $\mathfrak{A}$. Define $\eta : \mathfrak{P}(X) \to [0, \infty]$ to be

$$\eta(A) = \left\{ \begin{array}{ll} 0 & \text{if } A \text{ is upward enclosable } [\alpha < \infty] \\ \infty & \text{if not} \end{array} \right\}.$$

Then η is a ccontent. 2) Let $\alpha : \mathfrak{A} \to [0, \infty]$ be a cmeasure on a σ algebra $\mathfrak{A}$. Define $\eta : \mathfrak{P}(X) \to [0, \infty]$ to be

$$\eta(A) = \left\{ \begin{array}{ll} 0 & \text{if } A \text{ is upward enclosable } [\alpha < \infty]^\sigma \\ \infty & \text{if not} \end{array} \right\}.$$

Then η is a cmeasure.

New versus Conventional Contents and Measures

The present subsection is of more theoretical interest and will not be used before chapter VIII. We shall establish a correspondence between the new and the conventional contents and measures which is almost one-to-one. This correspondence expresses the new entities in terms of the conventional ones. Also it makes clear how to transfer to them certain notions for which the transfer is not obvious, for example the notion of null sets.

Let $\mathfrak{S}$ be an oval in X. The treatment will be based on the last subsection of section 1. We recall from 1.24 the ring $a(\mathfrak{S}) \subset R(\mathfrak{S}) \subset A(\mathfrak{S})$ and its properties.

2.14. PROPOSITION. *Let $\varphi : \mathfrak{S} \to \overline{\mathbb{R}}$ be a content $\dotplus$ on the oval $\mathfrak{S}$. Fix $P \in \mathfrak{S}$ with $\varphi(P) \in \mathbb{R}$ and define $\varphi^\wedge : A(\mathfrak{S}) \to [0, \infty]$ to be*

$$\varphi^\wedge(A) = \left\{ \begin{array}{ll} \varphi(P \cup A) + \left(-\varphi(P \cap A')\right) & \text{if } A \in a(\mathfrak{S}) \\ \infty & \text{if } A \notin a(\mathfrak{S}) \end{array} \right\};$$

note that in case $A \in a(\mathfrak{S})$ we have $P \cup A, P \cap A' \in \mathfrak{S}$ by 1.24.3), and $\varphi(P \cup A) \geqq \varphi(P) > -\infty$ and $-\varphi(P \cap A') \geqq -\varphi(P) > -\infty$, so that the above sum is defined and in $[0, \infty]$. *Then $\varphi^\wedge$ is a ccontent on the algebra* $A(\mathfrak{S})$ *and independent of P.*

If in particular $\mathfrak{S}$ is a ring and $\varphi : \mathfrak{S} \to [0, \infty]$ is a ccontent, then we can take $P := \varnothing$ and obtain $\varphi^\wedge(A) = \varphi(A)$ if $A \in a(\mathfrak{S}) = \mathfrak{S}$ and $\varphi^\wedge(A) = \infty$ if $A \notin \mathfrak{S}$.

Proof. i) It is obvious that $\varphi^\wedge(\varnothing) = 0$ and that $\varphi^\wedge$ is isotone. ii) In order to prove that φ is modular it suffices to consider $A, B \in a(\mathfrak{S})$ since otherwise both sides of the assertion are $= \infty$. Then $A \cup B, A \cap B \in a(\mathfrak{S})$ as well. We obtain

$$
\begin{aligned}
\varphi^\wedge(A) + \varphi^\wedge(B) &= \varphi(P \cup A) + \big(-\varphi(P \cap A')\big) \\
&+ \varphi(P \cup B) + \big(-\varphi(P \cap B')\big) \\
&= \varphi(P \cup (A \cup B)) + \varphi(P \cup (A \cap B)) \\
&+ \Big(-\varphi(P \cap (A \cup B)') - \varphi(P \cap (A \cap B)')\Big) \\
&= \varphi^\wedge(A \cup B) + \varphi^\wedge(A \cap B),
\end{aligned}
$$

where we made repeated use of the fact noted above that all terms are $> -\infty$. iii) For the last assertion fix $P, Q \in \mathfrak{S}$ with $\varphi(P), \varphi(Q) \in \mathbb{R}$. To be shown is

$$
\varphi(P \cup A) + \big(-\varphi(P \cap A')\big) = \varphi(Q \cup A) + \big(-\varphi(Q \cap A')\big) \text{ for } A \in a(\mathfrak{S}).
$$

Now we have

$$
\varphi(P \cup A) \dotplus \varphi(Q \cap A') = \varphi(P \cup Q \cup A) \dotplus \varphi(P \cap Q \cap A'),
$$

because $(P \cup A) \cup (Q \cap A') = P \cup Q \cup A$ and $(P \cup A) \cap (Q \cap A') = P \cap Q \cap A'$. Since the right side is symmetric in P and Q it follows that

$$
\varphi(P \cup A) \dotplus \varphi(Q \cap A') = \varphi(Q \cup A) \dotplus \varphi(P \cap A').
$$

In case $\dotplus$ we deduce that $\varphi(P \cup A) = \infty \Leftrightarrow \varphi(Q \cup A) = \infty$, and then the assertion is clear. If otherwise $\varphi(P \cup A), \varphi(Q \cup A) \in \mathbb{R}$ then it is clear as well. In case $\dotplus$ we deduce that $\varphi(Q \cap A') = -\infty \Leftrightarrow \varphi(P \cap A') = -\infty$, and then the assertion is clear. If otherwise $\varphi(Q \cap A'), \varphi(P \cap A') \in \mathbb{R}$ then it is clear as well. The proof is complete.

We next define a collection of maps with the opposite direction.

2.15. REMARK. *Let $\phi : A(\mathfrak{S}) \to [0, \infty]$ be a ccontent. Fix $P \in \mathfrak{S}$ and define $\phi^P : \mathfrak{S} \to \overline{\mathbb{R}}$ to be*

$$
\phi^P(A) = \phi(A \cap P') \dotplus \big(-\phi(A' \cap P)\big) \quad \text{for } A \in \mathfrak{S}.
$$

Then ϕ^P is a content $\dotplus$ on $\mathfrak{S}$ with $\phi^P(P) = 0$.

Proof. i) It is obvious that $\phi^P(P) = 0$ and that ϕ^P is isotone. ii) To prove that ϕ^P is modular $\dotplus$ let $A, B \in \mathfrak{S}$. Then

$$
\begin{aligned}
\phi^P(A) \dotplus \phi^P(B) &= \phi(A \cap P') \dotplus \phi(B \cap P') \\
&\dotplus \left(-\phi(A' \cap P) \right) \dotplus \left(-\phi(B' \cap P) \right) \\
&= \left(\phi(A \cap P') + \phi(B \cap P') \right) \\
&\dotplus \left(-\phi(A' \cap P) - \phi(B' \cap P) \right) \\
&= \left(\phi((A \cup B) \cap P') + \phi((A \cap B) \cap P') \right) \\
&\dotplus \left(-\phi((A \cup B)' \cap P) - \phi((A \cap B)' \cap P) \right) \\
&= \phi((A \cup B) \cap P') \dotplus \phi((A \cap B) \cap P') \\
&\dotplus \left(-\phi((A \cup B)' \cap P) \right) \dotplus \left(-\phi((A \cap B)' \cap P) \right) \\
&= \phi^P(A \cup B) \dotplus \phi^P(A \cap B).
\end{aligned}
$$

It turns out that the two processes defined above are inverse to each other.

2.16. THEOREM. *Let $\mathfrak{S}$ be an oval.* 1) *Let $\varphi : \mathfrak{S} \to \overline{\mathbb{R}}$ be a content $\dotplus$; thus $\varphi^\wedge : A(\mathfrak{S}) \to [0, \infty]$ is a ccontent with $\varphi^\wedge(A) = \infty$ when $A \notin a(\mathfrak{S})$. For each $P \in \mathfrak{S}$ with $\varphi(P) \in \mathbb{R}$ then $(\varphi^\wedge)^P = \varphi - \varphi(P)$.* 2) *Let $\phi : A(\mathfrak{S}) \to [0, \infty]$ be a ccontent with $\phi(A) = \infty$ when $A \notin a(\mathfrak{S})$; thus for each $P \in \mathfrak{S}$ the set function $\phi^P : \mathfrak{S} \to \overline{\mathbb{R}}$ is a content $\dotplus$ with $\phi^P(P) = 0$. Then $(\phi^P)^\wedge = \phi$.*

Proof. 1) For $A \in \mathfrak{S}$ we have

$$
(\varphi^\wedge)^P(A) = \varphi^\wedge(A \cap P') \dotplus \left(-\varphi^\wedge(A' \cap P) \right).
$$

By definition $A \cap P', A' \cap P \in a(\mathfrak{S})$ and hence

$$
\begin{aligned}
\varphi^\wedge(A \cap P') &= \varphi(P \cup (A \cap P')) + \left(-\varphi(P \cap (A' \cup P)) \right) \\
&= \varphi(P \cup A) - \varphi(P), \\
\varphi^\wedge(A' \cap P) &= \varphi(P \cup (A' \cap P)) + \left(-\varphi(P \cap (A \cup P')) \right) \\
&= \varphi(P) - \varphi(P \cap A).
\end{aligned}
$$

It follows that

$$
\begin{aligned}
(\varphi^\wedge)^P(A) &= \left(\varphi(P \cup A) - \varphi(P) \right) \dotplus \left(\varphi(P \cap A) - \varphi(P) \right) \\
&= \varphi(P \cup A) \dotplus \varphi(P \cap A) - 2\varphi(P) \\
&= \varphi(P) \dotplus \varphi(A) - 2\varphi(P) = \varphi(A) - \varphi(P).
\end{aligned}
$$

2) We have to prove that $(\phi^P)^\wedge(A) = \phi(A)$ for $A \in a(\mathfrak{S})$. In view of $\phi^P(P) = 0$ we have $(\phi^P)^\wedge(A) = \phi^P(P \cup A) + (-\phi^P(P \cap A'))$. Here

$$
\begin{aligned}
\phi^P(P \cup A) &= \phi((P \cup A) \cap P') \dotplus \left(-\phi((P' \cap A') \cap P) \right) \\
&= \phi(A \cap P'), \\
\phi^P(P \cap A') &= \phi((P \cap A') \cap P') \dotplus \left(-\phi((P' \cup A) \cap P) \right) \\
&= -\phi(A \cap P).
\end{aligned}
$$

It follows that $(\phi^P)^\wedge(A) = \phi(A \cap P') + \phi(A \cap P) = \phi(A)$.

2.17. CONSEQUENCE. *Let $\mathfrak{S}$ be an oval and $P \in \mathfrak{S}$. There is a one-to-one correspondence between the set functions $\varphi : \mathfrak{S} \to \overline{\mathbb{R}}$ which are contents $\dotplus$ with $\varphi(P) = 0$, and the set functions $\phi : A(\mathfrak{S}) \to [0, \infty]$ which are ccontents with $\phi(A) = \infty$ when $A \notin a(\mathfrak{S})$. The correspondence is $\phi = \varphi^\wedge$ and $\varphi = \phi^P$.*

We continue with the discussion of null sets. If $\varphi : \mathfrak{S} \to [0, \infty]$ is a ccontent on a ring $\mathfrak{S}$ then it is common to define the null sets for φ to be the sets $A \in \mathfrak{S}$ with $\varphi(A) = 0$. If $\varphi : \mathfrak{S} \to \overline{\mathbb{R}}$ is a content $\dotplus$ on an oval $\mathfrak{S}$ then this definition does not make sense. After the above results it is reasonable to define the **null sets** for φ to be the sets $A \in A(\mathfrak{S})$ with $\varphi^\wedge(A) = 0$. Of course then $A \in a(\mathfrak{S})$. There are some useful reformulations.

2.18. REMARK. *For a subset $A \in a(\mathfrak{S})$ the following are equivalent. 0) $\varphi^\wedge(A) = 0$. 1) $\varphi(S \cup A) = \varphi(S)$ for all $S \in \mathfrak{S}$. 2) $\varphi(S \cap A') = \varphi(S)$ for all $S \in \mathfrak{S}$. Note that 1)2) make sense in view of 1.24.3).*

Proof. Fix $P \in \mathfrak{S}$ with $\varphi(P) \in \mathbb{R}$. Then

$$
\varphi^\wedge(A) = \left(\varphi(P \cup A) - \varphi(P) \right) + \left(\varphi(P) - \varphi(P \cap A') \right),
$$

so that $\varphi^\wedge(A) = 0$ is equivalent to $\varphi(P \cup A) = \varphi(P) = \varphi(P \cap A')$.
1)$\Rightarrow$0) For $S := P \cap A' \in \mathfrak{S}$ we obtain $\varphi(P \cup A) = \varphi(P) = \varphi(P \cap A')$.
2)$\Rightarrow$0) For $S := P \cup A \in \mathfrak{S}$ we obtain $\varphi(P \cap A') = \varphi(P \cup A)$.
0)$\Rightarrow$1)2) For $U, V \in \mathfrak{S}$ we have

$$
\begin{aligned}
\varphi(U \cup A) \;&\dotplus\; \varphi(V \cap A') \\
&= \varphi((U \cup A) \cup (V \cap A')) \dotplus \varphi((U \cup A) \cap (V \cap A')) \\
&= \varphi(U \cup V \cup A) \dotplus \varphi(U \cap V \cap A').
\end{aligned}
$$

Since the right side is symmetric in U and V it follows that

$$
\varphi(U \cup A) \dotplus \varphi(V \cap A') = \varphi(V \cup A) \dotplus \varphi(U \cap A').
$$

Now put $V := P$ and note that $\varphi(P \cap A') = \varphi(P) = \varphi(P \cup A)$. It follows that $\varphi(U \cup A) = \varphi(U \cap A')$ for all $U \in \mathfrak{S}$. This implies 1)2).

It remains to establish the connection between the new and the conventional measures. The next two assertions are routine and will be left as exercises.

2.19. EXERCISE. Let $\varphi : \mathfrak{S} \to \overline{\mathbb{R}}$ be a content $\dotplus$ on the oval $\mathfrak{S}$ which is almost upward σ continuous and almost downward $\dot\sigma$ continuous. 1) $\varphi^\wedge | a(\mathfrak{S})$ is upward σ continuous. 2) $\varphi^\wedge$ is upward σ continuous whenever $a(\mathfrak{S})$ is a σ ring. 3) $\varphi^\wedge$ need not be upward σ continuous.

2.20. EXERCISE. Let $\phi : A(\mathfrak{S}) \to [0, \infty]$ be a ccontent such that $\phi | a(\mathfrak{S})$ is upward σ continuous. For each $P \in \mathfrak{S}$ then $\phi^P : \mathfrak{S} \to \overline{\mathbb{R}}$ is almost upward σ continuous and almost downward σ continuous.

2.21. CONSEQUENCE. *Let $\mathfrak{S}$ be a σ oval, so that $A(\mathfrak{S})$ is a σ algebra by 1.25, and let $P \in \mathfrak{S}$. Then there is a one-to-one correspondence between the set functions $\varphi : \mathfrak{S} \to \overline{\mathbb{R}}$ which are measures $\dotplus$ with $\varphi(P) = 0$, and the set functions $\phi : A(\mathfrak{S}) \to [0, \infty]$ which are cmeasures with $\phi(A) = \infty$ when $A \notin a(\mathfrak{S})$. The correspondence is $\phi = \varphi^\wedge$ and $\varphi = \phi^P$.*

Proof. Combine the former correspondence 2.17 with 2.19 and 2.20.

The Main Example: The Volume in $\mathbb{R}^n$

Let $\mathfrak{K} = \mathfrak{K}^n = \mathrm{Comp}(\mathbb{R}^n)$ denote the system of the compact subsets of $\mathbb{R}^n$. We shall define and explore the set function $\lambda = \lambda^n : \mathfrak{K} \to [0, \infty[$ which reflects the naive notion of volume in $\mathbb{R}^n$. It will later become the basis for the Lebesgue measure. We emphasize that its domain $\mathfrak{K} = \mathfrak{K}^n$ is a lattice which is by far not a ring, and which appears to be the most natural domain for the naive volume, at best besides the lattice of the finite unions of compact intervals. Thus the present procedure is in accordance with the systematic theories of the next chapter, which start from set functions defined on lattices.

We define $\mathfrak{W}_s (s = 0, 1, 2, \cdots)$ to consist of those compact cubes $Q \subset \mathbb{R}^n$ which arise under s consecutive midpoint subdivisions from the compact cubes with corners in $\mathbb{Z}^n$, that is of the subsets

$$Q(p) := \left\{ x = (x_1, \cdots, x_n) \in \mathbb{R}^n : \frac{1}{2^s}(p_l - 1) \leqq x_l \leqq \frac{1}{2^s} p_l \, (l = 1, \cdots, n) \right\}$$

for $p = (p_1, \cdots, p_n) \in \mathbb{Z}^n$. Let now $K \in \mathfrak{K}$. We define

$$Z_s(K) := \#(\{Q \in \mathfrak{W}_s : Q \cap K \neq \varnothing\}) \, (s = 0, 1, 2, \cdots).$$

Thus the naive interpretation of $\frac{1}{2^{ns}} Z_s(K)$ is the sum of the volumes of the $Q \in \mathfrak{W}_s$ which meet K. It follows that

$$Z_{s+1}(K) \leqq 2^n Z_s(K) \text{ and hence } \frac{1}{2^{n(s+1)}} Z_{s+1}(K) \leqq \frac{1}{2^{ns}} Z_s(K).$$

In fact, under midpoint subdivision each $Q \in \mathfrak{W}_s$ produces 2^n cubes in $\mathfrak{W}_{s+1}$, and $Q \cap K \neq \varnothing$ iff at least one of these subcubes meets K. We define

$$\lambda(K) = \lambda^n(K) := \lim_{s \to \infty} \frac{1}{2^{ns}} Z_s(K) = \inf \left\{ \frac{1}{2^{ns}} Z_s(K) : s \geqq 0 \right\}.$$

It is not hard to explore the basic properties of this set function. It is obvious that $\lambda(\varnothing) = 0$.

2.22. REMARK. *Let $A \in \mathfrak{K}^m$ and $B \in \mathfrak{K}^n$ and hence $A \times B \in \mathfrak{K}^{m+n}$. Then $\lambda^{m+n}(A \times B) = \lambda^m(A)\lambda^n(B)$.*

Proof. In obvious notation the $Q \in \mathfrak{W}_s^{m+n}$ are the products $Q = U \times V$ of $U \in \mathfrak{W}_s^m$ and $V \in \mathfrak{W}_s^n$. We have $Q \cap (A \times B) \neq \varnothing \Leftrightarrow U \cap A \neq \varnothing$ and $V \cap B \neq \varnothing$. It follows that $Z_s^{m+n}(A \times B) = Z_s^m(A)Z_s^n(B)$. The assertion is now obvious.

2.23. REMARK. *For a compact interval*

$$K = [a, b] := \{x \in \mathbb{R}^n : a_l \leqq x_l \leqq b_l \ (l = 1, \cdots, n)\},$$

where $a, b \in \mathbb{R}^n$ with $a_l \leqq b_l \ (l = 1, \cdots, n)$, we have

$$\lambda(K) = (b_1 - a_1) \cdots (b_n - a_n).$$

Proof. In view of 2.22 we can restrict ourselves to the case $n = 1$. i) We first prove the assertion: For an interval $[a, b] \subset \mathbb{R}$ the number $N := \#(\mathbb{Z} \cap [a, b])$ fulfils $(b - a) - 1 < N \leqq (b - a) + 1$. In fact, take the $m, n \in \mathbb{Z}$ with $m - 1 < a \leqq m$ and $n \leqq b < n + 1$. Then $N = n - m + 1$. But the above inequalities can be written $-a < -m + 1 \leqq -a + 1$ and $b - 1 < n \leqq b$. The assertion follows by addition. ii) Fix $s \geqq 0$. For $p \in \mathbb{Z}$ then $Q(p) \cap [a, b] \neq \varnothing$ is equivalent to $a \leqq \frac{1}{2^s}p$ and $\frac{1}{2^s}(p - 1) \leqq b$, hence to $2^s a \leqq p \leqq 2^s b + 1$. By i) we have $2^s(b - a) < Z_s(K) \leqq 2^s(b - a) + 2$. The assertion follows.

2.24. REMARK. *λ is isotone.* This is obvious.

2.25. PROPOSITION. *λ is downward τ continuous.*

Proof. Let $\mathfrak{M} \subset \mathfrak{K}$ be a paving with $\mathfrak{M} \downarrow A$, so that $A \in \mathfrak{K}$ as well. We have $\inf\{\lambda(M) : M \in \mathfrak{M}\} =: R \geqq \lambda(A)$; to be shown is $R \leqq \lambda(A)$. 1) Let $V \subset \mathbb{R}^n$ be open with $A \subset V$. Then $M \subset V$ for some $M \in \mathfrak{M}$. In fact, from $\{M \cap V' : M \in \mathfrak{M}\} \downarrow A \cap V' = \varnothing$ it follows that $M \cap V' = \varnothing$ for some $M \in \mathfrak{M}$. 2) Fix integers $0 \leqq s \leqq t$. We pass from $Q = Q(p) \in \mathfrak{W}_s$ to a larger compact cube $Q^\sim$, in that we extend each coordinate interval by $\frac{1}{2^t}$ on either side. Thus

$$
\begin{aligned}
Q^\sim : \frac{1}{2^s}(p_l - 1) - \frac{1}{2^t} &= \frac{1}{2^t}\left(2^{t-s}(p_l - 1) - 1\right) \leqq x_l \\
&\leqq \frac{1}{2^s}p_l + \frac{1}{2^t} = \frac{1}{2^t}\left(2^{t-s}p_l + 1\right) \ (l = 1, \cdots, n).
\end{aligned}
$$

We have $Q \subset \mathrm{Int}Q^\sim \subset Q^\sim$; and $Q^\sim$ consists of

$$\prod_{l=1}^n \left(\left(2^{t-s}p_l + 1\right) - \left(2^{t-s}(p_l - 1) - 1\right)\right) = \left(2^{t-s} + 2\right)^n \text{ cubes } \in \mathfrak{W}_t.$$

Now from

$$A \subset \bigcup_{Q \in \mathfrak{W}_s, Q \cap A \neq \varnothing} Q \subset \bigcup_{Q \in \mathfrak{W}_s, Q \cap A \neq \varnothing} \mathrm{Int}Q^\sim$$

and from 1) we obtain an $M \in \mathfrak{M}$ such that

$$M \subset \bigcup_{Q \in \mathfrak{W}_s, Q \cap A \neq \varnothing} \mathrm{Int}Q^\sim \subset \bigcup_{Q \in \mathfrak{W}_s, Q \cap A \neq \varnothing} Q^\sim.$$

The set on the right is a union of cubes from $\mathfrak{W}_t$, the number of which is $\leqq (2^{t-s} + 2)^n Z_s(A)$. It follows that

$$
\begin{aligned}
R \leqq \lambda(M) \leqq \frac{1}{2^{nt}} Z_t(M) &\leqq \frac{1}{2^{nt}} (2^{t-s} + 2)^n Z_s(A) \\
&= \left(\frac{1}{2^s} + \frac{2}{2^t}\right)^n Z_s(A).
\end{aligned}
$$

This holds true for all $0 \leqq s \leqq t$. We let $t \to \infty$ and obtain $R \leqq \frac{1}{2^{ns}} Z_s(A)$ for all $s \geqq 0$. It follows that $R \leqq \lambda(A)$.

2.26. PROPOSITION. λ is modular.

For nonvoid $K \in \mathfrak{K}$ and $\delta > 0$ we form $K(\delta) := \{x \in \mathbb{R}^n : \mathrm{dist}(x, K) \leqq \delta\}$. Thus $K(\delta) \in \mathfrak{K}$ and $K \subset \mathrm{Int}K(\delta)$. Furthermore $K(\delta) \downarrow K$ for $\delta \downarrow 0$.

Proof. Fix nonvoid $A, B \in \mathfrak{K}$. For $s \geqq 0$ we have $Z_s(A \cup B) + N_s = Z_s(A) + Z_s(B)$ with

$$
N_s := \#(\{Q \in \mathfrak{W}_s : Q \cap A, Q \cap B \neq \varnothing\}).
$$

Therefore

$$
\frac{1}{2^{ns}} N_s \to \lambda(A) + \lambda(B) - \lambda(A \cup B) =: D \quad \text{for } s \to \infty.
$$

Now we have $Z_s(A \cap B) \leqq N_s \leqq Z_s(A(\delta) \cap B)$ for all $\delta \geqq \frac{1}{2^s} \sqrt{n}$. We let $s \to \infty$ and obtain $\lambda(A \cap B) \leqq D \leqq \lambda(A(\delta) \cap B)$ for all $\delta > 0$. In view of 2.25 we have $\lambda(A(\delta) \cap B) \downarrow \lambda(A \cap B)$ for $\delta \downarrow 0$. Hence $D = \lambda(A \cap B)$, and the assertion follows.

2.27. PROPOSITION. λ is upward σ continuous.

Proof. Let $(A_l)_l$ be a sequence in $\mathfrak{K}$ with $A_l \uparrow A \in \mathfrak{K}$. We have $\lambda(A_l) \uparrow$ some $R \leqq \lambda(A)$; to be shown is $R \geqq \lambda(A)$. Let us fix $\varepsilon > 0$. 1) There exists a sequence $(B_l)_l$ in $\mathfrak{K}$ with the properties

i) $A_l \subset \mathrm{Int}B_l \subset B_l$;
ii) $\lambda(B_l) < \lambda(A_l) + \varepsilon\left(1 - \frac{1}{2^l}\right)$;
iii) $B_l \uparrow$.

First note that by 2.25 there exists $\delta_l > 0$ such that $\lambda(A_l(\delta_l)) < \lambda(A_l) + \frac{\varepsilon}{2^l}$. We put $B_l := A_1(\delta_1) \cup \cdots \cup A_l(\delta_l) \in \mathfrak{K}$. Then i)iii) are clear. We prove ii) via induction. For $l = 1$ the assertion is clear. To see $1 \leqq l \Rightarrow l + 1$ note that $B_{l+1} = B_l \cup A_{l+1}(\delta_{l+1})$. From 2.26 and the induction hypothesis we see that

$$
\begin{aligned}
&\lambda(B_{l+1}) - \lambda(A_{l+1}) \\
={}& \lambda(B_l) + \lambda(A_{l+1}(\delta_{l+1})) - \lambda(B_l \cap A_{l+1}(\delta_{l+1})) - \lambda(A_{l+1}) \\
\leqq{}& \left(\lambda(B_l) - \lambda(A_l)\right) + \lambda(A_{l+1}(\delta_{l+1})) - \lambda(A_{l+1}) \\
<{}& \varepsilon\left(1 - \frac{1}{2^l}\right) + \frac{\varepsilon}{2^{l+1}} = \varepsilon\left(1 - \frac{1}{2^{l+1}}\right),
\end{aligned}
$$

where $\leqq$ follows from $B_l \cap A_{l+1}(\delta_{l+1}) \supset A_l$. 2) Now from $A = \bigcup\limits_{l=1}^{\infty} A_l \subset$ $\bigcup\limits_{l=1}^{\infty} \mathrm{Int}(B_l)$ we obtain an $l \in \mathbb{N}$ such that $A \subset \mathrm{Int}B_l \subset B_l$. It follows that $\lambda(A) \leqq \lambda(B_l) < \lambda(A_l) + \varepsilon \leqq R + \varepsilon$. This holds true for all $\varepsilon > 0$. For $\varepsilon \downarrow 0$ we obtain the assertion.

3. Some Classical Extension Theorems for Set Functions

The fundamental extension theorems of the next chapter will be based on characteristic combinations of regularity and continuity requirements. The classical extension theorems of the present section are important results as well, but of different kind because regularity is not involved. The section consists of three independent subsections. We assume that X is a nonvoid set.

The Classical Uniqueness Theorem

The classical uniqueness theorem will be the first substantial application of the transporter theorem 1.16. It will be obtained in $\star$ and σ versions.

3.1. THEOREM. $\star$) *Let* $\varphi, \psi : \mathfrak{A} \to [0, \infty]$ *be ccontents on a ring* $\mathfrak{A}$ *in* X. *Let* $\mathfrak{S} \subset \mathfrak{A}$ *be a paving such that* $\mathfrak{S} \subset \mathrm{A}(\mathfrak{S}\top)$ *(this weakens the requirement* $\mathfrak{S} \subset \mathfrak{S}\top$ *which means that* $\mathfrak{S}$ *has* $\cap$ *). If* $\varphi(S) = \psi(S) < \infty$ *for all* $S \in \mathfrak{S}$ *then* $\varphi = \psi$ *on* $\mathrm{R}(\mathfrak{S})$.

σ) *Let* $\varphi, \psi : \mathfrak{A} \to [0, \infty]$ *be cmeasures on a* σ *ring* $\mathfrak{A}$ *in* X. *Let* $\mathfrak{S} \subset \mathfrak{A}$ *be a paving such that* $\mathfrak{S} \subset \mathrm{A}\sigma(\mathfrak{S}\top)$ *(this weakens the requirement* $\mathfrak{S} \subset \mathfrak{S}\top$ *which means that* $\mathfrak{S}$ *has* $\cap$ *). If* $\varphi(S) = \psi(S) < \infty$ *for all* $S \in \mathfrak{S}$ *then* $\varphi = \psi$ *on* $\mathrm{R}\sigma(\mathfrak{S})$.

Proof. Define $\mathfrak{N}$ to consist of all subsets $A \in \mathfrak{A}$ with $\varphi(A) = \psi(A) < \infty$. Then $\mathfrak{N}$ has $\setminus$, and in case σ) also $\downarrow \sigma$. By assumption $\mathfrak{S} \subset \mathfrak{N}$, so that $\mathfrak{N}$ is nonvoid. Thus the transporter theorem 1.16 furnishes

in case $\star$): $\mathrm{A}(\mathfrak{S}) \subset \mathrm{A}(\mathfrak{S}\top) \subset \mathfrak{S}\top\mathfrak{N}$;
in case σ): $\mathrm{A}\sigma(\mathfrak{S}) \subset \mathrm{A}\sigma(\mathfrak{S}\top) \subset \mathfrak{S}\top\mathfrak{N}$.

We now continue with σ); the case $\star$) is similar and simpler. i) For $A \in \mathrm{A}\sigma(\mathfrak{S})$ and $S_1, \cdots, S_n \in \mathfrak{S}$ we have $A \cap S_1 \cap \cdots \cap S_n \in \mathrm{A}\sigma(\mathfrak{S}) \subset \mathfrak{S}\top\mathfrak{N}$ and hence $A \cap S_1 \cdots \cap S_n = (A \cap S_1 \cdots \cap S_n) \cap S_n \in \mathfrak{N}$. ii) Let now $A \in \mathrm{R}\sigma(\mathfrak{S})$. By $1.8.2\sigma$) A is upward enclosable $\mathfrak{S}^\sigma$; thus there exists a sequence $(S_l)_l$ in $\mathfrak{S}$ such that $A \subset \bigcup\limits_{l=1}^{\infty} S_l$. Hence the sequence of the subsets $V_n := S_1 \cup \cdots \cup S_n \in \mathfrak{A}$ satisfies $A \cap V_n \uparrow A$. iii) We apply exercise 2.5.1) to the restrictions of φ and ψ to $[\varphi < \infty] \cap [\psi < \infty]$ and to the subsets $A \cap S_1, \cdots, A \cap S_n \in \mathfrak{N}$. By i) all their intersections of nonvoid subfamilies are in $\mathfrak{N}$ as well, and their union is $A \cap V_n$. It follows that $\varphi(A \cap V_n) = \psi(A \cap V_n)$. For $n \to \infty$ we obtain $\varphi(A) = \psi(A)$ from ii).

3.2. EXERCISE. The conclusions of 3.1.$\star$) and 3.1.σ) do not persist when one deletes $< \infty$ from the assumptions, even when $\mathfrak{S}$ is a lattice with $\varnothing \in \mathfrak{S}$. Hint: Let $X = \mathbb{N}$ and $\mathfrak{S}$ consist of $\varnothing$ and of the cofinite subsets.

3.3. EXERCISE. The conclusion of 3.1.σ) does not persist when one deletes $< \infty$ from the assumption, even when $\mathfrak{S}$ is a ring. Hint: 1) Construct a ring $\mathfrak{S}$ in a suitable X such that i) all nonvoid $S \in \mathfrak{S}$ are uncountable, but ii) there are nonvoid countable $S \in R\sigma(\mathfrak{S})$. 2) On $\mathfrak{A} := R\sigma(\mathfrak{S})$ then define $\varphi, \psi : \mathfrak{A} \to [0, \infty]$ to be

$$\varphi(A) = \left\{ \begin{array}{ll} 0 & \text{for } A = \varnothing \\ \infty & \text{for } A \neq \varnothing \end{array} \right\} \quad \text{and} \quad \psi(A) = \left\{ \begin{array}{ll} 0 & \text{for } A \text{ countable} \\ \infty & \text{for } A \text{ uncountable} \end{array} \right\}.$$

The Smiley-Horn-Tarski Extension Theorem

3.4. THEOREM. *Let $\varphi : \mathfrak{S} \to \mathbb{R}$ be a modular set function on a lattice $\mathfrak{S}$ in X. Then there exists a unique modular set function $\phi : O(\mathfrak{S}) \to \mathbb{R}$ which extends φ. If φ is isotone then ϕ is isotone as well.*

The classical Smiley-Horn-Tarski theorem is for lattices $\mathfrak{S}$ with $\varnothing \in \mathfrak{S}$ and hence $O(\mathfrak{S}) = R(\mathfrak{S})$; see for example Rao-Rao [1983] chapter 3. We shall comment on the usefulness of these results at the end of the subsection. The proof requires an elaborate construction. The first lemma below is obvious.

3.5. LEMMA. *Assume that $A(1), \cdots, A(r) \subset X$. For the nonvoid index sets $T \subset \{1, \cdots, r\}$ we form*

$$D(T) := \bigcap_{l \in T} A(l) \cap \bigcap_{l \notin T} (A(l))'.$$

Then the $D(T)$ are pairwise disjoint. Furthermore

$$\bigcup_{l=1}^{r} A(l) = \bigcup_{T} D(T) \text{ and } A(l) = \bigcup_{T \ni l} D(T) \ (l = 1, \cdots, r).$$

3.6. PROPOSITION. *Let $\mathfrak{S}$ be a paving with $\cap$. 1) $R(\mathfrak{S})$ consists of all $A \subset X$ such that*

$$\chi_A = \sum_{l=1}^{r} a_l \chi_{A(l)} \quad \text{with } A(1), \cdots, A(r) \in \mathfrak{S} \text{ and } a_1, \cdots, a_r \in \mathbb{Z}.$$

2) $O(\mathfrak{S})$ *consists of all $A \subset X$ such that*

$$\chi_A = \sum_{l=1}^{r} a_l \chi_{A(l)} \quad \text{as above with } \sum_{l=1}^{r} a_l = 1.$$

Proof. Define $\mathfrak{M}$ to consist of all $A \subset X$ such that χ_A can be represented as above. i) Assume that $A \in \mathfrak{M}$. By 3.5 then

$$\chi_A = \sum_{l=1}^{r} a_l \sum_{T \ni l} \chi_{D(T)} = \sum_{T} \left(\sum_{l \in T} a_l \right) \chi_{D(T)}.$$

Now $D(T) \in \mathrm{R}(\mathfrak{S})$ since $T \neq \varnothing$, and A is the disjoint union of some of the $D(T)$. Hence $A \in \mathrm{R}(\mathfrak{S})$. Now in case 2) consider $R := \{1, \cdots, r\}$. We have

$$\sum_{l \in R} a_l = \sum_{l=1}^{r} a_l = 1 \text{ and hence } A \supset D(R).$$

Furthermore $D(R) \in \mathfrak{S}$ by definition. From 1.17.$\star$) and 1.23.$\star$) we see that A is in $\mathrm{R}(\mathfrak{S}) \cap (\sqsupset \mathfrak{S}) = \mathrm{A}(\mathfrak{S}) \cap (\mathfrak{S} \sqsubset \mathfrak{S}^{\star}) = \mathrm{O}(\mathfrak{S})$. Thus we have $\mathfrak{M} \subset \mathrm{R}(\mathfrak{S})$ in case 1) and $\mathfrak{M} \subset \mathrm{O}(\mathfrak{S})$ in case 2). ii) For the converse note that $\mathfrak{S} \subset \mathfrak{M}$. Thus we have to show that $\mathfrak{M}$ is a ring in case 1) and an oval in case 2). This follows in case 1) from

$$\chi_{B \cap A'} = \chi_B - \chi_{B \cap A} = \chi_B - \chi_B \chi_A,$$
$$\chi_{A \cup B} = \chi_A + \chi_{B \cap A'} = \chi_A + \chi_B - \chi_B \chi_A \text{ for } A, B \subset X,$$

and in case 2) from

$$\chi_{U|A|V} = \chi_{U \cap A'} + \chi_{V \cap A} = \chi_U + \left(\chi_V - \chi_U \right) \chi_A \text{ for } U, V, A \subset X,$$

both times by the assumption that $\mathfrak{S}$ has $\cap$.

For the next step we need an auxiliary formula.

3.7. LEMMA. Let $A(0), A(1), \cdots, A(r) \subset X$ with $r \geq 1$, and form $U(p) := \bigcup_{l=0}^{p} A(l)$ for $0 \leq p \leq r$. Then

$$\sum_{l=0}^{r} \chi_{A(l)} = \sum_{l=1}^{r} \chi_{U(l-1) \cap A(l)} + \chi_{U(r)}.$$

Proof. The case $r = 1$: Here we have $U(0) = A(0)$ and $U(1) = A(0) \cup A(1)$. Thus the assertion is clear. The induction step $1 \leq r \Rightarrow r + 1$: Let $A(0), A(1), \cdots, A(r+1) \subset X$ with the $U(p)$ for $0 \leq p \leq r + 1$. By the induction hypothesis

$$\sum_{l=0}^{r+1} \chi_{A(l)} = \sum_{l=0}^{r} \chi_{A(l)} + \chi_{A(r+1)} = \sum_{l=1}^{r} \chi_{U(l-1) \cap A(l)} + \chi_{U(r)} + \chi_{A(r+1)},$$

and furthermore

$$\chi_{U(r)} + \chi_{A(r+1)} = \chi_{U(r) \cap A(r+1)} + \chi_{U(r+1)}.$$

The assertion follows.

3.8. PROPOSITION. *Let $\varphi : \mathfrak{S} \to \mathbb{R}$ be a modular set function on a lattice* $\mathfrak{S}$. 1) *For* $A(1), \cdots, A(r), B(1), \cdots, B(r) \in \mathfrak{S}$ *we have*

$$\sum_{l=1}^{r} \chi_{A(l)} = \sum_{l=1}^{r} \chi_{B(l)} \Rightarrow \sum_{l=1}^{r} \varphi(A(l)) = \sum_{l=1}^{r} \varphi(B(l)).$$

2) *Assume that* φ *is isotone. For* $A(1), \cdots, A(r), B(1), \cdots, B(r) \in \mathfrak{S}$ *then*

$$\sum_{l=1}^{r} \chi_{A(l)} \leqq \sum_{l=1}^{r} \chi_{B(l)} \Rightarrow \sum_{l=1}^{r} \varphi(A(l)) \leqq \sum_{l=1}^{r} \varphi(B(l)).$$

Proof. Both times the case $r = 1$ is obvious; thus it remains the induction step $1 \leqq r \Rightarrow r + 1$. 1) Let $A(0), A(1), \cdots, A(r), B(0), B(1), \cdots, B(r) \in \mathfrak{S}$ with $\sum_{l=0}^{r} \chi_{A(l)} = \sum_{l=0}^{r} \chi_{B(l)}$, and form the $U(p), V(p) \in \mathfrak{S}$ for $0 \leqq p \leqq r$ after 3.7. Then first of all $U(r) = V(r)$. By 3.7 we obtain

$$\sum_{l=1}^{r} \chi_{U(l-1) \cap A(l)} = \sum_{l=1}^{r} \chi_{V(l-1) \cap B(l)},$$

and hence by the induction hypothesis

$$\sum_{l=1}^{r} \varphi(U(l-1) \cap A(l)) = \sum_{l=1}^{r} \varphi(V(l-1) \cap B(l)).$$

Here the left side is

$$\sum_{l=1}^{r} \Big(\varphi(U(l-1)) + \varphi(A(l)) - \varphi(U(l)) \Big)$$
$$= \sum_{l=1}^{r} \varphi(A(l)) + \varphi(U(0)) - \varphi(U(r)) = \sum_{l=0}^{r} \varphi(A(l)) - \varphi(U(r)).$$

The same applies to the right side. The assertion follows. 2) Let $A(0), A(1),$ $\cdots, A(r), B(0), B(1), \cdots, B(r) \in \mathfrak{S}$ with $\sum_{l=0}^{r} \chi_{A(l)} \leqq \sum_{l=0}^{r} \chi_{B(l)}$, and form the $U(p), V(p) \in \mathfrak{S}$ for $0 \leqq p \leqq r$ as above. Then first of all $U(r) \subset V(r)$. Now pass to the subsets $b(l) := B(l) \cap U(r) \in \mathfrak{S}$ for $0 \leqq l \leqq r$ and once more form the $v(p) \in \mathfrak{S}$ for $0 \leqq p \leqq r$ as above. Then $v(p) = V(p) \cap U(r)$, hence in particular $v(r) = U(r)$. The assumption implies that $\sum_{l=0}^{r} \chi_{A(l)} \leqq \sum_{l=0}^{r} \chi_{b(l)}$, which is obvious both on $U(r)$ and outside of $U(r)$. By 3.7 we obtain

$$\sum_{l=1}^{r} \chi_{U(l-1) \cap A(l)} \leqq \sum_{l=1}^{r} \chi_{v(l-1) \cap b(l)},$$

and hence by the induction hypothesis

$$\sum_{l=1}^{r} \varphi(U(l-1) \cap A(l)) \leqq \sum_{l=1}^{r} \varphi(v(l-1) \cap b(l)) \leqq \sum_{l=1}^{r} \varphi(V(l-1) \cap B(l)).$$

As in the proof of 1) it follows that

$$\sum_{l=0}^{r}\varphi(A(l)) - \varphi(U(r)) \leqq \sum_{l=0}^{r}\varphi(B(l)) - \varphi(V(r)).$$

We add $\varphi(U(r)) \leqq \varphi(V(r))$ and obtain the assertion.

3.9. REFORMULATION. *Let* $\varphi : \mathfrak{S} \to \mathbb{R}$ *be a modular set function on a lattice* $\mathfrak{S}$. 1) *If* $A(1), \cdots , A(r) \in \mathfrak{S}$ *and* $a_1, \cdots , a_r \in \mathbb{Z}$ *with*

$$\sum_{l=1}^{r} a_l = 0 \text{ and } \sum_{l=1}^{r} a_l \chi_{A(l)} = 0 \text{ then } \sum_{l=1}^{r} a_l \varphi(A(l)) = 0.$$

2) *Assume that* φ *is isotone. If* $A(1), \cdots , A(r) \in \mathfrak{S}$ *and* $a_1, \cdots , a_r \in \mathbb{Z}$ *with*

$$\sum_{l=1}^{r} a_l = 0 \text{ and } \sum_{l=1}^{r} a_l \chi_{A(l)} \leqq 0 \text{ then } \sum_{l=1}^{r} a_l \varphi(A(l)) \leqq 0.$$

Proof. Put $a_l = p_l - q_l$ with $p_l, q_l \in \mathbb{N}$ for $1 \leqq l \leqq r$. Then $\sum_{l=1}^{r} p_l = \sum_{l=1}^{r} q_l =: n \in \mathbb{N}$. Thus 3.8 can be applied in the obvious manner.

Proof of 3.4. Let $\varphi : \mathfrak{S} \to \mathbb{R}$ be a modular set function on a lattice $\mathfrak{S}$. i) Existence of an extension: Let $A \in O(\mathfrak{S})$. By 3.9.1) then all representations of χ_A after 3.6.2) produce the same value

$$\sum_{l=1}^{r} a_l \varphi(A(l)) =: \phi(A).$$

Thus we obtain a set function $\phi : O(\mathfrak{S}) \to \mathbb{R}$ which is an extension of φ. In order to see that ϕ is modular let $A, B \in O(\mathfrak{S})$ and fix representations

$$\chi_A = \sum_{k=1}^{r} a_k \chi_{A(k)} \text{ and } \chi_B = \sum_{l=1}^{s} b_l \chi_{B(l)}$$

after 3.6.2). Then

$$\chi_{A \cap B} = \chi_A \chi_B = \sum_{k=1}^{r}\sum_{l=1}^{s} a_k b_l \chi_{A(k) \cap B(l)} \text{ with } \sum_{k=1}^{r}\sum_{l=1}^{s} a_k b_l = 1$$

is a representation for $A \cap B \in O(\mathfrak{S})$ of the same kind, and then $\chi_{A \cup B} = \chi_A + \chi_B - \chi_{A \cap B}$ with the above representations on the right likewise produces a representation for $A \cup B \in O(\mathfrak{S})$ of the same kind. Now it is obvious that $\phi(A \cup B) = \phi(A) + \phi(B) - \phi(A \cap B)$. ii) Assume that φ is isotone. Then by 3.9.2) the extension $\phi : O(\mathfrak{S}) \to \mathbb{R}$ of φ obtained in i) is isotone as well. iii) Uniqueness of the extension: Let $\phi : O(\mathfrak{S}) \to \mathbb{R}$ be a modular set function which is an extension of φ. Fix $A \in O(\mathfrak{S})$ and a representation of χ_A after 3.6.2). Then apply 3.9.1) to $\phi : O(\mathfrak{S}) \to \mathbb{R}$ and

$$\sum_{l=1}^{r} a_l \chi_{A(l)} + (-1)\chi_A = 0.$$

It follows that

$$\sum_{l=1}^{r} a_l \phi(A(l)) + (-1)\phi(A) = 0 \text{ or } \phi(A) = \sum_{l=1}^{r} a_l \varphi(A(l)).$$

The proof is complete.

3.10. EXERCISE. *Let* $\varphi : \mathfrak{S} \to \mathbb{R}$ *be a modular set function on a lattice* $\mathfrak{S}$ *with* $\varnothing \notin \mathfrak{S}$. *1) For each* $c \in \mathbb{R}$ *there exists a unique modular set function* $\phi : R(\mathfrak{S}) \to \mathbb{R}$ *with* $\phi|\mathfrak{S} = \varphi$ *and* $\phi(\varnothing) = c$. *2) Assume that* φ *is isotone. Then the extension* $\phi : R(\mathfrak{S}) \to \mathbb{R}$ *of* φ *obtained in 1) is isotone iff* φ *is bounded below and* $c \leqq \inf \varphi$.

Thus in the terms of section 2 each modular and isotone set function $\varphi : \mathfrak{S} \to \mathbb{R}$ on a lattice $\mathfrak{S}$ has a unique extension $\phi : O(\mathfrak{S}) \to \mathbb{R}$ which is a content on $O(\mathfrak{S})$. In particular each modular and isotone set function $\varphi : \mathfrak{S} \to [0, \infty[$ on a lattice $\mathfrak{S}$ such that $\varnothing \in \mathfrak{S}$ and $\varphi(\varnothing) = 0$ has a unique extension $\phi : R(\mathfrak{S}) \to [0, \infty[$ which is a ccontent on $R(\mathfrak{S})$. This result looks as beautiful and powerful as one could hope for. However, it is burdened with a disastrous defect: *The extension procedure can destroy sequential continuity.* Therefore it becomes useless as soon as one wants to pass from contents to measures. In fact, we shall construct an example of a modular and isotone set function $\varphi : \mathfrak{S} \to \{0, 1\}$ on a lattice $\mathfrak{S}$ with $\varnothing, X \in \mathfrak{S}$ and $\varphi(\varnothing) = 0$ which is upward and downward σ continuous, whereas the ccontent $\phi : R(\mathfrak{S}) \to \{0, 1\}$ has values $\phi(A_l) = 1$ for some sequence $(A_l)_l$ in $R(\mathfrak{S})$ such that $A_l \downarrow \varnothing$.

3.11. EXAMPLE. Let X be an infinite set, and let $T \subset X$ be such that both T and T' are infinite. Then form sequences of subsets

$(P_l)_l$ in X with $P_l \uparrow T$ but $P_l \neq T \; \forall l \in \mathbb{N}$, and $P_1 = \varnothing$;
$(Q_l)_l$ in X with $Q_l \downarrow T$ but $Q_l \neq T \; \forall l \in \mathbb{N}$, and $Q_1 = X$.

The paving $\mathfrak{S} := \{P_l, Q_l : l \in \mathbb{N}\}$ is a lattice in X by 1.2.7), and $\varphi(P_l) = 0$ and $\varphi(Q_l) = 1$ defines a modular and isotone set function $\varphi : \mathfrak{S} \to \{0, 1\}$ with $\varphi(\varnothing) = 0$ by 2.3.2). It is obvious that φ is upward and downward σ an even τ continuous. However, the sequence of the difference sets $A_l := Q_l - P_l \in R(\mathfrak{S})$ fulfils $A_l \downarrow \varnothing$ and $\phi(A_l) = \varphi(Q_l) - \varphi(P_l) = 1$ for all $l \in \mathbb{N}$.

Therefore essential new ideas are required for the extension of set functions from lattices to the level of ovals and rings. Such an idea is regularity. It will dominate the procedures of the next chapter.

Extensions of Set Functions to Lattices

The main extension theorems in this text start from set functions defined on pavings which are at least lattices. The exception is the present subsection, where we complement those theorems with results how to obtain well-behaved set functions on lattices from more primitive ones.

Let $\mathfrak{U}$ be a paving with $\cap$ in the nonvoid set X. We know from 1.2.10) that $\mathfrak{V} := L(\mathfrak{U})$ consists of the unions $A_1 \cup \cdots \cup A_r$ for all finite sequences

$A_1, \cdots, A_r \in \mathfrak{U}$. We fix a set function $\varphi : \mathfrak{U} \to \mathbb{R}$. The question whether and when it admits modular extensions $\phi : \mathfrak{V} \to \mathbb{R}$ receives an immediate hint from 2.5.1).

3.12. REMARK. *The set function* $\varphi : \mathfrak{U} \to \mathbb{R}$ *admits at most one modular extension* $\phi : \mathfrak{V} \to \mathbb{R}$. *If it exists then*

$$\phi(A_1 \cup \cdots \cup A_r) = \sum_{\varnothing \neq T \subset \{1, \cdots, r\}} (-1)^{\#(T)-1} \varphi(A_T) \quad \text{for } A_1, \cdots, A_r \in \mathfrak{U},$$

with the abbreviation $A_T := \bigcap_{l \in T} A_l \in \mathfrak{U}$ *for the nonvoid* $T \subset \{1, \cdots, r\}$.

We are thus led to extend $\varphi : \mathfrak{U} \to \mathbb{R}$ to all finite sequences $A_1, \cdots, A_r \in \mathfrak{U}$ by the definition

$$\varphi(A_1, \cdots, A_r) = \sum_{\varnothing \neq T \subset \{1, \cdots, r\}} (-1)^{\#(T)-1} \varphi(A_T).$$

It is obvious that $\varphi(\cdot, \cdots, \cdot)$ is a symmetric function of its arguments. Also $\varphi(A_1, \cdots, A_r) = \varphi(A_1 \cup \cdots \cup A_r)$ in case $\mathfrak{U}$ is a lattice and φ is modular. We proceed to collect further properties which will be needed for the main theorems.

3.13. PROPERTIES. 1) *For* $A_0, A_1, \cdots, A_r \in \mathfrak{U}$ *we have the recursion formula*

$$\varphi(A_0, A_1, \cdots, A_r) = \varphi(A_0) + \varphi(A_1, \cdots, A_r) - \varphi(A_0 \cap A_1, \cdots, A_0 \cap A_r).$$

2) *If* $A_0, A_1, \cdots, A_r \in \mathfrak{U}$ *are such that* $A_0 \subset A_l$ *for some* $l \in \{1, \cdots, r\}$ *then* $\varphi(A_0, A_1, \cdots, A_r) = \varphi(A_1, \cdots, A_r)$. 3) *For* $A_1, \cdots, A_r, B_1, \cdots, B_s \in \mathfrak{U}$ *we have*

$$\varphi(A_1, \cdots, A_r, B_1, \cdots B_s) + \varphi(A_1 \cap B_1, \cdots, A_r \cap B_s)$$
$$= \varphi(A_1, \cdots, A_r) + \varphi(B_1, \cdots, B_s),$$

where the second argument on the left consists of all intersections $A_k \cap B_l$ *with* $k \in \{1, \cdots, r\}$ *and* $l \in \{1, \cdots, s\}$.

Proof. 1) The assertion follows when one splits the sum in

$$\varphi(A_0, A_1, \cdots, A_r) = \sum_{\varnothing \neq T \subset \{0,1,\cdots,r\}} (-1)^{\#(T)-1} \varphi(A_T)$$

into the three partial sums which consist of the term $T = \{0\}$ alone, of the terms with $0 \notin T$, and of the terms with $0 \in T$ except $T = \{0\}$. 2) We can assume that $A_0 \subset A_1$. By 1) then in case $r = 1$

$$\varphi(A_0, A_1) - \varphi(A_1) = \varphi(A_0) - \varphi(A_0 \cap A_1) = \varphi(A_0) - \varphi(A_0) = 0,$$

and in case $r \geqq 2$

$$\varphi(A_0, A_1, \cdots, A_r) - \varphi(A_1, \cdots, A_r) = \varphi(A_0) - \varphi(A_0 \cap A_1, \cdots, A_0 \cap A_r)$$
$$= \varphi(A_0) - \varphi(A_0, A_0 \cap A_2, \cdots, A_0 \cap A_r),$$

which is $= 0$ once more by 1). 3) The proof is by induction. The case $r = 1$ is clear by 1). We turn to the induction step $1 \leqq r \Rightarrow r + 1$. For $A_0, A_1, \cdots, A_r, B_1, \cdots, B_s \in \mathfrak{U}$ we have by 1)

$$\varphi(A_0 \cap B_1, \cdots, A_r \cap B_s) = \varphi(A_0, A_0 \cap B_1, \cdots, A_r \cap B_s)$$
$$-\varphi(A_0) + \varphi(A_0 \cap A_0 \cap B_1, \cdots, A_0 \cap A_r \cap B_s).$$

In view of 2) we can omit in the first term on the right all places $A_0 \cap B_l$ with $l \in \{1, \cdots, s\}$, and in the last term all places $A_0 \cap A_k \cap B_l$ with $k \in \{1, \cdots, r\}$ and $l \in \{1, \cdots, s\}$. Thus the expression is

$$= \varphi(A_0, A_1 \cap B_1, \cdots, A_r \cap B_s) - \varphi(A_0) + \varphi(A_0 \cap B_1, \cdots, A_0 \cap B_s)$$
$$= \varphi(A_1 \cap B_1, \cdots, A_r \cap B_s) - \varphi(A_0 \cap A_1 \cap B_1, \cdots, A_0 \cap A_r \cap B_s)$$
$$+ \varphi(A_0 \cap B_1, \cdots, A_0 \cap B_s),$$

once more by 1). Now we apply the induction hypothesis to the first two terms on the right, and then 1) two times. The expression becomes

$$= \varphi(A_1, \cdots, A_r) + \varphi(B_1, \cdots, B_s) - \varphi(A_1, \cdots, A_r, B_1, \cdots, B_s)$$
$$- \varphi(A_0 \cap A_1, \ldots, A_0 \cap A_r)$$
$$+ \varphi(A_0 \cap A_1, \cdots, A_0 \cap A_r, A_0 \cap B_1, \cdots, A_0 \cap B_s)$$
$$= \big(\varphi(A_0, A_1, \cdots, A_r) - \varphi(A_0)\big)$$
$$- \big(\varphi(A_0, A_1, \cdots, A_r, B_1, \cdots, B_s) - \varphi(A_0)\big) + \varphi(B_1, \cdots, B_s)$$
$$= \varphi(A_0, A_1, \cdots, A_r) + \varphi(B_1, \cdots, B_s) - \varphi(A_0, A_1, \cdots, A_r, B_1, \cdots, B_s).$$

This is the assertion.

3.14. EXERCISE. For $A_1, \cdots, A_r, B_1, \cdots, B_s \in \mathfrak{U}$ we have

$$\varphi(A_1, \cdots, A_r, B_1, \cdots B_s) = \sum_{\varnothing \neq T \subset \{1, \cdots, r\}} (-1)^{\#(T)-1} \varphi(A_T, B_1, \cdots, B_s).$$

3.15. THEOREM. *The set function $\varphi : \mathfrak{U} \to \mathbb{R}$ admits a modular extension $\phi : \mathfrak{V} \to \mathbb{R}$ (and hence a unique one) iff it satisfies*

(mod) $A_1, \cdots, A_r, A \in \mathfrak{U}$ *with* $A_1 \cup \cdots \cup A_r = A \Rightarrow \varphi(A_1, \cdots, A_r) = \varphi(A)$.

Then $\phi(A) = \varphi(A_1, \cdots, A_r)$ for $A_1, \cdots, A_r \in \mathfrak{U}$ with $A_1 \cup \cdots \cup A_r = A \in \mathfrak{V}$. In this case the extension ϕ is isotone iff

(isot) $A_1, \cdots, A_r, A \in \mathfrak{U}$ *with* $A_1 \cup \cdots \cup A_r \subset A \Rightarrow \varphi(A_1, \cdots, A_r) \leqq \varphi(A)$.

Proof. i) If $\phi : \mathfrak{V} \to \mathbb{R}$ is a modular extension of φ then 3.12 says that $\phi(A) = \varphi(A_1, \cdots, A_r)$ for $A_1, \cdots, A_r \in \mathfrak{U}$ with $A_1 \cup \cdots \cup A_r = A \in \mathfrak{V}$. In particular we have (mod). ii) Assume that (mod) is satisfied. Then for $A_0, A_1, \cdots, A_r \in \mathfrak{U}$ with $A_0 \subset A_1 \cup \cdots \cup A_r$ we have

$$\varphi(A_0, A_1, \cdots, A_r) - \varphi(A_1, \cdots, A_r)$$
$$= \varphi(A_0) - \varphi(A_0 \cap A_1, \cdots, A_0 \cap A_r) = 0,$$

thus $\varphi(A_0, A_1, \cdots, A_r) = \varphi(A_1, \cdots, A_r)$. This implies that for $A_1, \cdots, A_r, B_1, \cdots, B_s \in \mathfrak{U}$ with $A_1 \cup \cdots \cup A_r = B_1 \cup \cdots \cup B_s$ we have

$$\varphi(A_1, \cdots, A_r) = \varphi(A_1, \cdots, A_r, B_1, \cdots, B_s) = \varphi(B_1, \cdots, B_s).$$

Therefore there is a well-defined function $\phi : \mathfrak{V} \to \mathbb{R}$ such that $\phi(A) = \varphi(A_1, \cdots, A_r)$ for $A_1, \cdots, A_r \in \mathfrak{U}$ with $A_1 \cup \cdots \cup A_r = A \in \mathfrak{V}$. This set function extends φ, and by 3.13.3) it is modular. iii) If ϕ is isotone then (isot) is clear. It on the other hand (isot) is satisfied then for $A_1, \cdots, A_r \in \mathfrak{U}$ we have

$$\varphi(A_0, A_1, \cdots, A_r) - \varphi(A_1, \cdots, A_r)$$
$$= \varphi(A_0) - \varphi(A_0 \cap A_1, \cdots, A_0 \cap A_r) \geqq 0.$$

This implies that ϕ is isotone. The proof is complete.

3.16. THEOREM. *The set function $\varphi : \mathfrak{U} \to \mathbb{R}$ admits a modular extension $\phi : \mathfrak{V} \to \mathbb{R}$ such that*

$$\phi(V_l) \to \phi(V) \text{ for all sequences } (V_l)_l \text{ in } \mathfrak{V} \text{ with } V_l \uparrow V \in \mathfrak{V}$$

(note that ϕ need not be isotone!) iff it satisfies

$$(\text{mod}\sigma) \quad (A_l)_l \text{ in } \mathfrak{U} \text{ with } \bigcup_{l=1}^{\infty} A_l = A \in \mathfrak{U} \Rightarrow \varphi(A_1, \cdots, A_r) \to \varphi(A).$$

Proof. i) If $\phi : \mathfrak{V} \to \mathbb{R}$ is as required then $(\text{mod}\sigma)$ follows from the assumption applied to the sequence of the $V_r := A_1 \cup \cdots \cup A_r \in \mathfrak{V}$. ii) Assume now that $(\text{mod}\sigma)$ is satisfied. Then first of all (mod) is fulfilled since $\varphi(A_1, \cdots, A_r) = \varphi(A_1, \cdots, A_r, \cdots, A_r)$ for $A_1, \cdots, A_r \in \mathfrak{U}$ by 3.13.2). iii) Consider a sequence $(V_r)_r$ in $\mathfrak{V}$ with $V_r \uparrow V \in \mathfrak{V}$. Then there is a sequence $(A_l)_l$ in $\mathfrak{U}$ such that $(V_r)_r$ is a subsequence of $(A_1 \cup \cdots \cup A_r)_r$. Thus we can assume that $V_r = A_1 \cup \cdots \cup A_r$ for $r \in \mathbb{N}$. On the other hand $V = B_1 \cup \cdots \cup B_s$ for some $B_1, \cdots, B_s \in \mathfrak{U}$. Thus for each nonvoid $T \subset \{1, \cdots, s\}$ we have $\bigcup_{l=1}^{\infty} (A_l \cap B_T) = B_T$ and hence by $(\text{mod}\sigma)$

$$\varphi(B_T, A_1, \cdots, A_r) - \varphi(A_1, \cdots, A_r)$$
$$= \varphi(B_T) - \varphi(B_T \cap A_1, \cdots, B_T \cap A_r) \to 0.$$

Thus exercise 3.14 implies that

$$\phi(V) - \phi(V_r) = \varphi(B_1, \cdots, B_s) - \varphi(A_1, \cdots, A_r)$$
$$= \varphi(B_1, \cdots, B_s, A_1, \cdots, A_r) - \varphi(A_1, \cdots, A_r)$$
$$= \sum_{T} (-1)^{\#(T)-1} \big(\varphi(B_T, A_1, \cdots, A_r) - \varphi(A_1, \cdots, A_r) \big) \to 0.$$

The proof is complete.

We conclude with the specialization to an additional condition on the paving $\mathfrak{U}$ which is frequent in applications.

3.17. SPECIAL CASE. *Let $\mathfrak{U}$ be a paving with $\cap$ and such that*

$$A, B \in \mathfrak{U} \text{ with } A, B \subset \text{ some } U \in \mathfrak{U} \Rightarrow A \cup B \in \mathfrak{U}.$$

Then a set function $\varphi : \mathfrak{U} \to \mathbb{R}$ admits a modular extension $\phi : \mathfrak{V} \to \mathbb{R}$ (and hence a unique one) iff it satisfies

$$(\star) \quad \varphi(A \cup B) + \varphi(A \cap B) = \varphi(A) + \varphi(B) \text{ for all } A, B \in \mathfrak{U} \text{ with } A \cup B \in \mathfrak{U}.$$

In this case the extension ϕ is isotone iff φ is isotone. Furthermore ϕ satisfies

$$\phi(V_l) \to \phi(V) \text{ for all sequences } (V_l)_l \text{ in } \mathfrak{V} \text{ with } V_l \uparrow V \in \mathfrak{V}$$

iff φ does the same on $\mathfrak{U}$.

Proof. For fixed $U \in \mathfrak{U}$ the subpaving $\mathfrak{U}(U) := \{A \in \mathfrak{U} : A \subset U\}$ of $\mathfrak{U}$ is a lattice. Under $(\star)$ the restriction $\varphi|\mathfrak{U}(U)$ is modular. It follows that $\varphi(A_1, \cdots, A_r) = \varphi(A_1 \cup \cdots \cup A_r)$ for $A_1, \cdots, A_r \in \mathfrak{U}(U)$. Therefore (mod) is satisfied. Furthermore we have (modσ) as soon as φ is as required. Thus we obtain all assertions.

CHAPTER II

The Extension Theories Based on Regularity

The theme of the present chapter is the construction of contents and measures from more primitive set functions. The construction is based on interrelated regularity and continuity conditions. These conditions are either both of outer or both of inner type. We want to demonstrate that the outer and inner theories are identical. To achieve this we have to work with the unconventional notions introduced in the first chapter, with set systems which avoid the empty set like the entire set, and with isotone set functions which take values in $\mathbb{R}$ or $\overline{\mathbb{R}}$. We start with the complete development of the outer extension theory. Then the upside-down transform method initiated in the first chapter will transform the outer into the inner extension theory. The chapter concludes with a detailed bibliographical annex.

4. The Outer Extension Theory: Concepts and Instruments

The Basic Definition

Let $\mathfrak{S}$ be a lattice in a nonvoid set X. We start with the basic definition which describes the final aim of the outer enterprise.

DEFINITION. Let $\varphi : \mathfrak{S} \to]-\infty, \infty]$ be an isotone set function $\not\equiv \infty$. For $\bullet = \star\sigma\tau$ we define an **outer** $\bullet$ **extension** of φ to be an extension of φ which is a $\dotplus$ content $\alpha : \mathfrak{A} \to \overline{\mathbb{R}}$ on an oval $\mathfrak{A}$, such that also $\mathfrak{S}^{\bullet} \subset \mathfrak{A}$ and that

α is outer regular $\mathfrak{S}^{\bullet}$, and

$\alpha|\mathfrak{S}^{\bullet}$ is upward $\bullet$ continuous; in this connection note that $\alpha|\mathfrak{S}^{\bullet} > -\infty$.

We define φ to be an **outer** $\bullet$ **premeasure** iff it admits outer $\bullet$ extensions. Thus an outer $\bullet$ premeasure is modular and upward $\bullet$ continuous.

The principal aim is to characterize those φ which are outer $\bullet$ premeasures, and then to describe all outer $\bullet$ extensions of φ. We shall obtain a beautiful answer in natural terms. Our approach will be based on two formations due to Carathéodory: On the one hand the so-called outer measure, and on the other hand the so-called measurable sets. Both of them need substantial reformulation. These two tasks will be attacked in the present section.

The restriction $\varphi > -\infty$ imposed in the definition will be justified by success. Without it the presentation would be burdened, at least in the cases $\bullet = \sigma\tau$, with useless and unpleasant complications. We shall not pursue this point.

The Outer Envelopes

Let $\varphi : \mathfrak{S} \to \overline{\mathbb{R}}$ be an isotone set function on a lattice $\mathfrak{S}$ in X. It is natural to form its **crude outer envelope** $\varphi^\star : \mathfrak{P}(X) \to \overline{\mathbb{R}}$, defined to be

$$\varphi^\star(A) = \inf\{\varphi(S) : S \in \mathfrak{S} \text{ with } S \supset A\} \quad \text{for } A \subset X.$$

However, this set function does not allow an adequate treatment of our outer $\bullet$ extension problem for $\bullet = \sigma\tau$. The decisive idea is to form for $\bullet = \sigma$ the set function $\varphi^\sigma : \mathfrak{P}(X) \to \overline{\mathbb{R}}$, defined to be

$$\varphi^\sigma(A) = \inf\{\lim_{l\to\infty} \varphi(S_l) : (S_l)_l \text{ in } \mathfrak{S} \text{ with } S_l \uparrow \supset A\} \quad \text{for } A \subset X.$$

It is a variant of the traditional Carathéodory outer measure which itself will not be used below. One of the benefits of φ^σ is that it has an immediate nonsequential counterpart. This is the set function $\varphi^\tau : \mathfrak{P}(X) \to \overline{\mathbb{R}}$, defined to be

$$\varphi^\tau(A) = \inf\{\sup_{S\in\mathfrak{M}} \varphi(S) : \mathfrak{M} \text{ paving } \subset \mathfrak{S} \text{ with } \mathfrak{M} \uparrow \supset A\} \quad \text{for } A \subset X.$$

These are the three **outer envelopes** $\varphi^\bullet : \mathfrak{P}(X) \to \overline{\mathbb{R}}$ of φ for $\bullet = \star\sigma\tau$ which will dominate the outer extension theory. From 1.3 we obtain the common formula

$$\varphi^\bullet(A) = \inf\{\sup_{S\in\mathfrak{M}} \varphi(S) : \mathfrak{M} \text{ paving } \subset \mathfrak{S} \text{ of type } \bullet \text{ with } \mathfrak{M} \uparrow \supset A\}.$$

We turn to the basic properties of these formations.

4.1. PROPERTIES. 1) $\varphi^\star | \mathfrak{S} = \varphi$. 2) $\varphi^\star \geqq \varphi^\sigma \geqq \varphi^\tau$. 3) $\varphi^\bullet$ *is isotone.* 4) $\varphi^\bullet$ *is outer regular* $[\varphi^\bullet | \mathfrak{S}^\bullet < \infty] \subset \mathfrak{S}^\bullet$. 5) *Assume that* φ *is submodular* $\dotplus$. *Then* $\varphi^\star$ *is submodular* $\dotplus$, *and* $\varphi^\bullet$ *for* $\bullet = \sigma\tau$ *is submodular* $\dotplus$ *when either* $\varphi > -\infty$ *or* $\varphi^\bullet < \infty$.

Proof. 1)2)3) are obvious. 4) Fix $A \subset X$ with $\varphi^\bullet(A) < \infty$. For fixed real $c > \varphi^\bullet(A)$ there exists a paving $\mathfrak{M} \subset \mathfrak{S}$ of type $\bullet$ such that $\mathfrak{M} \uparrow$ some $M \supset A$ and $\sup_{S\in\mathfrak{M}} \varphi(S) \leqq c$. Then $M \in \mathfrak{S}^\bullet$, and by definition $\varphi^\bullet(M) \leqq c$. The assertion follows. 5) Fix $A, B \subset X$. We can assume that $\varphi^\bullet(A), \varphi^\bullet(B) < \infty$. For fixed real $a > \varphi^\bullet(A)$ and $b > \varphi^\bullet(B)$ there exist pavings $\mathfrak{M}, \mathfrak{N} \subset \mathfrak{S}$ of type $\bullet$ such that

$$\mathfrak{M} \uparrow \text{ some } M \supset A \text{ and } \sup_{S\in\mathfrak{M}} \varphi(S) \leqq a \,, \; \mathfrak{N} \uparrow \text{ some } N \supset B \text{ and } \sup_{S\in\mathfrak{N}}(T) \leqq b.$$

From them we have the pavings

$$\{S \cup T : S \in \mathfrak{M} \text{ and } T \in \mathfrak{N}\} \quad \uparrow \quad M \cup N \supset A \cup B,$$
$$\{S \cap T : S \in \mathfrak{M} \text{ and } T \in \mathfrak{N}\} \quad \uparrow \quad M \cap N \supset A \cap B.$$

Now we start with $\bullet = \star$. Here $M \in \mathfrak{M}$ and $N \in \mathfrak{N}$. Thus we have $\varphi^\star(A \cup B) \leqq \varphi(M \cup N)$ and $\varphi^\star(A \cap B) \leqq \varphi(M \cap N)$. It follows that

$$\varphi^\star(A \cup B) \dotplus \varphi^\star(A \cap B) \leqq \varphi(M \cup N) \dotplus \varphi(M \cap N) \leqq \varphi(M) \dotplus \varphi(N) \leqq a + b,$$

and hence the assertion. We turn to the cases $\bullet = \sigma\tau$. We fix

$P, Q \in \mathfrak{M}$ and then $S \in \mathfrak{M}$ with $P, Q \subset S$,
$U, V \in \mathfrak{N}$ and then $T \in \mathfrak{N}$ with $U, V \subset T$,

and obtain

$$\varphi(P \cup U) \dotplus \varphi(Q \cap V) \leqq \varphi(S \cup T) \dotplus \varphi(S \cap T) \leqq \varphi(S) \dotplus \varphi(T) \leqq a + b.$$

Therefore $\varphi(P \cup U), \varphi(Q \cap V) < \infty$. If some $\varphi(Q \cap V)$ is $\in \mathbb{R}$ then

$$\varphi^\bullet(A \cup B) \;\leqq\; \sup\{\varphi(P \cup U) : P \in \mathfrak{M} \text{ and } U \in \mathfrak{N}\} \in \mathbb{R},$$
$$\varphi^\bullet(A \cap B) \;\leqq\; \sup\{\varphi(Q \cap V) : Q \in \mathfrak{M} \text{ and } V \in \mathfrak{N}\} \in \mathbb{R},$$

and hence $\varphi^\bullet(A \cup B) \dotplus \varphi^\bullet(A \cap B) \leqq a + b$. If not then $\varphi^\bullet(A \cap B) = -\infty$, and by assumption $\varphi^\bullet(A \cup B) < \infty$. Both times the assertion follows.

We shall later need a counterpart of the first assertion in 4.1.5) for supermodular $\dotplus$.

4.2. REMARK. Let φ be supermodular $\dotplus$. Assume that $A, B \subset X$ are **separated** $\mathfrak{S}$ in the sense that

for each $M \in \mathfrak{S}$ with $A \cap B \subset M$
there exist $S, T \in \mathfrak{S}$ with $A \subset S$ and $B \subset T$ such that $S \cap T \subset M$.
Then $\varphi^\star(A \cup B) \dotplus \varphi^\star(A \cap B) \geqq \varphi^\star(A) \dotplus \varphi^\star(B)$.

Proof. We can assume that $\varphi^\star(A \cup B) < \infty$ and hence all other values $\varphi^\star(\cdot) < \infty$ as well. Fix $M, N \in \mathfrak{S}$ with

$$M \supset A \cap B \text{ and } \varphi(M) < \infty, \quad N \supset A \cup B \text{ and } \varphi(N) < \infty.$$

Then choose $S, T \in \mathfrak{S}$ as assumed. It follows that

$$\begin{aligned}
\varphi(N) + \varphi(M) \;&\geqq\; \varphi\big(N \cap (S \cup T)\big) + \varphi\big(N \cap (S \cap T)\big) \\
&=\; \varphi\big((N \cap S) \cup (N \cap T)\big) + \varphi\big((N \cap S) \cap (N \cap T)\big) \\
&\geqq\; \varphi(N \cap S) + \varphi(N \cap T) \geqq \varphi^\star(A) + \varphi^\star(B).
\end{aligned}$$

This implies the assertion.

4.3. EXERCISE. Let $\varnothing \in \mathfrak{S}$ and $\varphi(\varnothing) = 0$ and φ be superadditive. Assume that $A, B \subset X$ with $A \cap B = \varnothing$ are separated $\mathfrak{S}$, that is

there exist $S, T \in \mathfrak{S}$ with $A \subset S$ and $B \subset T$ such that $S \cap T = \varnothing$.
Then $\varphi^\star(A \cup B) \geqq \varphi^\star(A) + \varphi^\star(B)$.

4.4. EXERCISE. The second assertion in 4.1.5) becomes false without additional assumptions. Hint for an example: Let X be the disjoint union of two infinite countable subsets U and V, and let $\mathfrak{S}$ consist of its finite subsets. Define $\varphi : \mathfrak{S} \to [-\infty, \infty[$ to be $\varphi(S) = \#(S)$ if S meets both U and V, and $\varphi(S) = -\infty$ otherwise.

In contrast to $\varphi^\star | \mathfrak{S} = \varphi$ the relation $\varphi^\bullet | \mathfrak{S} = \varphi$ need not be true for $\bullet = \sigma\tau$. It can be characterized as follows.

4.5. PROPOSITION. *For an isotone set function* $\varphi : \mathfrak{S} \to \overline{\mathbb{R}}$ *and* $\bullet = \sigma\tau$ *the following are equivalent.*

i) $\varphi^\bullet | \mathfrak{S} = \varphi$;
ii) φ *is upward* $\bullet$ *continuous.*

In this case we have furthermore

iii) $\varphi^\bullet | \mathfrak{S}^\bullet$ *is upward* $\bullet$ *continuous;*
iv) *if* $\{S \in \mathfrak{S}^\bullet : \varphi^\bullet(S) < \infty\} \subset \mathfrak{S}$ *then* $\varphi^\bullet = \varphi^\star$.

Proof. i) $\Rightarrow$ ii) Let $A \in \mathfrak{S}$ and $\mathfrak{M} \subset \mathfrak{S}$ be a paving of type $\bullet$ with $\mathfrak{M} \uparrow A$. By i) and the definition of $\varphi^\bullet$ then

$$\varphi(A) = \varphi^\bullet(A) \leqq \sup_{S \in \mathfrak{M}} \varphi(S) \text{ and hence } = \sup_{S \in \mathfrak{M}} \varphi(S).$$

ii) $\Rightarrow$ i) Let $A \in \mathfrak{S}$ and $\mathfrak{M} \subset \mathfrak{S}$ be a paving of type $\bullet$ with $\mathfrak{M} \uparrow \supset A$. Then $\{S \cap A : S \in \mathfrak{M}\}$ is a paving $\subset \mathfrak{S}$ of type $\bullet$ with $\uparrow A$. By ii) therefore

$$\varphi(A) = \sup_{S \in \mathfrak{M}} \varphi(S \cap A) \leqq \sup_{S \in \mathfrak{M}} \varphi(S).$$

It follows that $\varphi(A) \leqq \varphi^\bullet(A)$, and hence from $\varphi^\bullet(A) \leqq \varphi^\star(A) \leqq \varphi(A)$ the assertion. i) $\Rightarrow$ iv) Assume that this is false. Fix $A \subset X$ with $\varphi^\bullet(A) < \varphi^\star(A)$. By 4.1.4) there exists $S \in \mathfrak{S}^\bullet$ with $S \supset A$ and $\varphi^\bullet(S) < \varphi^\star(A)$. By assumption then $S \in \mathfrak{S}$ and $\varphi(S) = \varphi^\bullet(S) < \varphi^\star(A)$. This is a contradiction.

The most involved part of the proof is for the implication i) $\Rightarrow$ iii). We first prove a lemma.

4.6. LEMMA. *Let* $\mathfrak{M} \subset \mathfrak{S}^\bullet$ *be a paving of type* $\bullet$ *with* $\mathfrak{M} \uparrow A$. *Then of course* $A \in \mathfrak{S}^\bullet$. *Furthermore there exists a paving* $\mathfrak{N} \subset \mathfrak{S}$ *of type* $\bullet$ *with* $\mathfrak{N} \uparrow A$ *such that* $\mathfrak{N} \subset (\sqsubset \mathfrak{M})$.

Proof. Nontrivial are the cases $\bullet = \sigma\tau$. The case $\bullet = \sigma$: Choose a sequence $(M_n)_n$ in $\mathfrak{M}$ with $M_n \uparrow$ such that each member of $\mathfrak{M}$ is contained in some M_n. Then $M_n \uparrow A$. Now for each $n \in \mathbb{N}$ there exists a sequence $(S_n^l)_l$ in $\mathfrak{S}$ with $S_n^l \uparrow M_n$. We put $S_l := S_1^l \cup \cdots \cup S_l^l \in \mathfrak{S}$. Then $S_l \uparrow$ some $S \subset X$. We have on the one hand $S_l \subset M_1 \cup \cdots \cup M_l = M_l$, and on the other hand $S_l \supset S_n^l$ for $1 \leqq n \leqq l$. It follows that $S \subset A$ and $S \supset M_n$ for all $n \in \mathbb{N}$, so that $S = A$. Thus the paving $\mathfrak{N} := \{S_l : l \in \mathbb{N}\} \subset \mathfrak{S}$ is as required. The case $\bullet = \tau$: Define $\mathfrak{N} := \{S \in \mathfrak{S} : S \subset \text{ some } M \in \mathfrak{M}\} \subset \mathfrak{S}$. Then $\mathfrak{N}$ is nonvoid and has $\cup$ and is therefore upward directed. Furthermore

$$\bigcup_{S \in \mathfrak{N}} S = \bigcup_{M \in \mathfrak{M}} \bigcup_{S \in \mathfrak{S},\, S \subset M} S = \bigcup_{M \in \mathfrak{M}} M = A.$$

Thus $\mathfrak{N}$ is as required.

Proof of 4.5.i) $\Rightarrow$ iii). Consider a paving $\mathfrak{M} \subset \mathfrak{S}^\bullet$ of type $\bullet$ with $\mathfrak{M} \uparrow A \in \mathfrak{S}^\bullet$, and take $\mathfrak{N} \subset \mathfrak{S}$ as obtained in 4.6. By i) then

$$\varphi(S) = \varphi^\bullet(S) \leqq \sup_{M \in \mathfrak{M}} \varphi^\bullet(M) \quad \text{for each } S \in \mathfrak{N},$$

and therefore by definition

$$\varphi^\bullet(A) \leqq \sup_{S \in \mathfrak{N}} \varphi(S) \leqq \sup_{M \in \mathfrak{M}} \varphi^\bullet(M).$$

The assertion follows.

The most remarkable fact about the outer envelopes is the sequential continuity theorem which follows. Note that there is no continuity assumption on the set function φ itself.

4.7. THEOREM. *Assume that $\varphi : \mathfrak{S} \to \overline{\mathbb{R}}$ is isotone and submodular $\dotplus$. Then φ^σ and φ^τ are almost upward σ continuous.*

4.8. LEMMA. *Assume that $\varphi : \mathfrak{S} \to \overline{\mathbb{R}}$ is isotone and submodular $\dotplus$. For $P_1, \cdots, P_n, Q \in \mathfrak{S}$ with $\varphi(P_1), \cdots, \varphi(P_n), \varphi(Q) < \infty$ then $\varphi(P_1 \cup \cdots \cup P_n \cup Q) < \infty$ and*

$$\varphi(P_1 \cup \cdots \cup P_n \cup Q) + \sum_{l=1}^{n} \varphi(P_l \cap Q) \leqq \sum_{l=1}^{n} \varphi(P_l) + \varphi(Q).$$

Proof of 4.8. The case $n = 1$ is obvious. The induction step $1 \leqq n \Rightarrow n + 1$: Let $P_0, P_1, \cdots, P_n, Q \in \mathfrak{S}$ with $\varphi(P_0), \varphi(P_1), \cdots, \varphi(P_n), \varphi(Q) < \infty$. We know from 2.4 that $[\varphi < \infty]$ is a lattice. Thus from the induction hypothesis we obtain

$$\varphi(P_0 \cup P_1 \cup \cdots \cup P_n \cup Q) + \sum_{l=0}^{n} \varphi(P_l \cap Q)$$

$$= \varphi(P_1 \cup \cdots \cup P_n \cup (P_0 \cup Q)) + \sum_{l=1}^{n} \varphi(P_l \cap Q) + \varphi(P_0 \cap Q)$$

$$\leqq \sum_{l=1}^{n} \varphi(P_l) + \varphi(P_0 \cup Q) + \varphi(P_0 \cap Q) \leqq \sum_{l=0}^{n} \varphi(P_l) + \varphi(Q).$$

Proof of 4.7. We fix a sequence $(A_n)_n$ of subsets of X with $A_n \uparrow A$ and $\varphi^\bullet(A_n) > -\infty \; \forall n \in \mathbb{N}$. Then $\varphi^\bullet(A_n) \uparrow R \leqq \varphi^\bullet(A)$. To be shown is $\varphi^\bullet(A) \leqq R$. We can assume that $R < \infty$, so that the $\varphi^\bullet(A_n)$ and R are finite. We fix $\varepsilon > 0$, and then for each $n \in \mathbb{N}$ a paving $\mathfrak{M}(n) \subset \mathfrak{S}$ of type $\bullet$ such that $\mathfrak{M}(n) \uparrow$ some $M_n \supset A_n$ and

$$\sup_{S \in \mathfrak{M}(n)} \varphi(S) \leqq \varphi^\bullet(A_n) + \varepsilon 2^{-n-1}.$$

1) We claim that

$$\varphi(S_1 \cup \cdots \cup S_n) < R + \varepsilon \quad \text{for } S_l \in \mathfrak{M}(l) \; (l = 1, \cdots, n) \text{ and } n \in \mathbb{N}.$$

To see this fix $l \in \{1, \cdots, n\}$. Then $\{P \cap Q : P \in \mathfrak{M}(l) \text{ and } Q \in \mathfrak{M}(n+1)\}$ is a paving $\subset \mathfrak{S}$ of type $\bullet$ which $\uparrow M_l \cap M_{n+1} \supset A_l \cap A_{n+1} = A_l$. Hence there exist $P_l \in \mathfrak{M}(l)$ and $Q_l \in \mathfrak{M}(n+1)$ such that $\varphi(P_l \cap Q_l) \geq \varphi^\bullet(A_l) - \varepsilon 2^{-l-1}$. We can assume that $P_l \supset S_l$. Also there exists $Q \in \mathfrak{M}(n+1)$ with $Q \supset Q_1 \cup \cdots \cup Q_n$. It follows that

$$\varphi(P_l \cap Q) \geq \varphi^\bullet(A_l) - \varepsilon 2^{-l-1} \ (l = 1, \cdots, n).$$

Now from 4.8 we have $\varphi(P_1 \cup \cdots \cup P_n \cup Q) < \infty$ and

$$\varphi(P_1 \cup \cdots \cup P_n \cup Q) + \sum_{l=1}^{n} \varphi(P_l \cap Q) \leq \sum_{l=1}^{n} \varphi(P_l) + \varphi(Q).$$

From the above we see that in this formula

the left side is $\geq \varphi(S_1 \cup \cdots \cup S_n) + \sum_{l=1}^{n} (\varphi^\bullet(A_l) - \varepsilon 2^{-l-1}),$

the right side is $\leq \sum_{l=1}^{n} (\varphi^\bullet(A_l) + \varepsilon 2^{-l-1}) + (\varphi^\bullet(A_{n+1}) + \varepsilon 2^{-n-2}).$

It follows that

$$\varphi(S_1 \cup \cdots \cup S_n) < \varphi^\bullet(A_{n+1}) + \sum_{l=1}^{n+1} \varepsilon 2^{-l} < R + \varepsilon.$$

2) Let $\mathfrak{M}$ consist of all unions $S_1 \cup \cdots \cup S_n$ with $S_l \in \mathfrak{M}(l) \ (l = 1, \cdots, n)$ and $n \in \mathbb{N}$. Then $\mathfrak{M}$ is a paving $\subset \mathfrak{S}$ of type $\bullet$. It is clear that $\mathfrak{M}$ is upward directed and $\mathfrak{M} \uparrow \bigcup_{n=1}^{\infty} M_n \supset \bigcup_{n=1}^{\infty} A_n = A$. Thus we have $\varphi^\bullet(A) \leq \sup_{S \in \mathfrak{M}} \varphi(S)$. Combined with 1) we obtain $\varphi^\bullet(A) \leq R + \varepsilon$ for all $\varepsilon > 0$ and hence the assertion.

Complements for the Nonsequential Situation

The sequential continuity theorem 4.7 has no nonsequential counterpart. The present subsection is a short discussion of the complications which arise from this fact.

4.9. REMARK. Let $\mathfrak{S}$ be a lattice with $\varnothing \in \mathfrak{S}$ and $\varphi : \mathfrak{S} \to [0, \infty]$ be an isotone and modular set function with $\varphi(\varnothing) = 0$ which attains at least one finite positive value. Assume that φ is upward τ continuous and that $\mathfrak{S}^\tau = \mathfrak{S}$. Then 4.5 implies that $\varphi^\tau = \varphi^\sigma = \varphi^\star \geq 0$. In this situation it can happen that $\varphi^\tau = \varphi^\sigma = \varphi^\star$ is not upward τ continuous. For example this is obvious when $\varphi^\tau(F) = \varphi^\sigma(F) = \varphi^\star(F) = 0$ for all finite $F \subset X$. As the simplest example we anticipate from 5.14 the Lebesgue measure on $\mathbb{R}^n$ restricted to $\mathrm{Op}(\mathbb{R}^n)$.

However, the outer $\bullet$ main theorem requires a certain touch of upward $\bullet$ continuity which is trivial for $\bullet = \star$ and a consequence of 4.7 for $\bullet = \sigma$.

Let us define an isotone set function $\varphi : \mathfrak{S} \to \overline{\mathbb{R}}$ to be **upward • essential** iff

$$\varphi^\bullet(A) = \sup\{\varphi^\bullet(A \cap S) : S \in [\varphi < \infty]\} \quad \text{for all } A \subset X \text{ with}$$
$$\infty > \varphi^\bullet(A) \geqq \sup\{\varphi^\bullet(A \cap S) : S \in [\varphi < \infty]\} > -\infty.$$

Then we obtain what follows.

4.10. PROPOSITION. *Let $\varphi : \mathfrak{S} \to \overline{\mathbb{R}}$ be isotone. $\star$) φ is upward $\star$ essential. σ) If φ is submodular $\dotplus$ then it is upward σ essential. τ) Assume that φ is submodular $\dotplus$ and such that each $A \subset X$ with $\varphi^\tau(A) < \infty$ is upward enclosable $[\varphi < \infty]^\sigma$. Then φ is upward τ essential.*

Proof. $\star$) is obvious since $\varphi^\star(A) < \infty$ implies the existence of some $S \in [\varphi < \infty]$ with $S \supset A$ and hence $A \cap S = A$. $\sigma)\tau$) We prove for $\bullet = \sigma\tau$ and φ submodular $\dotplus$ an intermediate assertion which implies both results: If $A \subset X$ is upward enclosable $[\varphi < \infty]^\sigma$ and fulfils

$$\sup\{\varphi^\bullet(A \cap S) : S \in [\varphi < \infty]\} > -\infty \text{ then}$$
$$\varphi^\bullet(A) = \sup\{\varphi^\bullet(A \cap S) : S \in [\varphi < \infty]\}.$$

In fact, let $(S_l)_l$ be a sequence in $[\varphi < \infty]$ with $S_l \uparrow \supset A$, and let $T \in [\varphi < \infty]$ with $\varphi^\bullet(A \cap T) > -\infty$. Then $S_l \cup T \in [\varphi < \infty]$ since φ is submodular $\dotplus$. Furthermore $A \cap (S_l \cup T) \uparrow A$ and $\varphi^\bullet(A \cap (S_l \cup T)) > -\infty$. Thus we obtain $\varphi^\bullet(A \cap (S_l \cup T)) \uparrow \varphi^\bullet(A)$ from 4.7.

In view of these results an isotone and submodular $\dotplus$ set function $\varphi : \mathfrak{S} \to \overline{\mathbb{R}}$ will be called **upward essential** instead of upward τ essential.

We conclude with an example which will illuminate the outer τ main theorems in the next section.

4.11. EXAMPLE. We fix $\mathfrak{S}$ in X and $\varphi : \mathfrak{S} \to [0, \infty]$ as described in 4.9 above. Define Y to consist of two disjoint copies of X, that is $Y := X \times \{0, 1\}$. We write the subsets $A \subset Y$ in the form $A = A^0 \sqcup A^1$ with $A^0, A^1 \subset X$. Define $\mathfrak{T}$ to consist of the subsets $A = A^0 \sqcup A^1 \subset Y$ with $A^0 \in \mathfrak{S}$ and A^1 finite $\subset A^0$. Thus $\mathfrak{T}$ is a lattice in Y with $\varnothing \in \mathfrak{T}$. Furthermore $\mathfrak{T}^\tau$ consists of the subsets $A = A^0 \sqcup A^1 \subset Y$ with $A^0 \in \mathfrak{S}$ and $A^1 \subset A^0$. Then define $\psi : \mathfrak{T} \to [0, \infty]$ to be $\psi(S) = \varphi(S^0)$ for $S = S^0 \sqcup S^1 \in \mathfrak{T}$. Thus ψ is isotone and modular with $\psi(\varnothing) = 0$. Also ψ is upward τ continuous. We prove three assertions.

1) ψ has no outer τ extension.
2) $\psi^\tau(A) = \varphi^\tau(A^0 \cup A^1)$ for $A = A^0 \sqcup A^1 \subset Y$.
3) ψ is not upward τ essential.

For 3) we use that $\varphi^\tau(F) = 0$ for all finite $F \subset X$.

Proof of 1). Let $\alpha : \mathfrak{A} \to \overline{\mathbb{R}}$ be an outer τ extension of ψ. Then $\varnothing \in \mathfrak{A}$ and $\alpha(\varnothing) = 0$, so that α is a ccontent on a ring $\mathfrak{A}$. Furthermore $\alpha(A) = \psi(A) = \varphi(A^0)$ for $A = A^0 \sqcup A^1 \in \mathfrak{T}$, and hence $\alpha(A) = \varphi(A^0)$ for $A = A^0 \sqcup A^1 \in \mathfrak{T}^\tau$ since $\alpha|\mathfrak{T}^\tau$ is upward τ continuous. Now fix $E \in \mathfrak{S}$ with $c := \varphi(E) \in]0, \infty[$. Then on the one hand

$$E \sqcup \varnothing \in \mathfrak{T} \quad \text{with } \alpha(E \sqcup \varnothing) = \varphi(E) = c,$$
$$E \sqcup E \in \mathfrak{T}^\tau \text{ with } \alpha(E \sqcup E) = \varphi(E) = c.$$

On the other hand $\varnothing \sqcup E \in \mathfrak{A}$ since $\mathfrak{A}$ is a ring, and

$$
\begin{aligned}
\alpha(\varnothing \sqcup E) &= \inf\{\alpha(A) : A \in \mathfrak{T}^\tau \text{ with } A \supset \varnothing \sqcup E\} \\
&= \inf\{\varphi(A^0) : A^0 \in \mathfrak{S} \text{ and } A^1 \subset A^0 \text{ with } A^1 \supset E\} \\
&= \varphi(E) = c,
\end{aligned}
$$

since α is outer regular $\mathfrak{T}^\tau$. These values combine to contradict the fact that α is modular.

Proof of 2). Both directions $\leqq$ and $\geqq$ are routine verifications. Proof of 3). Fix as above $E \in \mathfrak{S}$ with $c := \varphi(E) \in]0, \infty[$. For $A := \varnothing \sqcup E \subset Y$ then $\psi^\tau(A) = \varphi^\tau(E) = \varphi(E) = c$. On the other hand we obtain for $S = S^0 \sqcup S^1 \in \mathfrak{T}$ that $\psi^\tau(A \cap S) = \psi^\tau(\varnothing \sqcup (E \cap S^1)) = \varphi^\tau(E \cap S^1) = 0$ since S^1 is finite. The assertion follows.

The Extended Carathéodory Construction

We turn to the second task of the present section. We consider a set function $\phi : \mathfrak{P}(X) \to H$, defined on the full power set $\mathfrak{P}(X)$ of a nonvoid set X, and with values in a nonvoid set H which carries an associative and commutative addition $+$. We shall define and explore the so-called Carathéodory class $\mathfrak{C}(\phi)$ of ϕ, a paving in X. The definition is classical in case that H has the neutral element 0 and $\phi(\varnothing) = 0$. But in the present context there is no restriction for $\phi(\varnothing)$, it can in particular be a non-cancellable element of H. Recall that $a \in H$ is named **cancellable** iff for each pair $u, v \in H$ the implication $u + a = v + a \Rightarrow u = v$ holds true. Thus $H = \overline{\mathbb{R}}$ with $\dot{+}$ or $\dot{+}$ has the non-cancellable elements $\pm\infty$.

The new situation requires a drastic modification of the classical definition. We define the **Carathéodory class $\mathfrak{C}(\phi)$** of ϕ to consist of those subsets $A \subset X$ which fulfil

$$\phi(U) + \phi(V) = \phi(U|A|V) + \phi(U|A'|V) \quad \text{for all } U, V \subset X.$$

We proceed to list its basic properties.

4.12. PROPERTIES. 1) $\varnothing, X \in \mathfrak{C}(\phi)$. *Also $\mathfrak{C}(\phi)$ has $\perp$.* 2) *Assume that* $E \subset X$ *has cancellable value* $\phi(E) \in H$. *Then $\mathfrak{C}(\phi)$ consists of the subsets* $A \subset X$ *which fulfil*

$$\phi(P) + \phi(E) = \phi(P|A|E) + \phi(P|A'|E) \quad \text{for all } P \subset X.$$

3) *In particular assume that $\phi(\varnothing) = 0$ is neutral in H. Then $\mathfrak{C}(\phi)$ consists of the subsets $A \subset X$ which fulfil*

$$\phi(P) = \phi(P \cap A') + \phi(P \cap A) \quad \text{for all } P \subset X.$$

Thus we come back to the traditional definition of the class $\mathfrak{C}(\phi)$. 4) *If* $A \in \mathfrak{C}(\phi)$ *then*

$$(+) \qquad \phi(P) + \phi(A) = \phi(P \cup A) + \phi(P \cap A) \quad \text{for all } P \subset X.$$

*On the other hand a subset $A \subset X$ which satisfies $(+)$ need not be in $\mathfrak{C}(\phi)$;
by* 2) *it is in $\mathfrak{C}(\phi)$ when $\phi(A) \in H$ is cancellable.* 5) *Assume that there exists
an $E \subset X$ such that $\phi(E) \in H$ is cancellable. Then $\mathfrak{C}(\phi)$ is an algebra.*

Proof. 1) is obvious. 2) Let $A \subset X$ be as described above. In order to
see that $A \in \mathfrak{C}(\phi)$ we fix $P, Q \subset X$ and form $U := P|A|Q$ and $V := P|A'|Q$.
By assumption

$$\big(\phi(U) + \phi(E)\big) + \big(\phi(V) + \phi(E)\big)$$
$$= \big(\phi(U|A|E) + \phi(U|A'|E)\big) + \big(\phi(V|A|E) + \phi(V|A'|E)\big)$$
$$= \big(\phi(P|A|E) + \phi(Q|A'|E)\big) + \big(\phi(Q|A|E) + \phi(P|A'|E)\big)$$
$$= \big(\phi(P|A|E) + \phi(P|A'|E)\big) + \big(\phi(Q|A|E) + \phi(Q|A'|E)\big)$$
$$= \big(\phi(P) + \phi(E)\big) + \big(\phi(Q) + \phi(E)\big),$$

and hence $\phi(U) + \phi(V) = \phi(P) + \phi(Q)$ since $\phi(E)$ is cancellable. This is the
assertion. 3) is an obvious special case of 2). 4) For $A \in \mathfrak{C}(\phi)$ the equation
$(+)$ is the definition with $V := A$. For the converse a counterexample will
be in exercise 4.13 below. The last assertion is obvious. 5) We have to prove
that $\mathfrak{C}(\phi)$ has $\cup$. Fix $A, B \in \mathfrak{C}(\phi)$. For $P \subset X$ we form $U := P|A \cup B|E$
and $V := P|(A \cup B)'|E$. With the notations $M := P|A|E$ and $N := P|A'|E$
one computes that

$$
\begin{aligned}
M|B|E &= (M \cap B') \cup (E \cap B) \\
&= (P \cap A' \cap B') \cup (E \cap A \cap B') \cup (E \cap B) \\
&= \big(P \cap (A \cup B)'\big) \cup \big(E \cap (A \cup B)\big) = P|A \cup B|E = U, \\
M|B'|E &= (M \cap B) \cup (E \cap B') \\
&= (P \cap A' \cap B) \cup (E \cap A \cap B) \cup (E \cap B') \\
&= (P \cap A' \cap B) \cup (E \cap A' \cap B') \cup (E \cap A) \\
&= (V \cap A' \cap B) \cup (V \cap A' \cap B') \cup (E \cap A) \\
&= (V \cap A') \cup (E \cap A) = V|A|E, \\
N &= P|A'|E = V|A'|E.
\end{aligned}
$$

Since $A, B \in \mathfrak{C}(\phi)$ it follows that

$$\phi(U) + \big(\phi(V) + \phi(E)\big) = \phi(U) + \big(\phi(V|A|E) + \phi(V|A'|E)\big)$$
$$= \big(\phi(M|B|E) + \phi(M|B'|E)\big) + \phi(N) = \big(\phi(M) + \phi(E)\big) + \phi(N)$$
$$= \big(\phi(P|A|E) + \phi(P|A'|E)\big) + \phi(E) = \big(\phi(P) + \phi(E)\big) + \phi(E),$$

and hence that $\phi(U) + \phi(V) = \phi(P) + \phi(E)$. By 2) this is the assertion.

4.13. EXERCISE. Construct an example $\phi : \mathfrak{P}(X) \to H$ such that there
exists a subset $A \subset X$ which fulfils $(+)$ but is not in $\mathfrak{C}(\phi)$, and that furthermore there exists a subset $E \subset X$ with cancellable value $\phi(E) \in H$.
Hint: Let $H := [0, \infty]$ with the usual addition, and let $X = Y \cup Z$ with
nonvoid disjoint Y and Z. Define $\phi : \mathfrak{P}(X) \to H$ to be $\phi(Y) = \phi(Z) = 0$
and $\phi(A) = \infty$ for all other $A \subset X$. Then proceed as follows. 0) For
$E \subset X : \phi(E) \in H$ is cancellable iff $E = Y$ or Z. 1) For $A \subset X : A$ fulfils
$(+)$ iff $A \neq Y, Z$. 2) For $A \subset X : A \in \mathfrak{C}(\phi)$ iff $A = \varnothing$ or X.

In the sequel we concentrate on the particular case that $H = \overline{\mathbb{R}}$ with one of the additions $\dotplus$ and $+$. For a set function $\phi : \mathfrak{P}(X) \to \overline{\mathbb{R}}$ we then write $\mathfrak{C}(\phi, \dotplus)$. No specification $\dotplus$ is needed when ϕ attains at most one of the values $\pm\infty$.

4.14. REMARK. $\mathfrak{C}(\phi, \dotplus)$ *is an algebra.*

Proof. By the above 4.12.5) it remains to consider the cases that the value set of ϕ is one of the singletons $\{\pm\infty\}$ or $\{-\infty, \infty\}$. Then it suffices to note that on $\{-\infty, \infty\}$ the element $-\infty$ is cancellable for $\dotplus$ and the element ∞ is cancellable for $+$.

4.15. EXERCISE. $\mathfrak{C}(\phi, +) = \mathfrak{C}(\phi\perp, \dotplus)$.

4.16. REMARK (Symmetrization). *Assume that $A \subset X$ satisfies*

$$\phi(P)\dotplus\phi(Q) \geqq \phi(P|A|Q)\dotplus\phi(P|A'|Q) \quad \text{for all } P, Q \subset X.$$

Then $A \in \mathfrak{C}(\phi, \dotplus)$.

Proof. We know from 1.1.5) that $U := P|A|Q$ and $V := P|A'|Q$ have $U|A|V = P$ and $U|A'|V = Q$. It follows that

$$\begin{aligned}\phi(P|A|Q)\dotplus\phi(P|A'|Q) &= \phi(U)\dotplus\phi(V) \\ &\geqq \phi(U|A|V)\dotplus\phi(U|A'|V) = \phi(P)\dotplus\phi(Q),\end{aligned}$$

and hence the assertion.

The Carathéodory Class in the Spirit of the Outer Theory

We proceed to consider the Carathéodory class $\mathfrak{C}(\phi, \dotplus)$ of an isotone set function $\phi : \mathfrak{P}(X) \to \overline{\mathbb{R}}$ under assumptions in the spirit of the outer theory. We shall see that the definition of $\mathfrak{C}(\phi, \dotplus)$ then admits substantial simplifications. We start with a simple remark.

4.17. REMARK. *Let $\mathfrak{T}$ be a paving in X and $\phi : \mathfrak{P}(X) \to \overline{\mathbb{R}}$ be isotone and outer regular $\mathfrak{T}$. If $A \subset X$ satisfies*

$$\phi(P)\dotplus\phi(Q) \geqq \phi(P|A|Q)\dotplus\phi(P|A'|Q) \quad \text{for all } P, Q \in \mathfrak{T},$$

then $A \in \mathfrak{C}(\phi, \dotplus)$.

Proof. By symmetrization it suffices to prove that

$$\phi(U)\dotplus\phi(V) \geqq \phi(U|A|V)\dotplus\phi(U|A'|V) \quad \text{for all } U, V \subset X.$$

We fix $U, V \subset X$ and can assume that $\phi(U), \phi(V) < \infty$. For fixed real $c > \phi(U) + \phi(V)$ there are $a, b \in \mathbb{R}$ with $c = a + b$ and with $\phi(U) < a$ and $\phi(V) < b$. By assumption there exist $P, Q \in \mathfrak{T}$ such that $P \supset U$ and $\phi(P) < a$, $Q \supset V$ and $\phi(Q) < b$. Hence

$$\phi(U|A|V)\dotplus\phi(U|A'|V) \leqq \phi(P|A|Q)\dotplus\phi(P|A'|Q) \leqq \phi(P)\dotplus\phi(Q) < a + b = c.$$

The assertion follows.

4.18. EXERCISE. *Let $\alpha : \mathfrak{A} \to \overline{\mathbb{R}}$ be a content $\dotplus$ on an oval $\mathfrak{A}$. Then $\mathfrak{A} \subset \mathfrak{C}(\alpha^\star, \dotplus)$.*

The main point is that the verification of $A \in \mathfrak{C}(\phi, \dotplus)$ can be reduced to certain pairs of subsets $P \subset Q$ of X. The basis is the fundamental lemma which follows. We use the abbreviation $P \sqsubset Q := \{P\} \sqsubset \{Q\} = \{A \subset X : P \subset A \subset Q\}$.

4.19. LEMMA. *Let* $P \subset Q \subset X$. *Assume that* $\varphi : P \sqsubset Q \to \mathbb{R}$ *is submodular. If* $A \subset X$ *satisfies*

$$\varphi(P) + \varphi(Q) \geqq \varphi(P|A|Q) + \varphi(P|A'|Q),$$

then

$$\varphi(U) + \varphi(V) = \varphi(U|A|V) + \varphi(U|A'|V) \quad \text{for all } U, V \in P \sqsubset Q.$$

Proof. By symmetrization it suffices to prove the assertion with $\geqq$. Fix $U, V \in P \sqsubset Q$. i) From the assumption we have

$$\varphi(P) + \varphi(Q) + 2\varphi(U) \geqq \big(\varphi(P|A|Q) + \varphi(U)\big) + \big(\varphi(P|A'|Q) + \varphi(U)\big).$$

Since φ is submodular and $P \subset U \subset Q$ this is

$$\geqq \big(\varphi(U|A|Q) + \varphi(P|A|U)\big) + \big(\varphi(U|A'|Q) + \varphi(P|A'|U)\big);$$

and when we use submodularity for the two first terms in the brackets and repeat the two second terms this is

$$\geqq \varphi(Q) + \varphi(U) + \varphi(P|A|U) + \varphi(P|A'|U).$$

Thus we have

$$\varphi(P) + \varphi(U) \geqq \varphi(P|A|U) + \varphi(P|A'|U).$$

Of course we have likewise

$$\varphi(P) + \varphi(V) \geqq \varphi(P|A|V) + \varphi(P|A'|V).$$

ii) We add the last two inequalities and use submodularity twice on the right for the two pairs of terms which were in crosswise position. Then we obtain

$$2\varphi(P) + \varphi(U) + \varphi(V) \geqq \big(\varphi(V|A|U) + \varphi(P)\big) + \big(\varphi(U|A|V) + \varphi(P)\big),$$

and hence the assertion.

From the lemma we deduce the next result which looks somewhat technical but will be a powerful tool.

4.20. PROPOSITION. *Assume that* $\phi : \mathfrak{P}(X) \to \bar{\mathbb{R}}$ *is an isotone set function. Let* $\mathfrak{P} \downarrow$ *and* $\mathfrak{Q} \uparrow$ *be pavings in* X *with nonvoid* $\mathfrak{P} \sqsubset \mathfrak{Q}$ *such that*

$\phi|\mathfrak{P}$ *and* $\phi|\mathfrak{Q}$ *are finite, and* $\phi|\mathfrak{P} \sqsubset \mathfrak{Q}$ *is submodular.*

Furthermore let $\mathfrak{H} \uparrow$ *be a paving in* X *with* $\mathfrak{Q} \subset \mathfrak{H}$ *such that*

ϕ *is outer regular* $\mathfrak{P} \sqsubset \mathfrak{H}$,

$\phi(T) = \sup\limits_{Q \in \mathfrak{Q}} \phi(T \cap Q)$ *for all* $T \in \mathfrak{P} \sqsubset \mathfrak{H}.$

If $A \subset X$ *satisfies*

$$\phi(P) + \phi(Q) \quad \geqq \quad \phi(P|A|Q) + \phi(P|A'|Q)$$
$$\text{for all } P \in \mathfrak{P} \text{ and } Q \in \mathfrak{Q} \text{ with } P \subset Q,$$

then $A \in \mathfrak{C}(\phi, \dotplus).$

Proof. i) We know from 1.8.1$\star$) that $\mathfrak{P} \sqsubset \mathfrak{Q}$ and $\mathfrak{P} \sqsubset \mathfrak{H}$ are ovals. ii) For each pair $P \in \mathfrak{P}$ and $Q \in \mathfrak{Q}$ there exists a pair $A \in \mathfrak{P}$ and $B \in \mathfrak{Q}$ with $A \subset P$ and $Q \subset B$ such that $A \subset B$. In fact, by assumption there exist $U \in \mathfrak{P}$ and $V \in \mathfrak{Q}$ with $U \subset V$. Then by directedness there are $A \in \mathfrak{P}$ with $A \subset P, U$ and $B \in \mathfrak{Q}$ with $B \supset Q, V$. It is obvious that A and B are as required. iii) For each pair $P \in \mathfrak{P}$ and $Q \in \mathfrak{Q}$ with $P \subset Q$ we see from 4.19 that

$$\phi(U) + \phi(V) = \phi(U|A|V) + \phi(U|A'|V) \quad \text{for all } U, V \in P \sqsubset Q.$$

By directedness it follows that

$$\phi(U) + \phi(V) = \phi(U|A|V) + \phi(U|A'|V) \quad \text{for all } U, V \in \mathfrak{P} \sqsubset \mathfrak{Q}.$$

Note that $U|A|V$ and $U|A'|V$ are in $\mathfrak{P} \sqsubset \mathfrak{Q}$ as well. iv) In view of 4.17 applied to $\mathfrak{T} := \mathfrak{P} \sqsubset \mathfrak{H}$ it suffices to prove that

$$\phi(U) + \phi(V) \geq \phi(U|A|V) + \phi(U|A'|V) \quad \text{for all } U, V \in \mathfrak{T}.$$

Note that $U|A|V$ and $U|A'|V$ are in $\mathfrak{T}$ as well, and that $\phi|\mathfrak{T} > -\infty$. v) Fix $U, V \in \mathfrak{T}$. Also fix $M, N \in \mathfrak{Q}$ and then $Q \in \mathfrak{Q}$ with $M, N \subset Q$. By ii) we can assume that Q is downward enclosable $\mathfrak{P}$. Thus $U \cap Q, V \cap Q \in \mathfrak{P} \sqsubset \mathfrak{Q}$. By iii) therefore

$$
\begin{aligned}
\phi(U) + \phi(V) \quad &\geq \quad \phi(U \cap Q) + \phi(V \cap Q) \\
&= \quad \phi(U \cap Q|A|V \cap Q) + \phi(U \cap Q|A'|V \cap Q) \\
&= \quad \phi\big((U|A|V) \cap Q\big) + \phi\big((U|A'|V) \cap Q\big) \\
&\geq \quad \phi\big((U|A|V) \cap M\big) \dot{+} \phi\big((U|A'|V) \cap N\big).
\end{aligned}
$$

Now the supremum over $M, N \in \mathfrak{Q}$ of the right side is $= \phi(U|A|V) + \phi(U|A'|V)$ since the two partial suprema are both $> -\infty$. By iv) the proof is complete.

4.21. ADDENDUM. *Assume in addition that $\phi|\mathfrak{P} \sqsubset \mathfrak{Q}$ is upward σ continuous. Then $\mathfrak{C}(\phi, \dot{+})$ is a σ algebra.*

Proof. Let $(A_l)_l$ be a sequence in $\mathfrak{C}(\phi, \dot{+})$ with $A_l \uparrow A$. To be shown is $A \in \mathfrak{C}(\phi, \dot{+})$. Fix $P \in \mathfrak{P}$ and $Q \in \mathfrak{Q}$ with $P \subset Q$. By assumption we have

$$
\begin{aligned}
\phi(P) + \phi(Q) \quad &= \quad \phi(P|A_l|Q) + \phi(P|A_l'|Q) \\
&= \quad \phi\big(P \cup (Q \cap A_l)\big) + \phi\big(P \cup (Q \cap A_l')\big) \\
&\geq \quad \phi\big(P \cup (Q \cap A_l)\big) + \phi\big(P \cup (Q \cap A')\big),
\end{aligned}
$$

since $A_l \subset A$ and hence $A_l' \supset A'$. Here all arguments are in $P \sqsubset Q$ and hence all values are finite. By assumption it follows that

$$
\begin{aligned}
\phi(P) + \phi(Q) \quad &\geq \quad \phi\big(P \cup (Q \cap A)\big) + \phi\big(P \cup (Q \cap A')\big) \\
&= \quad \phi(P|A|Q) + \phi(P|A'|Q).
\end{aligned}
$$

Thus from 4.20 we obtain $A \in \mathfrak{C}(\phi, \dot{+})$.

We include another addendum to 4.20 for the sake of chapter VI.

4.22. ADDENDUM. *Assume in addition that* $\theta : \mathfrak{P}(X) \to \overline{\mathbb{R}}$ *is an isotone set function with* $\theta \geqq \phi$ *such that*

$\theta|\mathfrak{P} = \phi|\mathfrak{P}$ *and* $\theta|\mathfrak{Q} = \phi|\mathfrak{Q}$,

$\theta(T) = \sup_{Q \in \mathfrak{Q}} \theta(T \cap Q)$ *for all* $T \in \mathfrak{P} \sqsubset \mathfrak{H}$.

Then $\phi|\mathfrak{C}(\phi, \dotplus)$ *is an extension of* $\theta|\mathfrak{C}(\theta, \dotplus)$.

Proof. Fix $A \in \mathfrak{C}(\theta, \dotplus)$. i) In order to prove that $A \in \mathfrak{C}(\phi, \dotplus)$ let $P \in \mathfrak{P}$ and $Q \in \mathfrak{Q}$ with $P \subset Q$. Then

$$\phi(P) + \phi(Q) = \theta(P) + \theta(Q) \;=\; \theta(P|A|Q) + \theta(P|A'|Q)$$
$$\geqq\; \phi(P|A|Q) + \phi(P|A'|Q).$$

From 4.20 the assertion follows. ii) We claim that $\theta(P|A|Q) = \phi(P|A|Q) \in \mathbb{R}$ for all $P \in \mathfrak{P}$ and $Q \in \mathfrak{Q}$. This follows at once from

$$\phi(P) + \phi(Q) = \theta(P) + \theta(Q) \;=\; \theta(P|A|Q) \dotplus \theta(P|A'|Q)$$
$$\geqq\; \phi(P|A|Q) \dotplus \phi(P|A'|Q) = \phi(P) + \phi(Q).$$

iii) It remains to prove that $\theta(A) \leqq \phi(A)$ and hence $\theta(A) = \phi(A)$. We fix $V \in \mathfrak{P} \sqsubset \mathfrak{H}$ with $V \supset A$ and have to show that $\theta(A) \leqq \phi(V)$. Let $P \in \mathfrak{P}$ with $P \subset V$ and $Q \in \mathfrak{Q}$. Then on the one hand

$$\theta(P|A|Q) \geqq \theta(P \cap Q|A|V \cap Q) = \theta\big((P|A|V) \cap Q\big),$$

so that from the assumption and $P|A|V \in \mathfrak{P} \sqsubset \mathfrak{H}$ and from $P|A|V \supset V \cap A = A$ we obtain

$$\sup_{Q \in \mathfrak{Q}} \theta(P|A|Q) \geqq \theta(P|A|V) \geqq \theta(A).$$

On the other hand we have $P|A|Q \subset P \cup A \subset V$ and hence

$$\sup_{Q \in \mathfrak{Q}} \phi(P|A|Q) \leqq \phi(V).$$

From ii) the assertion follows.

5. The Outer Extension Theory: The Main Theorem

The Outer Main Theorem

In the last section we have developed the concepts and instruments which we need in order to reach our principal aim as formulated after the basic definition. The first theorem below is a clear hint that these devices are adequate.

The present subsection is under the assumption that $\varphi : \mathfrak{S} \to]-\infty, \infty]$ is an isotone set function $\not\equiv \infty$ on a lattice $\mathfrak{S}$ in X.

5.1. THEOREM. *Assume that* $\alpha : \mathfrak{A} \to \overline{\mathbb{R}}$ *is an outer* • *extension of* φ. *Then* α *is a restriction of* $\varphi^{\bullet}|\mathfrak{C}(\varphi^{\bullet}, \dotplus)$.

Proof. i) $\varphi = \alpha|\mathfrak{S}$ is upward $\bullet$ continuous, and hence $\varphi = \varphi^\bullet|\mathfrak{S}$ by 4.5. Then $\alpha = \varphi = \varphi^\bullet$ on $\mathfrak{S}$ implies $\alpha = \varphi^\bullet$ on $\mathfrak{S}^\bullet$ by 4.3.iii), and hence $\alpha = \varphi^\bullet$ on $\mathfrak{A}$ by 4.1.4). ii) It remains to prove that $\mathfrak{A} \subset \mathfrak{C}(\varphi^\bullet, \dotplus)$. Fix $A \in \mathfrak{A}$. For $P, Q \in \mathfrak{S}^\bullet$ then

$$\alpha(P) \dotplus \alpha(Q) = \alpha(P|A|Q) \dotplus (P|A'|Q),$$

where all arguments are in $\mathfrak{A}$ since $\mathfrak{A}$ is an oval. In fact, since α is modular $\dotplus$ both sides are $= \alpha(P \cup Q) \dotplus \alpha(P \cap Q)$. Since now $\alpha = \varphi^\bullet$ on $\mathfrak{A}$ by i) we obtain $A \in \mathfrak{C}(\varphi^\bullet, \dotplus)$ from 4.17 applied to $\phi := \varphi^\bullet$ and $\mathfrak{T} := \mathfrak{S}^\bullet$.

We prepare the outer main theorem with the important next result.

5.2. PROPOSITION. *Let φ be submodular with $\varphi^\bullet|\mathfrak{S} > -\infty$, and upward essential in case $\bullet = \tau$. Fix pavings*

$\mathfrak{P} \subset [\varphi < \infty]$ *downward cofinal, that is such that $[\varphi < \infty] \subset (\sqsupset \mathfrak{P})$,*
$\mathfrak{Q} \subset [\varphi < \infty]$ *upward cofinal, that is such that $[\varphi < \infty] \subset (\sqsubset \mathfrak{Q})$.*

If $A \subset X$ satisfies

$$\varphi^\bullet(P) + \varphi^\bullet(Q) \;\geqq\; \varphi^\bullet(P|A|Q) + \varphi^\bullet(P|A'|Q)$$
$$\text{for all } P \in \mathfrak{P} \text{ and } Q \in \mathfrak{Q} \text{ with } P \subset Q,$$

then $A \in \mathfrak{C}(\varphi^\bullet, \dotplus)$.

5.3. ADDENDUM. *For $\bullet = \sigma\tau$ the class $\mathfrak{C}(\varphi^\bullet, \dotplus)$ is a σ algebra.*

Proof of 5.2 and 5.3. We deduce the assertions from 4.20 and 4.21. i) $[\varphi < \infty]$ is a lattice by 2.4 and nonvoid by assumption. Therefore $\mathfrak{P} \downarrow$ and $\mathfrak{Q} \uparrow$, and $\mathfrak{P} \sqsubset \mathfrak{Q} \supset [\varphi < \infty]$ is nonvoid. ii) $\phi := \varphi^\bullet$ is isotone and submodular $\dotplus$ by 4.1.3)5). By assumption and 4.1.1)2) ϕ is finite on $[\varphi < \infty]$ and hence on $\mathfrak{P} \sqsubset \mathfrak{Q}$. iii) We define $\mathfrak{H} := [\varphi^\bullet|\mathfrak{S}^\bullet < \infty]$ and note that $\varphi^\bullet|\mathfrak{S}^\bullet > -\infty$. Then $[\varphi < \infty] \subset \mathfrak{H}$ and hence $\mathfrak{Q} \subset \mathfrak{H}$. By ii) and 2.4 $\mathfrak{H}$ is a lattice and hence $\mathfrak{H} \uparrow$. iv) $\sqsupset \mathfrak{P}$ contains $[\varphi < \infty]$ and hence $\mathfrak{S}$, therefore $\mathfrak{S}^\bullet$ and in particular $\mathfrak{H}$. Thus $\mathfrak{H} \subset \mathfrak{P} \sqsubset \mathfrak{H}$. By 4.1.4) therefore ϕ is outer regular $\mathfrak{P} \sqsubset \mathfrak{H}$. v) For $T \in \mathfrak{P} \sqsubset \mathfrak{H}$ we have by definition $\varphi^\bullet(T) < \infty$ and $\sup\{\varphi^\bullet(T \cap S) : S \in [\varphi < \infty]\} > -\infty$. Since φ is upward $\bullet$ essential by 4.10.$\star$)σ) and by assumption we have

$$\phi(T) = \sup\{\phi(T \cap S) : S \in [\varphi < \infty]\} = \sup\{\phi(T \cap S) : S \in \mathfrak{Q}\}.$$

vi) After this the assertions follow from 4.20 and 4.21.

5.4. CONSEQUENCE. *Let φ be submodular with $\varphi^\bullet|\mathfrak{S} > -\infty$, and upward essential in case $\bullet = \tau$. Assume that $\mathfrak{Q} \subset [\varphi < \infty]$ is upward cofinal. Then*

$$\mathfrak{Q}\top\mathfrak{C}(\varphi^\bullet, \dotplus) \subset \mathfrak{C}(\varphi^\bullet, \dotplus).$$

Proof. We put $\mathfrak{P} := [\varphi < \infty]$. Fix $A \in \mathfrak{Q}\top\mathfrak{C}(\varphi^\bullet, \dotplus)$. For $P \in \mathfrak{P}$ and $Q \in \mathfrak{Q}$ we have $A \cap Q \in \mathfrak{C}(\varphi^\bullet, \dotplus)$ and hence

$$\varphi^\bullet(P) + \varphi^\bullet(Q) = \varphi^\bullet(P|A \cap Q|Q) + \varphi^\bullet(P|(A \cap Q)'|Q).$$

In case $P \subset Q$ the right side is $= \varphi^\bullet(P|A|Q) + \varphi^\bullet(P|A'|Q)$. Thus from 5.2 we obtain $A \in \mathfrak{C}(\varphi^\bullet, \dotplus)$.

We come to the central result of the present chapter.

5.5. THEOREM (Outer Main Theorem). *Let* $\varphi : \mathfrak{S} \to]-\infty, \infty]$ *be an isotone and submodular set function* $\not\equiv \infty$ *on a lattice* $\mathfrak{S}$. *Fix pavings*

$\mathfrak{P} \subset [\varphi < \infty]$ *downward cofinal, and*
$\mathfrak{Q} \subset [\varphi < \infty]$ *upward cofinal.*

Then the following are equivalent.

1) *There exist outer* • *extensions of* φ, *that is* φ *is an outer* • *premeasure.*

2) $\varphi^\bullet | \mathfrak{C}(\varphi^\bullet, \dotplus)$ *is an outer* • *extension of* φ. *Furthermore*

$$if \ \bullet = \star \ : \ \ \varphi^\bullet | \mathfrak{C}(\varphi^\bullet, \dotplus) \ is \ a \ content \ \dotplus \ on \ the \ algebra \ \mathfrak{C}(\varphi^\bullet, \dotplus),$$
$$if \ \bullet = \sigma\tau \ : \ \ \varphi^\bullet | \mathfrak{C}(\varphi^\bullet, \dotplus) \ is \ a \ measure \ \dotplus \ on \ the \ \sigma \ algebra \ \mathfrak{C}(\varphi^\bullet, \dotplus).$$

3) $\varphi^\bullet | \mathfrak{C}(\varphi^\bullet, \dotplus)$ *is an extension of* φ *in the crude sense, that is* $\mathfrak{S} \subset \mathfrak{C}(\varphi^\bullet, \dotplus)$ *and* $\varphi = \varphi^\bullet | \mathfrak{S}$.

4) $\varphi(U) + \varphi(V) = \varphi(M) \dotplus \varphi^\bullet(U|M'|V)$ *for all* $U \subset M \subset V$ *in* $\mathfrak{S}$; *note that* $M = U|M|V$. *In case* $\bullet = \tau$ *furthermore* φ *is upward essential.*

5) $\varphi = \varphi^\bullet | \mathfrak{S}$; *and* $\varphi(P) + \varphi(Q) \geq \varphi(M) + \varphi^\bullet(P|M'|Q)$ *for all* $P \subset M \subset Q$ *with* $P \in \mathfrak{P}, Q \in \mathfrak{Q}$, *and* $M \in \mathfrak{S}$ *and hence* $\in [\varphi < \infty]$. *In case* $\bullet = \tau$ *furthermore* φ *is upward essential.*

Note that 5.4 then implies $\mathfrak{Q}\top\mathfrak{S}^\bullet \subset \mathfrak{C}(\varphi^\bullet, \dotplus)$.

A posteriori it turns out that condition 5) is independent of the pavings $\mathfrak{P}$ and $\mathfrak{Q}$. But the present formulation is important for later specializations.

Proof. We prove 2)⇒1)⇒3)⇒4)⇒5)⇒2). The implication 2)⇒1) is obvious, and 1)⇒3) follows from 5.1. 3)⇒4) The first assertion follows from the definition of $\mathfrak{C}(\varphi^\bullet, \dotplus)$. It remains to show in case $\bullet = \tau$ that φ is upward τ essential. If not, then there exists $A \subset X$ such that

$$\infty > \varphi^\tau(A) > \sup \{\varphi^\tau(A \cap S) : S \in [\varphi < \infty]\} > -\infty.$$

Let $\varepsilon := \varphi^\tau(A) - \sup \{\varphi^\tau(A \cap S) : S \in [\varphi < \infty]\} > 0$. For $S \in [\varphi < \infty]$ then $\varphi(S) = \varphi^\tau(S) \in \mathbb{R}$ and $S \in \mathfrak{C}(\varphi^\tau, \dotplus)$. Therefore

$$\varphi^\tau(S) \dotplus \varphi^\tau(A) = \varphi^\tau(S|S|A) \dotplus \varphi^\tau(S|S'|A) = \varphi^\tau(A \cap S) \dotplus \varphi^\tau(A \cup S),$$

with all terms finite, and hence

$$\varphi(S) + \varphi^\tau(A) \leq \varphi^\tau(A) - \varepsilon + \varphi^\tau(A \cup S), \ or \ \varphi(S) + \varepsilon \leq \varphi^\tau(A \cup S).$$

Fix a paving $\mathfrak{M} \subset [\varphi < \infty]$ with $\mathfrak{M} \uparrow M \supset A$ and $\sup\limits_{S \in \mathfrak{M}} \varphi(S) < \infty$. By 4.5.iii) then $M \in \mathfrak{S}^\tau$ and $\sup\limits_{S \in \mathfrak{M}} (S) = \varphi^\tau(M) \in \mathbb{R}$. It follows that $\varphi^\tau(M) + \varepsilon \leq \varphi^\tau(M)$ and thus a contradiction. 4)⇒5) For $V \in \mathfrak{S}$ and $U = M \in [\varphi < \infty]$ contained in V we obtain from 4) that $\varphi(V) = \varphi^\bullet(V)$.

It remains to prove the implication 5)⇒2). For the remainder of the proof we assume 5). Then the assumptions of 5.2 and 5.3 are fulfilled. i) $\mathfrak{C}(\varphi^\bullet, \dotplus)$ is an algebra by 4.14, and $\alpha := \varphi^\bullet | \mathfrak{C}(\varphi^\bullet, \dotplus)$ is isotone and modular $\dotplus$ by 4.12.4). For $\bullet = \sigma\tau$ the class $\mathfrak{C}(\varphi^\bullet, \dotplus)$ is a σ algebra by 5.3.

ii) We conclude from 5.2 that $\mathfrak{S}^\bullet \subset \mathfrak{C}(\varphi^\bullet, \dot{+})$. Fix $A \in \mathfrak{S}^\bullet$. Then let $\mathfrak{M} \subset \mathfrak{S}$ be a paving of type $\bullet$ with $\mathfrak{M} \uparrow A$. Furthermore fix $P \in \mathfrak{P}$ and $Q \in \mathfrak{Q}$ with $P \subset Q$. For $S \in \mathfrak{M}$ we form

$$M := P|S|Q = P \cup (Q \cap S) \in \mathfrak{S} \text{ with } P \subset M \subset Q,$$

so that from 5) we obtain $\varphi(P) + \varphi(Q) \geq \varphi(M) + \varphi^\bullet(P|M'|Q)$. Here

$$
\begin{aligned}
P|M'|Q &= P \cup (Q \cap M') = P \cup (Q \cap P' \cap (Q' \cup S')) \\
&= P \cup (Q \cap P' \cap S') = P \cup (Q \cap S') \supset P \cup (Q \cap A') = P|A'|Q,
\end{aligned}
$$

so that

$$\varphi(P) + \varphi(Q) \geq \varphi\big(P \cup (Q \cap S)\big) + \varphi^\bullet(P|A'|Q).$$

Now $\{P \cup (Q \cap S) : S \in \mathfrak{M}\} \uparrow P \cup (Q \cap A) = P|A|Q$. From 4.5.iii) it follows that

$$\varphi(P) + \varphi(Q) \geq \varphi^\bullet(P|A|Q) + \varphi^\bullet(P|A'|Q).$$

Thus 5.2 implies that $A \in \mathfrak{C}(\varphi^\bullet, \dot{+})$ as claimed.

iii) In particular $\mathfrak{S} \subset \mathfrak{C}(\varphi^\bullet, \dot{+})$, so that α is an extension of φ. Thus α attains at least one finite value and hence is a content $\dot{+}$. Furthermore α is outer regular $\mathfrak{S}^\bullet$ by 4.1.4), and $\alpha|\mathfrak{S}^\bullet$ is upward $\bullet$ continuous by 4.5.iii). Therefore α is an outer $\bullet$ extension of φ. iv) It remains to prove that α is a measure $\dot{+}$ when $\bullet = \sigma\tau$. By 4.7 α is almost upward σ continuous. Thus by 2.11 α is almost downward σ continuous as well, provided that it is semifinite below, that is outer regular $[\alpha > -\infty]$. But we know from ii) that $\mathfrak{S}^\bullet$ is in $\mathfrak{C}(\varphi^\bullet, \dot{+})$ and hence in $[\alpha > -\infty]$. Thus α is outer regular $[\alpha > -\infty]$. The proof is complete.

The above outer main theorem fulfils the promise made after the basic definition: Conditions 4) and 5) characterize the outer $\bullet$ premeasures. Combined with 5.1 we see that for an outer $\bullet$ premeasure φ all outer $\bullet$ extensions are restrictions of a unique maximal one, which is $\varphi^\bullet|\mathfrak{C}(\varphi^\bullet, \dot{+})$. Thus we arrive at a natural and simple situation, and our concepts and instruments prove to be adequate. We want to put particular emphasis on the role of the Carathéodory class, because its initial creation was not at all connected with regularity.

5.6. REMARK. The outer main theorem would be false in case $\bullet = \tau$ if in 4)5) the condition that φ be upward essential would be omitted. To see this we return to example 4.11 and adopt the former notations. The set function $\psi : \mathfrak{T} \to [0, \infty]$ then violates 1) as shown in 4.11.1). On the other hand we anticipate from 5.14 that we could have started from a set function $\varphi : \mathfrak{S} \to [0, \infty]$ as described in 4.9 with $\varphi^\tau(F) = 0$ for all finite $F \subset X$ which has an outer τ extension and thus is an outer τ premeasure. Then the set function $\psi : \mathfrak{T} \to [0, \infty]$ fulfils the first part of 4). In fact, we see from 4.11.2) and since φ^τ is submodular that this condition reads

$$\varphi(U) + \varphi(V) = \varphi(M) + \varphi^\tau(U|M'|V) \quad \text{for all } U \subset M \subset V \text{ in } \mathfrak{S},$$

and therefore is satisfied.

5.7. SPECIAL CASE (Traditional Type). *Assume that $\mathfrak{S}$ is an oval. Then condition 5) simplifies to*

5o) $\varphi = \varphi^\bullet | \mathfrak{S}$; *and φ is supermodular. In case $\bullet = \tau$ furthermore φ is upward essential.*

Proof of 5o)$\Rightarrow$5). If $P \subset M \subset Q$ are as in 5) then $P|M'|Q = Q|M|P =:$ $N \in \mathfrak{S}$ and hence $\in [\varphi < \infty]$. We have $M \cap N = P$ and $M \cup N = Q$, and hence $\varphi(M) + \varphi^\bullet(N) = \varphi(M) + \varphi(N) \leq \varphi(P) + \varphi(Q)$. Proof of 5)$\Rightarrow$5o). One notes that 5)$\Rightarrow$1) $\Rightarrow$ φ is supermodular, or that 5) for $\mathfrak{P} = \mathfrak{Q} = [\varphi < \infty] \Rightarrow$ φ is supermodular.

We shall soon turn to the most important special cases. But first we want to terminate the present context with a short comparison of the three cases $\bullet = \star\sigma\tau$.

Comparison of the three Outer Theories

In the present subsection we assume that $\varphi : \mathfrak{S} \to]-\infty, \infty]$ is an isotone and submodular set function $\not\equiv \infty$ on a lattice $\mathfrak{S}$.

5.8. PROPOSITION. σ) *In case $\varphi = \varphi^\sigma | \mathfrak{S}$ we have $\mathfrak{C}(\varphi^\star, \dotplus) \subset \mathfrak{C}(\varphi^\sigma, \dotplus)$. τ) In case $\varphi = \varphi^\tau | \mathfrak{S}$ and φ upward essential we have $\mathfrak{C}(\varphi^\sigma, \dotplus) \subset \mathfrak{C}(\varphi^\tau, \dotplus)$.*

Proof. Combine $\varphi^\star \geq \varphi^\sigma \geq \varphi^\tau$ with 5.2 for $\mathfrak{P} = \mathfrak{Q} = [\varphi < \infty]$.

5.9. PROPOSITION. *Assume that φ is modular. σ) In case $\varphi = \varphi^\sigma | \mathfrak{S}$ we have $\varphi^\star(A) = \varphi^\sigma(A)$ for all $A \in \mathfrak{C}(\varphi^\star, \dotplus)$ with $\varphi^\star(A) < \infty$. τ) In case $\varphi = \varphi^\tau | \mathfrak{S}$ we have $\varphi^\sigma(A) = \varphi^\tau(A)$ for all $A \in \mathfrak{C}(\varphi^\sigma, \dotplus)$ with $\varphi^\sigma(A) < \infty$.*

Proof. σ) Fix $A \in \mathfrak{C}(\varphi^\star, \dotplus)$ with $\varphi^\star(A) < \infty$. For $P, Q \in [\varphi < \infty]$ we have by 4.1.5)

$$
\begin{aligned}
\varphi(P) + \varphi(Q) &= \varphi^\star(P) + \varphi^\star(Q) = \varphi^\star(P|A|Q) + \varphi^\star(P|A'|Q) \\
&\geq \varphi^\sigma(P|A|Q) + \varphi^\sigma(P|A'|Q) \geq \varphi^\sigma(P \cup Q) + \varphi^\sigma(P \cap Q) \\
&= \varphi(P \cup Q) + \varphi(P \cap Q) = \varphi(P) + \varphi(Q),
\end{aligned}
$$

where all arguments are between $P \cap Q$ and $P \cup Q$ and hence all values are finite. It follows that

$$\varphi^\star(P|A|Q) = \varphi^\sigma(P|A|Q) \text{ for all } P, Q \in [\varphi < \infty].$$

Since $\varphi^\star(A) < \infty$ there exists $Q \in [\varphi < \infty]$ with $Q \supset A$. Then $\varphi^\star(P \cup A) = \varphi^\sigma(P \cup A)$ for all $P \in [\varphi < \infty]$. Now we have to prove that $\varphi^\star(A) \leq \varphi^\sigma(A)$ and can thus assume that $\varphi^\sigma(A) < \infty$. Consider $U \in \mathfrak{S}^\sigma$ with $U \supset A$, and recall that φ^σ is outer regular $\mathfrak{S}^\sigma$. There are subsets $P \in [\varphi < \infty]$ such that $P \subset U$. It follows that

$$\varphi^\star(A) \leq \varphi^\star(P \cup A) = \varphi^\sigma(P \cup A) \leq \varphi^\sigma(U),$$

and hence $\varphi^\star(A) \leq \varphi^\sigma(A)$.

τ) Fix $A \in \mathfrak{C}(\varphi^\sigma, \dotplus)$ with $\varphi^\sigma(A) < \infty$. For $P, Q \in [\varphi < \infty]$ we have as above

$$
\begin{aligned}
\varphi(P) + \varphi(Q) &= \varphi^\sigma(P) + \varphi^\sigma(Q) = \varphi^\sigma(P|A|Q) + \varphi^\sigma(P|A'|Q) \\
&\geq \varphi^\tau(P|A|Q) + \varphi^\tau(P|A'|Q) \geq \varphi^\tau(P \cup Q) + \varphi^\tau(P \cap Q) \\
&= \varphi(P \cup Q) + \varphi(P \cap Q) = \varphi(P) + \varphi(Q),
\end{aligned}
$$

where all values are finite. It follows that

$$\varphi^\sigma(P|A|Q) = \varphi^\tau(P|A|Q) \text{ for all } P, Q \in [\varphi < \infty].$$

Since φ^σ and φ^τ are upward σ continuous by 4.7 this implies that

$$\varphi^\sigma(P|A|Q) = \varphi^\tau(P|A|Q) \text{ for all } P, Q \in [\varphi < \infty]^\sigma.$$

Since $\varphi^\sigma(A) < \infty$ there exists $Q \in [\varphi < \infty]^\sigma$ with $Q \supset A$. Then $\varphi^\sigma(P \cup A) = \varphi^\tau(P \cup A)$ for all $P \in [\varphi < \infty]^\sigma$. Now we have to prove that $\varphi^\sigma(A) \leqq \varphi^\tau(A)$ and can thus assume that $\varphi^\tau(A) < \infty$. Consider $U \in \mathfrak{S}^\tau$ with $U \supset A$, and recall that φ^τ is outer regular $\mathfrak{S}^\tau$. There are subsets $P \in [\varphi < \infty]$ such that $P \subset U$. It follows that

$$\varphi^\sigma(A) \leqq \varphi^\sigma(P \cup A) = \varphi^\tau(P \cup A) \leqq \varphi^\tau(U),$$

and hence $\varphi^\sigma(A) \leqq \varphi^\tau(A)$.

5.10. EXERCISE. σ) Construct an example which shows that 5.9.σ) becomes false without the condition $\varphi^\star(A) < \infty$. Hint: Let $\mathfrak{S}$ consist of the finite subsets of an infinite countable X, and let $\varphi = 0$. Determine $\mathfrak{C}(\varphi^\star, \dotplus)$ with the aid of 5.2. τ) Do the same for 5.9.τ).

However, we shall see that the three properties of φ to be an outer • premeasure for • $= \star \sigma \tau$ are independent, except that as a consequence of 5.5.5) the combination $+ - +$ of these properties cannot occur. The independence is plausible after 5.5.5): This condition can be subdivided into two partial ones, such that the one increases and the other decreases with • $= \star \sigma \tau$. We shall come back to this point in 5.15 in the frame of the conventional outer situation.

The Conventional Outer Situation

The above central theorem of the chapter will be most important in two particular cases. These are the specializations

$$\varnothing \in \mathfrak{S} \text{ and } \varphi(\varnothing) = 0, \text{ and}$$
$$X \in \mathfrak{S} \text{ and } \varphi(X) = 0.$$

The first one is called the conventional outer situation. It will be the theme of the present subsection. This specialization contains, and unifies and clarifies those earlier extension procedures which were in visible or invisible manner based on outer regularity. The other one is what later will become the conventional inner situation. It will achieve the same for the earlier extension procedures based on inner regularity. It is obvious that this specialization should be treated via the upside-down transform method. However, it seems more natural to perform the upside-down procedure for the entire development, and then to specialize to the case $\varnothing \in \mathfrak{S}$ and $\varphi(\varnothing) = 0$ as before. This will be done in the next section.

For the present we consider a lattice $\mathfrak{S}$ with $\varnothing \in \mathfrak{S}$ and an isotone set function $\varphi : \mathfrak{S} \to [0, \infty]$ with $\varphi(\varnothing) = 0$. There are certain immediate simplifications: An outer $\bullet$ extension of φ is an extension of φ which is a ccontent $\alpha : \mathfrak{A} \to [0, \infty]$ on a ring $\mathfrak{A}$, with the further properties as above. Furthermore we have $\varphi^\bullet : \mathfrak{P}(X) \to [0, \infty]$ with $\varphi^\bullet(\varnothing) = 0$. Thus we can write $\mathfrak{C}(\varphi^\bullet)$ instead of $\mathfrak{C}(\varphi^\bullet, \dotplus)$. Also the definition of upward essential simplifies in an obvious manner. It is natural to specialize 5.2 and 5.5 to $\mathfrak{P} = \{\varnothing\}$, and for simplicity we take $\mathfrak{Q} = [\varphi < \infty]$. Let us then rewrite the outer main theorem with these simplifications.

5.11. THEOREM (Conventional Outer Main Theorem). *Let $\mathfrak{S}$ be a lattice with $\varnothing \in \mathfrak{S}$, and $\varphi : \mathfrak{S} \to [0, \infty]$ be an isotone and submodular set function with $\varphi(\varnothing) = 0$. Then the following are equivalent.*

1) *There exist outer $\bullet$ extensions of φ, that is φ is an outer $\bullet$ premeasure.*

2) *$\varphi^\bullet | \mathfrak{C}(\varphi^\bullet)$ is an outer $\bullet$ extension of φ. Furthermore*

$$\text{if } \bullet = \star \ : \ \varphi^\bullet | \mathfrak{C}(\varphi^\bullet) \text{ is a ccontent on the algebra } \mathfrak{C}(\varphi^\bullet),$$

$$\text{if } \bullet = \sigma\tau \ : \ \varphi^\bullet | \mathfrak{C}(\varphi^\bullet) \text{ is a cmeasure on the } \sigma \text{ algebra } \mathfrak{C}(\varphi^\bullet).$$

3) *$\varphi^\bullet | \mathfrak{C}(\varphi^\bullet)$ is an extension of φ in the crude sense, that is $\mathfrak{S} \subset \mathfrak{C}(\varphi^\bullet)$ and $\varphi = \varphi^\bullet | \mathfrak{S}$.*

4) *$\varphi(B) = \varphi(A) + \varphi^\bullet(B \setminus A)$ for all $A \subset B$ in $\mathfrak{S}$. In case $\bullet = \tau$ furthermore φ is upward essential.*

5) *$\varphi = \varphi^\bullet | \mathfrak{S}$; and $\varphi(B) \geqq \varphi(A) + \varphi^\bullet(B \setminus A)$ for all $A \subset B$ in $[\varphi < \infty]$. In case $\bullet = \tau$ furthermore φ is upward essential.*

Note that 5.4 then implies $[\varphi < \infty] \top \mathfrak{S}^\bullet \subset \mathfrak{C}(\varphi^\bullet)$.

Assume that $\mathfrak{S}$ is a lattice with $\varnothing \in \mathfrak{S}$. We define an isotone set function $\varphi : \mathfrak{S} \to [0, \infty]$ with $\varphi(\varnothing) = 0$ to be **outer** $\bullet$ **tight** iff it fulfils

$$\varphi(B) \geqq \varphi(A) + \varphi^\bullet(B \setminus A) \quad \text{for all } A \subset B \text{ in } \mathfrak{S},$$

as it appears in condition 5) above. It is obvious that

$$\text{outer } \star \text{ tight } \Rightarrow \text{ outer } \sigma \text{ tight } \Rightarrow \text{ outer } \tau \text{ tight } .$$

We show on the spot that both converses $\Leftarrow$ are false. The counterexamples will be isotone and modular set functions $\varphi : \mathfrak{S} \to [0, \infty[$ with $\varphi(\varnothing) = 0$ which are upward τ continuous.

5.12. EXERCISE. We recall from 2.3.1) for $a \in X$ the Dirac set functions $\delta_a : \mathfrak{P}(X) \to \{0, 1\}$. δ_a is a cmeasure and upward and downward τ continuous. Now assume that X is a Hausdorff topological space. We consider the set functions $\varphi := \delta_a | \mathrm{Op}(X)$ and $\psi := \delta_a | \mathrm{Cl}(X)$. 1) We have $\varphi^\bullet = \delta_a$ for all $\bullet = \star\sigma\tau$. Therefore φ is outer $\bullet$ tight and an outer $\bullet$ premeasure. 2) We have the equivalences

$$\psi \text{ outer } \bullet \text{ tight } \Longleftrightarrow \psi^\bullet(\{a\}') = 0 \Longleftrightarrow \{a\} \in (\mathrm{Op}(X))_\bullet.$$

The condition on the right side can be different for $\bullet = \star\sigma\tau$: For $\bullet = \star$ it means that a is an isolated point of X. For $\bullet = \sigma$ it means that in classical notation $\{a\}$ is a G_δ set. For $\bullet = \tau$ it is always fulfilled. Thus we obtain obvious counterexamples as announced above. Furthermore $\psi^\tau = \delta_a$; therefore ψ is upward essential and hence an outer τ premeasure.

5.13. SPECIAL CASE (Traditional Type). *Assume that $\mathfrak{S}$ is a ring. Then condition 5) simplifies to*
5o) $\varphi = \varphi^\bullet|\mathfrak{S}$; *and φ is supermodular. In case $\bullet = \tau$ furthermore φ is upward essential.*

The conventional outer main theorem will henceforth be one of our systematic tools. A fundamental achievement will be the extension 7.12.1) of the last special case. For the present it will be applied to the former main example $\lambda : \mathfrak{K} = \mathrm{Comp}(\mathbb{R}^n) \to [0, \infty[$ in order to obtain the Lebesgue measure on $\mathbb{R}^n$ and its basic properties in the spirit of the outer theory.

The decisive fact follows from a simple observation which will be systematized below: For each pair $A \subset B$ in $\mathfrak{K}$ we have $B \setminus A \in \mathfrak{K}^\sigma$, that is there exists a sequence $(K_l)_l$ in $\mathfrak{K}$ such that $K_l \uparrow B \setminus A$. Then $A \cap K_l = \varnothing$ and $A \cup K_l \uparrow B$. We conclude from 2.26 that $\lambda(A) + \lambda(K_l) = \lambda(A \cup K_l) \leqq \lambda(B)$ and hence

$$\lambda^\sigma(B \setminus A) \leqq \lim_{l \to \infty} \lambda(K_l) \leqq \lambda(B) - \lambda(A);$$

note that 2.27 even implies that $\lambda(A \cup K_l) \uparrow \lambda(B)$ and hence $\lambda(K_l) \uparrow \lambda(B) - \lambda(A)$. Thus λ is outer σ tight. From 2.27 it follows that λ is an outer σ premeasure. The achievement of the conventional outer main theorem is then the cmeasure

$$\Lambda := \lambda^\sigma|\mathfrak{C}(\lambda^\sigma) = \lambda^\sigma|\mathfrak{L} \quad \text{on} \quad \mathfrak{L} := \mathfrak{C}(\lambda^\sigma),$$

defined to be the **Lebesgue measure** on $\mathbb{R}^n$. The last assertion in 5.11 furnishes

$$\mathrm{Cl}(\mathbb{R}^n) \subset \mathfrak{K}\top\mathfrak{K} \subset \mathfrak{K}\top\mathfrak{K}^\sigma \subset \mathfrak{C}(\lambda^\sigma) = \mathfrak{L} \quad \text{and hence} \quad \mathrm{Bor}(\mathbb{R}^n) \subset \mathfrak{L}.$$

The restriction $\Lambda|\mathrm{Bor}(\mathbb{R}^n)$ is called the **Borel-Lebesgue measure** on $\mathbb{R}^n$. All this is the first statement in the comprehensive theorem which follows.

5.14. THEOREM. 1) $\lambda : \mathfrak{K} = \mathrm{Comp}(\mathbb{R}^n) \to [0, \infty[$ *is an outer σ premeasure. The Lebesgue measure $\Lambda := \lambda^\sigma|\mathfrak{C}(\lambda^\sigma)$ has the domain $\mathfrak{L} := \mathfrak{C}(\lambda^\sigma) \supset \mathrm{Bor}(\mathbb{R}^n)$. 2) λ is not upward τ continuous and hence not an outer τ premeasure. 3) λ is not outer $\star$ tight and hence not an outer $\star$ premeasure.*

4) λ^σ *and hence Λ are outer regular $\mathrm{Op}(\mathbb{R}^n)$. 5) Λ is inner regular $\mathfrak{K} = \mathrm{Comp}(\mathbb{R}^n)$. 6) $\Lambda|\mathrm{Op}(\mathbb{R}^n)$ is upward τ continuous. 7) $\Lambda|\mathrm{Op}(\mathbb{R}^n) =: \omega$ is an outer $\bullet$ premeasure for all $\bullet = \star\sigma\tau$. It satisfies $\omega^\bullet = \lambda^\sigma$ and hence $\omega^\bullet|\mathfrak{C}(\omega^\bullet) = \Lambda$.*

Proof. 1) has been proved above. 2) is obvious since $\lambda(F) = 0$ for all finite $F \subset \mathbb{R}^n$. The next proofs require the preparations which follow. i) For $K \in \mathfrak{K}$ and $\varepsilon > 0$ there exists an open $U \supset K$ with $\Lambda(U) \leqq \lambda(K) + \varepsilon$. In fact, for $K \neq \varnothing$ and $K(\delta) := \{x \in \mathbb{R}^n : \mathrm{dist}(x, K) \leqq \delta\}$ we see from 2.24 that $\Lambda(\mathrm{Int}K(\delta)) \downarrow \lambda(K)$ for $\delta \downarrow 0$. ii) For $A \in \mathfrak{K}^\sigma$ with $\Lambda(A) < \infty$

and $\varepsilon > 0$ there exists an open $U \supset A$ with $\Lambda(U) \leqq \Lambda(A) + \varepsilon$. To see this choose a sequence $(K_l)_l$ in $\mathfrak{K}$ with $K_l \uparrow A$ and open subsets $U_l \supset K_l$ with $\Lambda(U_l) \leqq \lambda(K_l) + \varepsilon 2^{-l}$. For the $V_l := U_1 \cup \cdots \cup U_l$ one obtains via induction $\Lambda(V_l) \leqq \lambda(K_l) + \varepsilon(1 - 2^{-l})$. Thus $V_l \uparrow V$ furnishes an open $V \supset A$ with $\Lambda(V) \leqq \Lambda(A) + \varepsilon$.

3) Let $B := \{x \in \mathbb{R}^n : 0 \leqq x_1, \cdots, x_n \leqq 1\}$ be the unit cube of $\mathbb{R}^n$ and $D \subset \text{Int}B$ be a countable dense subset. From $\Lambda(D) = 0$ and ii) we obtain an open subset $U \supset D$ of $\text{Int}B$ with $\Lambda(U) < 1$. Note that $\lambda^\star(U) = 1$. Thus for the compact $A := B \setminus U \subset B$ we have $U = B \setminus A$ and

$$\lambda(B) = \lambda(A) + \Lambda(U) < \lambda(A) + 1 = \lambda(A) + \lambda^\star(B \setminus A).$$

It follows that λ is not outer $\star$ tight. 4) follows from 4.1.4) and the above ii). 5) Since $\mathbb{R}^n$ is in $\mathfrak{K}^\sigma$ we can restrict ourselves to $A \in \mathfrak{L}$ such that $A \subset$ some $K \in \mathfrak{K}$. Fix $\varepsilon > 0$. By 4) there exists an open $U \supset K \cap A'$ with $\Lambda(U) \leqq \Lambda(K \cap A') + \varepsilon$. Then $K \cap U'$ is compact $\subset A$, and we have

$$\begin{aligned}
\Lambda(A) + \Lambda(K \cap U) &\leqq \Lambda(A) + \Lambda(U) \leqq \Lambda(A) + \Lambda(K \cap A') + \varepsilon \\
&= \lambda(K) + \varepsilon = \lambda(K \cap U') + \Lambda(K \cap U) + \varepsilon,
\end{aligned}$$

and hence $\Lambda(A) \leqq \lambda(K \cap U') + \varepsilon$. 6) Let $A \in \text{Op}(\mathbb{R}^n)$, and $\mathfrak{M} \subset \text{Op}(\mathbb{R}^n)$ be a paving with $\mathfrak{M} \uparrow A$. For real $c < \Lambda(A)$ we obtain from 5) a compact $K \subset A$ with $c < \lambda(K)$. Now $K \subset$ some $M \in \mathfrak{M}$, and hence $c < \sup\{\Lambda(M) : M \in \mathfrak{M}\}$. The assertion follows. 7) We see from 4)6) that Λ is an outer $\bullet$ extension of ω, so that ω is an outer $\bullet$ premeasure. Now $\omega^\bullet$ and λ^σ coincide on $\text{Op}(\mathbb{R}^n)$ and are both outer regular $\text{Op}(\mathbb{R}^n)$. Therefore $\omega^\bullet = \lambda^\sigma$. The proof is complete.

5.15. EXERCISE. We can now prove that for the isotone and modular set functions $\varphi : \mathfrak{S} \to [0, \infty[$ with $\varphi(\varnothing) = 0$ the three properties to be an outer $\bullet$ premeasure for $\bullet = \star\sigma\tau$ are independent, as announced at the end of the last subsection. There are $2^3 = 8$ combinations of these properties. We know that the combination $+ - +$ cannot occur. 1) Deduce examples for $- - +$ and $- + +$ from 5.12.2). 2) Deduce examples for $- - -$ and $+ - -$ from 3.11. 3) Deduce examples for $- + -$ and $+ + +$ and also for $+ + -$ from 5.14.

6. The Inner Extension Theory

The basic part of the present section obtains the inner extension theory as a mere transcription of the outer extension theory via the upside-down transform method. Thus the two extension theories are in fact identical. We shall add a subsection on further results in the τ case, which in practice is much more important in the inner than in the outer situation. Then we specialize to the conventional inner situation as announced.

The Basic Definition

Let as before $\mathfrak{S}$ be a lattice in a nonvoid set X.

DEFINITION. Let $\varphi : \mathfrak{S} \to [-\infty, \infty[$ be an isotone set function $\neq -\infty$. For $\bullet = \star \sigma \tau$ we define an **inner $\bullet$ extension** of φ to be an extension of φ which is a $\dot{+}$ content $\alpha : \mathfrak{A} \to \overline{\mathbb{R}}$ on an oval $\mathfrak{A}$, such that also $\mathfrak{S}_\bullet \subset \mathfrak{A}$ and that

  α is inner regular $\mathfrak{S}_\bullet$, and
  $\alpha|\mathfrak{S}_\bullet$ is downward $\bullet$ continuous; in this context note that $\alpha|\mathfrak{S}_\bullet < \infty$.

We define φ to be an **inner $\bullet$ premeasure** iff it admits inner $\bullet$ extensions. Thus an inner $\bullet$ premeasure is modular and downward $\bullet$ continuous.

As before the principal aim is to characterize those φ which are inner $\bullet$ premeasures, and then to describe all inner $\bullet$ extensions of φ.

6.1. EXERCISE. Let $\varphi : \mathfrak{S} \to [-\infty, \infty[$ be an isotone set function $\neq -\infty$, and hence $\varphi\bot : \mathfrak{S}\bot \to]-\infty, \infty]$ an isotone set function $\neq \infty$. Then a set function $\alpha : \mathfrak{A} \to \overline{\mathbb{R}}$ is an inner $\bullet$ extension of φ iff the set function $\alpha\bot : \mathfrak{A}\bot \to \overline{\mathbb{R}}$ is an outer $\bullet$ extension of $\varphi\bot$.

We also refer to the instructive exercise 9.21 below. It is ab-ovo and could have been placed here, but will be postponed until it will be needed.

The Inner Envelopes

Let $\varphi : \mathfrak{S} \to \overline{\mathbb{R}}$ be an isotone set function on a lattice $\mathfrak{S}$. As before we define its **crude inner envelope** $\varphi_\star : \mathfrak{P}(X) \to \overline{\mathbb{R}}$ to be

$$\varphi_\star(A) = \sup\{\varphi(S) : S \in \mathfrak{S} \text{ with } S \subset A\} \quad \text{for } A \subset X.$$

Likewise we define the **inner envelopes** $\varphi_\sigma, \varphi_\tau : \mathfrak{P}(X) \to \overline{\mathbb{R}}$ as the counterparts of the respective outer formations to be

$$\varphi_\sigma(A) = \sup\{\lim_{l \to \infty} \varphi(S_l) : (S_l)_l \text{ in } \mathfrak{S} \text{ with } S_l \downarrow \subset A\} \quad \text{for } A \subset X,$$

$$\varphi_\tau(A) = \sup\{\inf_{S \in \mathfrak{M}} \varphi(S) : \mathfrak{M} \text{ paving } \subset \mathfrak{S} \text{ with } \mathfrak{M} \downarrow \subset A\} \quad \text{for } A \subset X.$$

As before we have for $\bullet = \star \sigma \tau$ the common formula

$$\varphi_\bullet(A) = \sup\{\inf_{S \in \mathfrak{M}} \varphi(S) : \mathfrak{M} \text{ paving } \subset \mathfrak{S} \text{ of type } \bullet \text{ with } \mathfrak{M} \downarrow \subset A\}.$$

6.2. EXERCISE. $(\varphi_\bullet)\bot = (\varphi\bot)^\bullet$ for $\bullet = \star \sigma \tau$.

The upside-down transform method thus furnishes the inner counterparts of the respective properties proved in the outer situation. For convenience we list the basic ones.

6.3. PROPERTIES. 1) $\varphi_\star|\mathfrak{S} = \varphi$. 2) $\varphi_\star \leqq \varphi_\sigma \leqq \varphi_\tau$. 3) $\varphi_\bullet$ *is isotone.* 4) $\varphi_\bullet$ *is inner regular* $[\varphi_\bullet|\mathfrak{S}_\bullet > -\infty] \subset \mathfrak{S}_\bullet$. 5) *Assume that* φ *is supermodular* $\dot{+}$. *Then* $\varphi_\star$ *is supermodular* $\dot{+}$, *and* $\varphi_\bullet$ *for* $\bullet = \sigma \tau$ *is supermodular* $\dot{+}$ *when either* $\varphi < \infty$ *or* $\varphi_\bullet > -\infty$.

6.4. EXERCISE. *Let φ be submodular $\overset{.}{+}$. Assume that $A, B \subset X$ are* **coseparated** $\mathfrak{S}$ *in the sense that*

for each $M \in \mathfrak{S}$ with $M \subset A \cup B$
there exist $S, T \in \mathfrak{S}$ with $S \subset A$ and $T \subset B$ such that $M \subset S \cup T$.

Then $\varphi_\star(A \cup B) \overset{.}{+} \varphi_\star(A \cap B) \leqq \varphi_\star(A) \overset{.}{+} \varphi_\star(B)$.

6.5. PROPOSITION. *For an isotone set function $\varphi : \mathfrak{S} \to \overline{\mathbb{R}}$ and $\bullet = \sigma\tau$ the following are equivalent.*

i) $\varphi_\bullet | \mathfrak{S} = \varphi$;
ii) φ *is downward $\bullet$ continuous.*

In this case we have furthermore

iii) $\varphi_\bullet | \mathfrak{S}_\bullet$ *is downward $\bullet$ continuous;*
iv) *if* $\{S \in \mathfrak{S}_\bullet : \varphi_\bullet(S) > -\infty\} \subset \mathfrak{S}$ *then* $\varphi_\bullet = \varphi_\star$.

6.6. LEMMA. *Let $\mathfrak{M} \subset \mathfrak{S}_\bullet$ be a paving of type $\bullet$ with $\mathfrak{M} \downarrow A$. Then of course $A \in \mathfrak{S}_\bullet$. Furthermore there exists a paving $\mathfrak{N} \subset \mathfrak{S}$ of type $\bullet$ with $\mathfrak{N} \downarrow A$ and $\mathfrak{N} \subset (\sqsupset \mathfrak{M})$.*

6.7. THEOREM. *Assume that $\varphi : \mathfrak{S} \to \overline{\mathbb{R}}$ is isotone and supermodular $\overset{.}{+}$. Then φ_σ and φ_τ are almost downward σ continuous.*

6.8. LEMMA. *Assume that $\varphi : \mathfrak{S} \to \overline{\mathbb{R}}$ is isotone and supermodular $\overset{.}{+}$. For $P_1, \cdots, P_n, Q \in \mathfrak{S}$ with $\varphi(P_1), \cdots, \varphi(P_n), \varphi(Q) > -\infty$ then $\varphi(P_1 \cap \cdots \cap P_n \cap Q) > -\infty$ and*

$$\varphi(P_1 \cap \cdots \cap P_n \cap Q) + \sum_{l=1}^{n} \varphi(P_l \cup Q) \geqq \sum_{l=1}^{n} \varphi(P_l) + \varphi(Q).$$

Next we define an isotone set function $\varphi : \mathfrak{S} \to \overline{\mathbb{R}}$ to be **downward $\bullet$ essential** iff its upside-down transform $\varphi\bot : \mathfrak{S}\bot \to \overline{\mathbb{R}}$ is upward $\bullet$ essential. One verifies that this means that

$$\varphi_\bullet(A) = \inf\{\varphi_\bullet(A \cup S) : S \in [\varphi > -\infty]\} \quad \text{for all } A \subset X \text{ with}$$
$$-\infty < \varphi_\bullet(A) \leqq \inf\{\varphi_\bullet(A \cup S) : S \in [\varphi > -\infty]\} < \infty.$$

We obtain the counterpart of the former result.

6.9. PROPOSITION. *Let $\varphi : \mathfrak{S} \to \overline{\mathbb{R}}$ be isotone.* $\star$) φ *is downward $\star$ essential.* σ) *If φ is supermodular $\overset{.}{+}$ then it is downward σ essential.* τ) *Assume that φ is supermodular $\overset{.}{+}$ and such that each $A \subset X$ with $\varphi_\tau(A) > -\infty$ is downward enclosable $[\varphi > -\infty]_\sigma$. Then φ is downward τ essential.*

Therefore an isotone and supermodular $\overset{.}{+}$ set function $\varphi : \mathfrak{S} \to \overline{\mathbb{R}}$ will be called **downward essential** instead of downward τ essential.

After these transcriptions we conclude the subsection with some simple but important relations which involve envelopes of both kinds.

6.10. PROPOSITION. *1) We have $\varphi^\bullet \leqq \varphi_\star$ on $\mathfrak{S}^\bullet$. Therefore $\varphi^\bullet \leqq (\varphi_\star | \mathfrak{S}^\bullet)^\star$. 2) The following are equivalent. i) $\varphi = \varphi^\bullet | \mathfrak{S}$. ii) $\varphi^\bullet \geqq \varphi_\star$ on $\mathfrak{S}^\bullet$ and hence $\varphi^\bullet = \varphi_\star$ on $\mathfrak{S}^\bullet$. iii) $\varphi^\bullet \geqq \varphi_\star$. iv) $\varphi^\bullet \geqq (\varphi_\star | \mathfrak{S}^\bullet)^\star$ and hence $\varphi^\bullet = (\varphi_\star | \mathfrak{S}^\bullet)^\star$.*

Proof. 1) Let $A \in \mathfrak{S}^{\bullet}$ and $\mathfrak{M} \subset \mathfrak{S}$ be a paving of type $\bullet$ with $\mathfrak{M} \uparrow A$. By definition then $\varphi(S) \leqq \varphi_{\star}(A)\ \forall S \in \mathfrak{M}$ and hence

$$\varphi^{\bullet}(A) \leqq \sup_{S \in \mathfrak{M}} \varphi(S) \leqq \varphi_{\star}(A).$$

The second relation is then clear. 2) The implications iii)$\Rightarrow$iv)$\Rightarrow$ii)$\Rightarrow$i) are obvious. i)$\Rightarrow$iii) Let $A \subset X$. For $S \in \mathfrak{S}$ with $S \subset A$ then $\varphi(S) = \varphi^{\bullet}(S) \leqq \varphi^{\bullet}(A)$. Thus $\varphi_{\star}(A) \leqq \varphi^{\bullet}(A)$.

6.11. EXERCISE. 1) *We have* $\varphi_{\bullet} \geqq \varphi^{\star}$ *on* $\mathfrak{S}_{\bullet}$. *Therefore* $\varphi_{\bullet} \geqq (\varphi^{\star}|\mathfrak{S}_{\bullet})_{\star}$. 2) *The following are equivalent.* i) $\varphi = \varphi_{\bullet}|\mathfrak{S}$. ii) $\varphi_{\bullet} \leqq \varphi^{\star}$ *on* $\mathfrak{S}_{\bullet}$ *and hence* $\varphi_{\bullet} = \varphi^{\star}$ *on* $\mathfrak{S}_{\bullet}$. iii) $\varphi_{\bullet} \leqq \varphi^{\star}$. iv) $\varphi_{\bullet} \leqq (\varphi^{\star}|\mathfrak{S}_{\bullet})_{\star}$ *and hence* $\varphi_{\bullet} = (\varphi^{\star}|\mathfrak{S}_{\bullet})_{\star}$.

In some earlier versions of the outer and inner extension theories the above formations

$$\varphi^{(\bullet)} := (\varphi_{\star}|\mathfrak{S}^{\bullet})^{\star} \quad \text{and} \quad \varphi_{(\bullet)} := (\varphi^{\star}|\mathfrak{S}_{\bullet})_{\star}$$

have been used in more or less explicit manner, in places where in the present text the envelopes $\varphi^{\bullet}$ and $\varphi_{\bullet}$ are the natural means. For some details we refer to the bibliographical annex to the chapter. Of course $\varphi^{(\star)} = \varphi^{\star}$ and $\varphi_{(\star)} = \varphi_{\star}$. For $\bullet = \sigma\tau$ the formations $\varphi^{(\bullet)}$ and $\varphi_{(\bullet)}$ are much more complicated than $\varphi^{\bullet}$ and $\varphi_{\bullet}$. For example, it is unclear whether beyond 6.10.2) and 6.11.2) they preserve semimodularity in the appropriate sense.

The Carathéodory Class in the Spirit of the Inner Theory

The second tool in the outer extension theory was the Carathéodory class $\mathfrak{C}(\cdot)$. Its definition and basic properties were not related to outer/inner aspects. Thus there is no reason for transcription. However, there was a subsequent subsection on the Carathéodory class in the spirit of the outer theory. For the present context this subsection consisted of intermediate results which need not be transcribed, so that the transcription could proceed to the next section on the main theorem. But the transcribed versions of some former results will be needed later, and therefore will be inserted at this point. For the transcription we refer to the earlier 4.15.

6.12. REMARK. *Let* $\mathfrak{T}$ *be a paving in* X *and* $\phi : \mathfrak{P}(X) \to \overline{\mathbb{R}}$ *be isotone and inner regular* $\mathfrak{T}$. *If* $A \subset X$ *satisfies*

$$\phi(P)+\phi(Q) \leqq \phi(P|A|Q)+\phi(P|A'|Q) \quad \textit{for all } P, Q \in \mathfrak{T},$$

then $A \in \mathfrak{C}(\phi, +)$.

6.13. EXERCISE. *Let* $\alpha : \mathfrak{A} \to \overline{\mathbb{R}}$ *be a content* $\dotplus$ *on an oval* $\mathfrak{A}$. *Then* $\mathfrak{A} \subset \mathfrak{C}(\alpha_{\star}, \dotplus)$.

6.14. LEMMA. *Let* $P \subset Q \subset X$. *Assume that* $\varphi : P \sqsubset Q \to \mathbb{R}$ *is supermodular. If* $A \subset X$ *satisfies*

$$\varphi(P) + \varphi(Q) \leqq \varphi(P|A|Q) + \varphi(P|A'|Q),$$

then

$$\varphi(U) + \varphi(V) = \varphi(U|A|V) + \varphi(U|A'|V) \quad \textit{for all } U, V \in P \sqsubset Q.$$

6.15. PROPOSITION. *Assume that* $\phi : \mathfrak{P}(X) \to \overline{\mathbb{R}}$ *is an isotone set function. Let* $\mathfrak{P} \downarrow$ *and* $\mathfrak{Q} \uparrow$ *be pavings in* X *with nonvoid* $\mathfrak{P} \sqsubset \mathfrak{Q}$ *such that*

$\phi|\mathfrak{P}$ *and* $\phi|\mathfrak{Q}$ *are finite, and* $\phi|\mathfrak{P} \sqsubset \mathfrak{Q}$ *is supermodular.*

Furthermore let $\mathfrak{H} \downarrow$ *be a paving in* X *with* $\mathfrak{P} \subset \mathfrak{H}$ *such that*

ϕ *is inner regular* $\mathfrak{H} \sqsubset \mathfrak{Q}$,

$$\phi(T) = \inf_{P \in \mathfrak{P}} \phi(T \cup P) \quad \text{for all } T \in \mathfrak{H} \sqsubset \mathfrak{Q}.$$

If $A \subset X$ *satisfies*

$$\phi(P) + \phi(Q) \;\leqq\; \phi(P|A|Q) + \phi(P|A'|Q)$$
$$\text{for all } P \in \mathfrak{P} \text{ and } Q \in \mathfrak{Q} \text{ with } P \subset Q,$$

then $A \in \mathfrak{C}(\phi, \dotplus)$.

As before we have two addenda.

6.16. ADDENDUM. *Assume in addition that* $\phi|\mathfrak{P} \sqsubset \mathfrak{Q}$ *is downward* σ *continuous. Then* $\mathfrak{C}(\phi, \dotplus)$ *is a* σ *algebra.*

6.17. ADDENDUM. *Assume in addition that* $\theta : \mathfrak{P}(X) \to \overline{\mathbb{R}}$ *is an isotone set function with* $\theta \leqq \phi$ *such that*

$\theta|\mathfrak{P} = \phi|\mathfrak{P}$ *and* $\theta|\mathfrak{Q} = \phi|\mathfrak{Q}$,

$$\theta(T) = \inf_{P \in \mathfrak{P}} \theta(T \cup P) \quad \text{for all } T \in \mathfrak{H} \sqsubset \mathfrak{Q}.$$

Then $\phi|\mathfrak{C}(\phi, \dotplus)$ *is an extension of* $\theta|\mathfrak{C}(\theta, \dotplus)$.

The Inner Main Theorem

We assume that $\varphi : \mathfrak{S} \to [-\infty, \infty[$ is an isotone set function $\not\equiv -\infty$ on a lattice $\mathfrak{S}$ in X. The assertions which follow are immediate consequences of their outer counterparts via the upside-down transform method. We recall 6.1 and 6.2, and once more the earlier 4.15.

6.18. THEOREM. *Assume that* $\alpha : \mathfrak{A} \to \overline{\mathbb{R}}$ *is an inner* $\bullet$ *extension of* φ. *Then* α *is a restriction of* $\varphi_\bullet|\mathfrak{C}(\varphi_\bullet, +)$.

6.19. PROPOSITION. *Let* φ *be supermodular with* $\varphi_\bullet|\mathfrak{S} < \infty$, *and downward essential in case* $\bullet = \tau$. *Fix pavings*

$\mathfrak{P} \subset [\varphi > -\infty]$ *downward cofinal, and*
$\mathfrak{Q} \subset [\varphi > -\infty]$ *upward cofinal.*

If $A \subset X$ *satisfies*

$$\varphi_\bullet(P) + \varphi_\bullet(Q) \;\leqq\; \varphi_\bullet(P|A|Q) + \varphi_\bullet(P|A'|Q)$$
$$\text{for all } P \in \mathfrak{P} \text{ and } Q \in \mathfrak{Q} \text{ with } P \subset Q,$$

then $A \in \mathfrak{C}(\varphi_\bullet, \dotplus)$.

6.20. ADDENDUM. *For* $\bullet = \sigma\tau$ *the class* $\mathfrak{C}(\varphi_\bullet, \dotplus)$ *is a* σ *algebra.*

The next assertion does not result from the upside-down technique but is an immediate consequence of 6.19 as before.

6.21. EXERCISE. *Let φ be supermodular with $\varphi_\bullet|\mathfrak{S} < \infty$, and downward essential in case $\bullet = \tau$. Assume that $\mathfrak{Q} \subset [\varphi > -\infty]$ is upward cofinal. Then*

$$\mathfrak{Q}\top\mathfrak{C}(\varphi_\bullet, +) \subset \mathfrak{C}(\varphi_\bullet, +).$$

After all the upside-down transform method furnishes the inner main theorem.

6.22. THEOREM (Inner Main Theorem). *Let $\varphi : \mathfrak{S} \to [-\infty, \infty[$ be an isotone and supermodular set function $\not\equiv -\infty$ on a lattice $\mathfrak{S}$. Fix pavings*

$\mathfrak{P} \subset [\varphi > -\infty]$ *downward cofinal, and*
$\mathfrak{Q} \subset [\varphi > -\infty]$ *upward cofinal.*

Then the following are equivalent.

1) *There exist inner $\bullet$ extensions of φ, that is φ is an inner $\bullet$ premeasure.*

2) *$\varphi_\bullet|\mathfrak{C}(\varphi_\bullet, +)$ is an inner $\bullet$ extension of φ. Furthermore*

if $\bullet = \star$: $\varphi_\bullet|\mathfrak{C}(\varphi_\bullet, +)$ is a content $+$ on the algebra $\mathfrak{C}(\varphi_\bullet, +)$,

if $\bullet = \sigma\tau$: $\varphi_\bullet|\mathfrak{C}(\varphi_\bullet, +)$ is a measure $+$ on the σ algebra $\mathfrak{C}(\varphi_\bullet, +)$.

3) *$\varphi_\bullet|\mathfrak{C}(\varphi_\bullet, +)$ is an extension of φ in the crude sense, that is $\mathfrak{S} \subset \mathfrak{C}(\varphi_\bullet, +)$ and $\varphi = \varphi_\bullet|\mathfrak{S}$.*

4) *$\varphi(U) + \varphi(V) = \varphi(M) + \varphi_\bullet(U|M'|V)$ for all $U \subset M \subset V$ in $\mathfrak{S}$; note that $M = U|M|V$. In case $\bullet = \tau$ furthermore φ is downward essential.*

5) *$\varphi = \varphi_\bullet|\mathfrak{S}$; and $\varphi(P) + \varphi(Q) \leqq \varphi(M) + \varphi_\bullet(P|M'|Q)$ for all $P \subset M \subset Q$ with $P \in \mathfrak{P}$, $Q \in \mathfrak{Q}$, and $M \in \mathfrak{S}$ and hence $\in [\varphi > -\infty]$. In case $\bullet = \tau$ furthermore φ is downward essential.*

Note that 6.21 then implies $\mathfrak{Q}\top\mathfrak{S}_\bullet \subset \mathfrak{C}(\varphi_\bullet, +)$.

6.23. SPECIAL CASE (Traditional Type). *Assume that $\mathfrak{S}$ is an oval. Then condition 5) simplifies to*

5o) *$\varphi = \varphi_\bullet|\mathfrak{S}$; and φ is submodular. In case $\bullet = \tau$ furthermore φ is downward essential.*

Comparison of the three Inner Theories

We obtain the inner counterparts of the respective outer results. In the present subsection let $\varphi : \mathfrak{S} \to [-\infty, \infty[$ be an isotone and supermodular set function $\not\equiv -\infty$ on a lattice $\mathfrak{S}$.

6.24. PROPOSITION. *σ) In case $\varphi = \varphi_\sigma|\mathfrak{S}$ we have $\mathfrak{C}(\varphi_\star, +) \subset \mathfrak{C}(\varphi_\sigma, +)$. τ) In case $\varphi = \varphi_\tau|\mathfrak{S}$ we have $\mathfrak{C}(\varphi_\sigma, +) \subset \mathfrak{C}(\varphi_\tau, +)$.*

6.25. PROPOSITION. *Assume that φ is modular. σ) In case $\varphi = \varphi_\sigma|\mathfrak{S}$ we have $\varphi_\star(A) = \varphi_\sigma(A)$ for all $A \in \mathfrak{C}(\varphi_\star, +)$ with $\varphi_\star(A) > -\infty$. τ) In case $\varphi = \varphi_\tau|\mathfrak{S}$ we have $\varphi_\sigma(A) = \varphi_\tau(A)$ for all $A \in \mathfrak{C}(\varphi_\sigma, +)$ with $\varphi_\sigma(A) > -\infty$.*

However, as before the three properties of φ to be an inner $\bullet$ premeasure for $\bullet = \star\sigma\tau$ are independent, except that as a consequence of 6.22.5) the combination $+ - +$ of these properties cannot occur.

Further Results on Nonsequential Continuity

In contrast to the ubiquitous σ continuity of measures the occurrence of τ continuity is restricted to particular and foremost situations and bound to severe limitations. The simplest illustration is as follows.

6.26. EXAMPLE. Let $\mathfrak{S}$ be a lattice in X which contains the finite subsets of X, and let $\varphi : \mathfrak{S} \to \overline{\mathbb{R}}$ be isotone with $\varphi(S) = 0$ for all finite $S \subset X$. If φ is upward τ continuous then $\varphi = 0$.

In our extension theories the τ extensions are defined to possess a certain τ continuity, but it is restricted to the direct descendants of the initial domain $\mathfrak{S}$. In the sequel we deduce two further results on τ continuity. We restrict this topic to the inner situation where it is much more important. In the present subsection we assume $\varphi : \mathfrak{S} \to [-\infty, \infty[$ to be an isotone and supermodular set function $\not\equiv -\infty$ on a lattice $\mathfrak{S}$.

The main feature is the occurrence of the transporter $\mathfrak{S} \top \mathfrak{S}_\tau$, which of course is $= \mathfrak{S}_\tau \top \mathfrak{S}_\tau = (\mathfrak{S}_\tau)\top$ as well. As a rule the members of $\mathfrak{S} \top \mathfrak{S}_\tau$ can be much larger subsets of X than those of $\mathfrak{S}_\tau$. We note that for an inner τ premeasure φ we have

$$\mathfrak{S} \top \mathfrak{S}_\tau \subset [\varphi > -\infty] \top \mathfrak{S}_\tau \subset \mathfrak{C}(\varphi_\tau, +),$$

and hence $(\mathfrak{S} \top \mathfrak{S}_\tau) \bot \subset \mathfrak{C}(\varphi_\tau, +)$ as well.

6.27. PROPOSITION. *Assume that $\varphi = \varphi_\tau | \mathfrak{S}$. Then the restriction $\varphi_\tau | \mathfrak{S} \top \mathfrak{S}_\tau$ is almost downward τ continuous.*

6.28. PROPOSITION. *Assume that φ is an inner τ premeasure. Then the restriction $\varphi_\tau | (\mathfrak{S} \top \mathfrak{S}_\tau) \bot$ is almost upward τ continuous.*

Proof of 6.27. Let $\mathfrak{M} \subset \mathfrak{S} \top \mathfrak{S}_\tau$ be a paving with $\varphi_\tau(M) < \infty \; \forall M \in \mathfrak{M}$ such that $\mathfrak{M} \downarrow H$; thus $H \in \mathfrak{S} \top \mathfrak{S}_\tau$. To be shown is

$$c := \inf_{M \in \mathfrak{M}} \varphi_\tau(M) \leqq \varphi_\tau(H).$$

We can assume that $c > -\infty$ and hence $c \in \mathbb{R}$. Then $\varphi_\tau(M) \in \mathbb{R} \; \forall M \in \mathfrak{M}$, but a priori $\varphi_\tau(H) = -\infty$ is possible. Let us fix real $\varepsilon > 0$ and $\lambda > \varphi_\tau(H)$. i) Fix $P \in \mathfrak{M}$. By 6.3.4) there exists $S \in \mathfrak{S}_\tau$ with $S \subset P$ such that $\varphi_\tau(S) > \varphi_\tau(P) - \varepsilon$. Note that $\varphi_\tau(S) \leqq \varphi_\tau(P) < \infty$ and hence $\varphi_\tau(S) \in \mathbb{R}$. ii) By assumption $\{M \cap S : M \in \mathfrak{M}\}$ is a paving $\subset \mathfrak{S}_\tau$ with $\downarrow H \cap S \in \mathfrak{S}_\tau$. By 6.5.iii) therefore

$$\inf_{M \in \mathfrak{M}} \varphi_\tau(M \cap S) = \varphi_\tau(H \cap S) \leqq \varphi_\tau(H) < \lambda.$$

Thus there exists $Q \in \mathfrak{M}$ such that $\varphi_\tau(Q \cap S) < \lambda$. Since $\mathfrak{M} \downarrow$ we can assume that $Q \subset P$, so that $Q \cup S \subset P$. iii) By 6.3.5) we then have

$$\begin{aligned} c + \varphi_\tau(S) &\leqq \varphi_\tau(Q) + \varphi_\tau(S) \leqq \varphi_\tau(Q \cup S) + \varphi_\tau(Q \cap S) \\ &\leqq \varphi_\tau(P) + \varphi_\tau(Q \cap S) < \varphi_\tau(S) + \varepsilon + \lambda, \end{aligned}$$

and hence $c < \varepsilon + \lambda$. The assertion follows.

Proof of 6.28. Let $\mathfrak{M} \subset (\mathfrak{S} \top \mathfrak{S}_\tau) \bot$ be a paving with $\varphi_\tau(M) > -\infty \; \forall M \in \mathfrak{M}$ such that $\mathfrak{M} \uparrow H$; thus $H \in (\mathfrak{S} \top \mathfrak{S}_\tau) \bot$. To be shown is

$$c := \sup_{M \in \mathfrak{M}} \varphi_\tau(M) \geqq \varphi_\tau(H).$$

We can assume that $c < \infty$ and hence $c \in \mathbb{R}$. Then $\varphi_\tau(M) \in \mathbb{R} \; \forall M \in \mathfrak{M}$, but a priori $\varphi_\tau(H) = \infty$ is possible. Let us fix real $\varepsilon > 0$ and $\lambda < \varphi_\tau(H)$. i) Fix $P \in \mathfrak{M}$. By 6.3.4) there exists $S \in \mathfrak{S}_\tau$ with $S \subset P$ such that $\varphi_\tau(S) > -\infty$ and hence $\varphi_\tau(S) \in \mathbb{R}$. ii) Once more by 6.3.4) there exists $T \in \mathfrak{S}_\tau$ with $T \subset H$ such that $\varphi_\tau(T) > \lambda$ and hence $\varphi_\tau(T) \in \mathbb{R}$. In view of $S \subset P \subset H$ we can assume that $S \subset T$. iii) By assumption $\{M' \cap T : M \in \mathfrak{M}\}$ is a paving $\subset \mathfrak{S}_\tau$ with $\downarrow H' \cap T = \varnothing$. Hence $\{(M' \cap T) \cup S : M \in \mathfrak{M}\}$ is a paving $\subset \mathfrak{S}_\tau$ with $\downarrow S \in \mathfrak{S}_\tau$. By 6.5.iii) therefore

$$\inf_{M \in \mathfrak{M}} \varphi_\tau((M' \cap T) \cup S) = \varphi_\tau(S).$$

Thus there exists $Q \in \mathfrak{M}$ such that $\varphi_\tau((Q' \cap T) \cup S) < \varphi_\tau(S) + \varepsilon$. Since $\mathfrak{M} \uparrow$ we can assume that $Q \supset P$. iv) We have $Q \in \mathfrak{C}(\varphi_\tau, +)$ and hence

$$\begin{aligned}
\varphi_\tau(S) + \lambda \; &< \; \varphi_\tau(S) + \varphi_\tau(T) = \varphi_\tau(S|Q|T) + \varphi_\tau(S|Q'|T) \\
&= \; \varphi_\tau(Q \cap T) + \varphi_\tau((Q' \cap T) \cup S) \\
&< \; \varphi_\tau(Q) + \varphi_\tau(S) + \varepsilon \leqq c + \varphi_\tau(S) + \varepsilon,
\end{aligned}$$

and therefore $\lambda < c + \varepsilon$. The assertion follows.

The Conventional Inner Situation

The conventional inner situation is defined to be the specialization that $\varnothing \in \mathfrak{S}$ and $\varphi(\varnothing) = 0$. Thus we consider a lattice $\mathfrak{S}$ with $\varnothing \in \mathfrak{S}$ and an isotone set function $\varphi : \mathfrak{S} \to [0, \infty[$ with $\varphi(\varnothing) = 0$. Although the full inner situation is known to be identical with the full outer situation, there are characteristic discrepancies between the two conventional situations, as it must be expected from traditional measure theory. Thus we have to assume this time that $\varphi < \infty$. As in the outer situation there are certain immediate simplifications: An inner $\bullet$ extension of φ is a ccontent $\alpha : \mathfrak{A} \to [0, \infty]$ on a ring $\mathfrak{A}$, with the further properties as above. Furthermore we have $\varphi_\bullet : \mathfrak{P}(X) \to [0, \infty]$ with $0 = \varphi(\varnothing) = \varphi_\star(\varnothing) \leqq \varphi_\sigma(\varnothing) \leqq \varphi_\tau(\varnothing)$. Thus as before we can write $\mathfrak{C}(\varphi_\bullet)$ instead of $\mathfrak{C}(\varphi_\bullet, +)$.

But there are two essential deviations from the conventional outer situation. One deviation is that this time all supermodular φ are downward essential. This is obvious from the definition. Therefore the respective condition can be deleted from the conventional inner results.

The other deviation from the conventional outer situation is that $\varphi_\bullet(\varnothing) = 0$ is a nontrivial condition when $\bullet = \sigma\tau$. This condition will be explored in the course of the present subsection. We shall see that it is much weaker than the full condition $\varphi_\bullet | \mathfrak{S} = \varphi$, and that its verification can be much easier and sometimes even trivial. Therefore it is desirable to have the conventional inner main theorem with a variant of condition 5) in which $\varphi_\bullet(\varnothing) = 0$ occurs instead of $\varphi_\bullet|\mathfrak{S} = \varphi$. Of course then the subsequent partial condition in 5) has to be fortified. For this purpose we need the so-called satellites of the envelopes $\varphi_\bullet$ which will be defined next.

For fixed $\bullet = \star\sigma\tau$ and $B \in \mathfrak{S}$ we define the **satellite inner $\bullet$ envelopes** $\varphi_\bullet^B : \mathfrak{P}(X) \to [0, \infty[$ to be

$$\varphi_\bullet^B(A) = \sup\{\inf_{S \in \mathfrak{M}} \varphi(S) \; : \; \mathfrak{M} \text{ paving } \subset \mathfrak{S} \text{ of type } \bullet \text{ with}$$
$$S \subset B \,\forall S \in \mathfrak{M} \text{ and } \mathfrak{M} \downarrow \subset A\} \text{ for } A \subset X.$$

We list the basic properties of these satellites.

6.29. PROPERTIES. 1) $\varphi_\bullet^B \leqq \varphi(B) < \infty$. 2) $\varphi_\bullet^B$ is isotone. 3) If φ is supermodular then $\varphi_\bullet^B$ is supermodular. 4) We have

$$\varphi_\bullet(A) = \sup\{\varphi_\bullet^B(A) : B \in \mathfrak{S}\} \quad \text{for } A \subset X.$$

5) Assume that $\varphi = \varphi_\bullet|\mathfrak{S}$. Then $\varphi_\bullet(A) = \varphi_\bullet^B(A)$ for $A \subset B \in \mathfrak{S}$.

Proof. 1) and 2) are obvious, and 3) follows from 6.3.5) when one notes that $\varphi_\bullet^B = (\varphi|\{S \in \mathfrak{S} : S \subset B\})_\bullet$. 4) We have to prove $\leqq$. Fix $A \subset X$, and let $\mathfrak{M}$ be a paving $\subset \mathfrak{S}$ of type $\bullet$ such that $\mathfrak{M} \downarrow \subset A$. For fixed $B \in \mathfrak{M}$ then $\mathfrak{N} := \{S \in \mathfrak{M} : S \subset B\}$ is a paving $\subset \mathfrak{S}$ of type $\bullet$ with $\mathfrak{N} \downarrow \subset A$ as well, and we have

$$\inf_{S \in \mathfrak{M}} \varphi(S) = \inf_{S \in \mathfrak{N}} \varphi(S) \leqq \varphi_\bullet^B(A).$$

The assertion follows. 5) Fix $A \subset B \in \mathfrak{S}$. We have to prove $\leqq$. Let $\mathfrak{M}$ be a paving $\subset \mathfrak{S}$ of type $\bullet$ such that $\mathfrak{M} \downarrow M \subset A$. For fixed $H \in \mathfrak{M}$ then $\{S \cup (H \cap B) : S \in \mathfrak{M}\}$ is a paving $\subset \mathfrak{S}$ of type $\bullet$ with $\downarrow M \cup (H \cap B) = H \cap B \in \mathfrak{S}$. Hence by assumption and 6.5

$$\inf_{S \in \mathfrak{M}} \varphi(S) \leqq \inf_{S \in \mathfrak{M}} \varphi(S \cup (H \cap B)) = \varphi(H \cap B).$$

Now $\mathfrak{N} := \{H \cap B : H \in \mathfrak{M}\}$ is a paving $\subset \mathfrak{S}$ of type $\bullet$ with $\mathfrak{N} \downarrow M \cap B = M$ as well. It follows that

$$\inf_{S \in \mathfrak{M}} \varphi(S) \leqq \inf_{S \in \mathfrak{N}} \varphi(S) \leqq \varphi_\bullet^B(A),$$

and hence the assertion.

The decisive fact on the satellite inner $\bullet$ envelopes is the next lemma.

6.30. LEMMA. Let φ be supermodular. Assume that $\varphi_\bullet(\varnothing) = 0$ and that

$$\varphi(B) \leqq \varphi(A) + \varphi_\bullet^B(B \setminus A) \quad \text{for all } A \in \mathfrak{S} \text{ and } B \in \mathfrak{Q} \text{ with } A \subset B,$$

where $\mathfrak{Q} \subset \mathfrak{S}$ is upward cofinal. Then $\varphi = \varphi_\bullet|\mathfrak{S}$.

Proof. Fix $A \in \mathfrak{S}$. For $B \in \mathfrak{Q}$ with $B \supset A$ we combine the assumptions with 6.29.3)1) to obtain

$$\begin{aligned}
\varphi_\bullet^B(A) + \varphi(B) &\leqq \varphi_\bullet^B(A) + \varphi(A) + \varphi_\bullet^B(B \setminus A) \\
&\leqq \varphi(A) + \varphi_\bullet^B(B) + \varphi_\bullet^B(\varnothing) \leqq \varphi(A) + \varphi(B),
\end{aligned}$$

and hence $\varphi_\bullet^B(A) \leqq \varphi(A)$. Since $\varphi_\bullet^B(\cdot)$ is isotone in $B \in \mathfrak{S}$ it follows from 6.29.4) that $\varphi_\bullet(A) \leqq \varphi(A)$.

As before it is natural to specialize 6.19 and 6.22 to $\mathfrak{P} = \{\varnothing\}$ and $\mathfrak{Q} = \mathfrak{S}$. We then obtain the conventional inner main theorem which follows.

6.31. THEOREM (Conventional Inner Main Theorem). *Let $\mathfrak{S}$ be a lattice with $\varnothing \in \mathfrak{S}$, and $\varphi : \mathfrak{S} \to [0, \infty[$ be an isotone and supermodular set function with $\varphi(\varnothing) = 0$. Then the following are equivalent.*

1) *There exist inner $\bullet$ extensions of φ, that is φ is an inner $\bullet$ premeasure.*

2) *$\varphi_\bullet|\mathfrak{C}(\varphi_\bullet)$ is an inner $\bullet$ extension of φ. Furthermore*

$$if \bullet = \star \; : \; \varphi_\bullet|\mathfrak{C}(\varphi_\bullet) \text{ is a ccontent on the algebra } \mathfrak{C}(\varphi_\bullet),$$

$$if \bullet = \sigma\tau \; : \; \varphi_\bullet|\mathfrak{C}(\varphi_\bullet) \text{ is a cmeasure on the } \sigma \text{ algebra } \mathfrak{C}(\varphi_\bullet).$$

3) *$\varphi_\bullet|\mathfrak{C}(\varphi_\bullet)$ is an extension of φ in the crude sense, that is $\mathfrak{S} \subset \mathfrak{C}(\varphi_\bullet)$ and $\varphi = \varphi_\bullet|\mathfrak{S}$.*

4) *$\varphi(B) = \varphi(A) + \varphi_\bullet(B \setminus A)$ for all $A \subset B$ in $\mathfrak{S}$.*

5) *$\varphi = \varphi_\bullet|\mathfrak{S}$; and $\varphi(B) \leqq \varphi(A) + \varphi_\bullet(B \setminus A)$ for all $A \subset B$ in $\mathfrak{S}$.*

5') *$\varphi_\bullet(\varnothing) = 0$; and $\varphi(B) \leqq \varphi(A) + \varphi_\bullet^B(B \setminus A)$ for all $A \subset B$ in $\mathfrak{S}$.*

Note that 6.21 then implies $\mathfrak{S}\top\mathfrak{S}_\bullet \subset \mathfrak{C}(\varphi_\bullet)$.

Assume that $\mathfrak{S}$ is a lattice with $\varnothing \in \mathfrak{S}$. We define an isotone set function $\varphi : \mathfrak{S} \to [0, \infty[$ with $\varphi(\varnothing) = 0$ to be **inner $\bullet$ tight** iff it fulfils

$$\varphi(B) \leqq \varphi(A) + \varphi_\bullet^B(B \setminus A) \quad \text{for all } A \subset B \text{ in } \mathfrak{S},$$

as it appears in condition 5') above. In case $\bullet = \star$ this means that

$$\varphi(B) \leqq \varphi(A) + \varphi_\star(B \setminus A) \quad \text{for all } A \subset B \text{ in } \mathfrak{S}.$$

It is obvious that

$$\text{inner } \star \text{ tight} \Rightarrow \text{inner } \sigma \text{ tight} \Rightarrow \text{inner } \tau \text{ tight}.$$

As before we show on the spot that both converses $\Leftarrow$ are false. The counterexamples will be isotone and modular set functions $\varphi : \mathfrak{S} \to [0, \infty[$ with $\varphi(\varnothing) = 0$ which are downward τ continuous.

6.32. EXERCISE. Consider the situation of exercise 5.12. We have the equivalences

$$\varphi \text{ inner } \bullet \text{ tight } \Longleftrightarrow \varphi_\bullet(\{a\}) = 1 \Longleftrightarrow \{a\} \in (\mathrm{Op}(X))_\bullet.$$

Thus we obtain obvious counterexamples as announced above.

6.33. SPECIAL CASE (Traditional Type). *Assume that $\mathfrak{S}$ is a ring. Then conditions 5)5') simplify to*

5o) *$\varphi = \varphi_\bullet|\mathfrak{S}$; and φ is submodular.*

5'o) *$\varphi_\bullet(\varnothing) = 0$; and φ is submodular.*

We next consider the weakened condition $\varphi_\bullet(\varnothing) = 0$ which occurs in 6.30 and in the conventional inner main theorem. For this discussion we assume an isotone set function $\varphi : \mathfrak{S} \to [0, \infty]$ with $\varphi(\varnothing) = 0$ on a lattice $\mathfrak{S}$ with $\varnothing \in \mathfrak{S}$. We define φ to be $\bullet$ **continuous at** $\varnothing$ iff

$$\inf_{S \in \mathfrak{M}} \varphi(S) = 0 \quad \text{for each paving } \mathfrak{M} \subset \mathfrak{S} \text{ of type } \bullet \text{ with } \mathfrak{M} \downarrow \varnothing,$$

and to be **almost** $\bullet$ **continuous at** $\varnothing$ iff this holds true whenever $\varphi(S) < \infty \; \forall S \in \mathfrak{M}$. It is clear that φ is $\bullet$ continuous at $\varnothing$ iff $\varphi_\bullet(\varnothing) = 0$. In order to obtain an obvious but famous criterion we define $\mathfrak{S}$ to be $\bullet$ **compact** iff each paving $\mathfrak{M} \subset \mathfrak{S}$ of type $\bullet$ with $\mathfrak{M} \downarrow \varnothing$ satisfies $\varnothing \in \mathfrak{M}$. The reason for this notion is obvious: In each Hausdorff topological space X the lattice $\mathrm{Comp}(X)$ of its compact subsets is τ compact. The next remark is then clear.

6.34. REMARK. *If $\mathfrak{S}$ is $\bullet$ compact then each isotone set function $\varphi :$ $\mathfrak{S} \to [0, \infty]$ with $\varphi(\varnothing) = 0$ is $\bullet$ continuous at $\varnothing$.*

It turns out that for $\bullet = \sigma\tau$ the condition $\varphi_\bullet(\varnothing) = 0$ is much weaker than $\varphi = \varphi_\bullet | \mathfrak{S}$. We shall present a dramatic example at the end of the subsection. The example will show in particular that the conventional inner main theorem becomes false when instead of 5)5') one forms the weaker condition which combines the first part of 5') with the second part of 5).

The most important and simplest nontrivial example for the conventional inner situation is the familiar set function $\lambda : \mathfrak{K} = \mathrm{Comp}(\mathbb{R}^n) \to [0, \infty[$. It has been an example for the conventional outer situation in 5.14.

6.35. EXAMPLE. 1) λ is inner $\star$ tight and hence inner $\bullet$ tight for $\bullet = \star\sigma\tau$. In fact, we know that for $A \subset B$ in $\mathfrak{K}$ there exists a sequence $(K_l)_l$ in $\mathfrak{K}$ with $K_l \uparrow B \setminus A$ and hence $\lambda(K_l) \uparrow \lambda(B) - \lambda(A)$. This implies that $\lambda_\star(B \setminus A) = \lambda(B) - \lambda(A)$. 2) λ is $\bullet$ continuous at $\varnothing$ in view of 6.34; we even know from 2.25 that λ is downward $\bullet$ continuous. 3) Therefore the conventional inner main theorem shows that λ is an inner $\bullet$ premeasure for $\bullet = \star\sigma\tau$. 4) From 6.5.iv) we have $\lambda_\star = \lambda_\sigma = \lambda_\tau$. We shall see in 7.5 below that the common maximal inner $\bullet$ extension $\lambda_\bullet | \mathfrak{C}(\lambda_\bullet)$ coincides with $\Lambda := \lambda^\sigma | \mathfrak{C}(\lambda^\sigma)$.

We conclude with the example announced above. It is quite complicated.

6.36. EXERCISE. Construct an example of an isotone and modular set function $\varphi : \mathfrak{S} \to [0, \infty[$ on a lattice $\mathfrak{S}$ with $\varnothing \in \mathfrak{S}$ and $X = \bigcup_{S \in \mathfrak{S}} S \neq \varnothing$ such that

   i) $\mathfrak{S}$ is σ compact and hence $\varphi_\sigma(\varnothing) = 0$, but
   ii) $\varphi_\sigma(A) = \infty$ for all nonvoid $A \subset X$.

One can proceed as follows. 1) Let E be an infinite set and $X := E \times \mathbb{N}$. A subset $A \subset X$ is described via its sections $A(s) := \{m \in \mathbb{N} : (s, m) \in A\} \subset \mathbb{N}$ for $s \in E$. Define $\mathfrak{E}$ to consist of the subsets $A \subset X$ such that

   1.i) $A(s) \subset \mathbb{N}$ is finite or cofinite for all $s \in E$;

1.ii) $A(s) = \varnothing$ for all $s \in E$ except a finite subset.

Then $\mathfrak{E}$ is a lattice with $\varnothing \in \mathfrak{E}$. Define $\phi : \mathfrak{E} \to [0, \infty[$ to be

$$\phi(A) = \#(\{s \in E : A(s) \subset \mathbb{N} \text{ is cofinite}\}) \quad \text{for } A \in \mathfrak{E}.$$

Then ϕ is isotone and modular with $\phi(\varnothing) = 0$. 2) If E is countable then there exists a function $\theta : X \to \mathfrak{P}(E)$ such that

2.i) $\theta(x) \subset E$ is infinite for all $x \in X$;

2.ii) for $u \neq v$ in X we have $\theta(u) \cap \theta(v) = \varnothing$;

2.iii) $E = \bigcup_{x \in X} \theta(x)$;

2.iv) for $x = (s, m) \in X$ we have $s \notin \theta(x)$.

Hint: We can assume that $E = \mathbb{N} \cup (-\mathbb{N})$. Let $I : \mathbb{N} \times \mathbb{N} \to \mathbb{N}$ be a bijection. Define $\theta : X \to \mathfrak{P}(E)$ to be

$$\begin{aligned} \theta(x) \quad &= \quad \{-\varepsilon I(n, I(p, q)) : n \in \mathbb{N}\} \subset E \\ &\quad \text{for } x = (\varepsilon p, q) \in X \text{ with } p, q \in \mathbb{N} \text{ and } \varepsilon \in \{-1, 1\}. \end{aligned}$$

Then θ is as required. For the sequel we fix E and θ. 3) Define $\mathfrak{S}$ to consist of the subsets $S \in \mathfrak{E}$ such that

$$x \in X \setminus S \Rightarrow S(s) \subset \mathbb{N} \text{ is finite for all } s \in \theta(x).$$

Then $\mathfrak{S}$ is a lattice with $\varnothing \in \mathfrak{S}$. Define $\varphi := \phi | \mathfrak{S}$. 4) $\mathfrak{S}$ is σ compact. Hint: Let $(S_l)_l$ be a sequence in $\mathfrak{S}$ with $S_l \downarrow \varnothing$. Fix a nonvoid finite subset $F \subset E$ such that $S_1(s) = \varnothing$ for all $s \in E \setminus F$. Then let $x_1, \cdots, x_r \in X$ such that $F \subset \bigcup_{k=1}^{r} \theta(x_k)$. If $x_1, \cdots, x_r \notin S_l$, which is true for almost all $l \geq 1$, then S_l is finite. This implies that $S_l = \varnothing$ for almost all $l \geq 1$. 5) Fix $a = (p, q) \in X$ and a nonvoid finite $F \subset \theta(a) \subset E$, and note that $p \notin \theta(a)$ and hence $p \notin F$. For $T \subset \mathbb{N}$ cofinite we define $S \subset X$ to be

5.i) $S(p) := \{q\}$;

5.ii) $S(s) := T$ for all $s \in F$;

5.iii) $S(s) := \varnothing$ for all other $s \in E$.

Then $S \in \mathfrak{S}$ with $a \in S$, and $\varphi(S) = \#(F)$. Deduce that $\varphi_\sigma(\{a\}) \geq \#(F)$. 6) It follows from 5) that $\varphi_\sigma(\{a\}) = \infty$ for all $a \in X$ and hence $\varphi_\sigma(A) = \infty$ for all nonvoid $A \subset X$.

7. Complements to the Extension Theories

The present section has two independent themes. The first one is to compare the outer and inner extension theories. The other theme is to exhibit certain classes of lattices with $\varnothing$ on which the relevant tightness conditions will be automatic facts like on rings, but which are much more natural initial domains than rings. The idea has been used for the Lebesgue measure in both the outer and the inner situation. Its systematization will lead to an essential increase of the frame of applications. The section concludes with a bibliographical annex.

Comparison of the Outer and Inner Extension Theories

The main result of the present subsection is for a set function $\varphi : \mathfrak{S} \to \mathbb{R}$ on a lattice $\mathfrak{S}$ which is both an outer and an inner $\bullet$ premeasure. We restrict ourselves to $\bullet = \star \sigma$ since the case $\bullet = \tau$ is unrealistic. The result will be that the two maximal extensions $\varphi^\bullet | \mathfrak{C}(\varphi^\bullet, \dotplus)$ and $\varphi_\bullet | \mathfrak{C}(\varphi_\bullet, \dotplus)$ coincide to the extent which can be expected in view of the classical uniqueness theorem 3.1.

7.1. LEMMA. *Assume that* $\varphi : \mathfrak{S} \to \overline{\mathbb{R}}$ *and* $\alpha : \mathfrak{A} \to \overline{\mathbb{R}}$ *are isotone set functions on lattices* $\mathfrak{S}$ *and* $\mathfrak{A}$, *and that* α *extends* φ.

$\star$) $\varphi_\star \leqq \alpha \leqq \varphi^\star$ *on* $\mathfrak{A}$, *and* $\varphi_\star \leqq \varphi^\star$ *on* $\mathfrak{P}(X)$.

σ) *If* $\mathfrak{A}$ *is a* σ *lattice and* α *is almost upward and downward* σ *continuous then*

$$\varphi_\sigma(A) \leqq \alpha(A) \quad \text{for } A \in \mathfrak{A} \text{ when } (+) \; \varphi < \infty \text{ or } \varphi_\sigma(A) < \infty;$$
$$\alpha(A) \leqq \varphi^\sigma(A) \quad \text{for } A \in \mathfrak{A} \text{ when } (-) - \infty < \varphi \text{ or } -\infty < \varphi^\sigma(A);$$
$$\varphi_\sigma(A) \leqq \varphi^\sigma(A) \quad \text{for } A \subset X \text{ when } (+) \text{ and } (-).$$

Proof. $\star$) is clear from the definitions. In σ) we can for the first assertion assume that $\varphi_\sigma(A) > -\infty$. Let $(S_l)_l$ be a sequence in $\mathfrak{S}$ with $S_l \downarrow$ some $U \subset A$. Then by definition $\lim_{l\to\infty} \varphi(S_l) \leqq \varphi_\sigma(A)$, so that both times we can assume that $\varphi(S_l) < \infty \; \forall l \in \mathbb{N}$. It follows that $U \in \mathfrak{S}_\sigma \subset \mathfrak{A}$ and $\varphi(S_l) = \alpha(S_l) \downarrow \alpha(U) \leqq \alpha(A)$. Therefore $\varphi_\sigma(A) \leqq \alpha(A)$. The second assertion is proved in the same manner. In order to prove the third assertion assume that $\varphi^\sigma(A) < \varphi_\sigma(A)$ and fix a real c with $\varphi^\sigma(A) < c < \varphi_\sigma(A)$. By 4.1.4) and 6.3.4) there are

$$V \in \mathfrak{S}^\sigma \subset \mathfrak{A} \qquad \text{with } V \supset A \text{ and } \varphi^\sigma(V) < c,$$
$$U \in \mathfrak{S}_\sigma \subset \mathfrak{A} \qquad \text{with } U \subset A \text{ and } \varphi_\sigma(U) > c.$$

We can apply the first assertion to U to obtain $\varphi_\sigma(U) \leqq \alpha(U)$, and the second assertion to V to obtain $\alpha(V) \leqq \varphi^\sigma(V)$. It follows that $\alpha(V) \leqq \varphi^\sigma(V) < c < \varphi_\sigma(U) \leqq \alpha(U)$ and hence a contradiction.

7.2. LEMMA. *Assume that* $\varphi : \mathfrak{S} \to \mathbb{R}$ *is an isotone set function on a lattice* $\mathfrak{S}$, *and that* $\alpha : \mathfrak{A} \to \overline{\mathbb{R}}$ *is an extension of* φ *which is*

$$\text{for } \bullet = \star \; : \quad \text{a content } \dotplus \text{ on an oval } \mathfrak{A};$$
$$\text{for } \bullet = \sigma \; : \quad \text{a measure } \dotplus \text{ on a } \sigma \text{ oval } \mathfrak{A}.$$

Then for $S, T \in \mathfrak{S}$ *and* $A \in \mathfrak{A}$ *we have*

$$\alpha(S|A|T) = \varphi^\bullet(S|A|T) \qquad \text{when } A \in \mathfrak{C}(\varphi^\bullet, \dotplus),$$
$$\alpha(S|A|T) = \varphi_\bullet(S|A|T) \qquad \text{when } A \in \mathfrak{C}(\varphi_\bullet, \dotplus).$$

Proof. Fix $S, T \in \mathfrak{S}$ and $A \in \mathfrak{A}$. Then $S|A|T \in \mathfrak{A}$ and $S|A'|T = T|A|S \in \mathfrak{A}$. From 7.1 we obtain

$$\varphi_\bullet(S|A|T) \leqq \alpha(S|A|T) \leqq \varphi^\bullet(S|A|T),$$

and the same for A'. Furthermore note that $S|A|T$ is between $S \cap T$ and $S \cup T$, and that in case $\bullet = \sigma$ the function φ is upward and downward σ continuous. Thus

$$\varphi^\bullet(S|A|T) \leqq \varphi^\bullet(S \cup T) = \varphi(S \cup T) < \infty,$$
$$\varphi_\bullet(S|A|T) \geqq \varphi_\bullet(S \cap T) = \varphi(S \cap T) > -\infty,$$

so that the above values are all finite. The same is true for A'. Now we have on the one hand

$$\alpha(S|A|T) + \alpha(S|A'|T) = \alpha(S) + \alpha(T) = \varphi(S) + \varphi(T),$$

since by the modularity $\dotplus$ of α both sides are $= \alpha(S \cup T) + \alpha(S \cap T)$. On the other hand we have

for $A \in \mathfrak{C}(\varphi^\bullet, \dotplus) : \varphi^\bullet(S|A|T) + \varphi^\bullet(S|A'|T) = \varphi^\bullet(S) + \varphi^\bullet(T) = \varphi(S) + \varphi(T)$,
for $A \in \mathfrak{C}(\varphi_\bullet, \dotplus) : \varphi_\bullet(S|A|T) + \varphi_\bullet(S|A'|T) = \varphi_\bullet(S) + \varphi_\bullet(T) = \varphi(S) + \varphi(T)$.

The combination furnishes the assertions.

The next result says that in a certain sense regularity can be turned around at the Carathéodory class.

7.3. PROPOSITION. *Assume that* $\phi : \mathfrak{P}(X) \to \overline{\mathbb{R}}$ *is isotone and submodular* $\dotplus$. *Let* $\mathfrak{T}$ *be a paving in* X *such that* ϕ *is outer regular* $\mathfrak{T}$. *If* $A \in \mathfrak{C}(\phi, \dotplus)$ *is such that there exists* $T \in \mathfrak{T}$ *with* $T \subset A$ *and* $-\infty < \phi(T) \leqq \phi(A) < \infty$ *then*

$$\phi(A) = \sup\{\phi(P) : P \in O(\mathfrak{T}) \text{ with } P \subset A\}.$$

Proof. Fix $\varepsilon > 0$. i) By assumption there exists $S \in \mathfrak{T}$ with $S \supset A$ such that $\phi(S) \leqq \phi(A) + \varepsilon$. Thus $\phi(S) \in \mathbb{R}$. Also fix $T \in \mathfrak{T}$ as described above. From $A \in \mathfrak{C}(\phi, \dotplus)$ we conclude that

$$\phi(T) \dotplus \phi(S) = \phi(T|A|S) \dotplus \phi(T|A'|S) = \phi(A) \dotplus \phi(S|A|T).$$

Therefore $\phi(S|A|T)$ is finite and $\leqq \phi(T) + \varepsilon$. ii) By assumption there exists $H \in \mathfrak{T}$ with $H \supset S|A|T$ such that

$$\phi(H) \leqq \phi(S|A|T) + \varepsilon \leqq \phi(T) + 2\varepsilon.$$

Thus $\phi(H) \in \mathbb{R}$. Note that

$$H \supset S|A|T \supset T \cap A = T \text{ since } T \subset A,$$
$$S \cap H' \subset S \cap (S'|A|T') = S \cap T' \cap A \subset A.$$

iii) Now define $P := S|H|T \in O(\mathfrak{T})$. The last inclusions show that $T \subset P \subset A$. Furthermore

$$P \cap H = T \cap H = T \text{ and } P \cup H = (S \cap H') \cup H = S \cup H \supset S \supset A.$$

Since ϕ is submodular $\dotplus$ this implies that

$$\phi(A) \dotplus \phi(T) \leqq \phi(P \cup H) \dotplus \phi(P \cap H) \leqq \phi(P) \dotplus \phi(H),$$

and we know that all terms are finite. From this and from ii) it follows that $\phi(A) + \phi(T) \leqq \phi(P) + \phi(T) + 2\varepsilon$ or $\phi(P) \geqq \phi(A) - 2\varepsilon$. Thus we have the assertion.

We shall also need the upside-down counterpart.

7.4. EXERCISE. *Assume that $\phi : \mathfrak{P}(X) \to \overline{\mathbb{R}}$ is isotone and supermodular $\dotplus$. Let $\mathfrak{T}$ be a paving in X such that ϕ is inner regular $\mathfrak{T}$. If $A \in \mathfrak{C}(\phi, \dotplus)$ is such that there exists $T \in \mathfrak{T}$ with $A \subset T$ and $-\infty < \phi(A) \leqq \phi(T) < \infty$ then*

$$\phi(A) = \inf\{\phi(P) : P \in O(\mathfrak{T}) \text{ with } P \supset A\}.$$

We can now obtain the desired comparison theorem.

7.5. THEOREM. *Let $\mathfrak{S}$ be a lattice and $\bullet = \star\sigma$. Assume that the set function $\varphi : \mathfrak{S} \to \mathbb{R}$ is both an outer and an inner $\bullet$ premeasure. Then $\mathfrak{C}(\varphi^\bullet, \dotplus) = \mathfrak{C}(\varphi_\bullet, \dotplus) =: \mathfrak{C}$. Furthermore $\varphi^\bullet(A) = \varphi_\bullet(A)$ for all $A \in \mathfrak{C} \cap (\mathfrak{S}_\bullet \sqsubset \mathfrak{S}^\bullet)$, except that in case $\bullet = \sigma$ one has also to admit that $\varphi^\sigma(A) = \infty$ and $\varphi_\sigma(A) = -\infty$.*

Note that in the conventional situation $\varnothing \in \mathfrak{S}$ and $\varphi(\varnothing) = 0$ the latter exceptional case cannot occur.

Proof. 1) φ is isotone and modular and fulfils $\varphi = \varphi^\bullet|\mathfrak{S} = \varphi_\bullet|\mathfrak{S}$. Thus 7.1 and 7.2 can be applied to both $\alpha := \varphi^\bullet|\mathfrak{C}(\varphi^\bullet, \dotplus)$ and $\alpha := \varphi_\bullet|\mathfrak{C}(\varphi_\bullet, \dotplus)$. Each time it follows that $\varphi_\bullet \leqq \varphi^\bullet$ on $\mathfrak{P}(X)$, and that

$$\varphi^\bullet(S|A|T) = \varphi_\bullet(S|A|T) \quad \text{for } S, T \in \mathfrak{S} \text{ and } A \in \mathfrak{C}(\varphi^\bullet, \dotplus) \cap \mathfrak{C}(\varphi_\bullet, \dotplus).$$

In view of $S \cap T \subset S|A|T \subset S \cup T$ the common value is finite. 2) We claim that

$$\varphi^\bullet(S|A|T) = \varphi_\bullet(S|A|T) \quad \text{for } S, T \in \mathfrak{S} \text{ and } A \in \mathfrak{C}(\varphi^\bullet, \dotplus).$$

To see this one applies 7.3 to $\phi := \varphi^\bullet$ and $\mathfrak{T} := \mathfrak{S}^\bullet$, and to the subset $S|A|T \in \mathfrak{C}(\varphi^\bullet, \dotplus)$; note that $S \cap T \in \mathfrak{S} \subset \mathfrak{S}^\bullet$ is as required in 7.3. It follows that

$$\begin{aligned} \varphi^\bullet(S|A|T) &= \sup\{\varphi^\bullet(P) : P \in O(\mathfrak{S}^\bullet) \text{ with } P \subset S|A|T\} \\ &= \sup\{\varphi^\bullet(P) : P \in O(\mathfrak{S}^\bullet) \text{ with } S \cap T \subset P \subset S|A|T\}. \end{aligned}$$

Now for the $P \in O(\mathfrak{S}^\bullet)$ of the last kind $P = S \cap T|P|S \cup T$. Furthermore we have $\mathfrak{S}^\bullet \subset \mathfrak{C}(\varphi^\bullet, \dotplus) \cap \mathfrak{C}(\varphi_\bullet, \dotplus)$ and hence $P \in O(\mathfrak{S}^\bullet) \subset \mathfrak{C}(\varphi^\bullet, \dotplus) \cap \mathfrak{C}(\varphi_\bullet, \dotplus)$. Thus 1) implies that

$$\varphi^\bullet(P) = \varphi^\bullet(S \cap T|P|S \cup T) = \varphi_\bullet(S \cap T|P|S \cup T) = \varphi_\bullet(P).$$

It follows that $\varphi^\bullet(S|A|T) \leqq \varphi_\bullet(S|A|T)$ and hence $= \varphi_\bullet(S|A|T)$. Note that the common value is finite as before. 2') Likewise we have

$$\varphi^\bullet(S|A|T) = \varphi_\bullet(S|A|T) \quad \text{for } S, T \in \mathfrak{S} \text{ and } A \in \mathfrak{C}(\varphi_\bullet, \dotplus).$$

The proof is as in 2), but with 7.4 instead of 7.3.

3) We next prove that $\mathfrak{C}(\varphi^\bullet, \dotplus) \subset \mathfrak{C}(\varphi_\bullet, \dotplus)$. In fact, let $A \in \mathfrak{C}(\varphi^\bullet, \dotplus)$. For $S, T \in \mathfrak{S}$ we obtain from 2)

$$\begin{aligned} \varphi(S) + \varphi(T) = \varphi^\bullet(S) + \varphi^\bullet(T) &= \varphi^\bullet(S|A|T) + \varphi^\bullet(S|A'|T) \\ &= \varphi_\bullet(S|A|T) + \varphi_\bullet(S|A'|T). \end{aligned}$$

Thus 6.19 implies that $A \in \mathfrak{C}(\varphi_\bullet, +)$. 3') We obtain $\mathfrak{C}(\varphi_\bullet, +) \subset \mathfrak{C}(\varphi^\bullet, \dotplus)$ as in 3), but based on 2') and 5.2 instead of 2) and 6.19. 4) So far we have proved that $\mathfrak{C}(\varphi^\bullet, \dotplus) = \mathfrak{C}(\varphi_\bullet, +) =: \mathfrak{C}$, and furthermore that

$$\varphi^\bullet(S|A|T) = \varphi_\bullet(S|A|T) \in \mathbb{R} \quad \text{for } S, T \in \mathfrak{S} \text{ and } A \in \mathfrak{C}.$$

5) We finish the case $\bullet = \star$. If $A \in \mathfrak{C} \cap (\mathfrak{S}_\star \sqsubset \mathfrak{S}^\star) = \mathfrak{C} \cap (\mathfrak{S} \sqsubset \mathfrak{S})$ then $S \subset A \subset T$ for some $S, T \in \mathfrak{S}$. It follows that $S|A|T = A$ and hence $\varphi^\star(A) = \varphi_\star(A)$.

6) We turn to the case $\bullet = \sigma$. Fix $A \in \mathfrak{C} \cap (\mathfrak{S}_\sigma \sqsubset \mathfrak{S}^\sigma)$. Thus $P \subset A \subset Q$ where $P_l \downarrow P$ and $Q_l \uparrow Q$ for some sequences $(P_l)_l$ and $(Q_l)_l$ in $\mathfrak{S}$. Furthermore fix $S \in \mathfrak{S}$. From 4) we obtain

$$P_l|A|S \downarrow P|A|S = S \cap A \quad \text{and hence} \quad \varphi^\sigma(S \cap A) = \varphi_\sigma(S \cap A) < \infty,$$
$$S|A|Q_l \uparrow S|A|Q = S \cup A \quad \text{and hence} \quad \varphi^\sigma(S \cup A) = \varphi_\sigma(S \cup A) > -\infty.$$

Now 4.12.4) implies that

$$\begin{aligned}
\varphi(S) + \varphi^\sigma(A) &= \varphi^\sigma(S \cup A) \dotplus \varphi^\sigma(S \cap A), \\
\varphi(S) + \varphi_\sigma(A) &= \varphi_\sigma(S \cup A) \dotplus \varphi_\sigma(S \cap A).
\end{aligned}$$

Thus if $\varphi^\sigma(A) \neq \varphi_\sigma(A)$ then we must have $\varphi^\sigma(S \cup A) = \varphi_\sigma(S \cup A) = \infty$ and $\varphi^\sigma(S \cap A) = \varphi_\sigma(S \cap A) = -\infty$, and hence $\varphi^\sigma(A) = \infty$ and $\varphi_\sigma(A) = -\infty$. We shall see in exercise 7.7 below that this indeed can happen.

7.6. EXAMPLE. For $\lambda : \mathfrak{K} = \mathrm{Comp}(\mathbb{R}^n) \to [0, \infty[$ the maximal inner σ extension $\lambda_\sigma | \mathfrak{C}(\lambda_\sigma)$, which is the common maximal inner $\bullet$ extension $\lambda_\bullet | \mathfrak{C}(\lambda_\bullet)$, coincides with $\Lambda := \lambda^\sigma | \mathfrak{C}(\lambda^\sigma)$. This fact has been announced in 6.35.

We conclude with the example announced in connection with 7.5.

7.7. EXERCISE. Construct an example which shows that in 7.5 it can happen that $\varphi^\sigma(A) = \infty$ and $\varphi_\sigma(A) = -\infty$. Hint: On $X = \mathbb{R}$ define the paving $\mathfrak{S}$ to consist of all $S \in \mathrm{Bor}(X)$ such that S is bounded above and S' is bounded below. Note that $\mathfrak{S}$ is an oval. We write $R := [0, \infty[$ and $L :=]-\infty, 0]$, and define $\varphi : \mathfrak{S} \to \mathbb{R}$ to be

$$\varphi(S) = \Lambda(S \cap R) - \Lambda(S' \cap L) \text{ for } S \in \mathfrak{S}.$$

Then show that $\varphi^\sigma(R) = \infty$ and $\varphi_\sigma(R) = -\infty$.

Lattices of Ringlike Types

We restrict ourselves to the conventional outer and inner situations. The most unfamiliar notion considered so far is that of tightness. Therefore it is desirable to have transparent assumptions which ensure the relevant tightness conditions. The simplest assumption of this type is that the initial domain be a ring. However, the previous theories allow to work with certain weaker assumptions which are much more realistic.

Let $\mathfrak{S}$ be a lattice in a nonvoid set X and $\bullet = \star\sigma\tau$. We define $\mathfrak{S}$ to be

upward $\bullet$ full iff $B \setminus A \in \mathfrak{S}^\bullet$ for each pair $A \subset B$ in $\mathfrak{S}$,
downward $\bullet$ full iff $B \setminus A \in \mathfrak{S}_\bullet$ for each pair $A \subset B$ in $\mathfrak{S}$.

Thus we have

$$\mathfrak{S} \text{ ring } \Leftrightarrow \mathfrak{S} \text{ upward } \star \text{ full}$$
$$\Rightarrow \mathfrak{S} \text{ upward } \sigma \text{ full } \Rightarrow \mathfrak{S} \text{ upward } \tau \text{ full};$$
$$\mathfrak{S} \text{ ring } \Leftrightarrow \mathfrak{S} \text{ downward } \star \text{ full}$$
$$\Rightarrow \mathfrak{S} \text{ downward } \sigma \text{ full } \Rightarrow \mathfrak{S} \text{ downward } \tau \text{ full}.$$

If $\mathfrak{S}$ is upward $\bullet$ full then $\varnothing \in \mathfrak{S}$, but trivial examples show that this need not be true if $\mathfrak{S}$ is downward $\bullet$ full.

7.8. EXAMPLES (for $\bullet = \sigma$). 1) Let X be a topological space. For the sublattices $\mathrm{CCl}(X)$ and $\mathrm{COp}(X)$ of $\mathrm{Baire}(X)$ defined in 1.6.4) we refer to 8.1 below. 2) Let X be a semimetrizable topological space. One verifies that $\mathrm{Cl}(X) \subset (\mathrm{Op}(X))_\sigma$ and hence that $\mathrm{Op}(X) \subset (\mathrm{Cl}(X))^\sigma$. It follows that $\mathrm{Cl}(X)$ is upward σ full and $\mathrm{Op}(X)$ is downward σ full. 3) Let X be a metrizable topological space. Then 2) implies that $\mathrm{Comp}(X)$ is upward σ full. This has been used for $X = \mathbb{R}^n$ in 5.14.1) and 6.35.

7.9. EXERCISE. Let X be a Hausdorff topological space. Prove that

$\mathrm{Cl}(X)$ is always upward τ full, and downward τ full iff X is discrete;
$\mathrm{Op}(X)$ is always downward τ full, and upward τ full iff X is discrete.

This makes clear that the case $\bullet = \tau$ is much less important than the cases $\bullet = \star\sigma$.

We come to the decisive point. We start with the upward fullness conditions.

7.10. PROPOSITION. *Assume that $\mathfrak{S}$ is upward $\bullet$ full. Let $\varphi : \mathfrak{S} \to [0,\infty]$ be isotone and modular with $\varphi(\varnothing) = 0$. 1) φ is outer $\bullet$ tight. 2) If $\varphi < \infty$ and $\varphi = \varphi^\bullet | \mathfrak{S}$ then φ is inner $\star$ tight.*

Proof. Fix $A \subset B$ in $\mathfrak{S}$. By 1.4.1) there exists a paving $\mathfrak{M} \subset \mathfrak{S}$ of type $\bullet$ with $\mathfrak{M} \uparrow B \setminus A$. 1) To be shown is $\varphi(A) + \varphi^\bullet(B \setminus A) \leqq \varphi(B)$. We can assume that $\varphi(B) < \infty$. For $S \in \mathfrak{M}$ we have $\varphi(A) + \varphi(S) = \varphi(A \cup S) \leqq \varphi(B)$. It follows that

$$\varphi(A) + \varphi^\bullet(B \setminus A) \leqq \varphi(A) + \sup_{S \in \mathfrak{M}} \varphi(S) \leqq \varphi(B).$$

2) To be shown is $\varphi(B) \leqq \varphi(A) + \varphi_*(B \setminus A)$. Now $\{A \cup S : S \in \mathfrak{M}\} \subset \mathfrak{S}$ is a paving of type $\bullet$ with $\uparrow B$, so that by assumption $\sup_{S \in \mathfrak{M}} \varphi(A \cup S) = \varphi(B)$. From $\varphi(A \cup S) = \varphi(A) + \varphi(S)$ for $S \in \mathfrak{M}$ we obtain

$$\varphi(B) = \varphi(A) + \sup_{S \in \mathfrak{M}} \varphi(S) \leqq \varphi(A) + \varphi_*(B \setminus A).$$

7.11. REMARK. We emphasize that

σ) if $\mathfrak{S}$ is upward σ full then φ need not be outer $\star$ tight;

τ) if $\mathfrak{S}$ is upward τ full then φ need not be outer σ tight.

Thus assertion 1) cannot be improved in this respect. For counterexamples we refer to 5.12: In a Hausdorff topological space X let $a \in X$ and $\psi := \delta_a | \mathrm{Cl}(X)$. σ) If X is metrizable then $\mathrm{Cl}(X)$ is upward σ full by 7.8.2). But if a is not an isolated point of X, that is if $\{a\} \notin \mathrm{Op}(X)$, then ψ is not outer $\star$ tight by 5.12.2). τ) $\mathrm{Cl}(X)$ is always upward τ full by 7.9. But if $\{a\} \notin (\mathrm{Op}(X))_\sigma$ then ψ is not outer σ tight by 5.12.2).

The main consequence which follows will be restricted to the case $\bullet = \sigma$. The case $\bullet = \star$ would be contained in the earlier 5.13 and 6.33, and the case $\bullet = \tau$ would be more involved and seems to be without substantial applications.

7.12. THEOREM. *Assume that* $\mathfrak{S}$ *is upward* σ *full. Let* $\varphi : \mathfrak{S} \to [0, \infty]$ *be isotone and modular with* $\varphi(\varnothing) = 0$ *and* $\varphi = \varphi^\sigma | \mathfrak{S}$. *1)* φ *is an outer* σ *premeasure. 2) If* $\varphi < \infty$ *then* φ *is an inner* σ *premeasure.*

Let us add at once that for $\varphi < \infty$ it follows from 7.5 that $\mathfrak{C}(\varphi^\sigma) = \mathfrak{C}(\varphi_\sigma) =: \mathfrak{C}$ and $\varphi^\sigma(A) = \varphi_\sigma(A)$ for all $A \in \mathfrak{C} \cap (\sqsubset \mathfrak{S}^\sigma)$.

Proof. 1) is clear from 7.10.1) and the conventional outer main theorem 5.11. 2) By 7.10.2) and the conventional inner main theorem 6.31 we have to prove that $\varphi_\sigma(\varnothing) = 0$. By 1) φ is an outer σ premeasure. Now consider a countable paving $\mathfrak{M} \subset \mathfrak{S}$ with $\mathfrak{M} \downarrow \varnothing$. To be shown is $\inf_{S \in \mathfrak{M}} \varphi(S) = 0$. We fix $E \in \mathfrak{M}$. Then likewise $\mathfrak{M}(E) := \{S \in \mathfrak{M} : S \subset E\} \subset \mathfrak{S}$ is a countable paving with $\downarrow \varnothing$. To be shown is of course $\inf_{S \in \mathfrak{M}(E)} \varphi(S) = 0$. By assumption we have $\{E \setminus S : S \in \mathfrak{M}(E)\} \subset \mathfrak{S}^\sigma$, and this is a countable paving with $\uparrow E \in \mathfrak{S} \subset \mathfrak{S}^\sigma$. Thus we have

$$\sup_{S \in \mathfrak{M}(E)} \varphi^\sigma(E \setminus S) = \varphi^\sigma(E) = \varphi(E).$$

In view of $\varphi^\sigma(E \setminus S) = \varphi(E) - \varphi(S)$ for $S \in \mathfrak{M}(E)$ this is the assertion.

7.13. EXERCISE. The above theorem becomes false in both parts when instead of $\varphi = \varphi^\sigma | \mathfrak{S}$ one assumes that $\varphi_\sigma(\varnothing) = 0$. In fact, we shall construct set functions $\varphi : \mathfrak{S} \to [0, \infty[$ on upward σ full lattices which are isotone and modular with $\varphi(\varnothing) = \varphi_\sigma(\varnothing) = 0$ but do not fulfil $\varphi = \varphi^\sigma | \mathfrak{S}$.

Let X be a Hausdorff topological space and $a \in X$ such that $\{a\}$ is not open but $\in (\mathrm{Op}(X))_\sigma$. 1) Construct a set function $\varphi : \mathrm{Cl}(X) \to [0, \infty[$ which is isotone and modular with $\varphi(\varnothing) = 0$ but not upward σ continuous. Hint: Consider on the real vector space $\mathrm{B}(X, \mathbb{R})$ of the bounded functions $X \to \mathbb{R}$ the sublinear functional $\vartheta : \mathrm{B}(X, \mathbb{R}) \to \mathbb{R}$, defined to be

$$\vartheta(f) = \limsup_{x \to a} f(x) := \inf\{\sup(f | U \setminus \{a\}) : a \in U \text{ open } \subset X\}.$$

Let after Hahn-Banach $\phi : \mathrm{B}(X, \mathbb{R}) \to \mathbb{R}$ be a linear functional $\leq \vartheta$. Note that $\phi \leq \vartheta \leq \sup$. Then define $\varphi : \mathrm{Cl}(X) \to [0, \infty[$ to be $\varphi(A) = \phi(\chi_A)$ for

$A \in \mathrm{Cl}(X)$. Consider for a sequence of open $U_l \downarrow \{a\}$ the sequence of the closed subsets $U_l' \cup \{a\}$. 2) If in particular X is compact then φ must be σ and even τ continuous at $\varnothing$.

Let us reformulate the last theorem in order that it looks like the classical Carathéodory extension theorem. The latter theorem is the upper closed path under the assumption that $\mathfrak{S}$ be a ring, and likewise the earlier 5.13 for $\bullet = \sigma$.

7.14. REFORMULATION. *Assume that $\mathfrak{S}$ is upward σ full. Let $\varphi : \mathfrak{S} \to [0, \infty]$ be isotone and modular with $\varphi(\varnothing) = 0$. Then we have the implications as shown below* (the simple arrows are obvious implications).

φ can be extended to cmeasure on σ algebra which is outer regular $\mathfrak{S}^\sigma$

$\Uparrow$ $\qquad\qquad\qquad\qquad\qquad\qquad\qquad\qquad\qquad\qquad\qquad$ $\downarrow$

φ is upward σ continuous $\leftarrow$ φ can be extended to cmeasure on σ algebra

$\Downarrow \varphi < \infty$ $\qquad\qquad\qquad\qquad\qquad\qquad\qquad\qquad\qquad\qquad$ $\uparrow$

φ can be extended to cmeasure on σ algebra which is inner regular $\mathfrak{S}_\sigma$.

We turn to the counterparts for the downward fullness conditions.

7.15. PROPOSITION. *Assume that $\mathfrak{S}$ is downward $\bullet$ full with $\varnothing \in \mathfrak{S}$. Let $\varphi : \mathfrak{S} \to [0, \infty]$ be isotone and modular with $\varphi(\varnothing) = 0$. 1) If $\varphi < \infty$ then φ is inner $\bullet$ tight. 2) If φ is almost $\bullet$ continuous at $\varnothing$ then φ is outer $\star$ tight.*

Proof. Fix $A \subset B$ in $\mathfrak{S}$. By 1.4.1) there exists a paving $\mathfrak{M} \subset \mathfrak{S}$ of type $\bullet$ with $\mathfrak{M} \downarrow B \setminus A$. We can assume that $S \subset B \; \forall S \in \mathfrak{M}$. 1) To be shown is $\varphi(B) \leqq \varphi(A) + \varphi_\bullet^B(B \setminus A)$. For $S \in \mathfrak{M}$ we have $A \cup S = B$ and hence $\varphi(B) \leqq \varphi(B) + \varphi(A \cap S) = \varphi(A \cup S) + \varphi(A \cap S) = \varphi(A) + \varphi(S)$. It follows that

$$\varphi(B) \leqq \varphi(A) + \inf_{S \in \mathfrak{M}} \varphi(S) \leqq \varphi(A) + \varphi_\bullet^B(B \setminus A).$$

2) To be shown is $\varphi(A) + \varphi^\star(B \setminus A) \leqq \varphi(B)$. We can assume that $\varphi(B) < \infty$. Then $\{A \cap S : S \in \mathfrak{M}\} \subset \mathfrak{S}$ is a paving of type $\bullet$ with $\downarrow \varnothing$, and all its members have $\varphi(\cdot) < \infty$. Hence by assumption $\inf_{S \in \mathfrak{M}} \varphi(A \cap S) = 0$. For $S \in \mathfrak{M}$ now

$$\begin{aligned} \varphi(B) + \varphi(A \cap S) &= \varphi(A \cup S) + \varphi(A \cap S) \\ &= \varphi(A) + \varphi(S) \geqq \varphi(A) + \varphi^\star(B \setminus A). \end{aligned}$$

The assertion follows.

7.16. THEOREM. *Assume that $\mathfrak{S}$ is downward σ full with $\varnothing \in \mathfrak{S}$. Let $\varphi : \mathfrak{S} \to [0, \infty]$ be isotone and modular with $\varphi(\varnothing) = 0$, and almost σ continuous at $\varnothing$. 1) If $\varphi < \infty$ then φ is an inner σ premeasure. 2) If φ is semifinite above then φ is an outer σ premeasure.*

Let us add as before that for $\varphi < \infty$ we obtain from 7.5 that $\mathfrak{C}(\varphi^\sigma) = \mathfrak{C}(\varphi_\sigma) =: \mathfrak{C}$ and $\varphi^\sigma(A) = \varphi_\sigma(A)$ for all $A \in \mathfrak{C} \cap (\sqsubset \mathfrak{S}^\sigma)$.

Proof. 1) In view of $\varphi < \infty$ we have $\varphi_\sigma(\varnothing) = 0$. Hence the assertion is clear from 7.15.1) and the conventional inner main theorem 6.31. 2) By

7.15.2) and the conventional outer main theorem 5.11 we have to prove that $\varphi = \varphi^\sigma | \mathfrak{S}$. Since $\varphi^\sigma | \mathfrak{S} \leqq \varphi$ by 4.1.1)2) we have to show that $\varphi \leqq \varphi^\sigma | \mathfrak{S}$; and since φ is assumed to be semifinite above it suffices to show that $\varphi(A) \leqq \varphi^\sigma(A)$ for all $A \in \mathfrak{S}$ with $\varphi(A) < \infty$. To achieve this we pass from $\mathfrak{S}$ to $\mathfrak{T} := [\varphi < \infty] \subset \mathfrak{S}$ which is a lattice and downward σ full with $\varnothing \in \mathfrak{T}$ as well. Also $\psi := \varphi | \mathfrak{T} < \infty$ is isotone and modular with $\psi(\varnothing) = 0$, and σ continuous at $\varnothing$. By 1) therefore ψ is an inner σ premeasure. Now fix $A \in \mathfrak{S}$ with $\varphi(A) < \infty$, that is $A \in \mathfrak{T}$. We have to show that $\sup_{S \in \mathfrak{M}} \varphi(S) = \sup_{S \in \mathfrak{M}} \psi(S)$ is $\geqq \varphi(A) = \psi(A)$ for each countable paving $\mathfrak{M} \subset \mathfrak{S}$ with $\mathfrak{M} \uparrow A$, which implies that $\mathfrak{M} \subset \mathfrak{T}$. By assumption we have $\{A \setminus S : S \in \mathfrak{M}\} \subset \mathfrak{T}_\sigma$, and this is a countable paving with $\downarrow \varnothing$. Hence $\inf_{S \in \mathfrak{M}} \psi_\sigma(A \setminus S) = 0$. In view of $\psi_\sigma(A \setminus S) = \psi(A) - \psi(S)$ for $S \in \mathfrak{M}$ this means that $\sup_{S \in \mathfrak{M}} \psi(S) = \psi(A)$. The proof is complete.

7.17. EXERCISE. Assertion 2) becomes false without the assumption that φ be semifinite above, even if $\mathfrak{S}$ is a ring. Hint for a counterexample: Let X be an infinite countable set, and let $\mathfrak{S}$ consist of its finite and cofinite subsets. Define $\varphi : \mathfrak{S} \to [0, \infty]$ to be $\varphi(A) = 0$ if A is finite and $\varphi(A) = \infty$ if A is cofinite.

As before we conclude with an obvious but useful reformulation.

7.18. REFORMULATION. *Assume that $\mathfrak{S}$ is downward σ full with $\varnothing \in \mathfrak{S}$. Let $\varphi : \mathfrak{S} \to [0, \infty]$ be isotone and modular with $\varphi(\varnothing) = 0$. Then we have the implications as shown below (the simple arrows are obvious implications).*

φ *can be extended to cmeasure on σ algebra which is outer regular $\mathfrak{S}^\sigma$*

$\qquad \Uparrow \varphi$ *semifinite above* $\qquad\qquad\qquad\qquad \downarrow$

φ *is almost σ cont at $\varnothing$* $\quad \leftarrow \quad$ φ *can be extended to cmeasure on σ algebra*

$\qquad \Downarrow \varphi < \infty$ $\qquad\qquad\qquad\qquad\qquad\qquad \uparrow$

φ *can be extended to cmeasure on σ algebra which is inner regular $\mathfrak{S}_\sigma$.*

Bibliographical Annex

The present subsection attempts to describe the development of the extension theories for contents and measures on the basis of lattices and of outer and inner regularity. We shall restrict ourselves to the conventional outer and inner situations in the above sense, because we know of no prior work in the full situations of isotone set functions with values in $\mathbb{R}$ or $\overline{\mathbb{R}}$. To be sure, there has been extensive work devoted to set functions with values in complete abelian Hausdorff topological groups, after the model of Sion [1969]. But in these papers the words isotone and regular do not occur, or at least attain different characters. Therefore we consider this work to be a domain on its own, and specialize its results to isotone set functions with values in $[0, \infty[\subset \mathbb{R}$. In compensation, the results will be considered to include regularity in the relevant sense whenever this can be read from the context.

Most of the papers to be discussed fall into the frame of the outer and inner • extensions for • = $\star\sigma\tau$, as defined at the outset in sections 4 and 6 above. The exceptions are the paper of Pettis [1951] cited in the introduction, and the extension procedures which follow the traditional two-step model of topological measure theory, in short from compact subsets via open subsets to arbitrary subsets. These contributions culminate in the work of Sapounakis-Sion [1983][1987] which will be discussed hereafter.

At present we start to formulate a scheme in order to describe the results of the former papers. The scheme is shaped after the conventional outer and inner main theorems 5.11 and 6.31, except that their properties 5) and 5)5') will be dropped and incorporated into 4). Let $\mathfrak{S}$ be a lattice with $\varnothing \in \mathfrak{S}$. Assume that

in the outer situation (=:out): $\varphi : \mathfrak{S} \to [0,\infty]$ is isotone and submodular
$$\text{with } \varphi(\varnothing) = 0,$$
in the inner situation (=:inn): $\varphi : \mathfrak{S} \to [0,\infty[$ is isotone and supermodular
$$\text{with } \varphi(\varnothing) = 0.$$

For fixed out/inn and • = $\star\sigma\tau$ we consider the properties of φ which follow.

(1) φ is an outer/inner • premeasure, that is φ has outer/inner • extensions. It is equivalent to require that φ has an outer/inner • extension which is

$$\text{for } \bullet = \star \quad : \quad \text{a ccontent on an algebra,}$$
$$\text{for } \bullet = \sigma\tau \quad : \quad \text{a cmeasure on a } \sigma \text{ algebra.}$$

The other properties of φ are with respect to a further isotone set function $\phi : \mathfrak{P}(X) \to [0,\infty]$. The formation $\mathfrak{C}(\phi)$ is as defined above.

(2 for ϕ) $\phi|\mathfrak{C}(\phi)$ is an outer/inner • extension of φ which is

$$\text{for } \bullet = \star \quad : \quad \text{a ccontent on an algebra,}$$
$$\text{for } \bullet = \sigma\tau \quad : \quad \text{a cmeasure on a } \sigma \text{ algebra.}$$

(3 for ϕ) $\phi|\mathfrak{C}(\phi)$ is an extension of φ in the crude sense.

(4 for ϕ) $\varphi(B) = \varphi(A) + \phi(B \setminus A)$ for all $A \subset B$ in $\mathfrak{S}$.

We consider one more condition for φ with respect to ϕ.

(U for ϕ) Each outer/inner • extension of φ is a restriction of $\phi|\mathfrak{C}(\phi)$.

We note the obvious implications

$$
\begin{array}{ccccc}
(2 \text{ for } \phi) & \Longrightarrow & (1) & & \\
\| & & \Downarrow (\text{U for } \phi) & & \\
(2 \text{ for } \phi) & \Longrightarrow & (3 \text{ for } \phi) & \Longrightarrow & (4 \text{ for } \phi).
\end{array}
$$

The most important of the above properties for φ is of course (1). For a subordinate set function ϕ the most valuable properties are (2 for ϕ) and (U for ϕ), because their combination means that ϕ dominates the set function φ in the formation of extensions of the respective kind. On the other hand the most direct and simplest of the properties of φ relative to ϕ is of course (4 for ϕ). Therefore the most needed implications are (4 for ϕ) $\Rightarrow \cdots$, in order to obtain sufficient conditions for (1), and (1) $\Rightarrow \cdots$, in particular (U for ϕ), in order to have necessary conditions for (1).

Before we describe the historical development we recall that the present conventional outer main theorem 5.11 asserts that in the outer cases $\bullet = \star \sigma \tau$ the properties (1) and (2 for $\varphi^\bullet$), (3 for $\varphi^\bullet$), (4 for $\varphi^\bullet$) are equivalent, provided that in case $\bullet = \tau$ one adds to (4 for φ^τ) the requirement that φ be upward essential. Furthermore 5.1 says that (U for $\varphi^\bullet$) holds true. The present conventional inner main theorem 6.31 combined with 6.18 asserts the same in the inner cases $\bullet = \star \sigma \tau$ with respect to $\varphi_\bullet$, this time without addendum in case $\bullet = \tau$. A provisional announcement of these facts was in König [1992c]. We do not have to come back to the outer case $\bullet = \tau$, because it has not been treated before.

In the outer and inner cases $\bullet = \star$ the results have been in the literature for quite some time in more or less comprehensive versions. See for example Topsøe [1970b] theorem 4.1 and Adamski [1984b] section 2. But the author has not seen the complete formulations before König [1992b] theorem A13.

We turn to the outer and inner cases $\bullet = \sigma$. We have to restrict ourselves to the basic achievements of the individual papers, perhaps with small simplifications. As the earlierst paper we mention Choksi [1958], because it comprised several previous results. Its theorem 1 asserts that

inn: (4 for $\varphi_\star$) and $\mathfrak{S} \, \sigma$ compact $\Rightarrow$ (1).

The leap forward around 1970 started in Topsøe [1970a] theorem 1 and [1970b] section 2 (and notes to section 5) with the results

inn: (4 for $\varphi_\star$) and $\varphi \, \sigma$ continuous at $\varnothing \Rightarrow$ (2 for $\varphi_\star$) and (U for $\varphi_\star$), when $\mathfrak{S}$ fulfils $\cap \sigma$,

inn: (4 for $\varphi_\star$) and $\varphi \, \sigma$ continuous at $\varnothing \Rightarrow$ (2 for $\varphi_{(\sigma)}$),

with $\varphi_{(\sigma)}$ and its relatives as defined after 6.10 and 6.11. Kelley-Srinivasan [1971] proved in corollary 2 that

out: (4 for φ°) $\Rightarrow$ (2 for φ°) and hence $\Leftrightarrow$ (2 for φ°),

for the Carathéodory outer measure φ° as defined in the introduction. Thus of course (4 for φ°) $\Rightarrow$ (1). The authors claimed without proof that even (4 for φ°) $\Leftrightarrow$ (1), but the present author cannot see this. In propositions 8 and 9 they proved via φ° that

out: (4 for $\varphi^\star$) and φ upward σ continuous $\Leftrightarrow$ (1), when $\mathfrak{S}$ fulfils $\cup \sigma$,

inn: (4 for $\varphi_\star$) and $\varphi \, \sigma$ continuous at $\varnothing \Leftrightarrow$ (1), when $\mathfrak{S}$ fulfils $\cap \sigma$.

Ridder [1971][1973] proved the last implications $\Rightarrow$ under the assumption that $\mathfrak{S}$ fulfils both $\cup \sigma$ and $\cap \sigma$. Then Kelley-Nayak-Srinivasan [1973] obtained an independent proof of the result of Topsøe [1970b] that

inn: (4 for $\varphi_\star$) and $\varphi \, \sigma$ continuous at $\varnothing \Rightarrow$ (2 for $\varphi_{(\sigma)}$).

The conditions $\cup \sigma$ and $\cap \sigma$ for $\mathfrak{S}$ are of course severe restrictions which often are not fulfilled.

From the present text we know that beyond these restrictions the converses $\cdots \Rightarrow$ (4 for $\varphi^\star$) and $\cdots \Rightarrow$ (4 for $\varphi_\star$) of the above assertions are all false. From 5.14 we see that in the outer situation even $\lambda : \mathfrak{K} = \mathrm{Comp}(\mathbb{R}^n) \to [0, \infty[$ is a counterexample. This expresses the basic inadequacy of the formations $\varphi^\star$ and $\varphi_\star$ for the treatment of $\bullet = \sigma\tau$.

To this line of papers we add the work of Lipecki [1974], who in the frame of abstract-valued set functions as described above proved an extended version of the last-mentioned result.

At last we quote from Adamski [1982] the two results

out: (4 for $\varphi^\star$) and φ upward σ continuous $\Rightarrow$ (2 for φ°),
inn: (4 for $\varphi_\star$) and φ σ continuous at $\varnothing$ $\Rightarrow$ (2 for $\varphi_{(\sigma)}$),

declared as direct counterparts. These results are contained in the former ones, the first one since its hypothesis implies at once (4 for φ°). We quote the results in their combination as an example for the odd kind of monopoly which the Carathéodory outer measure held in the outer situation, in spite of what we have said in the introduction. Another example is a note in the recent book of Kelley-Srinivasan [1988] page 20 which says that, in a certain sense, the properties (4 for φ°) and (4 for $\varphi_\star$) are dual to each other.

We remain in the outer and inner cases $\bullet = \sigma$. The next papers were essential improvements, because of results in which $\varphi^\star$ and $\varphi_\star$ as well as φ° did no more occur. The main results in Fox-Morales [1983] theorems 3.16 and 3.10 were

out: (4 for $\varphi^{(\sigma)}$) and φ upward σ continuous $\Rightarrow$ (1),
inn: (4 for $\varphi_{(\sigma)}$) and φ downward σ continuous $\Rightarrow$ (1).

Then Găină [1986] proved

out: (4 for $\varphi^{(\sigma)}$) and φ upward σ continuous $\Leftrightarrow$ (2 for $\varphi^{(\sigma)}$),
inn: (4 for $\varphi_{(\sigma)}$) and φ downward σ continuous $\Leftrightarrow$ (2 for $\varphi_{(\sigma)}$).

Both papers were in the frame of abstract-valued set functions, the first one still based on Sion [1969]. The independent work of König [1985] theorems 3.3 with 3.1 and 3.4 with 3.2 obtained

out: (4 for φ^σ) $\Leftrightarrow$ (2 for φ^σ), and furthermore (U for φ^σ),
inn: (4 for φ_σ) $\Leftrightarrow$ (2 for φ_σ), and furthermore (U for φ_σ),

and hence the full results as in the present text. We emphasize that the important fortified counterparts 7.14 and 7.18 of the classical extension theorem can be deduced from the last three papers, but not from the earlier ones. Their essence is in König [1985] theorems 3.8 and 3.9.

In the outer and inner cases $\bullet = \sigma$ it remains to review the work of Glazkov [1988] which stands somewhat apart. It assumed an arbitrary paving $\mathfrak{S}$ with $\varnothing \in \mathfrak{S}$ and an arbitrary set function $\varphi : \mathfrak{S} \to [0, \infty]$ with $\varphi(\varnothing) = 0$, and defined besides φ° the somewhat brutal inner counterpart $\varphi_\circ : \mathfrak{P}(X) \to [0, \infty]$ to be

$$\varphi_\circ(A) = \sup\{\sum_{l=1}^{r} \varphi(S_l) : S_1, \cdots, S_r \in \mathfrak{S} \text{ pairwise disjoint } \subset A\}.$$

As far as the present author knows, this formation had been considered in earlier decades, but was later abandoned because of severe unsymmetries with φ°. Nevertheless the paper obtained some notable results, based on appropriate definitions of outer and inner tightness. The outer result says that $\varphi^\circ|\mathfrak{C}(\varphi^\circ)$, which is known to be a cmeasure on a σ algebra, is an extension of φ iff φ is outer tight. However, the inner counterpart on $\varphi_\circ|\mathfrak{C}(\varphi_\circ)$ is not an equivalence assertion but restricted to certain sufficient conditions which, except the requirement that φ be inner tight, do not look adequate for an equivalence assertion. Thus there is not much hope for symmetry based on the formations φ° and $\varphi_\circ$.

The review of the inner case $\bullet = \tau$ is short. Prior to the present text we quote the work of Topsøe [1970ab], also reproduced in Pollard-Topsøe [1975]. It was our model in that it aimed at a uniform treatment of the three cases $\bullet = \star\sigma\tau$. Thus Pollard-Topsøe [1975] theorem B asserts that

  inn: (4 for $\varphi_\star$) and $\varphi \bullet$ continuous at $\varnothing \Rightarrow$ (2 for $\varphi_{(\bullet)}$).

The converse $\Leftarrow$ is false for $\bullet = \tau$ as it has been for $\bullet = \sigma$. There are also parts of the present comparison theorems 6.24 and 6.25 in Topsøe [1970b] theorem 5.1, and of the present τ continuity theorem 6.27 in Topsøe [1970b] lemma 2.3. At last Topsøe [1970a] lemma 1 seems to be the ancestor of the results like the present lemma 6.30.

At the end of the subsection we want to discuss the work of Sapounakis-Sion [1983][1987] as announced above. The concern here is Sapounakis-Sion [1987] part I with the fundamental theorem 1.1 and its corollaries. We shall later comment on certain applications. The reproduction will be a free one in certain minor points.

The situation is that of a two-step extension procedure. Assume that $\mathfrak{S}$ and $\mathfrak{T}$ are lattices in X which contain $\varnothing$ and fulfil $\mathfrak{S} \subset \mathfrak{T}\top\bot$, and let $\varphi : \mathfrak{S} \to [0, \infty[$ be an isotone set function with $\varphi(\varnothing) = 0$. We form $\psi := \varphi_\star|\mathfrak{T}$, so that $\psi : \mathfrak{T} \to [0, \infty]$ is an isotone set function with $\psi(\varnothing) = 0$. The aim is to obtain a cmeasure $\alpha : \mathfrak{A} \to [0, \infty]$ on a σ algebra $\mathfrak{A}$ with the properties

  I) $\mathfrak{A} \supset \mathfrak{S}$, and $\mathfrak{A} \supset \mathfrak{T}$ and hence $\mathfrak{A} \supset \mathfrak{T}^\sigma$;

  II) $\alpha|\mathfrak{S} = \varphi$, and α is inner regular $\mathfrak{S}$ at $\mathfrak{T}$ and outer regular $\mathfrak{T}^\sigma$.

Although this task seems to be quite different from those in the present text, we shall see that it can be incorporated into our extension theories. We do this with the next theorem which is based on the main results of the present chapter.

7.19. THEOREM. *Assume that $\mathfrak{S}$ is upward enclosable* $[\psi < \infty] = [\varphi_\star|\mathfrak{T} < \infty]$. *Then there exists a cmeasure* $\alpha : \mathfrak{A} \to [0, \infty]$ *on a* σ *algebra* $\mathfrak{A}$ *with the above properties* I)II) *iff*

  i) φ *is supermodular and inner* $\star$ *tight, and*

  ii) $\psi = \varphi_\star|\mathfrak{T}$ *is submodular and upward* σ *continuous.*

In this case ψ *is an outer* σ *and* $\star$ *premeasure, and* $\psi^\sigma|\mathfrak{C}(\psi^\sigma)$ *is as required. Furthermore each cmeasure* α *which is as required is a restriction of* $\psi^\sigma|\mathfrak{C}(\psi^\sigma)$.

This theorem can serve as a substitute for Sapounakis-Sion [1987] theorem 1.1 and some of the subsequent results. The main differences are that these authors on the one hand postulate $\alpha := \psi^\circ | \mathfrak{C}(\psi^\circ)$ from the start, and on the other hand do not present equivalence theorems in concrete terms like the above one, but are content with sufficient conditions.

Proof of the theorem. We first assume that $\alpha : \mathfrak{A} \to [0, \infty]$ is a cmeasure on a σ algebra $\mathfrak{A}$ with the properties I)II). Then $\alpha | \mathfrak{T}$ is an outer σ premeasure with $\alpha | \mathfrak{T} = (\alpha | \mathfrak{S})_\star | \mathfrak{T} = \varphi_\star | \mathfrak{T} = \psi$. Thus ψ is an outer σ premeasure, and hence in particular fulfils ii). We see from 5.1 that α is a restriction of $\psi^\sigma | \mathfrak{C}(\psi^\sigma)$. Thus $\psi^\sigma | \mathfrak{C}(\psi^\sigma)$ fulfils I)II) as well. Now we have to prove i). Since $\varphi = \alpha | \mathfrak{S}$ is modular it remains to show that it is inner $\star$ tight. To see this fix $A \subset B$ in $\mathfrak{S}$, and then $T \in \mathfrak{T}$ with $\alpha(T) = \psi(T) < \infty$ such that $B \subset T$. In view of $\mathfrak{S} \subset \mathfrak{T}\mathsf{T}\bot$ we have $T \backslash A \in \mathfrak{T}$, of course with $\alpha(T \backslash A) < \infty$. We fix $\varepsilon > 0$ and then $K \in \mathfrak{S}$ with $K \subset T \backslash A$ and $\alpha(T \backslash A) \leqq \alpha(K) + \varepsilon$. Now

$$\alpha(B \backslash A) + \alpha(T \backslash B) = \alpha(T \backslash A) \leqq \alpha(K) + \varepsilon = \alpha(K \cap B) + \alpha(K \cap B') + \varepsilon,$$

with all terms finite. On the other hand

$$K \cap B \in \mathfrak{S} \text{ with } K \cap B \subset A' \cap B = B \backslash A,$$
$$K \cap B' \subset T \cap B' = T \backslash B.$$

It follows that $\alpha(B \backslash A) \leqq \alpha(K \cap B) + \varepsilon$. Therefore

$$\begin{aligned} \varphi(B) - \varphi(A) &= \alpha(B) - \alpha(A) = \alpha(B \backslash A) \leqq \alpha(K \cap B) + \varepsilon \\ &= \varphi(K \cap B) + \varepsilon \leqq \varphi_\star(B \backslash A) + \varepsilon, \end{aligned}$$

and hence the assertion.

We next assume that φ and ψ fulfil i)ii). We first prove

$$(0) \qquad \varphi_\star(B \cap A) + \psi^\star(B \cap A') \leqq \psi(B) = \varphi_\star(B) \quad \text{for } A \subset X \text{ and } B \in \mathfrak{T}.$$

We can assume that $\psi(B) = \varphi_\star(B) < \infty$ and hence $\varphi_\star(B \cap A) < \infty$. We fix $\varepsilon > 0$ and then $S \in \mathfrak{S}$ with $S \subset B \cap A$ such that $\varphi_\star(B \cap A) \leqq \varphi(S) + \varepsilon$. Now

$$\varphi_\star(B) = \varphi_\star(B) + \varphi_\star(\varnothing) \geqq \varphi_\star(B \cap S) + \varphi_\star(B \cap S'),$$

since $\varphi_\star$ is supermodular by 6.3.5). Here we have on the one hand

$$B \cap S = S \text{ and hence } \varphi_\star(B \cap S) = \varphi(S) \geqq \varphi_\star(B \cap A) - \varepsilon.$$

On the other hand $B \cap S' \in \mathfrak{T}$ in view of $\mathfrak{S} \subset \mathfrak{T}\mathsf{T}\bot$; furthermore

$$B \cap S' \supset B \cap A' \text{ and hence } \varphi_\star(B \cap S') = \psi(B \cap S') \geqq \psi^\star(B \cap A').$$

It follows that $\varphi_\star(B) \geqq \varphi_\star(B \cap A) - \varepsilon + \psi^\star(B \cap A')$. Thus (0) is proved.

Now (0) will be applied three times. 1) From (0) for $A \subset B$ in $\mathfrak{T}$ we see that ψ is inner $\star$ tight and hence inner σ tight. Therefore ψ is an inner $\star$ and σ premeasure. From 5.8.σ) we know that $\mathfrak{C}(\psi^\star) \subset \mathfrak{C}(\psi^\sigma)$, and from 5.9.$\sigma$) that $\psi^\star(M) = \psi^\sigma(M)$ for all $M \in \mathfrak{C}(\psi^\star)$ with $\psi^\star(M) < \infty$. Thus $\alpha := \psi^\sigma | \mathfrak{C}(\psi^\sigma)$ is a cmeasure on a σ algebra $\supset \mathfrak{T}$ which is outer regular $\mathfrak{T}^\sigma$.

Furthermore we know from 5.11 that $\mathfrak{S}\bot \subset \mathfrak{T}\top = \mathfrak{T}\top\mathfrak{T} \subset \mathfrak{C}(\psi^\star) \subset \mathfrak{C}(\psi^\sigma)$ and hence $\mathfrak{S} \subset \mathfrak{C}(\psi^\star) \subset \mathfrak{C}(\psi^\sigma)$.

It remains to prove $\alpha|\mathfrak{S} = \varphi$. In fact, then we have also $(\alpha|\mathfrak{S})_\star|\mathfrak{T} = \varphi_\star|\mathfrak{T} = \psi = \alpha|\mathfrak{T}$, that is α is inner regular $\mathfrak{S}$ at $\mathfrak{T}$. To see the assertion fix $A \in \mathfrak{S}$. By assumption there exists $B \in [\psi < \infty] \subset \mathfrak{T}$ with $A \subset B$. 2) From (0) and ii) with 4.1.5) we have

$$\begin{aligned}\varphi_\star(B \cap A) + \psi^\star(B \cap A') &\leqq \psi(B) = \psi^\star(B) + \psi^\star(\varnothing) \\ &\leqq \psi^\star(B \cap A) + \psi^\star(B \cap A'),\end{aligned}$$

so that $B \cap A = A$ furnishes $\varphi(A) \leqq \psi^\star(A)$. 3) From (0) applied to A' and B and from $A \in \mathfrak{S} \subset \mathfrak{C}(\varphi_\star)$ in view of i) we obtain

$$\varphi_\star(B \cap A') + \psi^\star(B \cap A) \leqq \varphi_\star(B) = \varphi_\star(B \cap A') + \varphi_\star(B \cap A),$$

with all terms finite. Thus $B \cap A = A$ furnishes $\psi^\star(A) \leqq \varphi(A)$. Therefore we have $\varphi(A) = \psi^\star(A) < \infty$ and hence $\varphi(A) = \psi^\sigma(A) = \alpha(A)$. The proof is complete.

We mention at last that in Sapounakis-Sion [1983][1987] one requires but $\cup$ for $\mathfrak{S}$ and $\cap$ for $\mathfrak{T}$, instead of $\mathfrak{S}$ and $\mathfrak{T}$ to be lattices as above. Maurice Sion has pointed out to the author that it is a benefit of the Carathéodory outer measure that it permits to start with pavings which fulfil $\cap$ but need not be lattices. This aspect can of course become relevant, but it is expected that the last subsection of the present section 3 will be able to take care of it as well.

CHAPTER III

Applications of the Extension Theories

The most important consequences of the extension theories of chapter II appear to be the representation theorems of Daniell-Stone and Riesz type which will be obtained in chapter V. The present chapter is devoted to a few direct applications of the extension theories to different areas in the domain of set functions. All these are central areas of rich tradition and extensive literature. We do not intend to compete with them in technical depth and sophistication. We rather want to show that certain basic ideas and results come as natural outflows from the above theories. We shall arrive at certain forms which are more comprehensive and simpler, and also more explicit than those found in the literature so far. An example is the extension of the capacitability theorem due to Choquet in section 10. In most but not all cases we shall be concerned with the *conventional* outer and inner situations.

8. Baire Measures

Let X be a topological space. We recall from 1.2.8) and 1.6.4) the lattices $\mathrm{COp}(X) \subset \mathrm{Op}(X)$ and $\mathrm{CCl}(X) \subset \mathrm{Cl}(X)$ and the σ algebras $\mathrm{Baire}(X) \subset \mathrm{Bor}(X)$ in X. One defines the

Borel measures on X to be the cmeasures $\alpha : \mathrm{Bor}(X) \to [0, \infty]$,
Baire measures on X to be the cmeasures $\alpha : \mathrm{Baire}(X) \to [0, \infty]$.

The two kinds of measures have much in common, but also distinctive peculiarities. The present section concentrates on certain fundamentals for Baire measures.

Basic Properties of Baire Measures

We collect some essential properties of the lattices $\mathrm{COp}(X)$ and $\mathrm{CCl}(X)$.

8.1. PROPERTIES. 1) $\mathrm{COp}(X)$ *and* $\mathrm{CCl}(X)$ *are lattices which contain* $\varnothing$ *and* X, *and* $\mathrm{CCl}(X) = (\mathrm{COp}(X))\bot$. 2) $\mathrm{COp}(X)$ *has* $\cup\sigma$, *and* $\mathrm{CCl}(X)$ *has* $\cap\sigma$. 3) $\mathrm{CCl}(X) \subset (\mathrm{COp}(X))_\sigma$ *and* $\mathrm{COp}(X) \subset (\mathrm{CCl}(X))^\sigma$. 4) $\mathrm{CCl}(X)$ *is upward* σ *full, and* $\mathrm{COp}(X)$ *is downward* σ *full.* 5) *Assume that* X *is completely regular. Then* $\mathrm{Cl}(X) = (\mathrm{CCl}(X))_\tau$ *and* $\mathrm{Op}(X) = (\mathrm{COp}(X))^\tau$. 6) *Assume that* X *is normal. Then for each pair* $B \in \mathrm{Cl}(X)$ *and* $V \in \mathrm{Op}(X)$ *with* $B \subset V$ *there exist* $A \in \mathrm{CCl}(X)$ *and* $U \in \mathrm{COp}(X)$ *with* $B \subset U \subset A \subset V$. 7) *Consider for* $A \subset X$ *the properties*

i) $A \in \mathrm{COp}(X)$;

ii) *there exists a sequence of continuous* $f_l : X \to [0, \infty[$ *with* $f_l \uparrow \chi_A$;

iii) $A \in \mathrm{Op}(X)$ *and* $A \in (\mathrm{Cl}(X))^\sigma$.

Then i)$\Rightarrow$ii)$\Rightarrow$iii); *and* iii)$\Rightarrow$i) *when* X *is normal.* 8) *Assume that* X *is semimetrizable. Then* $\mathrm{COp}(X) = \mathrm{Op}(X)$ *and* $\mathrm{CCl}(X) = \mathrm{Cl}(X)$, *and hence* $\mathrm{Baire}(X) = \mathrm{Bor}(X)$.

Proof. 1) is known from 1.6.4). 2) We prove the first assertion. Let $(A_l)_l$ be a sequence in $\mathrm{COp}(X)$. We can assume that $A_l = [f_l > 0]$ for some $f_l \in C(X, \mathbb{R})$ with $0 \le f_l \le 1$. Then

$$f := \sum_{l=1}^{\infty} \frac{1}{2^l} f_l \in C(X, \mathbb{R}) \quad \text{with} \quad \bigcup_{l=1}^{\infty} A_l = [f > 0] \in \mathrm{COp}(X).$$

3) Once more we prove the first assertion. The set $A \in \mathrm{CCl}(X)$ of the form $A = [f \le 0]$ for some $f \in C(X, \mathbb{R})$ is the intersection of the sets $A_l := [f < 1/l] \in \mathrm{COp}(X) \, \forall l \in \mathbb{N}$. 4) follows from 1)3). 5) The assumption means that for each pair $A \in \mathrm{Cl}(X)$ and $u \in A'$ there exists a function $f \in C(X, \mathbb{R})$ with $f|A \le 0$ and $f(u) > 0$. Thus $A \subset [f \le 0] \in \mathrm{CCl}(X)$ and $u \notin [f \le 0]$. This implies the assertion. 6) By the Urysohn lemma there exists $f \in C(X, \mathbb{R})$ with $B \subset [f \le 0] \subset [f < 1] \subset V$. Then $U := [f < 1/2]$ and $A := [f \le 1/2]$ are as required. 7) For i)$\Rightarrow$ii) let $A = [f \ne 0]$ for some $f \in C(X, \mathbb{R})$. Then the $f_l := \min(l|f|, 1) \, \forall l \in \mathbb{N}$ are as required. ii)$\Rightarrow$iii) We have

$$A = \bigcup_{l=1}^{\infty} [f_l > 0] \in \mathrm{Op}(X), \quad \text{and} \quad A = \bigcup_{l=1}^{\infty} [f_l \ge t] \in (\mathrm{Cl}(X))^\sigma \text{ for } 0 < t < 1.$$

iii)$\Rightarrow$i) follows from 6). 8) is known from 1.6.4).

The result below shows in particular that the well-known regularity properties of Baire measures are immediate consequences of the extension theories of chapter II.

8.2. PROPOSITION. *Let* $\alpha : \mathrm{Baire}(X) \to [0, \infty]$ *be a cmeasure.* 1) *Define* $\mathfrak{S} := \{A \in \mathrm{CCl}(X) : \alpha(A) < \infty\}$ *and* $\varphi := \alpha | \mathfrak{S}$. *Then*

i) $\mathfrak{S}$ *is a lattice with* $\cap \sigma$ *which is upward* σ *full.*

ii) φ *is an outer and an inner* σ *premeasure.*

iii) $\mathfrak{C}(\varphi^\sigma) = \mathfrak{C}(\varphi_\sigma) =: \mathfrak{C} \supset \mathrm{Baire}(X)$, *and* $\varphi^\sigma = \varphi_\sigma$ *on* $\mathfrak{C} \cap (\sqsubset \mathfrak{S}^\sigma)$.

iv) $\varphi_\star = \varphi_\sigma$.

v) $\varphi_\star = \varphi_\sigma \le \alpha \le \varphi^\sigma$ *on* $\mathrm{Baire}(X)$.

vi) $\varphi_\star = \varphi_\sigma = \alpha = \varphi^\sigma$ *on* $\mathrm{Baire}(X) \cap (\sqsubset \mathfrak{S}^\sigma)$.

In particular if $X \in \mathfrak{S}^\sigma$ *then* α *is inner regular* $\mathfrak{S}$.

2) *Define* $\mathfrak{T} := \{A \in \mathrm{COp}(X) : \alpha(A) < \infty\}$ *and* $\psi := \alpha | \mathfrak{T}$. *Then*

i) $\mathfrak{T}$ *is a lattice which is downward* σ *full.*

ii) ψ *is an outer and an inner* σ *premeasure.*

iii) $\mathfrak{C}(\psi^\sigma) = \mathfrak{C}(\psi_\sigma) =: \mathfrak{C} \supset \mathrm{Baire}(X)$, *and* $\psi^\sigma = \psi_\sigma$ *on* $\mathfrak{C} \cap (\sqsubset \mathfrak{T}^\sigma)$.

iv) $\psi^\sigma = \psi^\star$.

v) $\psi_\sigma \le \alpha \le \psi^\sigma = \psi^\star$ *on* $\mathrm{Baire}(X)$.

vi) $\psi_\sigma = \alpha = \psi^\sigma = \psi^\star$ on $\mathrm{Baire}(X) \cap (\sqsubset \mathfrak{T}^\sigma)$.

In particular if $X \in \mathfrak{T}^\sigma$ then α is outer regular $\mathfrak{T}$. Note that α is outer regular $\mathfrak{T}$ iff it is outer regular $\mathrm{COp}(X)$.

Proof of 1). i) follows from 8.1.1)2)4), and hence ii) from 7.12. iii) follows from 7.5 and from $\mathrm{CCl}(X) \subset \mathfrak{S} \top \mathfrak{S} \subset \mathfrak{C}(\varphi^\sigma) = \mathfrak{C}(\varphi_\sigma)$ in 5.11 and 6.31. iv) follows from 6.5.iv). Then v) results from 7.1.σ), and vi) from 7.5.

Proof of 2). i) follows from 8.1.1)2)4), and hence ii) from 7.16. iii) follows from 7.5 and from $\mathrm{COp}(X) \subset \mathfrak{T} \top \mathfrak{T} \subset \mathfrak{C}(\psi^\sigma) = \mathfrak{C}(\psi_\sigma)$ in 5.11 and 6.31. iv) follows from 4.5.iv). Then v) results from 7.1.σ), and vi) from 7.5.

8.3. EXERCISE. We have $\mathfrak{T}^\sigma \subset \mathfrak{S}^\sigma$, but the converse need not be true. It fact, it can happen that $X \in \mathfrak{S}^\sigma$ and α is not outer regular $\mathfrak{T}$, which implies that $X \notin \mathfrak{T}^\sigma$. One can even achieve that X is compact Hausdorff. Hint for an example: Let $X := \mathbb{N} \cup \{\infty\}$. Define $\mathrm{Op}(X)$ to consist of all subsets of $\mathbb{N}$ and of all cofinite subsets of X which contain ∞. Note that $\mathrm{CCl}(X) = \mathrm{Cl}(X)$ and that $\mathrm{Baire}(X) = \mathfrak{P}(X)$. Let $\alpha : \mathrm{Baire}(X) = \mathfrak{P}(X) \to [0, \infty]$ be the counting measure, that is $\alpha(A) = \#(A)$ for all $A \subset X$.

8.4. EXERCISE. Give an alternative proof of the two assertions

$$\varphi_\star(A) = \alpha(A) \quad \text{for all } A \in \mathrm{Baire}(X) \text{ upward enclosable } \mathfrak{S}^\sigma,$$
$$\psi^\star(A) = \alpha(A) \quad \text{for all } A \in \mathrm{Baire}(X) \text{ upward enclosable } \mathfrak{T}^\sigma,$$

which instead of 7.5 uses the classical uniqueness theorem 3.1.σ) (plus certain facts around the transporter theorem).

8.5. ADDENDUM. i) $\varphi_\star = \varphi_\sigma = \alpha$ *on* $[\alpha < \infty]^\sigma$.
ii) $\varphi_\star = \varphi_\sigma = \alpha$ *on* $\mathrm{Baire}(X)$ *iff α is semifinite above.*

Therefore α is inner regular $\mathfrak{S}$ iff it is semifinite above.

Proof. i) It suffices to prove that $\alpha(A) \leqq \varphi_\star(A)$ for $A \in [\alpha < \infty]$. To see this one applies 8.2.1) to the finite Baire measure $S \mapsto \alpha(S \cap A)$ and to A. Then

$$\alpha(A) = \sup\{\alpha(S \cap A) : S \in \mathrm{CCl}(X) \text{ with } S \subset A\}$$
$$= \sup\{\alpha(S) : S \in \mathfrak{S} \text{ with } S \subset A\} = \varphi_\star(A).$$

ii) It α is semifinite above then i) implies that $\alpha(A) \leqq \varphi_\star(A)$ for all $A \in \mathrm{Baire}(X)$. On the other hand if $\varphi_\star = \varphi_\sigma = \alpha$ on $\mathrm{Baire}(X)$ then α is semifinite above by the definition.

The most important point in the present subsection is the extension of set functions defined on $\mathrm{CCl}(X)$ and $\mathrm{COp}(X)$ to Baire measures. The positive results below are immediate consequences of the former theorems 7.12=7.14 and 7.16=7.18. Note that each version contains some construction of the Lebesgue measure.

8.6. THEOREM. *Let $\varphi : \mathrm{CCl}(X) \to [0, \infty]$ be isotone and modular with $\varphi(\varnothing) = 0$, and upward σ continuous.*

1) *There exists a unique cmeasure* $\alpha : \mathrm{Baire}(X) \to [0, \infty]$ *which extends* φ *and is outer regular* $(\mathrm{CCl}(X))^\sigma$. *This is* $\alpha = \varphi^\sigma | \mathrm{Baire}(X)$. *It need not be outer regular* $\mathrm{COp}(X)$.

2) *Assume that* $\varphi < \infty$. *Then there exists a unique cmeasure* $\alpha : \mathrm{Baire}(X) \to [0, \infty]$ *which extends* φ. *This is*

$$\alpha = \varphi_\star | \mathrm{Baire}(X) = \varphi_\sigma | \mathrm{Baire}(X) = \varphi^\sigma | \mathrm{Baire}(X).$$

It is $\alpha < \infty$ *and hence outer regular* $\mathrm{COp}(X)$ *and inner regular* $\mathrm{CCl}(X)$.

Proof. 1) The lattice $\mathrm{CCl}(X)$ is upward σ full by 8.1.4). Thus φ is an outer σ premeasure by 7.12.1). The assertion follows from the conventional outer main theorem 5.11. The example constructed in 8.3 shows that α need not be outer regular $\mathrm{COp}(X)$, even when X is compact Hausdorff. 2) The existence assertion follows from 1). Let now β be any Baire measure extension of φ. Then $\beta < \infty$. By 8.2.1) β is inner regular $\mathrm{CCl}(X)$ and hence unique, and we have $\beta = \varphi_\star | \mathrm{Baire}(X) = \varphi_\sigma | \mathrm{Baire}(X)$. Furthermore β must be the extension $\alpha = \varphi^\sigma | \mathrm{Baire}(X)$ obtained in 1). At last it is outer regular $\mathrm{COp}(X)$ by 8.2.2).

8.7. THEOREM. *Let* $\varphi : \mathrm{COp}(X) \to [0, \infty[$ *be isotone and modular with* $\varphi(\varnothing) = 0$, *and* σ *continuous at* $\varnothing$. *Then there exists a unique cmeasure* $\alpha : \mathrm{Baire}(X) \to [0, \infty]$ *which extends* φ. *This is*

$$\alpha = \varphi^\star | \mathrm{Baire}(X) = \varphi^\sigma | \mathrm{Baire}(X) = \varphi_\sigma | \mathrm{Baire}(X).$$

It is $\alpha < \infty$ *and hence inner regular* $\mathrm{CCl}(X)$ *and outer regular* $\mathrm{COp}(X)$.

Proof. The lattice $\mathrm{COp}(X)$ is downward σ full by 8.1.4). Thus φ is an inner σ premeasure by 7.16.1). From the conventional inner main theorem 6.31 it follows that $\alpha := \varphi_\sigma | \mathrm{Baire}(X)$ is a cmeasure which extends φ. Let now β be any Baire measure extension of φ. Then $\beta < \infty$. By 8.2.2) β is outer regular $\mathrm{COp}(X)$ and hence unique, and we have $\beta = \varphi^\star | \mathrm{Baire}(X) = \varphi^\sigma | \mathrm{Baire}(X)$. Furthermore β must be the extension $\alpha = \varphi_\sigma | \mathrm{Baire}(X)$ obtained above. At last it is inner regular $\mathrm{CCl}(X)$ by 8.2.1).

8.8. REMARK. Theorem 8.6 becomes false when one replaces the assumption that φ be upward σ continuous by the assumption of 8.7 that φ be $< \infty$ and σ continuous at $\varnothing$. In fact, the example constructed in 7.13 shows that φ then need not be upward σ continuous.

8.9. BIBLIOGRAPHICAL NOTE. The author found no traces of 8.6 and 8.7 in the literature prior to his [1985]. A posteriori one sees that part of 8.6 and 8.7 could have been obtained from Topsøe [1970a] and Kelley-Srinivasan [1971], though on different paths. We restrict ourselves to 8.6, and have to assume that $\varphi < \infty$. From the assumptions one deduces on the one hand that φ is inner $\star$ tight, as noted in 7.10.2), and on the other hand that φ is σ continuous at $\varnothing$, which requires some nontrivial manipulation. Then the results of the above papers imply that φ has a Baire measure extension α which is inner regular $\mathrm{CCl}(X)$. Thus one obtains 8.6.2) except that $\alpha = \varphi^\sigma | \mathrm{Baire}(X)$. Also 8.6.1) does not seem to be accessible.

Inner Regularity in Separable Metric Spaces

The subsequent result is restricted to metric spaces, where the Borel and Baire formations are known to be identical. It serves to prepare the famous theorem 9.9 in the next section.

Let X be a metric space with metric d. As usual we form for $a \in X$ and $\delta > 0$

the open ball $V(a, \delta) := \{x \in X : d(a, x) < \delta\}$, and
the closed ball $\nabla(a, \delta) := \{x \in X : d(a, x) \leqq \delta\}$.

We recall the relevant notions and results. A subset $M \subset X$ is called **dispersed** (or **verstreut**) iff there exists $\delta > 0$ such that $d(u, v) \geqq \delta$ for all pairs $u, v \in M$ with $u \neq v$. It is a fundamental fact that for a subset $A \subset X$ the three properties below are equivalent.

1) For each $\delta > 0$ there exists a finite $F \subset A$ such that $A \subset \bigcup\limits_{u \in F} V(u, \delta)$.
2) Each dispersed subset of A is finite.
3) Each sequence in A has a subsequence which is a Cauchy sequence.

In this case A is called **precompact** (or **totally bounded**). It follows that A is compact iff it is precompact and complete. But note that neither precompactness nor completeness is a topological notion. The theorem in question then reads as follows.

8.10. THEOREM. *Assume that X is a separable metric space. Let $\alpha :$* Baire$(X) = Bor(X) \to [0, \infty]$ *be a cmeasure which is semifinite above, that is inner regular*

$$\mathfrak{S} := \{A \in \mathrm{CCl}(X) = \mathrm{Cl}(X) : \alpha(A) < \infty\}.$$

Then α is inner regular

$$\{A \in \mathrm{CCl}(X) = \mathrm{Cl}(X) : A \text{ precompact with } \alpha(A) < \infty\}.$$

Proof. 1) We first assume $\alpha < \infty$ and prove that

$$\alpha(X) = \sup\{\alpha(A) : A \in \mathrm{Cl}(X) \text{ precompact }\}.$$

Let $\{u_l : l \in \mathbb{N}\}$ be a countable dense subset of X and $\delta > 0$. For fixed $n \in \mathbb{N}$ we have $X = \bigcup\limits_{l=1}^{\infty} \nabla(u_l, 1/n)$. Hence there exists $k(n) \in \mathbb{N}$ such that $B_n := \bigcup\limits_{l=1}^{k(n)} \nabla(u_l, 1/n)$ fulfils $\alpha(B_n) \geqq \alpha(X) - \delta/2^n$ or $\alpha(B'_n) \leqq \delta/2^n$. Let $A_n := B_1 \cap \cdots \cap B_n$. Then $A'_n = B'_1 \cup \cdots \cup B'_n$ and hence

$$\alpha(A'_n) \leqq \sum_{p=1}^{n} \alpha(B'_p) \leqq \delta \quad \text{or} \quad \alpha(A_n) \geqq \alpha(X) - \delta.$$

Now $A_n \downarrow A \in \mathrm{Cl}(X)$ and hence $\alpha(A) \geqq \alpha(X) - \delta$. But A is precompact since $A \subset A_n \subset B_n$ for all $n \in \mathbb{N}$. 2) We turn to the assertion of the theorem. Fix $B \in$ Baire$(X) = Bor(X)$ and a real $c < \alpha(B)$. By assumption there exists $P \in \mathrm{Cl}(X)$ with $P \subset B$ and $c < \alpha(P) < \infty$. We can apply 1) to the restriction $\alpha|\mathrm{Baire}(P) = \mathrm{Bor}(P)$, where $\mathrm{Bor}(P) = \{M \in \mathrm{Bor}(X) : M \subset P\}$ by 1.13.2). Thus there exists $A \subset P$ closed and precompact in P, and hence in X, such that $c < \alpha(A)$. The assertion follows.

Extension of Baire Measures to Borel Measures

It is natural to ask whether and when a Baire measure $\alpha : \mathrm{Baire}(X) \to [0, \infty]$ can be extended to Borel measures. This is known to be a hard problem. The present subsection obtains an extension theorem which is an immediate consequence of the extension theories of chapter II. It assumes a certain τ behaviour and X to be completely regular, and connects the Borel and Baire structures via 8.1.5). We note that there are extension theorems of different type which are not based upon τ theories. We shall postpone them until section 19.

8.11. THEOREM. *Assume that X is completely regular. Let the cmeasure* $\alpha : \mathrm{Baire}(X) \to [0, \infty]$ *be inner regular* $\mathfrak{S} := \{A \in \mathrm{CCl}(X) : \alpha(A) < \infty\}$, *that is semifinite above, and assume that* $\varphi := \alpha|\mathfrak{S}$ *is τ continuous at $\varnothing$. Then*

1) φ *is an inner τ premeasure with* $\mathfrak{C}(\varphi_\tau) \supset \mathrm{Bor}(X)$. *The cmeasure $\beta :=$* $\varphi_\tau|\mathrm{Bor}(X)$ *extends α and is inner regular* $\mathfrak{S}_\tau = \mathrm{Cl}(X) \cap (\sqsubset \mathfrak{S})$. *Furthermore* $\beta|\mathrm{Cl}(X)$ *is almost downward τ continuous.*

2) *Each cmeasure $\vartheta : \mathrm{Bor}(X) \to [0, \infty]$ which extends $\varphi = \alpha|\mathfrak{S}$ and is inner regular* $\mathfrak{S}_\tau$ *is* $= \beta$.

Proof of 8.11.1). By the assumptions φ is an inner $\star$ premeasure and τ continuous at $\varnothing$, and hence an inner $\bullet$ premeasure for $\bullet = \star\sigma\tau$ by the conventional inner main theorem 6.31. Besides $\mathrm{Baire}(X) \subset \mathfrak{C}(\varphi_\sigma)$ we see from 6.31 that $\mathrm{Cl}(X) = (\mathrm{CCl}(X))_\tau \subset \mathfrak{S}\top\mathfrak{S}_\tau \subset \mathfrak{C}(\varphi_\tau)$ and hence $\mathrm{Bor}(X) \subset \mathfrak{C}(\varphi_\tau)$. By 6.24.$\tau$) and 6.25.$\tau$) $\varphi_\tau|\mathfrak{C}(\varphi_\tau)$ is an extension of $\varphi_\sigma|\mathfrak{C}(\varphi_\sigma)$ and hence of $\alpha = \varphi_\star|\mathrm{Baire}(X) = \varphi_\sigma|\mathrm{Baire}(X)$. Therefore $\beta := \varphi_\tau|\mathrm{Bor}(X)$ is a cmeasure which extends α and is inner regular $\mathfrak{S}_\tau$. By 8.1.5) we have $\mathfrak{S}_\tau = \mathrm{Cl}(X) \cap (\sqsubset \mathfrak{S})$. The last assertion follows from 6.27.

The proof of the second part requires a simple lemma.

8.12. LEMMA. *Let $\mathfrak{S}$ be a lattice with $\varnothing \in \mathfrak{S}$, and $\phi : \mathfrak{S}_\bullet \to [0, \infty]$ be isotone with $\phi(\varnothing) = 0$. If $\phi|\mathfrak{S}$ is $\bullet$ continuous at $\varnothing$ then ϕ is $\bullet$ continuous at $\varnothing$ as well.*

Proof of 8.12. Let $\mathfrak{M} \subset \mathfrak{S}_\bullet$ be a paving of type $\bullet$ with $\mathfrak{M} \downarrow \varnothing$. By 6.6 there exists a paving $\mathfrak{N} \subset \mathfrak{S}$ of type $\bullet$ with $\mathfrak{N} \downarrow \varnothing$ and $\mathfrak{N} \subset (\sqsupset \mathfrak{M})$. Fix $N \in \mathfrak{N}$. There exists $M \in \mathfrak{M}$ with $M \subset N$ and hence $0 \leqq \phi(M) \leqq \phi(N)$. Therefore

$$0 \leqq \inf_{M \in \mathfrak{M}} \phi(M) \leqq \phi(N) \quad \text{and hence} \quad 0 \leqq \inf_{M \in \mathfrak{M}} \phi(M) \leqq \inf_{N \in \mathfrak{N}} \phi(N).$$

The assertion follows.

Proof of 8.11.2). Let $\vartheta : \mathrm{Bor}(X) \to [0, \infty]$ be a cmeasure which extends $\varphi = \alpha|\mathfrak{S}$ and is inner regular $\mathfrak{S}_\tau$. Then $\vartheta|\mathfrak{S}_\tau$ is an inner $\star$ premeasure, and

τ continuous at $\varnothing$ by 8.12. Thus $\vartheta|\mathfrak{S}_\tau$ is an inner τ premeasure and hence downward τ continuous. Therefore $\vartheta = \varphi = \alpha = \beta$ on $\mathfrak{S}$ implies that $\vartheta = \beta$ on $\mathfrak{S}_\tau$, and hence that $\vartheta = \beta$ since both sides are inner regular $\mathfrak{S}_\tau$. The proof is complete.

We emphasize that we have not only proved the existence and uniqueness of the desired Borel extension, but have also obtained a certain explicit description. We turn to the special case $\alpha < \infty$ which has a simpler formulation, and then to the special case of a compact Hausdorff space X. The latter case leads to Radon measures in the sense of the next section.

8.13. SPECIAL CASE. *Assume that X is completely regular. Let $\alpha :$ Baire$(X) \to [0, \infty[$ be a cmeasure such that $\varphi := \alpha|\mathrm{CCl}(X)$ is τ continuous at $\varnothing$. Then*

1) *φ is an inner τ premeasure with $\mathfrak{C}(\varphi_\tau) \supset \mathrm{Bor}(X)$. The finite cmeasure $\beta := \varphi_\tau|\mathrm{Bor}(X)$ extends α and is inner regular $\mathrm{Cl}(X)$. Furthermore $\beta|\mathrm{Cl}(X)$ is downward τ continuous.*

2) *Each cmeasure $\vartheta : \mathrm{Bor}(X) \to [0, \infty[$ which extends $\varphi = \alpha|\mathrm{CCl}(X)$ and is inner regular $\mathrm{Cl}(X)$ is $= \beta$.*

8.14. SPECIAL CASE. *Assume that X is a compact Hausdorff space. Let $\alpha : \mathrm{Baire}(X) \to [0, \infty[$ be a cmeasure. Then*

1) *$\varphi := \alpha|\mathrm{CCl}(X)$ is an inner τ premeasure with $\mathfrak{C}(\varphi_\tau) \supset \mathrm{Bor}(X)$. The finite cmeasure $\beta := \varphi_\tau|\mathrm{Bor}(X)$ extends α and is inner regular $\mathrm{Cl}(X)$. Furthermore $\beta|\mathrm{Cl}(X)$ is downward τ continuous.*

2) *Each cmeasure $\vartheta : \mathrm{Bor}(X) \to [0, \infty[$ which extends $\varphi = \alpha|\mathrm{CCl}(X)$ and is inner regular $\mathrm{Cl}(X)$ is $= \beta$.*

8.15. BIBLIOGRAPHICAL NOTE. The essence of the extension theorem 8.11 is in Topsøe [1970b] theorem 5.1. A similar result is in Sapounakis-Sion [1987] theorem 7.2, in the spirit of the two-step extension method of this work as described in the bibliographical annex to chapter II. It extends the initial result of Knowles [1967] which was for finite Baire measures. An ab-ovo proof of the compact special case 8.14 is in Dudley [1989] section 7.3.

For the extensive literature on Borel and Baire measures we refer to the survey articles of Gardner-Pfeffer [1984] and Wheeler [1983].

The Hewitt-Yosida Theorem

In the last subsection we return to the abstract situation of a nonvoid set X. We want to add another application of the notion of upward σ full lattices. We start with the classical decomposition theorem of Hewitt-Yosida [1952].

8.16. THEOREM (Hewitt-Yosida). *Let $\varphi : \mathfrak{S} \to [0, \infty[$ be a ccontent on a ring $\mathfrak{S}$. Then there exists a unique decomposition $\varphi = \xi + \eta$ of φ into ccontents $\xi, \eta : \mathfrak{S} \to [0, \infty[$ such that*

i) *ξ is upward σ continuous.*

ii) *η is σ discontinuous in the sense that there is no nonzero upward σ continuous ccontent $\vartheta : \mathfrak{S} \to [0, \infty[$ with $\vartheta \leqq \eta$.*

We shall deduce from the outer σ extension procedure of chapter II that there is an identical result when $\mathfrak{S}$ is an upward σ full lattice. This extended result will be a special case of a more comprehensive theorem. We need an assertion which extends part of the conventional outer main theorem 5.11 for $\bullet = \sigma$.

8.17. PROPOSITION. *Let $\mathfrak{S}$ be a lattice with $\varnothing \in \mathfrak{S}$, and $\varphi : \mathfrak{S} \to [0, \infty]$ be isotone and submodular with $\varphi(\varnothing) = 0$ as well as outer σ tight. Then $\mathfrak{S} \subset \mathfrak{C}(\varphi^\sigma)$.*

Proof. Fix $S \in \mathfrak{S}$. By 5.2 applied to $\mathfrak{P} := \{\varnothing\}$ and $\mathfrak{Q} := [\varphi < \infty]$ we have to show that

$$\varphi^\sigma(Q) \geq \varphi^\sigma(Q \cap S) + \varphi^\sigma(Q \cap S') \quad \text{for all } Q \in [\varphi < \infty];$$

here we used that $\varphi^\sigma(\varnothing) = 0$. So fix $Q \in [\varphi < \infty]$. Let $(S_l)_l$ be a sequence in $\mathfrak{S}$ with $S_l \uparrow$ some $V \supset Q$. Since φ is outer σ tight we have $\varphi(S_l) \geq \varphi(S_l \cap S) + \varphi^\sigma(S_l \cap S')$. Now on the one hand $(S_l \cap S)_l$ is a sequence in $\mathfrak{S}$ with $S_l \cap S \uparrow V \cap S \supset Q \cap S$, so that by definition $\lim\limits_{l \to \infty} \varphi(S_l \cap S) \geq \varphi^\sigma(Q \cap S)$. On the other hand $S_l \cap S' \uparrow V \cap S'$, so that 4.7 implies that $\lim\limits_{l \to \infty} \varphi^\sigma(S_l \cap S') = \varphi^\sigma(V \cap S') \geq \varphi^\sigma(Q \cap S')$. Together we obtain $\lim\limits_{l \to \infty} \varphi(S_l) \geq \varphi^\sigma(Q \cap S) + \varphi^\sigma(Q \cap S')$. The assertion follows.

8.18. THEOREM. *Let $\mathfrak{S}$ be a lattice with $\varnothing \in \mathfrak{S}$, and $\varphi : \mathfrak{S} \to [0, \infty]$ be isotone and submodular with $\varphi(\varnothing) = 0$ as well as outer σ tight. Then*

i) *$\psi := \varphi^\sigma | \mathfrak{S} : \mathfrak{S} \to [0, \infty]$ is an outer σ premeasure with $\psi \leq \varphi$ and $\psi^\sigma = \varphi^\sigma$.*

ii) *If $\vartheta : \mathfrak{S} \to [0, \infty]$ is isotone and upward σ continuous with $\vartheta \leq \varphi$ then $\vartheta \leq \psi$.*

Proof of 1). i) ψ is isotone by 4.1.3) and submodular by 4.1.5), and $\psi(\varnothing) = \varphi^\sigma(\varnothing) = 0$ and $\psi \leq \varphi$ by 4.1.1)2). By 4.7 ψ is upward σ continuous. It follows that $\varphi^\sigma = \psi = \psi^\sigma$ on $\mathfrak{S}$, hence $\varphi^\sigma = \psi^\sigma$ on $\mathfrak{S}^\sigma$ and therefore on $\mathfrak{P}(X)$ by 4.1.4). ii) ψ is upward σ continuous, and by the above 8.17 we have $\mathfrak{S} \subset \mathfrak{C}(\varphi^\sigma) = \mathfrak{C}(\psi^\sigma)$. Thus ψ is an outer σ premeasure. Proof of 2). From the assumptions we obtain $\vartheta = \vartheta^\sigma | \mathfrak{S} \leq \varphi^\sigma | \mathfrak{S} = \psi$.

The final assertion below is the announced extension of the Hewitt-Yosida theorem.

8.19. THEOREM. *Let $\mathfrak{S}$ be an upward σ full lattice, and $\varphi : \mathfrak{S} \to [0, \infty[$ be isotone and modular with $\varphi(\varnothing) = 0$. Then there exists a unique decomposition $\varphi = \xi + \eta$ of φ with $\xi, \eta : \mathfrak{S} \to [0, \infty[$ isotone and modular (and of course $\xi(\varnothing) = \eta(\varnothing) = 0$) such that*

i) *ξ is upward σ continuous.*

ii) *η is σ discontinuous in the sense that there is no nonzero upward σ continuous isotone and modular $\vartheta : \mathfrak{S} \to [0, \infty[$ with $\vartheta \leq \eta$.*

Here we have $\xi = \varphi^\sigma | \mathfrak{S}$.

Proof. 1) We have $\varphi = \xi + \eta$ with $\xi := \varphi^\sigma | \mathfrak{S} \leqq \varphi$ and $\eta := \varphi - \xi \geqq 0$. This decomposition is as required: i) φ is outer σ tight by 7.10.1), and hence ξ an outer σ premeasure by 8.18.i). In particular ξ and η are modular. Of course ξ is isotone, but it is nontrivial that η is isotone. To see this note that for $A \subset B$ in $\mathfrak{S}$ we have

$$\xi(B) - \xi(A) = \xi^\sigma(B \setminus A) = \varphi^\sigma(B \setminus A) \leqq \varphi(B) - \varphi(A),$$

so that $\eta(A) \leqq \eta(B)$. ii) It remains to show that η is σ discontinuous. Let $\vartheta : \mathfrak{S} \to [0, \infty[$ be as assumed. Then $\vartheta + \xi \leqq \eta + \xi = \varphi$ is isotone and upward σ continuous. From 8.18.ii) we obtain $\vartheta + \xi \leqq \xi$ and hence $\vartheta = 0$.

2) It remains to prove the uniqueness assertion. Assume that $\varphi = \xi + \eta$ is any decomposition of φ with the required properties. Also put $\psi := \varphi^\sigma | \mathfrak{S}$ as in 8.18. i) From 8.18.ii) applied to $\vartheta := \xi$ we obtain $\xi \leqq \psi$. ii) Therefore $\vartheta := \psi - \xi \geqq 0$ is modular with $\vartheta(\varnothing) = 0$. We claim that ϑ is isotone. Let $A \subset B$ in $\mathfrak{S}$, and consider a sequence $(S_l)_l$ in $\mathfrak{S}$ with $S_l \uparrow B \setminus A$. Then $A \cup S_l \uparrow B$ and hence

$$\psi(A) + \psi(S_l) = \psi(A \cup S_l) \uparrow \psi(B),$$
$$\xi(A) + \xi(S_l) = \xi(A \cup S_l) \uparrow \xi(B).$$

From $\xi(S_l) \leqq \psi(S_l)$ we obtain $\xi(B) - \xi(A) \leqq \psi(B) - \psi(A)$ or $\vartheta(A) \leqq \vartheta(B)$. iii) Now $\vartheta := \psi - \xi$ is upward σ continuous and satisfies $\vartheta \leqq \eta$ as well. By assumption it follows that $\vartheta = 0$ or $\xi = \psi$. The proof is complete.

8.20. BIBLIOGRAPHICAL NOTE. A result similar to 8.18 is in Sapouna-kis-Sion [1987] theorem 12.1. It is restricted to the case that $\mathfrak{S}$ has $\sqcap \sigma$ and that φ is bounded above, but on the other hand obtains a more refined decomposition of φ. It is an extension of the initial result due to Knowles [1967] theorem 4.3 which was in the context of a completely regular topological space X.

9. Radon Measures

Let X be a Hausdorff topological space. The present section centers around a particular class of cmeasures on X which has turned out to be the most fundamental one. It descends from the lattice $\mathfrak{K} := \mathrm{Comp}(X) \subset \mathrm{Cl}(X)$ of the compact subsets of X.

Radon Contents and Radon Measures

We define a **Radon content** on X to be a ccontent $\alpha : \mathfrak{A} \to [0, \infty]$ on some $\mathfrak{A} \subset \mathfrak{P}(X)$ with $\mathfrak{K} \subset \mathfrak{A}$ such that $\alpha | \mathfrak{K} < \infty$ and α is inner regular $\mathfrak{K}$. Thus the restriction $\varphi := \alpha | \mathfrak{K}$ reproduces $\alpha = \varphi_\star | \mathfrak{A}$. When α is a cmeasure it is called a **Radon measure** on X. It will be seen that Radon contents can be extended to Radon measures and thus could be dismissed in principle. In the literature the name Radon measure is often reserved for the Borel-Radon measures $\alpha : \mathrm{Bor}(X) \to [0, \infty]$.

The most natural problem is to characterize those set functions $\varphi : \mathfrak{K} \to [0, \infty[$ which can be extended to Radon contents/measures, and then to describe all these Radon contents/measures. It is plain that this problem fits in the frame of our chapter II, and in fact it has been the historical source of the entire development.

We recall at the start those properties of $\mathfrak{K}$ which are decisive for the present purpose. These properties are

I) $\mathfrak{K}$ is a lattice with $\varnothing \in \mathfrak{K}$ and $\mathfrak{K} = \mathfrak{K}_\star = \mathfrak{K}_\sigma = \mathfrak{K}_\tau$.

II) If $\mathfrak{M} \subset \mathfrak{K}$ is a paving such that $\mathfrak{M} \downarrow \varnothing$ then $\varnothing \in \mathfrak{M}$, that is $\mathfrak{K}$ is τ compact as defined before 6.34.

Then section 6 implies the characterization theorem which follows. Part of the enumeration has been borrowed from the conventional inner main theorem 6.31.

9.1. THEOREM. *Let $\varphi : \mathfrak{K} \to [0, \infty[$ be isotone and supermodular with $\varphi(\varnothing) = 0$. Then the nine conditions below are equivalent.*

i) φ *can be extended to a Radon content.*
ii) φ *can be extended to a Radon measure*
iii) φ *can be extended to a (unique) Borel-Radon measure.*
1•) φ *is an inner • premeasure ($• = \star\sigma\tau$).*
5'•) φ *is inner • tight ($• = \star\sigma\tau$).*

Under these conditions $\varphi_\star = \varphi_\sigma = \varphi_\tau$. Furthermore $\phi := \varphi_\bullet|\mathfrak{C}(\varphi_\bullet)$ is a Radon measure extension of φ with $\mathfrak{C}(\varphi_\bullet) \supset \mathrm{Bor}(X)$, and each Radon content extension of φ is a restriction of ϕ.

Proof. 1) The conditions 6.31.1) and 6.31.5') for $• = \star\sigma\tau$ will be called 1•) and 5'•). By the above I) then 1•) attains the form

1•) φ can be extended to a Radon content and is downward • continuous.

By the above II) and 6.34 we have $\varphi_\bullet(\varnothing) = 0$, so that 5'•) attains the form

5'•) φ is inner • tight.

It is obvious that $1\star) \Leftarrow 1\sigma) \Leftarrow 1\tau)$ and $5'\star) \Rightarrow 5'\sigma) \Rightarrow 5'\tau)$. Now 6.31 says that $1\bullet) \Leftrightarrow 5'\bullet)$ for each $• = \star\sigma\tau$. It follows that all these six conditions are equivalent, and also equivalent to i)=1$\star$). Furthermore it is obvious that iii)$\Rightarrow$ii)$\Rightarrow$i).

2) Let us assume i) and hence the seven equivalent conditions as above. From 6.5.iv) we obtain $\varphi_\star = \varphi_\sigma = \varphi_\tau$. Then the common maximal inner • extension $\phi := \varphi_\bullet|\mathfrak{C}(\varphi_\bullet)$ is a Radon measure, and we have $\mathrm{Cl}(X) \subset \mathfrak{K}\top\mathfrak{K} \subset \mathfrak{C}(\varphi_\bullet)$ and hence $\mathrm{Bor}(X) \subset \mathfrak{C}(\varphi_\bullet)$. Thus we have iii). The last assertion follows from 6.18. The proof is complete.

An isotone and supermodular set function $\varphi : \mathfrak{K} \to [0, \infty[$ with $\varphi(\varnothing) = 0$ which fulfils the equivalent conditions of the last theorem will be called a **Radon premeasure**. Thus if $\alpha : \mathfrak{A} \to [0, \infty]$ is a Radon content then $\varphi := \alpha|\mathfrak{K}$ is a Radon premeasure, and $\alpha = \varphi_\bullet|\mathfrak{A}$ with $\mathfrak{K} \subset \mathfrak{A} \subset \mathfrak{C}(\varphi_\bullet)$.

9.2. EXAMPLE. We have seen in 6.35 that $\lambda : \mathrm{Comp}(\mathbb{R}^n) \to [0, \infty[$ is a Radon premeasure. Then 7.6 says that its maximal Radon measure extension $\lambda_\bullet|\mathfrak{C}(\lambda_\bullet)$ is the Lebesgue measure $\Lambda = \lambda^\sigma|\mathfrak{C}(\lambda^\sigma) = \lambda^\sigma|\mathfrak{L}$. Thus $\Lambda|\mathrm{Bor}(\mathbb{R}^n)$ is its unique Borel-Radon measure extension.

9.3. THEOREM. *Let* $\varphi : \mathfrak{K} \to [0, \infty[$ *be a Radon premeasure and* $\phi := \varphi_\bullet|\mathfrak{C}(\varphi_\bullet)$. *Then* $\phi|\mathfrak{K}\top\mathfrak{K}$ *is almost downward* τ *continuous, and* $\phi|(\mathfrak{K}\top\mathfrak{K})\bot$ *is upward* τ *continuous.*

Proof. This follows from 6.27 and 6.28.

9.4. CONSEQUENCE. *Let* $\alpha : \mathrm{Bor}(X) \to [0, \infty]$ *be a Borel-Radon measure. Then* $\alpha|\mathrm{Cl}(X)$ *is almost downward* τ *continuous, and* $\alpha|\mathrm{Op}(X)$ *is upward* τ *continuous.*

It is remarkable that there is an assertion in the opposite direction.

9.5. LEMMA. *Assume that* $\varphi : \mathfrak{K} \to [0, \infty[$ *is isotone with* $\varphi(\varnothing) = 0$ *and fulfils* $\varphi(A \cup B) \leqq \varphi(A) + \varphi(B)$ $\forall A, B \in \mathfrak{K}$, *and is downward* τ *continuous. Then* φ *is inner* $\star$ *tight.*

Proof. Fix $A \subset B$ in $\mathfrak{K}$. To be shown is $\varphi(B) \leqq \varphi(A) + \varphi_\star(B \setminus A)$. i) Define $\mathfrak{M} \subset \mathfrak{K}$ to consist of all $S \in \mathfrak{K}$ such that $U \cap B \subset S \subset B$ for some open $U \supset A$. $\mathfrak{M}$ is nonvoid since $B \in \mathfrak{M}$, and it has $\cap$. We claim that $\mathfrak{M} \downarrow A$. To see this fix $v \in B \setminus A$. Since A is compact there exist disjoint open U and V with $A \subset U$ and $v \in V$. Thus $S := B \cap V' \in \mathfrak{K}$ satisfies $U \cap B \subset S$ and hence $S \in \mathfrak{M}$. Since $v \notin S$ the assertion follows. ii) Now fix $\varepsilon > 0$. By i) there exists $S \in \mathfrak{M}$ and hence an open $U \supset A$ with $U \cap B \subset S \subset B$ such that $\varphi(S) < \varphi(A) + \varepsilon$. Then $S, B \cap U' \in \mathfrak{K}$ with $S \cup (B \cap U') = B$, and thus by assumption

$$\varphi(B) \leqq \varphi(S) + \varphi(B \cap U') < \varphi(A) + \varepsilon + \varphi(B \cap U').$$

But $B \cap U' \subset B \cap A' = B \setminus A$ and hence $\varphi(B \cap U') \leqq \varphi_\star(B \setminus A)$. The assertion follows.

9.6. THEOREM. *Assume that* $\varphi : \mathfrak{K} \to [0, \infty[$ *is isotone and additive and fulfils* $\varphi(A \cup B) \leqq \varphi(A) + \varphi(B)$ $\forall A, B \in \mathfrak{K}$. *Then* φ *is a Radon premeasure iff it is downward* τ *continuous.*

9.7. EXERCISE. 1) Let X be compact Hausdorff. If $\alpha : \mathrm{Bor}(X) \to [0, \infty[$ is a finite cmeasure such that $\alpha|\mathrm{Cl}(X)$ is downward τ continuous, then α is a Radon measure. Hint: Compare α with the Borel-Radon measure $(\alpha|\mathrm{Cl}(X))_\bullet|\mathrm{Bor}(X)$. 2) Let X be locally compact Hausdorff. If $\alpha : \mathrm{Bor}(X) \to [0, \infty[$ is a finite cmeasure such that $\alpha|\mathrm{Op}(X)$ is upward τ continuous, then α is a Radon measure. Hint: Look at the paving of those open subsets of X which are contained in compact subsets of X.

It is a nontrivial question whether all reasonable Borel measures $\alpha : \mathrm{Bor}(X) \to [0, \infty]$ with $\alpha|\mathfrak{K} < \infty$ are Radon measures. The next subsection will reproduce the classical counterexample where X is a compact Hausdorff space. There are partial positive results in 9.7 above. The most famous partial positive result is 9.9.ii) below where X is a Polish space.

We come to the notion of local finiteness. A Borel measure $\alpha : \text{Bor}(X) \to [0, \infty]$ on a topological space X is called **locally finite** iff each point $u \in X$ has an open neighbourhood $U \subset X$ such that $\alpha(U) < \infty$. When X is Hausdorff this implies that $\alpha|\mathfrak{K} < \infty$. In several texts the definition of Borel-Radon measures includes the requirement of local finiteness. The reason is that for certain deeper results this restriction cannot be dispensed with. The present text does not follow this habit. One basic reason is that the Riesz representation theorem which will be obtained in chapter V involves all Borel-Radon measures and not only the locally finite ones. Let us present an example of a Borel-Radon measure which is not locally finite.

9.8. EXAMPLE. Let X be an infinite countable set with a Hausdorff topology which is not discrete but in which all compact subsets are finite. There are several simple constructions of this kind; see for example König [1993]. Then of course $\text{Bor}(X) = \mathfrak{P}(X)$, and the counting measure $\alpha : \text{Bor}(X) = \mathfrak{P}(X) \to [0, \infty]$ is a Radon measure. Let now $a \in X$ be a point such that $\{a\}$ is not open. Then all open neighbourhoods $U \subset X$ of a must be infinite and hence have $\alpha(U) = \infty$. Thus α is not locally finite.

Related to local finiteness is the famous partial positive result announced above. One defines a **Polish space** to be a separable topological space which is metrizable under a complete metric. This is a fundamental notion in topology.

9.9. THEOREM. *Let $\alpha : \text{Bor}(X) \to [0, \infty]$ be a cmeasure on a Polish space. i) If α is semifinite above then it is inner regular $\{A \in \mathfrak{K} : \alpha(A) < \infty\}$. ii) If α is locally finite then it is a Radon measure.*

Proof. i) is an immediate consequence of 8.10. ii) We recall from topology that a separable metrizable space X has a countable base and hence is Lindelöf, that is each open cover of X has a countable subcover. Thus if α is locally finite then there exists a sequence $(U_l)_l$ in $\text{Op}(X)$ such that $U_l \uparrow X$ and $\alpha(U_l) < \infty$ $\forall l$. Therefore α is semifinite above. The assertion follows from i).

The last theme of the present subsection is outer regularity. A Borel-Radon measure $\alpha : \text{Bor}(X) \to [0, \infty]$ need not be outer regular $\text{Op}(X)$, because outer regularity $\text{Op}(X)$ enforces local finiteness. The next exercise formulates a simple idea which has a natural proof.

9.10. EXERCISE. *Let $\alpha : \text{Bor}(X) \to [0, \infty]$ be a Borel-Radon measure. 1) If $A \in \text{Bor}(X)$ has open supersets $U \supset A$ with $\alpha(U) < \infty$ then $\alpha(A) = \inf\{\alpha(U) : U \text{ open} \supset A\}$. 2) α is outer regular $\text{Op}(X)$ iff for each $A \in \text{Bor}(X)$ with $\alpha(A) < \infty$ there exist open supersets $U \supset A$ with $\alpha(U) < \infty$.*

9.11. BIBLIOGRAPHICAL NOTE. In the fifties and sixties, under the influence of Bourbaki [1952][1956][1965][1967], the notion of a Radon measure was restricted to locally compact Hausdorff topological spaces X. In this frame they were *defined* as certain positive linear functionals, in fact as those which occur in the traditional Riesz representation theorem formulated in the introduction. For such a functional Bourbaki defined two set functions $\mathfrak{P}(X) \to [0, \infty]$, one of which produces a Borel-Radon measure in the present sense. For a short description we refer to Berg-Christensen-Ressel [1984] notes and remarks to chapter 2.

Radon measures on arbitrary Hausdorff spaces X appeared in Bourbaki [1969] and in Schwartz [1973], with basic ideas due to Choquet. These definitions required local finiteness. For a discussion of the different definitions we refer to Schwartz [1973] and once more to Berg-Christensen-Ressel [1984] notes and remarks to chapter 2. At about the same time, but without connection to the above work, the fundamental paper of Kisyński [1968] characterized those set functions $\varphi : \text{Comp}(X) \to [0, \infty[$ which can be extended to Borel-Radon measures, this time without local finiteness. Kisyński proved the decisive equivalence iii)⇔5'⋆) in the present theorem 9.1. Within a short time this work became the source of the extension theories based on regularity as described in the bibliographical annex to chapter II. For the role of local finiteness we refer to Fremlin [1975].

At last we mention the work of Anger-Portenier [1992a][1992b] which has the aim to revive the functional-analytic aspects of the theory of Radon measures. We shall come back to it in chapter V.

The Classical Example of a Non-Radon Borel Measure

The example in question is based on the order topology on a certain uncountable well-ordered set. Its usual presentation involves ordinal numbers; see for example Kelley-Srinivasan [1988] page 52. We want to offer an ab-ovo presentation, based on the definitions of total order and well-order and on the existence of an uncountable set which carries a well-order.

Let X be a nonvoid set equipped with a total order $\leqq$. We form as usual

for $u, v \in X$ with $u \leqq v$ the interval $[u, v] := \{x \in X : u \leqq x \leqq v\}$,

for $u, v \in X$ with $u < v$ the interval $]u, v[:= \{x \in X : u < x < v\}$,

and for $a \in X$

the upper ends $[a, \cdot := \{x \in X : a \leqq x\}$ and $]a, \cdot := \{x \in X : a < x\}$,

the lower ends $\cdot, a] := \{x \in X : x \leqq a\}$ and $\cdot, a[:= \{x \in X : x < a\}$.

Define $\mathfrak{B}(\leqq)$ to consist of all these intervals $]u, v[$ and all these ends $]a, \cdot$ and $\cdot, a[$, and of $\varnothing$ and X. Then $\mathfrak{B}(\leqq)$ has $\cap$ and is therefore the base of a unique topology $\mathfrak{T}(\leqq)$ on X, called the **order topology** on X. One verifies that $\mathfrak{T}(\leqq)$ is Hausdorff.

9.12. REMARK. *Each sequence in X has a monotone subsequence.*

The proof is mathematical folklore. Let $(x_l)_l$ be a sequence in X. We define $n \in \mathbb{N}$ to be a top index iff $x_l \leqq x_n$ for all $l \geqq n$. Then there are two cases. 1) There are at most finitely many top indices. Thus there exists an $N \in \mathbb{N}$ such that no $n \geqq N$ is a top index. Hence for each $n \geqq N$

there exists an $l > n$ such that $x_l > x_n$. Therefore we obtain a sequence of indices $N = n(1) < \cdots < n(p) < \cdots$ such that the subsequence $(x_{n(p)})_p$ is strictly increasing. 2) There are infinitely many top indices. Assume that the sequence of indices $1 \leq n(1) < \cdots < n(p) < \cdots$ consists of top indices. Then the subsequence $(x_{n(p)})_p$ is decreasing.

For the remainder of the subsection we fix a set E equipped with a well-order $\leq$ such that E is uncountable but for each $a \in E$ the lower end $\cdot, a]$ is countable. This situation can be produced as follows. Let X be an uncountable set equipped with a well-order $\leq$. If for each $a \in X$ the lower end $\cdot, a]$ is countable then we take $E := X$. Otherwise the nonvoid subset of those $a \in X$ for which $\cdot, a]$ is uncountable has a first element $e \in X$. Then $E := \cdot, e[$ is as required.

9.13. PROPERTIES. 1) *For each $a \in E$ the lower end $\cdot, a]$ is compact.* 2) *For each $a \in E$ the lower end $\cdot, a]$ is open.* 3) *For each nonvoid countable $A \subset E$ there is an $a \in E$ such that $A \subset \cdot, a[$.* 4) *Each nonvoid compact subset $A \subset E$ is countable.*

Proof. 1) Let $A := \cdot, a]$, and fix a family $(U(x))_{x \in A}$ of open neighbourhoods $U(x)$ of the points $x \in A$. Define $F \subset E$ to consist of those $u \in A$ for which $[u, a]$ can be covered by finitely many of these $U(x) \, \forall x \in A$. Then $F \neq \varnothing$ since $a \in F$. Thus there is a first element $c \in F$. If c is the first element of E then we have $[c, a] = \cdot, a] = A$ and hence the assertion. Otherwise there exists an element $u \in E$ with $u < c$, and hence an element $u \in E$ with $u < c$ such that $]u, c[\subset U(c)$. Then $[u, c]$ is covered by $U(c)$ and $U(u)$, so that $u \in F$, which contradicts the role of c. Therefore the second case cannot happen. 2) The nonvoid subset $]a, \cdot$ has a first element $b \in E$, and for this we have $\cdot, a] = \cdot, b[$. 3) Assume not. Then for each $a \in E$ there is an $x \in A$ with $a \leq x$ or $a \in \cdot, x]$. It follows that $E = \bigcup_{x \in A} \cdot, x]$, which is impossible since E is uncountable. 4) The lower ends $\cdot, x[\; \forall x \in E$ form an open cover of A. Therefore $A \subset \cdot, x[$ for some $x \in E$. Hence A is countable.

9.14. REMARK. 1) *Let $(x_l)_l$ be a decreasing sequence in E. Then there exists $c \in E$ such that $x_l = c$ for almost all $l \in \mathbb{N}$. Thus $x_l \to c$.* 2) *Let $(x_l)_l$ be an increasing sequence in E. If $c \in E$ is the first element of the subset $\{x \in X : x_l \leq x \; \forall l \in \mathbb{N}\}$ (which is nonvoid by 9.13.3) then $x_l \to c$.* 3) *E is sequentially compact, that is each sequence in E has a subsequence which converges to some element of E.*

Proof. 1) Let $c \in E$ be the first element of $\{x_l : l \in \mathbb{N}\}$. Then $c = x_l$ for some $l \in \mathbb{N}$ and hence for almost all $l \in \mathbb{N}$. 2) We can assume that c is not the first element of E. Let U be an open neighbourhood of c. Then there exists $u \in E$ with $u < c$ such that $]u, c[\subset U$. By the definition of c we have $u < x_p$ for some $p \in \mathbb{N}$. Once more by the definition of c it follows that $x_l \in U$ for all $l \geq p$. 3) Combine 1)2) with 9.12.

9.15. PROPOSITION. *Assume that the subsets $A_l \subset E$ are closed and uncountable $\forall l \in \mathbb{N}$. Then $A := \bigcap_{l=1}^{\infty} A_l$ is closed and uncountable as well.*

Proof. 1) To be shown is the second assertion. We first prove that A is nonvoid. We perform an inductive construction which is based on the fact that for each uncountable $M \subset E$ and each $u \in E$ there exists $x \in M$ with $u < x$. Thus we construct in step $n \in \mathbb{N}$ elements $x_l^n \in A_l$ for $l = 1, \cdots, n$ such that $x_1^1 < \cdots < x_1^n < \cdots < x_n^n \cdots$. For each fixed $k \in \mathbb{N}$ then

$$\{x \in E : x_k^n \leqq x \text{ for all } n \geqq k\} = \{x \in E : x_l^n \leqq x \, \forall n, l \in \mathbb{N} \text{ with } l \leqq n\}.$$

Let $c \in E$ be the first element of this common subset. For each fixed $k \in \mathbb{N}$ then $x_k^n \to c$ for $n \to \infty$ by 9.14.2). Thus $c \in A_k$ since A_k is closed. It follows that $c \in A$. 2) For each fixed $u \in E$ the subsets $A_l \cap [u, \cdot \subset E$ are closed and uncountable $\forall l \in \mathbb{N}$. Thus 1) shows that $A \cap [u, \cdot$ is nonvoid. By 9.13.3) this implies that A is uncountable.

9.16. PROPOSITION. *Let $A \in \mathrm{Bor}(E)$. Then exactly one of the subsets A and A' contains an uncountable member of $\mathrm{Cl}(E)$.*

Proof. By 9.15 we have to show that at least one of the subsets A and A' contains an uncountable member of $\mathrm{Cl}(E)$. Define $\mathfrak{N}$ to consist of those subsets $N \subset E$ such that either N or N' contains an uncountable member of $\mathrm{Cl}(E)$. To be shown is then that $\mathfrak{N}$ is a σ algebra and that $\mathrm{Cl}(E) \subset \mathfrak{N}$.

i) $\mathfrak{N}$ is nonvoid since $\varnothing, E \in \mathfrak{N}$. It is obvious that $\mathfrak{N}$ fulfils $\perp$. ii) To see that $\mathfrak{N}_\sigma \subset \mathfrak{N}$ fix a sequence $(N_l)_l$ in $\mathfrak{N}$ with intersection N. If each N_l contains an uncountable member of $\mathrm{Cl}(E)$ then by 9.15 the same holds true for N. Otherwise $N_l' \subset N' \, \forall l \in \mathbb{N}$ shows that it holds true for N'. Thus $N \in \mathfrak{N}$. Therefore $\mathfrak{N}$ is a σ algebra. iii) Now let $N \in \mathrm{Cl}(E)$. If N is uncountable then of course $N \in \mathfrak{N}$. If N is countable then by 9.13.3) there exists an $a \in E$ such that $N \subset \cdot, a[$ and hence $[a, \cdot \subset N'$. But $[a, \cdot$ is closed and uncountable, so that $N \in \mathfrak{N}$ as well. It follows that $\mathrm{Cl}(E) \subset \mathfrak{N}$.

So far we remainded within E. From 9.13.1)2) we know that E with $\mathfrak{T}(\leqq)$ is a locally compact Hausdorff space. Let now $X := E \cup \{\omega\}$ be the usual one-point compactification of E. That means that $\omega \notin E$, and that the topology $\mathfrak{S}$ of X consists of the subsets

$A \subset E$ such that $A \in \mathrm{Op}(E)$, and of the
$A \subset X$ with $\omega \in A$ such that $A' \in \mathrm{Comp}(E)$.

It is standard that $\mathfrak{S}$ is a compact Hausdorff topology on X with $\mathfrak{S}|E = \mathfrak{T}(\leqq)$. By 1.13.2) we have $\mathrm{Bor}(E) = \{A \in \mathrm{Bor}(X) : A \subset E\}$.

9.17. THEOREM. *Define $\alpha : \mathrm{Bor}(X) \to \{0, 1\}$ to be*

$$\alpha(A) = \left\{ \begin{array}{l} 1 \text{ if } A \text{ contains an uncountable member of } \mathrm{Cl}(E) \\ 0 \text{ if } A' \text{ contains an uncountable member of } \mathrm{Cl}(E) \end{array} \right\},$$

so that α is well-defined by 9.16. Then α is a Borel measure on X which is not a Radon measure.

Proof. i) We prove that α is a ccontent. By 2.10 it is to be shown that $\alpha(A \cup B) = \alpha(A) + \alpha(B)$ for all disjoint pairs $A, B \in \mathrm{Bor}(X)$. By 9.16 the values $\alpha(A)$ and $\alpha(B)$ cannot both be $= 1$. The assertion is clear when one of them is $= 1$. So assume that $\alpha(A) = \alpha(B) = 0$. Then there are uncountable members $P, Q \in \mathrm{Cl}(E)$ with $P \subset A'$ and $Q \subset B'$ and hence $P \cap Q \subset A' \cap B' = (A \cup B)'$. By 9.15 it follows that $\alpha(A \cup B) = 0$. ii) By 2.11 α is a cmeasure as soon as it is downward σ continuous. Let $(A_l)_l$ be a sequence in $\mathrm{Bor}(X)$ with $A_l \downarrow A$. We can assume that $\alpha(A_l) = 1 \forall l \in \mathbb{N}$. But then it is immediate from 9.15 that $\alpha(A) = 1$ as well. iii) We have $\alpha(E) = 1$. But for the $K \in \mathrm{Comp}(X)$ with $K \subset E$, that is for the $K \in \mathrm{Comp}(E)$, we have $\alpha(K) = 0$ since K is countable by 9.13.4). Therefore α is not inner regular $\mathrm{Comp}(X)$. Thus the desired example has been obtained.

The Notion of Support and the Decomposition Theorem

Let $\alpha : \mathrm{Bor}(X) \to [0, \infty]$ be a Borel-Radon measure on a Hausdorff topological space X. Then there exists a unique maximal open subset $U \subset X$ such that $\alpha(U) = 0$. In fact, the paving $\{A \in \mathrm{Op}(X) : \alpha(A) = 0\}$ is upward directed to its union $U \in \mathrm{Op}(X)$. From 9.4 it follows that $\alpha(U) = 0$. One defines the closed subset $U' =: \mathrm{Supp}(\alpha)$ to be the **closed support** of α. One notes with dissatisfaction that this concept does not reflect the basic character of a Radon measure, because the lattice $\mathfrak{K} := \mathrm{Comp}(X)$ does not occur at all.

In fact, the notion of support does not seem to be a simple one. At first an example will show that it cannot be defined in the previous manner even for a finite cmeasure $\alpha : \mathrm{Bor}(X) \to [0, \infty[$ on a locally compact Hausdorff space X, unless $\alpha|\mathrm{Op}(X)$ is upward τ continuous, that is by 9.4 and 9.7.2) unless α is a Radon measure. In the introduction of Schwartz [1973] this is one of the reasons which the author invoked in order to plead for the concentration on Borel-Radon measures.

9.18. EXAMPLE. We return to the notations of the last subsection. The restriction $\eta := \alpha|\mathrm{Bor}(E)$ of the cmeasure $\alpha : \mathrm{Bor}(X) \to \{0, 1\}$ defined in 9.17 is a cmeasure $\eta : \mathrm{Bor}(E) \to \{0, 1\}$ on the locally compact Hausdorff space E such that $\eta(E) = 1$. But each point $a \in E$ has an open neighbourhood $U(a) := \cdot, a] \subset E$ such that $\eta(U(a)) = 0$.

The example makes clear that the notion of support requires a certain τ behaviour. At this point we return to the abstract measure theory. It will be seen that the extension theories of chapter II in their conventional inner τ version lead to an adequate notion of support, provided that one accepts a certain shift which a closer look at the Borel-Radon measure situation reveals to be natural: The support will a priori be defined not for cmeasures with certain properties, but for conventional inner τ premeasures. For Borel-Radon measures the result then can be different from the former one, but perhaps reflects their basic character better than before.

The definition reads as follows. Let $\mathfrak{S}$ be a lattice in a nonvoid set X with $\varnothing \in \mathfrak{S}$, and $\varphi : \mathfrak{S} \to [0, \infty[$ be an inner τ premeasure with $\varphi(\varnothing) = 0$ and with maximal inner τ extension $\phi := \varphi_\tau | \mathfrak{C}(\varphi_\tau)$. We know that $\mathfrak{S} \top \mathfrak{S}_\tau \subset \mathfrak{C}(\varphi_\tau)$ and hence $(\mathfrak{S} \top \mathfrak{S}_\tau)\bot \subset \mathfrak{C}(\varphi_\tau)$, and from 6.28 that $\phi | (\mathfrak{S} \top \mathfrak{S}_\tau)\bot$ is upward τ continuous. As before this implies the existence of a unique maximal subset $V \in (\mathfrak{S} \top \mathfrak{S}_\tau)\bot$ such that $\phi(V) = 0$. We define its complement $V' =: \operatorname{supp}(\varphi) \in \mathfrak{S} \top \mathfrak{S}_\tau$ to be the **support** of φ.

9.19. COMPARISON. *Assume that $\alpha : \operatorname{Bor}(X) \to [0, \infty]$ is a Borel-Radon measure on a Hausdorff topological space X. Let $\varphi := \alpha | \mathfrak{K} < \infty$ be its Radon premeasure and $\phi := \varphi_\bullet | \mathfrak{C}(\varphi_\bullet)$, so that $\alpha = \phi | \operatorname{Bor}(X)$. Then the formations*

$$\operatorname{Supp}(\alpha) \quad := \quad U' \in \operatorname{Cl}(X),$$

$$\textit{where } U \in \operatorname{Op}(X) \textit{ is maximal with } \alpha(U) = 0,$$

$$\operatorname{supp}(\alpha | \mathfrak{K}) \quad := \quad V' \in \mathfrak{K} \top \mathfrak{K},$$

$$\textit{where } V \in (\mathfrak{K} \top \mathfrak{K})\bot \textit{ is maximal with } \phi(V) = 0$$

are linked through $\operatorname{Supp}(\alpha) = \overline{\operatorname{supp}(\alpha | \mathfrak{K})}$.

We note that $\operatorname{supp}(\alpha | \mathfrak{K})$ need not be in $\operatorname{Cl}(X)$ (and that we even do not know whether it must be in $\operatorname{Bor}(X)$). However, the two notions are identical in the so-called **k spaces**, defined to mean that $\mathfrak{K} \top \mathfrak{K} = \operatorname{Cl}(X)$. See for example Engelking [1989] pages 152-155.

Proof. We have $\operatorname{Cl}(X) \subset \mathfrak{K} \top \mathfrak{K}$ or $\operatorname{Op}(X) \subset (\mathfrak{K} \top \mathfrak{K})\bot$. i) Therefore $U \subset V$ or $\operatorname{supp}(\varphi) = V' \subset U' = \operatorname{Supp}(\alpha)$. ii) If $A \subset V$ is open then $\alpha(A) = \phi(A) = 0$ and hence $A \subset U$. Thus $\operatorname{Int}(V) \subset U$ or $\operatorname{Supp}(\alpha) = U' \subset (\operatorname{Int}V)' = \overline{V'} = \overline{\operatorname{supp}(\varphi)}$. It follows that $\operatorname{Supp}(\alpha) = \overline{\operatorname{supp}(\varphi)}$.

We continue on the spot with an example that the two supports $\operatorname{Supp}(\alpha)$ and $\operatorname{supp}(\alpha | \mathfrak{K})$ are different. It will be obvious that in the example $\operatorname{supp}(\alpha | \mathfrak{K})$ is the better choice.

9.20. EXAMPLE. Let X and $a \in X$ be as in example 9.8. Define $\alpha : \operatorname{Bor}(X) = \mathfrak{P}(X) \to [0, \infty]$ to be the counting measure for $X \setminus \{a\}$, that is $\alpha(A) = \#(A \cap \{a\}')$ for all $A \subset X$. Then α is a Radon measure, and of course $\alpha = \phi$. The subsets $A \subset X$ with $\alpha(A) = 0$ are precisely $A = \varnothing$ and $A = \{a\}$. Now on the one hand $\mathfrak{K} \top \mathfrak{K} = \mathfrak{P}(X)$ since $\mathfrak{K}$ consists of the finite subsets of X, and hence $V = \{a\}$ or $\operatorname{supp}(\alpha | \mathfrak{K}) = X \setminus \{a\}$. On the other hand $U = \varnothing$ or $\operatorname{Supp}(\alpha) = X$.

We shall see in exercise 11.23 below that the new support defined above fulfils certain natural requirements connected with the notion of support.

We want to pursue the present context for a moment, because this will lead us to the fundamental decomposition theorem. We start with the ab-ovo exercise which has been announced after the basic definition in section 6. It consists of simple verifications.

9.21. EXERCISE. *Let $\varphi : \mathfrak{S} \to [-\infty, \infty[$ be an inner $\bullet$ premeasure and $\alpha : \mathfrak{A} \to \overline{\mathbb{R}}$ be an inner $\bullet$ extension of φ. Fix an $H \in \mathfrak{A}$ with $\alpha(H) > -\infty$,*

and define $\alpha_H : \mathfrak{A} \to \overline{\mathbb{R}}$ *to be* $\alpha_H(A) = \alpha(A \cap H)$ *for* $A \in \mathfrak{A}$. *Then* $\varphi_H :$
$= \alpha_H|\mathfrak{S}$ *is an inner* • *premeasure and* α_H *is an inner* • *extension of* φ_H.

We return to the assumption that $\varphi : \mathfrak{S} \to [0, \infty[$ is an inner τ premeasure on the lattice $\mathfrak{S}$ with $\varnothing \in \mathfrak{S}$ and $\varphi(\varnothing) = 0$, and put $\phi := \varphi_\tau|\mathfrak{C}(\varphi_\tau)$ as before. Fix $A \in \mathfrak{C}(\varphi_\tau)$. We see from 9.21 that $\varphi_A := \phi(\cdot \cap A)|\mathfrak{S}$ is an inner τ premeasure and $\phi(\cdot \cap A) : \mathfrak{C}(\varphi_\tau) \to [0, \infty]$ is an inner τ extension of φ_A (perhaps not the maximal one). Thus we can form the subset $C(A) := \mathrm{supp}(\varphi_A) \in \mathfrak{S}\top\mathfrak{S}_\tau$. Its definition can be rephrased to mean that

$$\text{for } V \in (\mathfrak{S}\top\mathfrak{S}_\tau)\bot : \phi(V \cap A) = 0 \Leftrightarrow V \subset C(A)'.$$

We list some simple properties.

9.22. PROPERTIES. 1) *If* $A \in \mathfrak{S}\top\mathfrak{S}_\tau$ *then* $C(A) \subset A$. 2) *For* $A \subset B$ *in* $\mathfrak{C}(\varphi_\tau)$ *we have* $C(A) \subset C(B)$. 3) *For* $A \in \mathfrak{C}(\varphi_\tau)$ *we have* $C\big(A \cap C(A)\big) = C(A)$.

Proof. 1) follows from the definition applied to $V := A'$. 2) For $V := C(B)' \in (\mathfrak{S}\top\mathfrak{S}_\tau)\bot$ we have $\phi(V \cap B) = 0$ and hence $\phi(V \cap A) = 0$. Therefore $V \subset C(A)'$. 3) To be shown is $C\big(A \cap C(A)\big) \supset C(A)$. For $V := C\big(A \cap C(A)\big)' \in (\mathfrak{S}\top\mathfrak{S}_\tau)\bot$ we have $\phi(V \cap A \cap C(A)) = 0$. Furthermore $\phi(A \cap C(A)') = 0$ and hence $\phi(V \cap A \cap C(A)') = 0$. Therefore $\phi(V \cap A) = 0$. It follows that $V \subset C(A)'$ as claimed.

We define a subset $A \in \mathfrak{C}(\varphi_\tau)$ to be **full** iff $A \subset C(A)$ and $0 < \phi(A) < \infty$. The basic fact on these subsets is as follows.

9.23. REMARK. *Each* $A \subset X$ *with* $\varphi_\tau(A) > 0$ *contains a full subset* $K \in \mathfrak{S}_\tau$.

Proof. Since φ_τ is inner regular $\mathfrak{S}_\tau$ there exists $S \in \mathfrak{S}_\tau$ with $S \subset A$ and $\varphi_\tau(S) > 0$. Let $K := S \cap C(S) \in \mathfrak{S}_\tau$. From 9.22.3) we have $C(K) = C(S) \supset S \cap C(S) = K$, and hence from 9.22.1) even $C(K) = K$. Furthermore $\phi(S \cap C(S)') = 0$ implies that $0 < \phi(K) = \phi(S) < \infty$. Thus K is as required.

We come to the basic definition. Let $\alpha : \mathfrak{A} \to [0, \infty]$ be a nonzero cmeasure on a σ algebra. One defines a **decomposition** for α to be a paving $\mathfrak{M} \subset \mathfrak{A}$ of pairwise disjoint subsets $M \in \mathfrak{A}$ with $0 < \alpha(M) < \infty$, which is such that

$$\text{each } A \in \mathfrak{A} \text{ with } 0 < \alpha(A) < \infty \text{ fulfils } \alpha(A \cap M) > 0 \text{ for some } M \in \mathfrak{M}.$$

It is obvious that there exist decompositions for α whenever $X \in [\alpha < \infty]^\sigma$.

Our aim is to prove that for the present $\phi : \mathfrak{C}(\varphi_\tau) \to [0, \infty]$ there exist decompositions, and even decompositions with important further properties, provided that ϕ is outer regular $(\mathfrak{S}\top\mathfrak{S}_\tau)\bot$ at $\mathfrak{S}_\tau$. It is equivalent to require that each $K \in \mathfrak{S}_\tau$ be contained in some $U \in (\mathfrak{S}\top\mathfrak{S}_\tau)\bot$ with $\phi(U) < \infty$, which for short could be named local finiteness; compare exercise 9.10.

9.24. THEOREM. *Let $\varphi : \mathfrak{S} \to [0, \infty[$ be a nonzero inner τ premeasure with $\varphi(\varnothing) = 0$, and let $\phi := \varphi_\tau | \mathfrak{C}(\varphi_\tau)$ be outer regular $(\mathfrak{S} \top \mathfrak{S}_\tau) \bot$ at $\mathfrak{S}_\tau$. Assume that $\mathfrak{M} \subset \mathfrak{C}(\varphi_\tau)$ is a maximal paving of pairwise disjoint full members of $\mathfrak{C}(\varphi_\tau)$* (this holds true in particular from 9.23 when $\mathfrak{M} \subset \mathfrak{S}_\tau$ is a maximal paving of pairwise disjoint full members of $\mathfrak{S}_\tau$). *Then $\mathfrak{M}$ is a decomposition for ϕ. If $A \in \mathfrak{C}(\varphi_\tau)$ is contained in some $U \in (\mathfrak{S} \top \mathfrak{S}_\tau) \bot$ with $\phi(U) < \infty$, then $A \cap M \neq \varnothing$ for at most countably many $M \in \mathfrak{M}$.*

9.25. ADDENDUM. 1) *For each family $(P(M))_{M \in \mathfrak{M}}$ in $\mathfrak{C}(\varphi_\tau)$ with $P(M) \subset M \; \forall M \in \mathfrak{M}$ we have $\bigcup_{M \in \mathfrak{M}} P(M) \in \mathfrak{C}(\varphi_\tau)$. In particular $\bigcup_{M \in \mathfrak{M}} M \in \mathfrak{C}(\varphi_\tau)$.*
2) *The complement $D := \left(\bigcup_{M \in \mathfrak{M}} M \right)' \in \mathfrak{C}(\varphi_\tau)$ has $\phi(D) = 0$.*

Proof of 9.24 and 9.25. i) For $A \in \mathfrak{C}(\varphi_\tau)$ with $\phi(A) < \infty$ it is clear that $\phi(A \cap M) > 0$ for at most countably many $M \in \mathfrak{M}$. ii) If $U \in (\mathfrak{S} \top \mathfrak{S}_\tau) \bot$ then $\phi(U \cap M) = 0$ implies that $U \subset (C(M))' \subset M'$ or $U \cap M = \varnothing$. Thus for $U \in (\mathfrak{S} \top \mathfrak{S}_\tau) \bot$ with $\phi(U) < \infty$ we have $U \cap M \neq \varnothing$ for at most countably many $M \in \mathfrak{M}$. iii) By assumption each $K \in \mathfrak{S}_\tau$ is contained in some $U \in (\mathfrak{S} \top \mathfrak{S}_\tau) \bot$ with $\phi(U) < \infty$. By ii) hence $K \cap M \neq \varnothing$ for at most countably many $M \in \mathfrak{M}$.

We first prove 9.25. 1) Let $P := \bigcup_{M \in \mathfrak{M}} P(M)$. By 6.21 we have to prove that $S \cap P \in \mathfrak{C}(\varphi_\tau)$ for each $S \in \mathfrak{S}$. From iii) we have $S \cap M = \varnothing$ and hence $S \cap P(M) = \varnothing$ except for at most countably many $M \in \mathfrak{M}$. Thus the union in $S \cap P = \bigcup_{M \in \mathfrak{M}} S \cap P(M)$ is at most countable. This implies $S \cap P \in \mathfrak{C}(\varphi_\tau)$.
2) We see from 1) that $D \in \mathfrak{C}(\varphi_\tau)$. By 9.23 the maximality of $\mathfrak{M}$ enforces that $\phi(D) = 0$.

We turn to the proof of 9.24. Let $A \in \mathfrak{C}(\varphi_\tau)$ with $0 < \phi(A) < \infty$. Then there exists $K \in \mathfrak{S}_\tau$ with $K \subset A$ and $\phi(K) > 0$. By iii) the union in $K = (K \cap D) \cup \bigcup_{M \in \mathfrak{M}} (K \cap M)$ is at most countable. From 9.25.2) we obtain an $M \in \mathfrak{M}$ such that $\phi(K \cap M) > 0$ and hence $\phi(A \cap M) > 0$. The second assertion is obvious from ii).

We shall present in 13.39 an example where ϕ is not outer regular $(\mathfrak{S} \top \mathfrak{S}_\tau) \bot$ at $\mathfrak{S}_\tau$, but $X \in [\phi < \infty]^\sigma$, so that there exist decompositions for ϕ. It will be a continuous counterpart to the example in 9.8 and 9.20.

9.26. BIBLIOGRAPHICAL NOTE. For rich cmeasures like the present ones the existence of a decomposition is equivalent to several fundamental properties. This context has been developed in Kölzow [1968]. The present decomposition theorem 9.24 unifies the known versions of the former topological decomposition theorem named after Godement-Bourbaki and the abstract decomposition theorem due to Kölzow [1966]. The Godement-Bourbaki theorem has been raised in Schwartz [1973] section I.6 to the locally finite Borel-Radon measures on arbitrary Hausdorff topological spaces. The Kölzow theorem, for which we also refer to Floret [1981] appendix and

Leinert [1995] chapter 14, had no equivalent development so far. It is a non-trivial task to describe the precise class of cmeasures to which this theorem applies. The presentations known to the author are based on the so-called abstract Bourbaki integral, which is the nonsequential counterpart of the traditional Daniell-Stone elementary integral, and is likewise in the historical headwaters of the present development in chapter V below. One proved the existence of decompositions for the maximal cmeasures constructed in the outer sense from these abstract Bourbaki integrals. This class of cmeasures has been characterized in Kölzow [1965][1967], in complicated terms which could not replace their former descent.

For the sake of comparison we add a little anticipation of our chapter V. This chapter is in the inner sense, with the result that we come at once to the so-called essential measures. The class of elementary integrals will be much more comprehensive than in the traditional situation. The class of relevant cmeasures will have a simple characterization for both $\bullet = \sigma\tau$: We shall see in 14.25.1) that it consists of the maximal inner $\bullet$ extensions $\phi := \varphi_\bullet | \mathfrak{C}(\varphi_\bullet)$ of the inner $\bullet$ premeasures $\varphi : \mathfrak{S} \to [0, \infty[$ with $\varphi(\varnothing) = 0$. Moreover 14.25.3) will show that these ϕ are outer regular $(\mathfrak{S}\top\mathfrak{S}_\bullet)\bot$ at $\mathfrak{S}_\bullet$ whenever they come from the traditional situation of a Stonean lattice subspace. Thus the present decomposition theorem 9.24 contains both the Godement-Bourbaki theorem of Schwartz [1973] and the theorem of Kölzow [1966], and to see the latter fact does not require the characterization theorem due to Kölzow [1965][1967].

10. The Choquet Capacitability Theorem

The theme of this section is a central topic in the domain of the so-called non-additive set functions. The main result is an extension and also simplification of the famous capacitability theorem due to Choquet [1959]. The theorem itself does not need the main results of the extension theories of chapter II. It is more an application of the idea to form the new envelopes φ^σ and φ_σ for isotone set functions $\varphi : \mathfrak{S} \to \overline{\mathbb{R}}$ on lattices $\mathfrak{S}$. Then we combine the theorem with results of chapter II in order to obtain classical applications to measures in the spirit of the present text. Thus the section wants to contribute to the unification of the theories of additive and non-additive set functions.

Suslin and Co-Suslin Sets

We start with the usual notations. For $n \in \mathbb{N}$ as usual $\mathbb{N}^n$ consists of the sequences $\lambda = (\lambda_1, \cdots, \lambda_n)$ of natural numbers $\lambda_1, \cdots, \lambda_n$. Then $\mathbb{N}^\infty$ denotes the disjoint union of the $\mathbb{N}^n \, \forall n \in \mathbb{N}$. Thus $\mathbb{N}^\infty$ is countable. At last $\mathbb{N}^{\mathbb{N}}$ is defined to consist of the infinite sequences $\alpha = (\alpha_l)_l$ of natural numbers $\alpha_l \, \forall l \in \mathbb{N}$. Thus $\mathbb{N}^{\mathbb{N}}$ is uncountable. For $\alpha \in \mathbb{N}^{\mathbb{N}}$ we write $\alpha|n :=$ $(\alpha_1, \cdots, \alpha_n) \in \mathbb{N}^n \, \forall n \in \mathbb{N}$.

Next let X be a nonvoid set. For each family $(A(\lambda))_{\lambda \in \mathbb{N}^\infty}$ of subsets $A(\lambda) \subset X \; \forall \lambda \in \mathbb{N}^\infty$ we form

the kernel $\quad \bigwedge\limits_{\lambda \in \mathbb{N}^\infty} A(\lambda) := \bigcup\limits_{\alpha \in \mathbb{N}^\mathbb{N}} \bigcap\limits_{n=1}^{\infty} A(\alpha|n) \subset X$, and

the cokernel $\quad \bigvee\limits_{\lambda \in \mathbb{N}^\infty} A(\lambda) := \bigcap\limits_{\alpha \in \mathbb{N}^\mathbb{N}} \bigcup\limits_{n=1}^{\infty} A(\alpha|n) \subset X$.

If $\mathfrak{S}$ is a set system in X then

$\mathfrak{S}^{\#}$ consists of the kernels $\quad \bigwedge\limits_{\lambda \in \mathbb{N}^\infty} A(\lambda)$ of all families $(A(\lambda))_{\lambda \in \mathbb{N}^\infty}$ in $\mathfrak{S}$,

$\mathfrak{S}_{\#}$ consists of the cokernels $\quad \bigvee\limits_{\lambda \in \mathbb{N}^\infty} A(\lambda)$ of all families $(A(\lambda))_{\lambda \in \mathbb{N}^\infty}$ in $\mathfrak{S}$.

The subsets $A \in \mathfrak{S}^{\#}$ are called the **Suslin sets** for $\mathfrak{S}$, and the $A \in \mathfrak{S}_{\#}$ the **co-Suslin sets** for $\mathfrak{S}$. As in 1.5.2) one verifies that $(\mathfrak{S}^{\#})\bot = (\mathfrak{S}\bot)_{\#}$.

10.1. PROPERTIES. 1) $\mathfrak{S} \subset \mathfrak{S}^{\#}$. 2) $\mathfrak{S}^{\sigma} \subset \mathfrak{S}^{\#}$. 3) $\mathfrak{S}_{\sigma} \subset \mathfrak{S}^{\#}$. 4) $\mathfrak{S}^{\#} \subset (\mathfrak{S}_{\sigma} \sqsubset \mathfrak{S}^{\sigma})$. *Of course the same properties hold true for* $\mathfrak{S}_{\#}$.

Proof. 1) If $A \in \mathfrak{S}$ then take the family $(A(\lambda))_{\lambda \in \mathbb{N}^\infty}$ with $A(\lambda) := A$ for all $\lambda \in \mathbb{N}^\infty$. 2) If $(A_l)_l$ in $\mathfrak{S}$ then take the family $(A(\lambda))_{\lambda \in \mathbb{N}^\infty}$ with $A(\lambda) := A_{\lambda_1}$ for all $\lambda \in \mathbb{N}^\infty$. Then

$$\bigwedge\limits_{\lambda \in \mathbb{N}^\infty} A(\lambda) = \bigcup\limits_{\alpha \in \mathbb{N}^\mathbb{N}} A_{\alpha_1} = \bigcup\limits_{l=1}^{\infty} A_l.$$

3) If $(A_l)_l$ in $\mathfrak{S}$ then take the family $(A(\lambda))_{\lambda \in \mathbb{N}^\infty}$ with $A(\lambda) = A_n$ for all $\lambda \in \mathbb{N}^n$. Then

$$\bigwedge\limits_{\lambda \in \mathbb{N}^\infty} A(\lambda) = \bigcup\limits_{\alpha \in \mathbb{N}^\mathbb{N}} \bigcap\limits_{n=1}^{\infty} A_n = \bigcap\limits_{n=1}^{\infty} A_n.$$

4) is clear since $\mathbb{N}^\infty$ is countable.

The next assertion is much harder. For the sake of completeness we include a proof, although we do not need its full power.

10.2. PROPOSITION. $\mathfrak{S}^{\#\#} = \mathfrak{S}^{\#}$. *Hence of course* $\mathfrak{S}_{\#\#} = \mathfrak{S}_{\#}$.

10.3. CONSEQUENCE. $\mathfrak{S}^{\#}$ *and* $\mathfrak{S}_{\#}$ *are* σ *lattices.*

Proof of 10.2. Fix $A \in \mathfrak{S}^{\#\#}$. To be shown is $A \in \mathfrak{S}^{\#}$. By definition there is a family $(A(\sigma))_{\sigma \in \mathbb{N}^\infty}$ in $\mathfrak{S}^{\#}$ such that

$$A = \bigcup\limits_{\alpha \in \mathbb{N}^\mathbb{N}} \bigcap\limits_{p=1}^{\infty} A(\alpha|p).$$

Thus for each $\sigma \in \mathbb{N}^\infty$ there is a family $(A_{\lambda}^{\sigma})_{\lambda \in \mathbb{N}^\infty}$ in $\mathfrak{S}$ such that

$$A(\sigma) = \bigcup\limits_{\beta \in \mathbb{N}^\mathbb{N}} \bigcap\limits_{q=1}^{\infty} A_{\beta|q}^{\sigma}.$$

It follows that

$$A = \bigcup_{\alpha \in \mathbb{N}^{\mathbb{N}}} \bigcap_{p=1}^{\infty} \bigcup_{\beta \in \mathbb{N}^{\mathbb{N}}} \bigcap_{q=1}^{\infty} A_{\beta|q}^{\alpha|p}.$$

We have to transfer this expression into the required form. 1) We start with an auxiliary formula. Let $(M_\beta^p)_{p \in \mathbb{N}, \beta \in \mathbb{N}^{\mathbb{N}}}$ be a family of subsets of X. Then

$$\bigcap_{p=1}^{\infty} \bigcup_{\beta \in \mathbb{N}^{\mathbb{N}}} M_\beta^p = \bigcup_{\tau \in \mathbb{N}^{\mathbb{N} \times \mathbb{N}}} \bigcap_{p=1}^{\infty} M_{\tau(p,\cdot)}^p.$$

In fact, an element $x \in X$ is a member of the left side iff for each $p \in \mathbb{N}$ there exists a sequence $\tau(p, \cdot) \in \mathbb{N}^{\mathbb{N}}$ with $x \in M_{\tau(p,\cdot)}^p$, that is iff there exists a function $\tau \in \mathbb{N}^{\mathbb{N} \times \mathbb{N}}$ such that $x \in \bigcap_{p=1}^{\infty} M_{\tau(p,\cdot)}^p$. From this formula we obtain

$$A = \bigcup_{\alpha \in \mathbb{N}^{\mathbb{N}}} \bigcup_{\tau \in \mathbb{N}^{\mathbb{N} \times \mathbb{N}}} \bigcap_{p=1}^{\infty} \bigcap_{q=1}^{\infty} A_{(\tau(p,1),\cdots,\tau(p,q))}^{\alpha|p}.$$

2) Let $\theta : \mathbb{N} \times \mathbb{N} \to \mathbb{N}$ be a fixed bijection with the properties that

$\forall p \in \mathbb{N} : \theta(p, \cdot)$ is a strictly increasing function $\mathbb{N} \to \mathbb{N}$,
$\forall q \in \mathbb{N} : \theta(\cdot, q)$ is a strictly increasing function $\mathbb{N} \to \mathbb{N}$.

For example, we can enumerate the members of $\mathbb{N} \times \mathbb{N}$ in consecutive diagonals; for $(p, q), (p', q') \in \mathbb{N} \times \mathbb{N}$ we then have $p + q < p' + q' \Rightarrow \theta(p, q) < \theta(p', q')$, which implies the required properties. It follows that $p, q \leqq \theta(p, q)$ for $(p, q) \in \mathbb{N} \times \mathbb{N}$. We also need the inverse bijection $\vartheta := \theta^{-1} : \mathbb{N} \times \mathbb{N} \leftarrow \mathbb{N}$, which we write in the form $\vartheta(n) = (\xi(n), \eta(n)) \, \forall n \in \mathbb{N}$. 3) We next form the maps

$I : \mathbb{N}^{\mathbb{N}} \times \mathbb{N}^{\mathbb{N} \times \mathbb{N}} \to \mathbb{N}^{\mathbb{N}}$, defined to be
$I : (\alpha, \tau) \longmapsto I(\alpha, \tau) = \lambda : \lambda_n = \theta(\alpha_n, \tau(\vartheta(n)) \, \forall n \in \mathbb{N}$; and

$J : \mathbb{N}^{\mathbb{N}} \to \mathbb{N}^{\mathbb{N}} \times \mathbb{N}^{\mathbb{N} \times \mathbb{N}}$, defined to be
$J : \lambda \longmapsto J(\lambda) = (\alpha, \tau) : \alpha_n = \xi(\lambda_n) \, \forall n \in \mathbb{N}$,
$\qquad\qquad\qquad \tau(p, q) = \eta(\lambda_{\theta(p,q)}) \, \forall (p, q) \in \mathbb{N} \times \mathbb{N}.$

We claim that I and J are bijections which are inverse to each other. In fact, for $(\alpha, \tau) \in \mathbb{N}^{\mathbb{N}} \times \mathbb{N}^{\mathbb{N} \times \mathbb{N}}$ and $\lambda \in \mathbb{N}^{\mathbb{N}}$ the relation $I(\alpha, \tau) = \lambda$ means that $\theta(\alpha_n, \tau(\vartheta(n))) = \lambda_n \, \forall n \in \mathbb{N}$, and hence is equivalent to

$\alpha_n = \xi(\lambda_n) \, \forall n \in \mathbb{N}$ and
$\tau(\vartheta(n)) = \eta(\lambda_n) \, \forall n \in \mathbb{N}$ or $\tau(p, q) = \eta(\lambda_{\theta(p,q)}) \, \forall (p, q) \in \mathbb{N} \times \mathbb{N}$,

that is to $(\alpha, \tau) = J(\lambda)$.

4) We now define $E(\lambda, n) \in \mathfrak{S}$ for $\lambda \in \mathbb{N}^{\mathbb{N}}$ and $n \in \mathbb{N}$ as follows. If

$\lambda = I(\alpha, \tau)$ with $(\alpha, \tau) \in \mathbb{N}^{\mathbb{N}} \times \mathbb{N}^{\mathbb{N} \times \mathbb{N}}$,
$n = \theta(p, q)$ with $(p, q) \in \mathbb{N} \times \mathbb{N}$,

then $E(\lambda, n) := A^{\alpha|p}_{(\tau(p,1),\cdots,\tau(p,q))} \in \mathfrak{S}$. By the final formula of 1) it follows that

$$A = \bigcup_{\lambda \in \mathbb{N}^{\mathbb{N}}} \bigcap_{n=1}^{\infty} E(\lambda, n).$$

We next claim that for $n \in \mathbb{N}$ and $\lambda, \lambda' \in \mathbb{N}^{\mathbb{N}}$ with $\lambda_l = \lambda'_l \; \forall l = 1, \cdots, n$ we have $E(\lambda, n) = E(\lambda', n)$. In fact, let $(p, q) \in \mathbb{N} \times \mathbb{N}$ correspond to n, and $(\alpha, \tau), (\alpha', \tau') \in \mathbb{N}^{\mathbb{N}} \times \mathbb{N}^{\mathbb{N} \times \mathbb{N}}$ correspond to λ, λ'. Then

for $l \leq p : l \leq p \leq \theta(p, q) = n$ and hence $\alpha_l = \alpha'_l$;
for $l \leq q : \theta(p, l) \leq \theta(p, q) = n$ and hence $\tau(p, l) = \tau'(p, l)$;

therefore $E(\lambda, n) = E(\lambda', n)$ as claimed. 5) The last assertion allows us to define the family $(B(\sigma))_{\sigma \in \mathbb{N}^\infty}$ in $\mathfrak{S}$, in that for $n \in \mathbb{N}$ and $\sigma = (\sigma_1, \cdots, \sigma_n) \in \mathbb{N}^n$ we put

$$B(\sigma) := E((\sigma_1, \cdots, \sigma_n, \cdots), n) \text{ with an arbitrary continuation of } \sigma.$$

Thus $E(\lambda, n) = B(\lambda|n)$ for all $n \in \mathbb{N}$ and $\lambda \in \mathbb{N}^{\mathbb{N}}$. It follows that

$$A = \bigcup_{\lambda \in \mathbb{N}^{\mathbb{N}}} \bigcap_{n=1}^{\infty} B(\lambda|n) = \bigwedge_{\sigma \in \mathbb{N}^\infty} B(\sigma) \in \mathfrak{S}^{\#}.$$

The proof is complete.

The Extended Choquet Theorem

Let as above X be a nonvoid set. The present subsection considers an isotone set function $\phi : \mathfrak{P}(X) \to \overline{\mathbb{R}}$. There is also a lattice $\mathfrak{S}$ in X. We prepare the main theorem with some formal relations between ϕ and $\mathfrak{S}$.

On $\mathfrak{S}$ we have $\phi = (\phi|\mathfrak{S})_\star$, but $\phi \leq (\phi|\mathfrak{S})_\sigma$, with $=$ iff $\phi|\mathfrak{S}$ is downward σ continuous. On the full power set $\mathfrak{P}(X)$ we have $(\phi|\mathfrak{S})_\star \leq \phi$, but $(\phi|\mathfrak{S})_\sigma \leq \phi$ is likewise restricted to a special situation. It will be described below.

10.4. REMARK. *Let $\mathfrak{S}$ be a lattice in X. Then the following are equivalent.* 1) $(\phi|\mathfrak{S})_\sigma \leq \phi$. 2) $(\phi|\mathfrak{S})_\sigma \leq (\phi|\mathfrak{S}_\sigma)_\star$. *In this connection note the obvious relation* $(\phi|\mathfrak{S}_\sigma)_\star \leq \phi$. 3) *For each sequence $(S_l)_l$ in $\mathfrak{S}$ with $S_l \downarrow D$ (which need not be in $\mathfrak{S}$) we have* $\phi(S_l) \downarrow \phi(D)$.

Proof. 1)$\Rightarrow$3) Let $(S_l)_l$ be a sequence in $\mathfrak{S}$ with $S_l \downarrow D$. From the definition of $(\phi|\mathfrak{S})_\sigma(D)$ and from 1) we obtain

$$\phi(D) \leq \lim_{l \to \infty} \phi(S_l) \leq (\phi|\mathfrak{S})_\sigma(D) \leq \phi(D).$$

3)$\Rightarrow$2) Fix $A \subset X$ with $(\phi|\mathfrak{S})_\sigma(A) > -\infty$. For each sequence $(S_l)_l$ in $\mathfrak{S}$ with $S_l \downarrow$ some $D \subset A$ it follows from 3) and from $D \in \mathfrak{S}_\sigma$ that

$$\lim_{l \to \infty} \phi(S_l) = \phi(D) \leq (\phi|\mathfrak{S}_\sigma)_\star(A).$$

2)$\Rightarrow$1) is obvious.

We proceed to the main theorem.

10.5. THEOREM. *Assume that* $\phi : \mathfrak{P}(X) \to \overline{\mathbb{R}}$ *is isotone and upward* σ *continuous. Let* $\mathfrak{S}$ *be a lattice in* X. *Then*

$$\phi(A) \leqq (\phi|\mathfrak{S})_\sigma(A) \quad \text{for all } A \in \mathfrak{S}^\#.$$

This is an extension of the classical Choquet theorem [1959]. In fact, let $\mathfrak{S}$ be a lattice. A **Choquet capacity for** $\mathfrak{S}$ is defined to be an isotone set function $\phi : \mathfrak{P}(X) \to \overline{\mathbb{R}}$ which is upward σ continuous and satisfies the equivalent conditions of 10.4 for the lattice $\mathfrak{S}$. The above theorem then implies that

$$\phi(A) = (\phi|\mathfrak{S})_\sigma(A) = (\phi|\mathfrak{S}_\sigma)_\star(A) \quad \text{for all } A \in \mathfrak{S}^\#.$$

The classical Choquet theorem is $\phi(A) = (\phi|\mathfrak{S}_\sigma)_\star(A)$ for all $A \in \mathfrak{S}^\#$, under the additional restriction that $\varnothing \in \mathfrak{S}$. Thus the introduction of the function $(\phi|\mathfrak{S})_\sigma$ produces a more comprehensive and simpler result, above all because no connection between ϕ and $\mathfrak{S}$ has been assumed. The proof which follows is an adaptation of an old proof due to Choquet [1959].

10.6. CONSTRUCTION. *Let* $(A(\lambda))_{\lambda \in \mathbb{N}^\infty}$ *be a family of subsets of* X. *Define the family* $(B(\lambda))_{\lambda \in \mathbb{N}^\infty}$ *to be*

$$B(\lambda) := \bigcup_{\tau \in \mathbb{N}^p, \tau \leqq \lambda} \bigcap_{l=1}^{p} A(\tau_1, \cdots, \tau_l) \quad \text{for } \lambda \in \mathbb{N}^p \text{ with } p \in \mathbb{N}.$$

Then

$$0) \quad \bigwedge_{\lambda \in \mathbb{N}^\infty} B(\lambda) = \bigwedge_{\lambda \in \mathbb{N}^\infty} A(\lambda).$$

Furthermore $(B(\lambda))_{\lambda \in \mathbb{N}^\infty}$ *has the monotonicity properties*

1) $B(\lambda, s) \subset B(\lambda)$ *for* $\lambda \in \mathbb{N}^\infty$ *and* $s \in \mathbb{N}$;
2) $B(\sigma) \supset B(\tau)$ *for* $\sigma, \tau \in \mathbb{N}^p$ *with* $\sigma \geqq \tau \ \forall p \in \mathbb{N}$.

If $(A(\lambda))_{\lambda \in \mathbb{N}^\infty}$ *itself has these properties* 1)2) *then* $B(\lambda) = A(\lambda) \ \forall \lambda \in \mathbb{N}^\infty$.

Proof. 1)2) and the final assertion are obvious. Furthermore

$$\text{for } \alpha \in \mathbb{N}^\mathbb{N} \text{ and } p \in \mathbb{N} : B(\alpha|p) \supset \bigcap_{l=1}^{p} A(\alpha|l) \supset \bigcap_{l=1}^{\infty} A(\alpha|l),$$

which implies the inclusion $\supset$ in 0). It remains to prove $\subset$ in 0). For this purpose fix $x \in \bigwedge_{\lambda \in \mathbb{N}^\infty} B(\lambda)$. Thus by definition there exists $\alpha \in \mathbb{N}^\mathbb{N}$ such that $x \in \bigcap_{p=1}^{\infty} B(\alpha|p)$. Hence for each $p \in \mathbb{N}$ there exists $(t_1^p, \cdots, t_p^p) \in \mathbb{N}^p$ with $t_l^p \leqq \alpha_l \ \forall 1 \leqq l \leqq p$ and $x \in A(t_1^p, \cdots, t_l^p) \ \forall 1 \leqq l \leqq p$.

i) We construct via induction for $l \in \mathbb{N}$ an index $\tau_l \leqq \alpha_l$ and an infinite subset $N(l) \subset \{l, l+1, \cdots\}$ such that $\mathbb{N} \supset N(1) \supset \cdots \supset N(l) \supset \cdots$ and that $t_l^p = \tau_l$ for all $p \in N(l)$. The step $l = 1$: We have $t_1^p \leqq \alpha_1 \ \forall p \in \mathbb{N}$. Hence there exists an index $\tau_1 \leqq \alpha_1$ and an infinite $N(1) \subset \mathbb{N}$ such that $t_1^p = \tau_1 \ \forall p \in N(1)$. The step $1 \leqq l \Rightarrow l+1$: We have $t_{l+1}^p \leqq \alpha_{l+1} \ \forall p \geqq l+1$, hence $\forall p \in N(l) \cap \{l+1, \cdots\}$. Therefore there exists an index $\tau_{l+1} \leqq \alpha_{l+1}$ and an infinite $N(l+1) \subset N(l) \cap \{l+1, \cdots\}$ such that $t_{l+1}^p = \tau_{l+1} \; \forall p \in N(l+1)$. This completes the inductive construction.

ii) Now let $\tau = (\tau_l)_l \in \mathbb{N}^\mathbb{N}$. Fix $n \in \mathbb{N}$. For $p \in N(n)$ then on the one hand $p \geqq n$. On the other hand

$$\text{for } 1 \leq l \leq n : p \in N(l) \text{ and hence } t_l^p = \tau_l;$$

therefore $(t_1^p, \cdots, t_n^p) = (\tau_1, \cdots, \tau_n)$. We combine these facts to obtain $x \in A(\tau_1, \cdots, \tau_n) = A(\tau|n)$ for each $n \in \mathbb{N}$. Thus $x \in \bigcap_{n=1}^{\infty} A(\tau|n)$ and hence $x \in \bigwedge_{\lambda \in \mathbb{N}^\infty} A(\lambda)$. This is the assertion.

10.7. THEOREM. *Assume that* $\phi : \mathfrak{P}(X) \to \overline{\mathbb{R}}$ *is isotone and upward* σ *continuous. Let* $(A(\lambda))_{\lambda \in \mathbb{N}^\infty}$ *be a family of subsets of* X. *Consider the family* $(B(\lambda))_{\lambda \in \mathbb{N}^\infty}$ *after 10.6, so that*

$$A := \bigwedge_{\lambda \in \mathbb{N}^\infty} A(\lambda) = \bigwedge_{\lambda \in \mathbb{N}^\infty} B(\lambda), \quad \text{and } B(\alpha|l) \downarrow\subset A \text{ for each } \alpha \in \mathbb{N}^\mathbb{N}.$$

Then

$$\phi(A) \leqq \sup\{\lim_{l \to \infty} \phi(B(\alpha|l)) : \alpha \in \mathbb{N}^\mathbb{N}\}.$$

It is obvious that 10.7 implies 10.5, and in fact furnishes a sharpened version of 10.5: Let $A \in \mathfrak{S}^\#$. Fix any family $(A(\lambda))_{\lambda \in \mathbb{N}^\infty}$ in $\mathfrak{S}$ with $A = \bigwedge_{\lambda \in \mathbb{N}^\infty} A(\lambda)$, and let $(B(\lambda))_{\lambda \in \mathbb{N}^\infty}$ be as above. Then for each $\alpha \in \mathbb{N}^\mathbb{N}$ the sequence $(B(\alpha|l))_l$ is in $\mathfrak{S}$ and fulfils $B(\alpha|l) \downarrow\subset A$. Therefore the second member in the assertion of 10.7 is $\leqq (\phi|\mathfrak{S})_\sigma(A)$. It is shaped like the definition of $(\phi|\mathfrak{S})_\sigma(A)$, except that its supremum is not extended over all sequences $(S_l)_l$ in $\mathfrak{S}$ with $S_l \downarrow\subset A$, but is restricted to the particular sequences $(B(\alpha|l))_l$. This is an essential fortification of 10.5.

Proof of 10.7. i) We form

$$D(\alpha) := \bigcap_{l=1}^{\infty} B(\alpha|l) \quad \text{for } \alpha \in \mathbb{N}^\mathbb{N}.$$

From 10.6.2) then $D(\alpha) \supset D(\beta)$ for $\alpha, \beta \in \mathbb{N}^\mathbb{N}$ with $\alpha \geqq \beta$. Next we form

$$U(\lambda) := \bigcup_{\alpha \in \mathbb{N}^\mathbb{N}, \alpha|p = \lambda} D(\alpha) \quad \text{for } \lambda \in \mathbb{N}^p \text{ with } p \in \mathbb{N}.$$

It follows that $U(\sigma) \supset U(\tau)$ for $\sigma, \tau \in \mathbb{N}^p$ with $\sigma \geqq \tau$. Furthermore $U(\lambda) \subset B(\lambda)$ for all $\lambda \in \mathbb{N}^\infty$, because for $\lambda \in \mathbb{N}^p$ and $\alpha \in \mathbb{N}^\mathbb{N}$ with $\alpha|p = \lambda$ we have $D(\alpha) \subset B(\alpha|p) = B(\lambda)$. ii) By definition we have

$$U(\lambda) = \bigcup_{l=1}^{\infty} U(\lambda, l) \quad \text{for } \lambda \in \mathbb{N}^\infty,$$

and hence from i) $U(\lambda, l) \uparrow U(\lambda)$. Likewise we have

$$A = \bigcup_{\alpha \in \mathbb{N}^{\mathbb{N}}} D(\alpha) = \bigcup_{l=1}^{\infty} U(l),$$

and hence from i) $U(l) \uparrow A$. iii) To prove the assertion we can assume that $\phi(A) > -\infty$. We fix a real $c < \phi(A)$. In view of ii) we can find via induction numbers $\tau_l \in \mathbb{N}$ $\forall l \in \mathbb{N}$ such that $\phi\big(U(\tau_1, \cdots, \tau_p)\big) > c$ $\forall p \in \mathbb{N}$. Thus $\tau = (\tau_l)_l \in \mathbb{N}^{\mathbb{N}}$ satisfies $\phi\big(U(\tau|p)\big) > c$ $\forall p \in \mathbb{N}$. From i) it follows that $\phi\big(B(\tau|p)\big) > c$ $\forall p \in \mathbb{N}$ and hence $\lim_{l \to \infty} \phi\big(B(\tau|l)\big) \geqq c$. This proves the assertion.

We leave as an exercise the result which follows from 10.5 via the upside-down transform method.

10.8. EXERCISE. *Assume that $\phi : \mathfrak{P}(X) \to \overline{\mathbb{R}}$ is isotone and downward σ continuous. Let $\mathfrak{S}$ be a lattice in X. Then*

$$\phi(A) \geqq (\phi|\mathfrak{S})^\sigma(A) \quad \text{for all } A \in \mathfrak{S}_\#.$$

Combination with Basic Properties of the σ Envelopes

We start with a simple but important remark, and then turn to the fundamental combinations of the main theorem with the sequential continuity theorems 4.7 and 6.7.

10.9. REMARK. *Let $\phi : \mathfrak{P}(X) \to \overline{\mathbb{R}}$ be a Choquet capacity for the lattice $\mathfrak{S}$. Assume that $\phi|\mathfrak{S}$ is $< \infty$ and supermodular. Then $\phi|\mathfrak{S}^\#$ is supermodular $+$.*

Proof. By 6.3.5) the set function $(\phi|\mathfrak{S})_\sigma$ is supermodular $\dotplus$. Thus we obtain for $A, B \in \mathfrak{S}^\#$ from 10.5 and 10.4

$$\phi(A) \dotplus \phi(B) \leqq (\phi|\mathfrak{S})_\sigma(A) \dotplus (\phi|\mathfrak{S})_\sigma(B)$$
$$\leqq (\phi|\mathfrak{S})_\sigma(A \cup B) \dotplus (\phi|\mathfrak{S})_\sigma(A \cap B) \leqq \phi(A \cup B) \dotplus \phi(A \cap B).$$

10.10. THEOREM. *Let $\mathfrak{S}$ be a lattice and $\varphi : \mathfrak{S} \to \overline{\mathbb{R}}$ be isotone.*
1) *If φ is bounded below and submodular then $\varphi^\sigma \leqq \varphi_\sigma$ on $\mathfrak{S}^\#$.*
2) *If φ is bounded above and supermodular then $\varphi^\sigma \leqq \varphi_\sigma$ on $\mathfrak{S}_\#$.*

Proof of 1). The set function $\phi := \varphi^\sigma$ is $> -\infty$ and even bounded below, and hence upward σ continuous by 4.7. Thus we obtain for $A \in \mathfrak{S}^\#$ from 10.5 and 4.1.1)2)

$$\varphi^\sigma(A) = \phi(A) \leqq (\phi|\mathfrak{S})_\sigma(A) = (\varphi^\sigma|\mathfrak{S})_\sigma(A) \leqq \varphi_\sigma(A).$$

In this connection we want to mention a simple fact.

10.11. EXERCISE. *Let $\mathfrak{S}$ be a lattice and $\varphi : \mathfrak{S} \to \overline{\mathbb{R}}$ be isotone. Then $\varphi^\sigma > -\infty$ iff φ is bounded below.*

The Measurability of Suslin and Co-Suslin Sets

We continue with a classical theme of measure theory, adapted to the frame of the present text. We start with an ad-hoc extension of a classical definition. An isotone set function $\alpha : \mathfrak{A} \to \overline{\mathbb{R}}$ on a lattice $\mathfrak{A}$ is called **complete** iff $U \subset A \subset V$ with $U, V \in \mathfrak{A}$ and $\alpha(U) = \alpha(V) \in \mathbb{R}$ implies that $A \in \mathfrak{A}$.

10.12. THEOREM. *Assume that $\mathfrak{A}$ is a σ lattice and that $\alpha : \mathfrak{A} \to \overline{\mathbb{R}}$ is isotone, almost upward and almost downward σ continuous, and complete. Let $\mathfrak{S} \subset \mathfrak{A}$ be a lattice such that $\alpha|\mathfrak{S}$ is bounded below and submodular. If $A \in \mathfrak{S}^{\#}$ with $(\alpha|\mathfrak{S})_\sigma(A) < \infty$ or $\alpha|\mathfrak{S} < \infty$ then*

$$A \in \mathfrak{A} \quad and \quad \alpha(A) = (\alpha|\mathfrak{S})_\sigma(A) = (\alpha|\mathfrak{S})^\sigma(A).$$

Proof. By assumption the set function $\varphi := \alpha|\mathfrak{S}$ is bounded below and submodular. Let us fix $A \in \mathfrak{S}^{\#}$. i) We first assume that $\varphi_\sigma(A) < \infty$. By 10.10.1) then $-\infty < \varphi^\sigma(A) \leqq \varphi_\sigma(A) < \infty$; in particular both values are finite. By the regularity properties 6.3.4) and 4.1.4) there exist

$$U \in (\mathfrak{S}_\sigma)^\sigma \subset \mathfrak{A} \quad with \quad U \subset A \text{ and } \varphi_\sigma(U) = \varphi_\sigma(A);$$
$$V \in (\mathfrak{S}^\sigma)_\sigma \subset \mathfrak{A} \quad with \quad V \supset A \text{ and } \varphi^\sigma(V) = \varphi^\sigma(A).$$

Then from 7.1.σ) we obtain $\varphi_\sigma(U) \leqq \alpha(U)$ and $\alpha(V) \leqq \varphi^\sigma(V)$. It follows that

$$\alpha(U) \leqq \alpha(V) \leqq \varphi^\sigma(V) = \varphi^\sigma(A) \leqq \varphi_\sigma(A) = \varphi_\sigma(U) \leqq \alpha(U);$$

hence we have $=$ at all places, and the common value is finite. The assertion follows. ii) We now assume that $\varphi < \infty$. By 10.1.4) A is upward enclosable $\mathfrak{S}^\sigma$; thus there exists a sequence $(S_l)_l$ in $\mathfrak{S}$ with $S_l \uparrow \supset A$ and hence $S_l \cap A \uparrow A$. Now $S_l \cap A \in \mathfrak{S}^{\#}$ since $\mathfrak{S}^{\#}$ is a lattice by 10.3, and $\varphi_\sigma(S_l \cap A) \leqq \varphi_\sigma(S_l) = \varphi(S_l) < \infty$ by 6.5 since φ is downward σ continuous. It follows from i) that $S_l \cap A \in \mathfrak{A}$ and $\alpha(S_l \cap A) = \varphi_\sigma(S_l \cap A) = \varphi^\sigma(S_l \cap A) \in \mathbb{R}$. By assumption and 4.7 we conclude that $A \in \mathfrak{A}$ and

$$\alpha(A) = \varphi^\sigma(A) \leqq \varphi_\sigma(A) \in \,] - \infty, \infty].$$

Now 7.1.σ) with $\varphi < \infty$ implies that $\varphi_\sigma(A) \leqq \alpha(A)$. Thus we obtain the assertion.

10.13. EXERCISE. *Assume that $\mathfrak{A}$ is a σ lattice and that $\alpha : \mathfrak{A} \to \overline{\mathbb{R}}$ is isotone, almost upward and almost downward σ continuous, and complete. Let $\mathfrak{S} \subset \mathfrak{A}$ be a lattice such that $\alpha|\mathfrak{S}$ is bounded above and supermodular. If $A \in \mathfrak{S}_{\#}$ with $(\alpha|\mathfrak{S})^\sigma(A) > -\infty$ or $\alpha|\mathfrak{S} > -\infty$ then*

$$A \in \mathfrak{A} \quad and \quad \alpha(A) = (\alpha|\mathfrak{S})_\sigma(A) = (\alpha|\mathfrak{S})^\sigma(A).$$

The natural candidates for completeness are of course the maximal extensions obtained in the outer and inner main theorems 5.5 and 6.22. We conclude the subsection with the proof of these facts.

10.14. THEOREM. *Let $\varphi : \mathfrak{S} \to] - \infty, \infty]$ be an outer $\bullet$ premeasure on the lattice $\mathfrak{S}$. Then $\varphi^\bullet|\mathfrak{C}(\varphi^\bullet, \dot{+})$ is complete.*

10.15. LEMMA. *Assume that $\mathfrak{S}$ is a lattice and that $\varphi : \mathfrak{S} \to \overline{\mathbb{R}}$ is isotone and modular $\dotplus$. Let $U \subset V$ be in $\mathfrak{S}$ with $\varphi(U) = \varphi(V) \in \mathbb{R}$. For all $S \in \mathfrak{S}$ then $\varphi(S \cup U) = \varphi(S \cup V)$ and $\varphi(S \cap U) = \varphi(S \cap V)$.*

Proof of 10.15. i) Since φ is submodular $\dotplus$ we obtain

$$\begin{aligned} \varphi(S \cup U) \dotplus \varphi(V) &\geqq \varphi(S \cup V) \dotplus \varphi((S \cup U) \cap V) \\ &\geqq \varphi(S \cup V) \dotplus \varphi(U) = \varphi(S \cup V) \dotplus \varphi(V). \end{aligned}$$

Thus $\varphi(V) \in \mathbb{R}$ implies that $\varphi(S \cup U) \geqq \varphi(S \cup V)$ and hence that $\varphi(S \cup U) = \varphi(S \cup V)$. ii) Since φ is supermodular $\dotplus$ we obtain

$$\begin{aligned} \varphi(S \cap V) \dotplus \varphi(U) &\leqq \varphi((S \cap V) \cup U) \dotplus \varphi(S \cap U) \\ &\leqq \varphi(V) \dotplus \varphi(S \cap U) = \varphi(U) \dotplus \varphi(S \cap U). \end{aligned}$$

Thus $\varphi(U) \in \mathbb{R}$ implies that $\varphi(S \cap V) \leqq \varphi(S \cap U)$ and hence $\varphi(S \cap V) = \varphi(S \cap U)$.

Proof of 10.14. i) We form

$$\mathfrak{T} := \{ S \in \mathfrak{C}(\varphi^\bullet, \dotplus) : \varphi^\bullet(S) \in \mathbb{R} \} \subset \mathfrak{C}(\varphi^\bullet, \dotplus).$$

Then $\mathfrak{T}$ is a lattice since $\varphi^\bullet | \mathfrak{C}(\varphi^\bullet, \dotplus)$ is modular $\dotplus$, and $[\varphi < \infty] = [\varphi \in \mathbb{R}] \subset \mathfrak{T}$. Now fix $U \subset A \subset V$ with $U, V \in \mathfrak{C}(\varphi^\bullet, \dotplus)$ and $\varphi^\bullet(U) = \varphi^\bullet(V) \in \mathbb{R}$. To be shown is $A \in \mathfrak{C}(\varphi^\bullet, \dotplus)$. First note that $U, V \in \mathfrak{T}$. Now let $P, Q \in \mathfrak{T}$ and apply 10.15 twice to $\varphi^\bullet | \mathfrak{C}(\varphi^\bullet, \dotplus)$. Then

$$\begin{aligned} \varphi^\bullet(Q \cap U) &= \varphi^\bullet(Q \cap V), \text{ and this is in } \mathbb{R}; \\ \varphi^\bullet(P \cup (Q \cap U)) &= \varphi^\bullet(P \cup (Q \cap V)), \text{ and this is in } \mathbb{R}. \end{aligned}$$

ii) We fix $P, Q \in [\varphi < \infty] = [\varphi \in \mathbb{R}] \subset \mathfrak{T}$ with $P \subset Q$. We note that all arguments which occur below will be in $P \sqsubset Q$, and hence all values $\varphi^\bullet(\cdot)$ will be finite. We obtain

$$\begin{aligned} \varphi^\bullet(P|A|Q) + \varphi^\bullet(P|A'|Q) &= \varphi^\bullet(P \cup (Q \cap A)) + \varphi^\bullet(P \cup (Q \cap A')) \\ &\leqq \varphi^\bullet(P \cup (Q \cap V)) + \varphi^\bullet(P \cup (Q \cap U')) \\ &= \varphi^\bullet(P \cup (Q \cap U)) + \varphi^\bullet(P \cup (Q \cap U')) \\ &= \varphi^\bullet(P|U|Q) + \varphi^\bullet(P|U'|Q) = \varphi^\bullet(P) + \varphi^\bullet(Q). \end{aligned}$$

From 5.2 it follows that $A \in \mathfrak{C}(\varphi^\bullet, \dotplus)$.

10.16. EXERCISE. *Let $\varphi : \mathfrak{S} \to [-\infty, \infty[$ be an inner $\bullet$ premeasure on the lattice $\mathfrak{S}$. Then $\varphi_\bullet | \mathfrak{C}(\varphi_\bullet, \dotplus)$ is complete.*

We specialize 10.12 and 10.13 on the basis of 10.14 and 10.16. For the sake of a simple formulation we restrict ourselves to finite-valued set functions $\varphi : \mathfrak{S} \to \mathbb{R}$.

10.17. THEOREM. *Assume that the set function $\varphi : \mathfrak{S} \to \mathbb{R}$ on the lattice $\mathfrak{S}$ is*

> *either an outer σ premeasure with $\mathfrak{A} := \mathfrak{C}(\varphi^\sigma, \dotplus)$,*
> *or an inner σ premeasure with $\mathfrak{A} := \mathfrak{C}(\varphi_\sigma, \dotplus)$.*

Then

> *if φ is bounded below: $A \in \mathfrak{S}^{\#}$ implies $A \in \mathfrak{A}$ and $\varphi^{\sigma}(A) = \varphi_{\sigma}(A)$,*
> *if φ is bounded above: $A \in \mathfrak{S}_{\#}$ implies $A \in \mathfrak{A}$ and $\varphi^{\sigma}(A) = \varphi_{\sigma}(A)$.*

10.18. BIBLIOGRAPHICAL NOTE. The first version of the theory of capacities in Choquet [1953-54] was in the frame of Hausdorff topological spaces. The abstract version quoted in the present text is in Choquet [1959]. It is the proof of the capacitability theorem in this paper which could have been adapted to furnish the present extended theorem 10.5. Our procedure via 10.7 likewise furnishes the so-called precise capacitability theorem of Dellacherie-Meyer [1988] chapter XI. The need for a more comprehensive capacitability theorem like the present 10.5, where the set function ϕ and the lattice $\mathfrak{S}$ be not tied to each other, has been expressed much earlier in Sion [1963] page 87.

For the subsequent parts of the present section we refer to Meyer [1966] chapter III and Dellacherie-Meyer [1978] chapter III, and to Jacobs [1978] chapter XIII. Let $\mathfrak{S}$ be a lattice in a nonvoid set X with $\varnothing \in \mathfrak{S}$, and $\varphi : \mathfrak{S} \to [0, \infty]$ be an isotone and submodular set function. The above authors assumed φ to be upward σ continuous and considered the formation $\varphi^{(\sigma)} : \mathfrak{P}(X) \to [0, \infty]$ defined above behind 6.10 and 6.11. They proved that $\varphi^{(\sigma)}$ is submodular and upward σ continuous. These are of course special cases of the present 4.1.5) and 4.7, in that the formation φ^{σ} is always submodular and upward σ continuous, and $\varphi^{(\sigma)} = \varphi^{\sigma}$ iff φ is upward σ continuous by 6.10.2). The authors then used the capacitability theorem and their above-mentioned results to reprove a number of basic results in traditional measure theory, like the Carathéodory extension theorem and the Daniell-Stone representation theorem, theorem 9.9.ii) on Borel-Radon measures, and in particular the traditional version of the present measurability theorem 10.12. But they never went beyond the traditional frame of measure theory.

CHAPTER IV

The Integral

In the present chapter the integral will be understood as a formation which is based upon a certain set function. The opposite concept will be the topic of the next chapter.

The continuation of the present work requires the adequate concepts and facts in the domain of integration. In spite of the vast literature on integration we have not seen the presentation which we need. Therefore the present chapter will offer a short but complete account, without claim for essential innovation. The main point is that the next chapter will require the integral for functions with values in $[0, \infty[$ or $[0, \infty]$ instead of $\mathbb{R}$ or $\overline{\mathbb{R}}$, but in return for more comprehensive classes of functions than usual. Therefore we first concentrate on functions with values in $[0, \infty]$. For these functions the next two sections will define two different notions of an integral, called the horizontal and the vertical integral. These notions are based on the two respective natural ideas for the formation of an integral. The horizontal integral will be in the spirit of the so-called non-additive theory of integration. We shall see that the two integrals are equal whenever both make sense.

Then for the sake of fairness and completeness it will be natural to demonstrate how the two integrals combine and specialize in order to furnish the conventional integral for functions with values in $\mathbb{R}$ or $\overline{\mathbb{R}}$. This will be done in the final section of the chapter.

11. The Horizontal Integral

Upper and Lower Measurable Functions

Let $\mathfrak{S}$ be a lattice with $\varnothing \in \mathfrak{S}$ in a nonvoid set X. We define a function $f : X \to [0, \infty]$ to be

upper measurable $\mathfrak{S}$ iff $[f \geq t] \in \mathfrak{S}$ for all $t > 0$,
lower measurable $\mathfrak{S}$ iff $[f > t] \in \mathfrak{S}$ for all $t > 0$.

We define $\mathrm{UM}(\mathfrak{S})$ and $\mathrm{LM}(\mathfrak{S})$ to consist of all these functions.

In the sequel we write as usual $u \vee v := \max(u, v)$ and $u \wedge v := \min(u, v)$ for $u, v \in \overline{\mathbb{R}}$, and likewise for the pointwise combinations of functions with values in $\overline{\mathbb{R}}$. In connection with 11.1.2) below we note that $u = u \wedge t + (u-t)^+$ for $u \in \overline{\mathbb{R}}$ and $t \in \mathbb{R}$.

11.1. PROPERTIES. 1) *If* $f, g \in \mathrm{UM}(\mathfrak{S})$ *then* $f \vee g, f \wedge g \in \mathrm{UM}(\mathfrak{S})$; *the same for* $\mathrm{LM}(\mathfrak{S})$.

2) *If* $f \in \mathrm{UM}(\mathfrak{S})$ *then* $tf, f \wedge t, (f-t)^+ \in \mathrm{UM}(\mathfrak{S})$ *for all real* $t > 0$; *the same for* $\mathrm{LM}(\mathfrak{S})$.

3) *If* $\mathfrak{S} = \mathfrak{S}_\sigma$ *then* $f, g \in \mathrm{UM}(\mathfrak{S})$ *implies* $f + g \in \mathrm{UM}(\mathfrak{S})$. *If* $\mathfrak{S} = \mathfrak{S}^\sigma$ *then* $f, g \in \mathrm{LM}(\mathfrak{S})$ *implies* $f + g \in \mathrm{LM}(\mathfrak{S})$.

4) *In case* $\mathfrak{S} = \mathfrak{S}_\sigma$ *we have* $\mathrm{LM}(\mathfrak{S}) \subset \mathrm{UM}(\mathfrak{S})$. *In case* $\mathfrak{S} = \mathfrak{S}^\sigma$ *we have* $\mathrm{UM}(\mathfrak{S}) \subset \mathrm{LM}(\mathfrak{S})$. *Thus* $\mathrm{UM}(\mathfrak{S}) = \mathrm{LM}(\mathfrak{S}) =: \mathrm{M}(\mathfrak{S})$ *when* $\mathfrak{S}$ *is a* σ *lattice.*

Proof. 1) is obvious. 2) For $s > 0$ we have

$$[f \wedge t \geqq s] = \left\{ \begin{array}{ll} \varnothing & \text{when } t < s \\ [f \geqq s] & \text{when } t \geqq s \end{array} \right\} \text{ and } [(f-t)^+ \geqq s] = [f \geqq t + s],$$

and likewise in case $>$. 3) will be deduced from the next lemma. 4) For $t > 0$ we have $[f > t - 1/n] \downarrow [f \geqq t]$ and $[f \geqq t + 1/n] \uparrow [f > t]$.

11.2. LEMMA. *Assume that* $f, g : X \to]-\infty, \infty]$ *and* $t > 0$. *If* $D \subset \mathbb{R}$ *is dense then*

$$[f + g \geqq t] = \bigcap_{s \in D} [f \geqq s] \cup [g \geqq t - s] \text{ and}$$

$$[f + g > t] = \bigcup_{s \in D} [f > s] \cap [g > t - s].$$

Proof of the first relation. $\subset$) Assume that $x \in X$ is not in the set on the right. Thus there exists $s \in D$ with $f(x) < s$ and $g(x) < t - s$, so that $f(x) + g(x) < t$. Hence x is not in the set on the left. $\supset$) Assume that $x \in X$ is not in the set on the left. Thus $f(x) + g(x) < t$. It follows that $f(x)$ and $g(x)$ are both finite, and $f(x) < t - g(x)$. Choose $s \in D$ with $f(x) < s < t - g(x)$. Then $x \notin [f \geqq s] \cup [g \geqq t - s]$, so that x is not in the set on the right. Proof of the second relation. $\subset$) Assume that $x \in X$ is in the set on the left, that is $f(x) + g(x) > t$. If $f(x) = \infty$ or $g(x) = \infty$ then it is obvious that x is in the set on the right. So assume that $f(x)$ and $g(x)$ are both finite, and thus $f(x) > t - g(x)$. Choose $s \in D$ with $f(x) > s > t - g(x)$. Then $x \in [f > s] \cap [g > t - s]$, so that x is in the set on the right. $\supset$) Assume that $x \in X$ is in the set on the right. Thus there exists $s \in D$ with $f(x) > s$ and $g(x) > t - s$, so that $f(x) + g(x) > t$. Hence x is in the set on the left.

Proof of 11.1.3). i) Assume that $\mathfrak{S} = \mathfrak{S}_\sigma$ and $f, g \in \mathrm{UM}(\mathfrak{S})$, and let $t > 0$. We have $[f \geqq s] \cup [g \geqq t - s] \in \mathfrak{S}$ for all $s \in \mathbb{R}$ with $0 < s < t$; furthermore $[f \geqq s] \cup [g \geqq t - s] = X$ for $s \leqq 0$ and for $s \geqq t$. The first relation in 11.2 with countable D implies that $[f + g \geqq t] \in \mathfrak{S}$. ii) Assume that $\mathfrak{S} = \mathfrak{S}^\sigma$ and $f, g \in \mathrm{LM}(\mathfrak{S})$, and let $t > 0$. We have $[f > s] \cap [g > t - s] \in \mathfrak{S}$ for all $s \in \mathbb{R}$ with $s \neq 0, t$; to see this distinguish the cases $s < 0$ and $s > t$

and $0 < s < t$. The second relation in 11.2 with countable D and $0, t \notin D$ implies that $[f + g > t] \in \mathfrak{S}$.

11.3. EXERCISE. 1) Show by examples that $\mathrm{UM}(\mathfrak{S})$ and $\mathrm{LM}(\mathfrak{S})$ need not be stable under addition, even when $\mathfrak{S}$ is an algebra. 2) Show by examples that neither $\mathrm{UM}(\mathfrak{S})$ nor $\mathrm{LM}(\mathfrak{S})$ need contain the other function class, even when $\mathfrak{S}$ is an algebra. Hint: Let $X = \mathbb{N} \cup (-\mathbb{N})$, and let $\mathfrak{S}$ consist of the finite and cofinite subsets of X.

11.4. PROPOSITION. *Assume that* $f : X \to [0, \infty[$ *has a finite value set. Then the following are equivalent.* i) $f \in \mathrm{UM}(\mathfrak{S})$. ii) $f \in \mathrm{LM}(\mathfrak{S})$. iii) *There exist* $A(1), \cdots, A(r) \in \mathfrak{S}$ *and* $t_1, \cdots, t_r > 0$ *such that*

$$f = \sum_{l=1}^{r} t_l \chi_{A(l)}.$$

iv) *The same as* iii)*, but with the additional requirement that* $A(1) \supset \cdots \supset A(r)$.

We define $S(\mathfrak{S})$ to consist of the functions $f : X \to [0, \infty[$ which fulfil the equivalent properties in 11.4.

11.5. LEMMA. *Assume that* $f : X \to [0, \infty[$ *has a finite value set* $f(X) \subset \{t(0), t(1), \cdots, t(r)\}$ *with* $0 = t(0) < t(1) < \cdots < t(r) < \infty$. *Then*

$$f = \sum_{l=1}^{r} \big(t(l) - t(l-1)\big) \chi_{[f \geq t(l)]}.$$

Furthermore note that $[f \geq t(l)] = [f > t]$ *for* $t(l-1) \leq t < t(l)$ $(l = 1, \cdots, r)$.

Proof of 11.5. For $x \in X$ with $f(x) = 0 = t(0)$ the second member is $= 0$, and in case $f(x) = t(p)(p = 1, ..., r)$ it is $= \sum_{l=1}^{p} \big(t(l) - t(l-1)\big) = t(p)$.

Proof of 11.4. The implications i)$\Rightarrow$iv) and ii)$\Rightarrow$iv) follow from 11.5. iv)$\Rightarrow$iii) is obvious. In order to prove iii)$\Rightarrow$i) and iii)$\Rightarrow$ii) note that for $t > 0$

$[f \geq t]$ is the union of the $\bigcap_{l \in T} A(l)$ for the nonvoid $T \subset \{1, \cdots, r\}$

$$\text{with } \sum_{l \in T} t_l \geq t,$$

$[f > t]$ is the union of the $\bigcap_{l \in T} A(l)$ for the nonvoid $T \subset \{1, \cdots, r\}$

$$\text{with } \sum_{l \in T} t_l > t.$$

We conclude the subsection with an estimation of related type which will be a useful tool.

11.6. LEMMA. *For* $f : X \to \overline{\mathbb{R}}$ *and real numbers* $a = t(0) < t(1) < \cdots < t(r) = b$ *we have*

$$\sum_{l=1}^{r} (t(l) - t(l-1)) \chi_{[f \geqq t(l)]} \quad \leqq \quad (f-a)^+ \wedge (b-a)$$

$$\leqq \quad \sum_{l=1}^{r} (t(l) - t(l-1)) \chi_{[f \geqq t(l-1)]}.$$

The same holds true for $[f > \cdot]$ *instead of* $[f \geqq \cdot]$.

Proof. Fix $x \in X$. 1) The first relation has to be proved for $[f \geqq \cdot]$. We distinguish the cases

$f(x) < t(1)$: left $= 0 \leqq$ right;
$t(p) \leqq f(x) < t(p+1)$ with $1 \leqq p < r$: left $= t(p) - t(0) \leqq f(x) - a =$ right;
$f(x) \geqq t(r)$: left $= t(r) - t(0) = b - a =$ right.

2) The second relation has to be proved for $[f > \cdot]$. We distinguish the cases

$f(x) \leqq t(0)$: right $= 0 =$ left;
$t(p-1) < f(x) \leqq t(p)$ with $1 \leqq p < r$: right $= t(p) - t(0) \geqq f(x) - a \geqq$ left;
$f(x) > t(r-1)$: right $= t(r) - t(0) = b - a \geqq$ left.

The proof is complete.

The Horizontal Integral

The notion to be defined is based on the elementary Riemann integral. We recall this notion as follows. We assume for real $a < b$ the proper Riemann integral $\int_a^b F(t)dt$ of a function $F : [a, b] \to \mathbb{R}$ which is Riemann integrable and hence bounded. All monotone functions $F : [a, b] \to \mathbb{R}$ are Riemann integrable. We need for $-\infty \leqq a < b \leqq \infty$ the improper Riemann integral of a monotone function $F :]a, b[\to [0, \infty]$. It is defined to be

$$\int_{a \leftarrow}^{\to b} F(t)dt := \sup \left\{ \int_u^v F(t)dt : a < u < v < b \right\} \quad \text{if } F \text{ is finite-valued,}$$

and $= \infty$ if F attains the value ∞ at some point and hence on some non-degenerate subinterval of $]a, b[$. We shall make free use of the elementary properties of these integrals.

We come to the definition. Let $\mathfrak{S}$ be a lattice with $\varnothing \in \mathfrak{S}$ in a nonvoid set X. Assume that $\varphi : \mathfrak{S} \to [0, \infty]$ is isotone with $\varphi(\varnothing) = 0$. We define the **horizontal integral** with respect to φ

$$\text{for } f \in \text{UM}(\mathfrak{S}) \quad \text{to be} \quad \fint f d\varphi := \int_{0 \leftarrow}^{\to \infty} \varphi([f \geqq t])dt \in [0, \infty],$$

$$\text{for } f \in \text{LM}(\mathfrak{S}) \quad \text{to be} \quad \fint f d\varphi := \int_{0 \leftarrow}^{\to \infty} \varphi([f > t])dt \in [0, \infty].$$

Thus the two formations are well-defined.

11.7. REMARK. *For* $f \in \text{UM}(\mathfrak{S}) \cap \text{LM}(\mathfrak{S})$ *the two horizontal integrals* $\fint f d\varphi$ *defined above are equal.*

Proof. We form $P : P(t) = \varphi([f \geqq t])$ and $Q : Q(t) = \varphi([f > t])$ for $t > 0$. Thus $P, Q :]0, \infty[\to [0, \infty]$ are monotone decreasing. We claim that

$$Q(t+) = P(t+) \leqq Q(t) \leqq P(t) \leqq Q(t-) = P(t-) \text{ for } t > 0,$$

which of course implies the assertion. In fact, we have i) $Q(t) \leqq P(t)$ for $t > 0$ and hence $Q(t-) \leqq P(t-)$ and $Q(t+) \leqq P(t+)$ for $t > 0$. The monotonicity implies ii) $P(t+) \leqq P(t) \leqq P(t-)$ and $Q(t+) \leqq Q(t) \leqq Q(t-)$ for $t > 0$. At last iii) $P(v) \leqq Q(u)$ for $0 < u < v$ and hence $P(t-) \leqq Q(t-)$ and $P(t+) \leqq Q(t+)$ for $t > 0$. The assertion follows.

11.8. PROPERTIES. 1) *If* $f \in S(\mathfrak{S})$ *is represented in the form* 11.4.iv) *then*

$$\fint f d\varphi = \sum_{l=1}^{r} t_l \varphi(A(l)).$$

In particular $\fint \chi_A d\varphi = \varphi(A)$ *for* $A \in \mathfrak{S}$.
2) *The horizontal integral is isotone on* $\mathrm{UM}(\mathfrak{S})/\mathrm{LM}(\mathfrak{S})$.
3) *The horizontal integral is positive-homogeneous on* $\mathrm{UM}(\mathfrak{S})/\mathrm{LM}(\mathfrak{S})$, *that is* $\fint (cf) d\varphi = c \fint f d\varphi$ *for real* $c > 0$. *But of course one cannot expect any additivity property.*
4) *If* $f \mapsto \fint f d\varphi$ *is subadditive/superadditive on* $S(\mathfrak{S})$ *then the set function* φ *is submodular/supermodular.*
5) *For* $t > 0$ *we have*

$$t\varphi([f \geqq t]) \quad \leqq \quad \fint f d\varphi \quad \text{in case } f \in \mathrm{UM}(\mathfrak{S}),$$

$$t\varphi([f > t]) \quad \leqq \quad \fint f d\varphi \quad \text{in case } f \in \mathrm{LM}(\mathfrak{S}).$$

In both cases therefore $\fint f d\varphi < \infty$ *implies that* $[f > 0] \in [\varphi < \infty]^{\sigma}$.

Proof. 1) We put $\tau_0 := 0$ and $\tau_p := \sum_{l=1}^{p} t_l$ for $p = 1, \cdots, r$. Then f can at most attain the values $\tau_0, \tau_1, \cdots, \tau_r$, and for $x \in X$ and $1 \leqq p \leqq r$ we have $f(x) \geqq \tau_p \Leftrightarrow x \in A(p)$. It follows for $t > 0$ that

$$[f \geqq t] = \left\{ \begin{array}{ll} A(p) & \text{when } \tau_{p-1} < t \leqq \tau_p \text{ with } 1 \leqq p \leqq r \\ \varnothing & \text{when } t > \tau_r \end{array} \right\}.$$

Therefore we have by definition

$$\fint f d\varphi = \sum_{p=1}^{r} \int_{\tau_{p-1}\leftarrow}^{\tau_p} \varphi([f \geqq t]) dt = \sum_{p=1}^{r} t_p \varphi(A(p)).$$

2) and 3) are obvious. 4) We prove the sub assertion. For $A, B \in \mathfrak{S}$ we have by 1) and by assumption

$$\begin{aligned} \varphi(A \cup B) + \varphi(A \cap B) &= \fint (\chi_{A\cup B} + \chi_{A\cap B}) d\varphi = \fint (\chi_A + \chi_B) d\varphi \\ &\leqq \fint \chi_A d\varphi + \fint \chi_B d\varphi = \varphi(A) + \varphi(B). \end{aligned}$$

The super assertion has the same proof. 5) is obvious.

The fundamental fact is that 11.8.4) admits a fortified converse.

11.9. LEMMA. *For $f \in \mathrm{UM}(\mathfrak{S})/\mathrm{LM}(\mathfrak{S})$ and $A \in \mathfrak{S}$ we have $f + \chi_A \in$* $\mathrm{UM}(\mathfrak{S})/\mathrm{LM}(\mathfrak{S})$. *If φ is submodular then*

$$\fint (f + \chi_A) d\varphi \leqq \fint f d\varphi + \varphi(A), \quad and \geqq if \ \varphi \ is \ supermodular.$$

Proof. We restrict ourselves to the case $\mathrm{UM}(\mathfrak{S})$. We have

$$\text{for } 0 < t \leqq 1 \quad : \quad [f + \chi_A \geqq t] = A \cup [f \geqq t],$$
$$\text{for } t > 1 \quad : \quad [f + \chi_A \geqq t] = [f \geqq t] \cup \big(A \cap [f \geqq t - 1]\big),$$

so that $f + \chi_A \in \mathrm{UM}(\mathfrak{S})$. Now we treat the submodular/supermodular cases at the same time. We can assume that $\fint f d\varphi < \infty$ and $\varphi(A) < \infty$; thus $\varphi([f \geqq t]) < \infty \ \forall t > 0$. Then

$$\fint (f + \chi_A) d\varphi$$

$$= \quad \int_{0\leftarrow}^{1} \varphi(A \cup [f \geqq t]) dt + \int_{1\leftarrow}^{\rightarrow\infty} \varphi\big([f \geqq t] \cup (A \cap [f \geqq t - 1])\big) dt$$

$$= \quad \lim_{n\to\infty} \bigg(\int_{0\leftarrow}^{1} \varphi(A \cup [f \geqq t]) dt$$

$$+ \quad \sum_{l=2}^{n} \int_{(l-1)\leftarrow}^{l} \varphi\big([f \geqq t] \cup (A \cap [f \geqq t - 1])\big) dt \bigg)$$

$$\leqq/\geqq \quad \lim_{n\to\infty} \bigg(\int_{0\leftarrow}^{1} \big(\varphi(A) + \varphi([f \geqq t]) - \varphi(A \cap [f \geqq t])\big) dt$$

$$+ \quad \sum_{l=2}^{n} \int_{(l-1)\leftarrow}^{l} \big(\varphi([f \geqq t]) + \varphi(A \cap [f \geqq t - 1]) - \varphi(A \cap [f \geqq t])\big) dt \bigg)$$

$$= \quad \lim_{n\to\infty} \bigg(\varphi(A) + \int_{0\leftarrow}^{n} \varphi([f \geqq t]) dt - \int_{(n-1)\leftarrow}^{n} \varphi(A \cap [f \geqq t]) dt \bigg).$$

In view of $\fint f d\varphi < \infty$ the third integral on the right tends to 0 for $n \to \infty$. Thus the right side tends to $\varphi(A) + \fint f d\varphi$. The assertion follows.

11.10. CONSEQUENCE. *For $f \in \mathrm{UM}(\mathfrak{S})/\mathrm{LM}(\mathfrak{S})$ and $g \in \mathrm{S}(\mathfrak{S})$ we have* $f + g \in \mathrm{UM}(\mathfrak{S})/\mathrm{LM}(\mathfrak{S})$. *If φ is submodular then*

$$\fint (f + g) d\varphi \leqq \fint f d\varphi + \fint g d\varphi, \quad and \geqq if \ \varphi \ is \ supermodular.$$

Proof. Combine 11.9 with 11.8.1).

11.11. THEOREM. *Assume that $f, g \in \mathrm{UM}(\mathfrak{S})/\mathrm{LM}(\mathfrak{S})$ are such that $f +$* $g \in \mathrm{UM}(\mathfrak{S})/\mathrm{LM}(\mathfrak{S})$. *If φ is submodular then*

$$\fint (f + g) d\varphi \leqq \fint f d\varphi + \fint g d\varphi,$$

and the same with $\geqq$ if φ is supermodular.

Proof. As before we restrict ourselves to the case UM($\mathfrak{S}$). We can assume that $\int f d\varphi < \infty$ and $\int g d\varphi < \infty$, in the supermodular case also $\int (f+g)d\varphi < \infty$. But in the submodular case we have $\int (f+g)d\varphi < \infty$ as a consequence of the two previous assumptions, because for $t > 0$

$$[f+g \geq t] \subset [f \geq t/2] \cup [g \geq t/2],$$
$$\varphi([f+g \geq t]) \leq \varphi([f \geq t/2]) + \varphi([g \geq t/2]),$$

and hence $\int (f+g)d\varphi \leq 2\int f d\varphi + 2\int g d\varphi < \infty$. Therefore we can assume that $\int f d\varphi, \int g d\varphi < \infty$ and $\int (f+g)d\varphi < \infty$.

1) Let us fix t : $0 = t(0) < t(1) < \cdots < t(r) = b < \infty$ and put $\delta(t) := \max\{t(l) - t(l-1) : l = 1, \cdots, r\}$. We form the function

$$g_t := \sum_{l=1}^{r}(t(l) - t(l-1))\chi_{[g \geq t(l)]} \in S(\mathfrak{S}).$$

From 11.6 we know that $g_t \leq g$. Furthermore we note for $x \in X$ the implications which follow.

i) $g(x) \geq b \Rightarrow g_t(x) = b$.

ii) $0 \leq g(x) < b \Rightarrow g(x) < g_t(x) + \delta(t)$. In fact, if $t(p-1) \leq g(x) < t(p)$ with $1 \leq p \leq r$, then $g_t(x) = t(p-1)$ and hence $g(x) < t(p) = (t(p) - t(p-1)) + t(p-1) \leq \delta(t) + g_t(x)$.

iii) For $0 < t \leq b$ we have

$$g(x) \geq t + \delta(t) \Rightarrow g_t(x) \geq t,$$
$$f(x) + g(x) \geq t + \delta(t) \Rightarrow f(x) + g_t(x) \geq t.$$

In fact, assume in one of these implications that the right relation is false. Then $g_t(x) < t$, hence $g(x) < b$ by i), and hence $g(x) < g_t(x) + \delta(t)$ by ii). Therefore

in the first case: $g(x) < g_t(x) + \delta(t) < t + \delta(t)$,
in the second case: $f(x) + g(x) < f(x) + g_t(x) + \delta(t) < t + \delta(t)$,

so that each time the left relation must be false.

2) The first implication in iii) furnishes

$$\int g d\varphi - \int g_t d\varphi$$

$$\overset{\leqq}{} \int_{0\leftarrow}^{\rightarrow\infty} \varphi([g \geq t])dt - \int_{0\leftarrow}^{b} \varphi([g_t \geq t])dt$$

$$\overset{\leqq}{} \int_{0\leftarrow}^{\rightarrow\infty} \varphi([g \geq t])dt - \int_{0\leftarrow}^{b} \varphi([g \geq t + \delta(t)])dt$$

$$\overset{\leqq}{} \int_{0\leftarrow}^{\rightarrow\infty} \varphi([g \geq t])dt - \int_{\delta(t)}^{b} \varphi([g \geq t])dt$$

$$= \int_{b}^{\rightarrow\infty} \varphi([g \geq t])dt + \int_{0\leftarrow}^{\delta(t)} \varphi([g \geq t])dt.$$

Likewise the second implication in iii) furnishes

$$\fint (f+g)d\varphi - \fint (f+g_t)d\varphi$$

$$\leqq \int_b^{\rightarrow\infty} \varphi([f+g\geqq t])dt + \int_{0\leftarrow}^{\delta(t)} \varphi([f+g\geqq t])dt.$$

3) Now let $\varepsilon > 0$. Then we can choose the above t such that in the two inequalities in 2) the last members are both $\leqq \varepsilon$. It follows that

$$\fint g d\varphi \leqq \fint g_t d\varphi + \varepsilon \text{ and } \fint (f+g)d\varphi \leqq \fint (f+g_t)d\varphi + \varepsilon.$$

So if φ is submodular then 11.10 implies that

$$\fint (f+g)d\varphi \leqq \fint (f+g_t)d\varphi + \varepsilon \leqq \fint f d\varphi + \fint g_t d\varphi + \varepsilon$$

$$\leqq \fint f d\varphi + \fint g d\varphi + \varepsilon.$$

Likewise if φ is supermodular then 11.10 implies that

$$\fint (f+g)d\varphi \geqq \fint (f+g_t)d\varphi \geqq \fint f d\varphi + \fint g_t d\varphi$$

$$\geqq \fint f d\varphi + \fint g d\varphi - \varepsilon.$$

Thus we obtain both assertions.

We continue with two examples in order to illustrate the wide extent of the present notion.

11.12. EXERCISE. 1) We let $X = \mathbb{R}$ and $\mathfrak{S} = \mathfrak{P}(\mathbb{R})$, so that $\mathrm{UM}(\mathfrak{S}) = \mathrm{LM}(\mathfrak{S})$ consists of all functions $f : \mathbb{R} \to [0,\infty]$. Define $\varphi : \mathfrak{P}(X) \to \{0,1\}$ to be

$$\varphi(A) = \left\{ \begin{array}{ll} 1 & \text{if } A \text{ is unbounded above} \\ 0 & \text{if } A \text{ is bounded above} \end{array} \right\}.$$

Prove that

$$\fint f d\varphi = \limsup_{x\to\infty} f(x) \quad \text{for all } f : \mathbb{R} \to [0,\infty].$$

2) The above example can be extended as follows. Let $\mathfrak{H}$ be a paving of nonvoid subsets of a set X. Define $\varphi : \mathfrak{P}(X) \to \{0,1\}$ to be

$$\varphi(A) = \left\{ \begin{array}{ll} 1 & \text{if } A \cap H \neq \varnothing \text{ for all } H \in \mathfrak{H} \\ 0 & \text{if } A \cap H = \varnothing \text{ for some } H \in \mathfrak{H} \end{array} \right\}.$$

Prove that

$$\fint f d\varphi = \inf_{H\in\mathfrak{H}} \sup_{x\in H} f(x) =: \limsup_{x\uparrow\mathfrak{H}} f(x) \quad \text{for all } f : X \to [0,\infty].$$

11.13. EXERCISE. Let $B(X, \mathbb{R})$ denote the real vector space of all bounded functions $X \to \mathbb{R}$ on a nonvoid set X. Let $\phi : B(X, \mathbb{R}) \to \mathbb{R}$ be a positive linear functional, that is $\phi(f) \geq 0$ for $f \geq 0$; it follows of course that ϕ is isotone. Define $\varphi : \mathfrak{P}(X) \to [0, \infty[$ to be $\varphi(A) = \phi(\chi_A)$ for $A \subset X$. Thus φ is isotone and modular with $\varphi(\varnothing) = 0$. Prove that $\fint f d\varphi = \phi(f)$ for all $0 \leq f \in B(X, \mathbb{R})$. Hint: Use 11.6.

The next exercise is a remarkable theoretical application of the basic theorem 11.11. It is of the type of a sandwich theorem.

11.14. EXERCISE. 1) *Let X be a nonvoid set, and S be a nonvoid set of functions $X \to [0, \infty[$ which is stable under addition. Assume that*

$Q : S \to [0, \infty[$ *is subadditive, and*
$P : S \to [0, \infty[$ *is superadditive with $P \leq Q$.*

Then there exists an additive functional $\phi : S \to [0, \infty[$ such that $P \leq \phi \leq Q$. If Q is isotone then ϕ can be chosen to be isotone. This is a Hahn-Banach type result. It can either be proved ab-ovo with the usual method or deduced from the Hahn-Banach theorem due to Rodé. See Rodé [1978] or König [1987].

2) *Let $\mathfrak{S}$ be a lattice in X with $\varnothing \in \mathfrak{S}$, and consider set functions*

$\beta : \mathfrak{S} \to [0, \infty[$ *isotone and submodular with $\beta(\varnothing) = 0$,*
$\alpha : \mathfrak{S} \to [0, \infty[$ *isotone and supermodular with $\alpha(\varnothing) = 0$,*

such that $\alpha \leq \beta$. Then there exists $\varphi : \mathfrak{S} \to [0, \infty[$ isotone and modular such that $\alpha \leq \varphi \leq \beta$. Hint: Define the functionals $P, Q : S(\mathfrak{S}) \to [0, \infty[$ to be

$$P(f) = \fint f d\alpha \text{ and } Q(f) = \fint f d\beta \quad \text{for } f \in S(\mathfrak{S}),$$

amd combine 11.11 with 1).

11.15. REMARK. The horizontal integral has the pleasant property that it retains its value under extension of the underlying set function: Assume that $\psi : \mathfrak{T} \to [0, \infty]$ is an extension of $\varphi : \mathfrak{S} \to [0, \infty]$ of the same kind. Then $\mathrm{UM}(\mathfrak{S}) \subset \mathrm{UM}(\mathfrak{T})$ and $\mathrm{LM}(\mathfrak{S}) \subset \mathrm{LM}(\mathfrak{T})$, and it is obvious from the definition that

$$\fint f d\varphi = \fint f d\psi \quad \text{for all } f \in \mathrm{UM}(\mathfrak{S})/\mathrm{LM}(\mathfrak{S}).$$

This applies in particular to the extensions $\varphi^\star, \varphi_\star : \mathfrak{P}(X) \to [0, \infty]$ of φ, the relevant properties of which are collected in 4.1 and 6.3.

Regularity and Continuity of the Horizontal Integral

The last subsection presents some of the most important properties of the horizontal integral. We assume as before that $\mathfrak{S}$ is a lattice with $\varnothing \in \mathfrak{S}$ and that $\varphi : \mathfrak{S} \to [0, \infty]$ is isotone with $\varphi(\varnothing) = 0$.

11.16. THEOREM. $\fint f d\varphi_\star = \sup\{\fint u d\varphi : u \in S(\mathfrak{S}) \text{ with } u \leq f\}$ *for all* $f : X \to [0, \infty]$.

Proof. Let $M \in [0, \infty]$ denote the supremum on the right. To be shown is $\fint f d\varphi_\star \leqq M$. 1) We can assume that $\varphi_\star([f \geqq t]) < \infty \; \forall t > 0$. In fact, if $\varphi_\star([f \geqq t]) = \infty$ for some $t > 0$ then for each $R > 0$ there exists a subset $A \in \mathfrak{S}$ with $A \subset [f \geqq t]$ and $t\varphi(A) \geqq R$, which implies that $u := t\chi_A \in S(\mathfrak{S})$ fulfils $u \leqq f$ and $\fint u d\varphi = t\varphi(A) \geqq R$, and hence that $M \geqq R$. Therefore $M = \infty$. 2) We fix $0 < a < b < \infty$ and have to prove that $\int_a^b \varphi_\star([f \geqq t]) dt \leqq M$. Let $\varepsilon > 0$. By definition of the Riemann integral there exists a subdivision $a = t(0) < t(1) < \cdots < t(r) = b$ with

$$\int_a^b \varphi_\star([f \geqq t]) dt \leqq \sum_{l=1}^r \big(t(l) - t(l-1)\big)\varphi_\star([f \geqq t(l)]) dt + \varepsilon.$$

Then there exist subsets $A(l) \in \mathfrak{S}$ with $A(l) \subset [f \geqq t(l)]$ and

$$\varphi_\star([f \geqq t(l)]) \leqq \varphi(A(l)) + \frac{\varepsilon}{b-a} \; \forall l = 1, \cdots, r.$$

In view of $[f \geqq t(r)] \subset \cdots \subset [f \geqq t(1)]$ we can achieve that $A(r) \subset \cdots \subset A(1)$. It follows that

$$\int_a^b \varphi_\star([f \geqq t]) dt \leqq \sum_{l=1}^r \big(t(l) - t(l-1)\big)\varphi(A(l)) + 2\varepsilon.$$

Now we form

$$u := \sum_{l=1}^r \big(t(l) - t(l-1)\big)\chi_{A(l)} \in S(\mathfrak{S}).$$

By 11.6 then $u \leqq (f - a)^+ \wedge (b - a) \leqq f$, and hence from 11.8.1) we obtain $\int_a^b \varphi_\star([f \geqq t]) dt \leqq \fint u d\varphi + 2\varepsilon$. The assertion follows.

We turn to the theorems on downward and upward $\bullet$ continuity for $\bullet = \sigma\tau$. In case $\bullet = \sigma$ these are archetypes of the Beppo Levi theorem. It is fundamental that they are also true for $\bullet = \tau$. We present the proof in the downward situation which is somewhat more involved, and then add the upward one as an exercise.

11.17. THEOREM. Assume that φ is almost downward $\bullet$ continuous. Then the horizontal integral $f \mapsto \fint f d\varphi$ is almost downward $\bullet$ continuous on $\mathrm{UM}(\mathfrak{S})$ in the natural sense: If $M \subset \mathrm{UM}(\mathfrak{S})$ is nonvoid of type $\bullet$ and downward directed in the pointwise order with $M \downarrow F \in \mathrm{UM}(\mathfrak{S})$, and if $\fint f d\varphi < \infty$ for all $f \in M$, then $\inf\{\fint f d\varphi : f \in M\} = \fint F d\varphi$.

Proof. 1) For $f \in M$ we define $\hat{f} :]0, \infty[\to [0, \infty[$ to be $\hat{f}(t) = \varphi([f \geqq t]) < \infty \; \forall t > 0$. Thus $\hat{f}$ is monotone decreasing with $\fint f d\varphi = \int_{0\leftarrow}^{\to\infty} \hat{f}(t) dt < \infty$. Furthermore $f \leqq g$ implies that $\hat{f} \leqq \hat{g}$. 2) For fixed $t > 0$ we have $\{[f \geqq t] : f \in M\} \downarrow [F \geqq t]$. Thus by assumption $\inf\{\varphi([f \geqq t]) = \hat{f}(t) : f \in M\} = \varphi([F \geqq t]) =: \hat{F}(t)$, with $\hat{F} :]0, \infty[\to [0, \infty[$ as above. It follows that $\{\hat{f} : f \in M\} \downarrow \hat{F}$ pointwise. 3) We fix $P \in M$ and $\varepsilon > 0$, and then $0 < a < b < \infty$ with

$$\int_{0\leftarrow}^a \hat{P}(t) dt \leqq \varepsilon \quad \text{and} \quad \int_b^{\to\infty} \hat{P}(t) dt \leqq \varepsilon.$$

For $f \in M$ with $f \leqq P$ then

$$\oint f d\varphi = \int_{0\leftarrow}^{\rightarrow\infty} \hat{f}(t)dt \leqq \int_a^b \hat{f}(t)dt + 2\varepsilon.$$

4) By definition of the Riemann integral there exists a subdivision $a = t_0 < t_1 < \cdots < t_r = b$ with

$$\sum_{l=1}^r (t_l - t_{l-1})\hat{F}(t_{l-1}) \leqq \int_a^b \hat{F}(t)dt + \varepsilon.$$

Then by 2) there exist functions $f_l \in M \; \forall l = 1, \cdots, r$ such that $\hat{f}_l(t_{l-1}) \leqq \hat{F}(t_{l-1}) + \varepsilon/b - a$. Thus for the functions $f \in M$ with $f \leqq f_1, \cdots, f_r, P$ we obtain from 3)

$$\oint f d\varphi \;\leqq\; \int_a^b \hat{f}(t)dt + 2\varepsilon \leqq \sum_{l=1}^r (t_l - t_{l-1})\hat{f}(t_{l-1}) + 2\varepsilon$$

$$\leqq \sum_{l=1}^r (t_l - t_{l-1})\hat{f}_l(t_{l-1}) + 2\varepsilon \leqq \sum_{l=1}^r (t_l - t_{l-1})\hat{F}(t_{l-1}) + 3\varepsilon$$

$$\leqq \int_a^b \hat{F}(t)dt + 4\varepsilon \leqq \oint F d\varphi + 4\varepsilon.$$

The assertion follows.

11.18. EXERCISE. *Assume that φ is upward • continuous. Then the horizontal integral $f \mapsto \oint f d\varphi$ is upward • continuous on $\mathrm{LM}(\mathfrak{S})$ in the natural sense: If $M \subset \mathrm{LM}(\mathfrak{S})$ is nonvoid of type • and upward directed in the pointwise order with $M \uparrow F \in \mathrm{LM}(\mathfrak{S})$, then $\sup\{\oint f d\varphi : f \in M\} = \oint F d\varphi$.*

We continue with the important specialization to Borel-Radon measures.

11.19. REMARK. Let X be a topological space. Then $\mathrm{UM}(\mathrm{Cl}(X))$ is the class $\mathrm{USC}(X, [0, \infty])$ of the upper semicontinuous functions $f : X \to [0, \infty]$, and $\mathrm{LM}(\mathrm{Op}(X))$ is the class $\mathrm{LSC}(X, [0, \infty])$ of the lower semicontinuous functions $f : X \to [0, \infty]$.

11.20. CONSEQUENCE. *Let $\alpha : \mathrm{Bor}(X) \to [0, \infty]$ be a Borel-Radon measure on the Hausdorff topological space X. Then the horizontal integral $f \mapsto \oint f d\alpha$ is almost downward τ continuous on $\mathrm{USC}(X, [0, \infty])$ and upward τ continuous on $\mathrm{LSC}(X, [0, \infty])$.*

Proof. We know from 9.4 that $\alpha|\mathrm{Cl}(X)$ is almost downward τ continuous and $\alpha|\mathrm{Op}(X)$ is upward τ continuous. Therefore the assertions follow from 11.17 and 11.18.

11.21. EXERCISE. 1) Let X be a completely regular topological space. Assume that $f \in \mathrm{USC}(X, [-\infty, \infty[)$ is such that there exists some $u \in C(X, \mathbb{R})$ with $u \geqq f$. Then f is the pointwise infimum of $\{u \in C(X, \mathbb{R}) : u \geqq f\}$. 2) Let $\alpha : \mathrm{Bor}(X) \to [0, \infty[$ be a finite Borel-Radon measure on a completely regular Hausdorff space X. Assume that $f \in \mathrm{USC}(X, [0, \infty[)$ is bounded above. Then

$$\int f d\alpha = \inf\left\{\int u d\alpha : u \in \mathrm{C}(X, [0, \infty[) \text{ bounded with } u \geqq f\right\}.$$

11.22. EXERCISE. This is another variant of 11.17 whose proof is similar and simpler. *Assume that φ is almost $\bullet$ continuous at $\varnothing$. If $M \subset \mathrm{UM}(\mathfrak{S})$ is nonvoid of type $\bullet$ such that $M \downarrow 0$ in the pointwise order, and if $\int f d\varphi < \infty$ for all $f \in M$, then $\inf\{\int f d\varphi : f \in M\} = 0$.*

The last exercise continues the discussion of the two notions of support defined in section 9.

11.23. EXERCISE. Let $\mathfrak{S}$ be a lattice with $\varnothing \in \mathfrak{S}$ and $\varphi : \mathfrak{S} \to [0, \infty]$ be isotone with $\varphi(\varnothing) = 0$. By definition

$$\text{for } f \in \mathrm{UM}(\mathfrak{S}) \quad : \quad \int f d\varphi = 0 \Leftrightarrow \varphi([f \geqq t]) = 0 \; \forall t > 0,$$

$$\text{for } f \in \mathrm{LM}(\mathfrak{S}) \quad : \quad \int f d\varphi = 0 \Leftrightarrow \varphi([f > 0]) = 0 \; \forall t > 0.$$

It follows that $\int f d\varphi = 0$ does not enforce that $f = 0$. For example, for $A \in \mathfrak{S}$ with $\varphi(A) = 0$ we have $\int \chi_A d\varphi = \varphi(A) = 0$ by 11.8.1), but of course A need not be void. Moreover $\int f d\varphi = 0$ does not even enforce that $f = 0$ on that part of X on which φ is *concentrated* in some reasonable sense, except under additional assumptions on f. In this context the previous notions of support become relevant.

0) The Borel-Lebesgue measure $\alpha =: \Lambda | \mathrm{Bor}(\mathbb{R}^n)$ fulfils

$$\mathrm{Supp}(\alpha) = \mathrm{supp}(\alpha | \mathfrak{K}) = \mathbb{R}^n.$$

Hint: $\mathfrak{K} \top \mathfrak{K} = \mathrm{Cl}(\mathbb{R}^n)$.

1) Let $\alpha : \mathrm{Bor}(X) \to [0, \infty]$ be a Borel-Radon measure on a Hausdorff topological space X. Then

$$\text{for } f \in \mathrm{LSC}(X, [0, \infty]) = \mathrm{LM}(\mathrm{Op}(X)) : \int f d\alpha = 0 \Rightarrow f = 0 \text{ on } \mathrm{Supp}(\alpha).$$

The example above shows that the same assertion need not be true for $f \in \mathrm{USC}(X, [0, \infty]) = \mathrm{UM}(\mathrm{Cl}(X))$. These are well-known facts. The positive assertion is considered as an indication that the notion of support is a reasonable one. It is remarkable that it has a counterpart for the new notion of support.

2) Let $\mathfrak{S}$ be a lattice with $\varnothing \in \mathfrak{S}$ in the nonvoid set X, and let $\varphi : \mathfrak{S} \to [0, \infty[$ be an inner τ premeasure with $\varphi(\varnothing) = 0$ and $\phi := \varphi_\tau | \mathfrak{C}(\varphi_\tau)$. Then

$$\text{for } f \in \mathrm{LM}((\mathfrak{S} \top \mathfrak{S}_\tau) \bot) : \int f d\phi = 0 \Rightarrow f = 0 \text{ on } \mathrm{supp}(\varphi).$$

3) Deduce 1) from 2).

11.24. BIBLIOGRAPHICAL NOTE. The concept of the horizontal integral was the basic idea of Lebesgue for the formation of an integral, while the vertical integral of the next section was the basic idea of Riemann. See for example Stroock [1994] introduction to chapter II. In recent textbooks the horizontal integral occurs in case that $\varphi : \mathfrak{S} \to [0, \infty]$ is a σ finite cmeasure on a σ algebra, but without much systematization. See for example Bauer [1992] Satz 23.8. The full notion of the horizontal integral appeared in Choquet [1953-54], so that this work formed the start of the so-called non-additive theory of integration. The fundamental result 11.11 is from Choquet [1953-54], Topsøe [1978] section 8, and Kindler [1986]. For more details we refer to Denneberg [1994] chapter 6. For the sandwich type results in 11.14 and their development see for example Kindler [1986]. The results in the last subsection are basic facts for Borel-Radon measures; see for example Dellacherie-Meyer [1978] chapter III section 3. The more comprehensive versions of the present text will be needed for the main results of chapters V and VII below. There have been hints in this direction in the literature before, for example in Topsøe [1978] section 8.

12. The Vertical Integral

Definition and Main Properties

This time we assume that $\mathfrak{A}$ is a ring in a nonvoid set X and that $\alpha : \mathfrak{A} \to [0, \infty]$ is a ccontent. We define the **vertical integral** for $f : X \to [0, \infty]$ with respect to α to be

$$
\begin{aligned}
\int_* f d\alpha : \ &= \ \sup \Big\{ \sum_{l=1}^{r} t_l \alpha(A(l)) : A(1), \cdots, A(r) \in \mathfrak{A} \text{ pairwise disjoint} \\
&\qquad \text{and } t_1, \cdots, t_r > 0 \text{ with } f|A(l) \geqq t_l \ \forall l = 1, \cdots, r \Big\} \\
&= \ \sup \Big\{ \sum_{l=1}^{r} t_l \alpha(A(l)) : A(1), \cdots, A(r) \in \mathfrak{A} \text{ pairwise disjoint} \\
&\qquad \text{and } t_1, \cdots, t_r \geqq 0 \text{ with } f|A(l) \geqq t_l \ \forall l = 1, \cdots, r \Big\},
\end{aligned}
$$

where the second expression is under the usual convention $0\infty := 0$. The first expression is the supremum of a nonvoid subset of $[0, \infty]$ since $\varnothing \in \mathfrak{A}$ and $\alpha(\varnothing) = 0$. It is clear that the two expressions are equal and with value in $[0, \infty]$.

12.1. REMARK. *For $f : X \to [0, \infty]$ we have*

$$\int_* f d\alpha : \quad = \quad \sup \left\{ \sum_{l=1}^{r} t_l \alpha(A(l)) : A(1), \cdots, A(r) \in \mathfrak{A} \right.$$

$$\left. and \ t_1, \cdots, t_r > 0 \ with \ f \geqq \sum_{l=1}^{r} t_l \chi_{A(l)} \right\}$$

$$= \quad \sup \left\{ \sum_{l=1}^{r} t_l \alpha(A(l)) : A(1), \cdots, A(r) \in \mathfrak{A} \right.$$

$$\left. and \ t_1, \cdots, t_r \geqq 0 \ with \ f \geqq \sum_{l=1}^{r} t_l \chi_{A(l)} \right\}.$$

Proof. As above it is clear that the two expressions on the right are equal. Let $I \in [0, \infty]$ denote their common value. To be shown is $I \leqq \int_* f d\alpha$, since the reverse is obvious. We can assume that $\int_* f d\alpha < \infty$, so that $A \in \mathfrak{A}$ with $f|A \geqq t > 0$ implies that $\alpha(A) < \infty$. Now fix $A(1), \cdots, A(r) \in \mathfrak{A}$ and $t_1, \cdots, t_r > 0$ with $f \geqq \sum_{l=1}^{r} t_l \chi_{A(l)}$. Thus $\alpha(A(l)) < \infty \ \forall l = 1, \cdots, r$. As in the earlier obvious lemma 3.5 we form the subsets

$$D(T) := \bigcap_{l \in T} A(l) \cap \bigcap_{l \in T'} (A(l))' \in \mathfrak{A} \quad \text{for the nonvoid } T \subset \{1, \cdots, r\}.$$

These subsets are pairwise disjoint with $A(l) = \bigcup_{T \ni l} D(T)$ or

$$\chi_{A(l)} = \sum_{T \ni l} \chi_{D(T)} \quad \text{and hence} \quad \alpha(A(l)) = \sum_{T \ni l} \alpha(D(T)) \ \forall l = 1, \cdots, r.$$

We also put $\tau_T := \sum_{l \in T} t_l > 0$. Then on the one hand

$$\sum_{l=1}^{r} t_l \chi_{A(l)} = \sum_{l=1}^{r} \sum_{T \ni l} t_l \chi_{D(T)} = \sum_{T} \sum_{l \in T} t_l \chi_{D(T)} = \sum_{T} \tau_T \chi_{D(T)},$$

and on the other hand

$$\sum_{l=1}^{r} t_l \alpha(A(l)) = \sum_{l=1}^{r} \sum_{T \ni l} t_l \alpha(D(T)) = \sum_{T} \sum_{l \in T} t_l \alpha(D(T)) = \sum_{T} \tau_T \alpha(D(T)).$$

The first relation shows that $f|D(T) \geqq \tau_T$ for all T. Therefore the common value in the second relation is $\leqq \int_* f d\alpha$. This is the assertion.

12.2. PROPERTIES. 1) $\int_* \chi_A d\alpha = \alpha(A)$ for $A \in \mathfrak{A}$. 2) *The vertical integral is isotone.* 3) *Assume that α is semifinite above. Then*

$$\int_* f d\alpha = \sup \left\{ \int_* \chi_P f d\alpha : P \in \mathfrak{A} \text{ with } \alpha(P) < \infty \right\} \text{ for all } f : X \to [0, \infty].$$

4) *The vertical integral is positive-homogeneous. Furthermore*

$$\int_* (f+g)d\alpha \geq \int_* f d\alpha + \int_* g d\alpha \quad for\ f, g : X \to [0, \infty].$$

Proof. 1) On the one hand $\chi_A | A \geq 1$ implies by definition that $\int_* \chi_A d\alpha \geq \alpha(A)$. On the other hand let $A(1), \cdots, A(r) \in \mathfrak{A}$ be pairwise disjoint and $t_1, \cdots, t_r > 0$ with $\chi_A | A(l) \geq t_l \ \forall l = 1, \cdots, r$. This implies that $A(l) \subset A$, and that $A(l) = \varnothing$ when $t_l > 1$. It follows that

$$\sum_{l=1}^{r} t_l \alpha(A(l)) \leq \sum_{l=1}^{r} \alpha(A(l)) \leq \alpha(A).$$

Therefore $\int_* \chi_A d\alpha \leq \alpha(A)$. 2)3) and the first part of 4) are obvious. It remains to prove the second part of 4). Let

$A(1), \cdots, A(r) \in \mathfrak{A}$ be pairwise disjoint
and $s_1, \cdots, s_r \geq 0$ with $f | A(k) \geq s_k \ \forall k = 1, \cdots, r$,
$B(1), \cdots, B(s) \in \mathfrak{A}$ be pairwise disjoint
and $t_1, \cdots, t_s \geq 0$ with $g | B(l) \geq t_l \ \forall l = 1, \cdots, s$.

We can assume that $\bigcup_{k=1}^{r} A(k) = \bigcup_{l=1}^{s} B(l)$. Then $(f+g) | A(k) \cap B(l) \geq s_k + t_l$ implies that

$$\int_* (f+g)d\alpha \geq \sum_{k=1}^{r} \sum_{l=1}^{s} (s_k + t_l)\alpha(A(k) \cap B(l))$$

$$= \sum_{k=1}^{r} \sum_{l=1}^{s} s_k \alpha(A(k) \cap B(l)) + \sum_{k=1}^{r} \sum_{l=1}^{s} t_l \alpha(A(k) \cap B(l))$$

$$= \sum_{k=1}^{r} s_k \alpha(A(k)) + \sum_{l=1}^{s} t_l \alpha(B(l)),$$

and hence the assertion.

12.3. PROPOSITION. *For $f, g : X \to [0, \infty]$ we have*

$$\int_* (f+g)d\alpha = \int_* f d\alpha + \int_* g d\alpha,$$

provided that at least one of the two functions is in $\mathrm{UM}(\mathfrak{A}) \cup \mathrm{LM}(\mathfrak{A})$.

Proof. We assume that $f \in \mathrm{UM}(\mathfrak{A}) \cup \mathrm{LM}(\mathfrak{A})$. By 12.2.4) we have to prove the direction $\leq$. Fix pairwise disjoint $A(1), \cdots, A(r) \in \mathfrak{A}$ and $t_1, \cdots, t_r > 0$ with $(f+g) | A(l) \geq t_l \ \forall l = 1, \cdots, r$. Furthermore fix $n \in \mathbb{N}$. For $1 \leq l \leq r$ then form

$$A^k(l) := \left\{ \begin{array}{ll} A(l) \cap [f \geq (k/n)t_l] & \text{when } f \in \mathrm{UM}(\mathfrak{A}) \\ A(l) \cap [f > (k/n)t_l] & \text{when } f \in \mathrm{LM}(\mathfrak{A}) \end{array} \right\} \in \mathfrak{A} \ \forall k = 1, \cdots, n.$$

Thus we have $A^n(l) \subset \cdots \subset A^1(l) \subset A^0(l) := A(l)$, and hence the decomposition

$$A(l) = \bigcup_{k=1}^{n} \left(A^{k-1}(l) \setminus A^k(l) \right) \cup A^n(l).$$

In both cases we have

for $1 \leq k \leq n$ on $A^{k-1}(l) \setminus A^k(l) : f \geq \dfrac{k-1}{n} t_l$ and $g \geq \dfrac{n-k}{n} t_l$.

Therefore by definition

$$\int_* f d\alpha \;\geq\; \sum_{l=1}^{r} \left(\sum_{k=1}^{n} \frac{k-1}{n} t_l \alpha \left(A^{k-1}(l) \setminus A^k(l) \right) + t_l \alpha \left(A^n(l) \right) \right),$$

$$\int_* g d\alpha \;\geq\; \sum_{l=1}^{r} \left(\sum_{k=1}^{n} \frac{n-k}{n} t_l \alpha \left(A^{k-1}(l) \setminus A^k(l) \right) \right),$$

and hence

$$\int_* f d\alpha + \int_* g d\alpha \geq \frac{n-1}{n} \sum_{l=1}^{r} t_l \alpha(A(l)).$$

Now let $n \to \infty$. Then the assertion follows.

12.4. CONSEQUENCE. *For $f_1, \cdots, f_r : X \to [0, \infty]$ we have*

$$\int_* \left(\sum_{l=1}^{r} f_l \right) d\alpha = \sum_{l=1}^{r} \int_* f_l d\alpha,$$

provided that all these functions except at most one are in $\mathrm{UM}(\mathfrak{A}) \cup \mathrm{LM}(\mathfrak{A})$.

12.5. SPECIAL CASE. *If $f \in \mathrm{S}(\mathfrak{A})$ is represented in the form* 11.4.iii) *then*

$$\int_* f d\alpha = \sum_{l=1}^{r} t_l \alpha(A(l)).$$

We conclude with some useful remarks.

12.6. REMARK. *For each $f : X \to [0, \infty]$ we have* 1) $\int_* (f \wedge t) d\alpha \uparrow \int_* f d\alpha$ *for $t \uparrow \infty$ and* 2) $\int_* (f - t)^+ d\alpha \uparrow \int_* f d\alpha$ *for $t \downarrow 0$.*

Proof. We can assume that $\int_* f d\alpha > 0$. Let $\int_* f d\alpha > c > 0$. By definition there exist pairwise disjoint $A(1), \cdots, A(r) \in \mathfrak{A}$ and $t_1, \cdots, t_r > 0$ such that $f|A(l) \geq t_l \ \forall l = 1, \cdots, r$ and $\sum_{l=1}^{r} t_l \alpha(A(l)) \geq c.$ 1) For $t \geq t_1, \cdots, t_r$ we have $(f \wedge t)|A(l) \geq t_l \ \forall l = 1, \cdots, r$ and hence $\int_* (f \wedge t) d\alpha \geq c.$ 2) For $0 < t < \varepsilon := \min(t_1, \cdots, t_r)$ we have $(f - t)|A(l) \geq t_l - t \geq (1 - t/\varepsilon) t_l \ \forall l = 1, \cdots, r$ and hence $\int_* (f - t)^+ d\alpha \geq (1 - t/\varepsilon)c.$

12.7. REMARK. *Assume that* $f : X \to [0, \infty]$ *and* $A \in \mathfrak{A}$ *with* $[f > 0]$
$\subset A$. *Then* $\int_* f d\alpha \leqq (\sup f) \alpha(A)$.

Proof. By 12.6.1) we can assume that $\sup f < \infty$. But then the assertion follows from 12.2.1).

Regularity and Continuity of the Vertical Integral

The next assertion is an immediate consequence of the definition combined with 12.5.

12.8. PROPOSITION. $\int_* f d\alpha = \sup \left\{ \int_* u d\alpha : u \in S(\mathfrak{A}) \text{ with } u \leqq f \right\}$ *for all*
$f : X \to [0, \infty]$.

The continuity theorem will be this time on upward σ continuity and a consequence of 4.7. It is a version of the famous Fatou theorem.

12.9. THEOREM. *Assume that* α *is upward* σ *continuous. Let* $(f_n)_n$ *be a sequence in* $\mathrm{UM}(\mathfrak{A})$ *or in* $\mathrm{LM}(\mathfrak{A})$, *and* $f : X \to [0, \infty]$ *with* $f \leqq \liminf\limits_{n\to\infty} f_n$.
Then $\int_* f d\alpha \leqq \liminf\limits_{n\to\infty} \int_* f_n d\alpha$.

Proof. We can assume that $\int_* f d\alpha > 0$. Let $\int_* f d\alpha > c > 0$. By definition there exist pairwise disjoint $A(1), \cdots, A(r) \in \mathfrak{A}$ and $t_1, \cdots, t_r > 0$ such that $f|A(l) \geqq t_l \; \forall l = 1, \cdots, r$ and $\sum\limits_{l=1}^{r} t_l \alpha(A(l)) > c$. Then choose $0 < \delta < t_1, \cdots, t_r$ with $\sum\limits_{l=1}^{r} (t_l - \delta) \alpha(A(l)) > c$; this can be done in all cases. 1) For $1 \leq l \leq r$ we form

$$D_n^l := \bigcap_{p=n}^{\infty} \left(A(l) \cap [f_p > t_l - \delta] \right) \; \forall n \in \mathbb{N}.$$

Thus $D_n^l \uparrow$ in $n \in \mathbb{N}$. For $x \in A(l)$ we have $t_l \leq f(x) \leq \liminf\limits_{n\to\infty} f_n(x)$ and hence $t_l - \delta < f_p(x)$ for almost all $p \in \mathbb{N}$; thus $x \in D_n^l$ for almost all $n \in \mathbb{N}$. It follows that $D_n^l \uparrow A(l)$ for $n \to \infty$. From 4.7 we obtain $\alpha^\sigma(D_n^l) \uparrow \alpha^\sigma(A(l))$. Thus $D_n^l \subset A(l) \cap [f_n \geqq / > t_l - \delta] \subset A(l)$ with 4.5 implies that

$$\alpha \left(A(l) \cap [f_n \geqq / > t_l - \delta] \right) \to \alpha(A(l)) \text{ for } n \to \infty.$$

2) For $n \in \mathbb{N}$ we have by definition

$$\int_* f_n d\alpha \geqq \sum_{l=1}^{r} (t_l - \delta) \alpha \left(A(l) \cap [f_n \geqq / > t_l - \delta] \right).$$

We know from 1) that the right side is $> c$ for almost all $n \in \mathbb{N}$. It follows that $\liminf\limits_{n\to\infty} \int_* f_n d\alpha \geqq c$. This is the assertion.

As a consequence we obtain a special case of the Beppo Levi theorem.

12.10. CONSEQUENCE. *Assume that α is upward σ continuous. Let $(f_n)_n$ be a sequence in* $\mathrm{UM}(\mathfrak{A})$ *or in* $\mathrm{LM}(\mathfrak{A})$, *and* $f_n \uparrow f : X \to [0, \infty]$. *Then $\int_* f_n d\alpha \uparrow \int_* f d\alpha$ for $n \to \infty$.*

Proof. From 12.9 we obtain $\int_* f d\alpha \leq \lim_{n \to \infty} \int_* f_n d\alpha$. On the other hand $\int_* f_n d\alpha \leq \int_* f d\alpha$ for $n \in \mathbb{N}$ and hence $\lim_{n \to \infty} \int_* f_n d\alpha \leq \int_* f d\alpha$.

Comparison of the two Integrals

The next theorem says that the horizontal and the vertical integral are equal whenever they are both defined.

12.11. THEOREM. *Assume that $\alpha : \mathfrak{A} \to [0, \infty]$ is a ccontent on a ring $\mathfrak{A}$ in X. Then*

$$\oint f d\alpha = \int_* f d\alpha \quad \text{for all } f \in \mathrm{UM}(\mathfrak{A}) \cup \mathrm{LM}(\mathfrak{A}).$$

Proof. We define the function $F :\,]0, \infty[\, \to [0, \infty]$ to be

$$F(t) = \left\{ \begin{array}{ll} \alpha([f \geq t]) & \text{when } f \in \mathrm{UM}(\mathfrak{A}) \\ \alpha([f > t]) & \text{when } f \in \mathrm{LM}(\mathfrak{A}) \end{array} \right\} \quad \text{for } t > 0.$$

Thus F is monotone decreasing, and by definition $\oint f d\alpha = \int_{0\leftarrow}^{\to\infty} F(t) dt$. If $F(t) = \infty$ for some $t > 0$, then $\int_{o\leftarrow}^{\to\infty} F(t) dt = \infty$ by definition, and $\int_* f d\alpha = \infty$ by definition as well. Thus we can assume that $F(t) < \infty$ for all $t > 0$.

1) We first claim that

$$\int_* (f - u)^+ \wedge (v - u) d\alpha = \int_u^v F(t) dt \quad \text{for } 0 < u < v < \infty.$$

In fact, for each subdivision $u = t(0) < t(1) < \cdots < t(r) = v$ we have by 11.6

$$\sum_{l=1}^{r} \left(t(l) - t(l-1) \right) \chi_{[f \geq t(l)]} \quad \leqq \quad (f - u)^+ \wedge (v - u)$$

$$\leqq \quad \sum_{l=1}^{r} \left(t(l) - t(l-1) \right) \chi_{[f \geq t(l-1)]},$$

and the same for $[f > \cdot]$ instead of $[f \geq \cdot]$. Upon application of $\int_*$ it follows from 12.5 that in both cases

$$\sum_{l=1}^{r} \left(t(l) - t(l-1) \right) F(t(l)) \quad \leqq \quad \int_* (f - u)^+ \wedge (v - u) d\alpha$$

$$\leqq \quad \sum_{l=1}^{r} \left(t(l) - t(l-1) \right) F(t(l-1)).$$

By the definition of the Riemann integral the assertion follows. 2) For $0 < u < v < \infty$ we have $(f - u)^+ \wedge (v - u) = ((f \wedge v) - u)^+$, so that 1) says that

$$\int_* ((f \wedge v) - u)^+ d\alpha = \int_u^v F(t)dt.$$

We conclude from 12.6.2) for $u \downarrow 0$ that

$$\int_* (f \wedge v)d\alpha = \int_{0\leftarrow}^v F(t)dt \text{ for } v > 0,$$

and then from 12.6.1) for $v \uparrow \infty$ that $\int_* f d\alpha = \int_{0\leftarrow}^{\to\infty} F(t)dt$. This is the assertion.

12.12. CONSEQUENCE. *Assume that* $\alpha : X \to [0, \infty]$ *is a ccontent on a ring* $\mathfrak{A}$ *on* X. *Then*

$$\oint f d\alpha_\star = \int_* f d\alpha \quad \text{for all } f : X \to [0, \infty].$$

Proof. From 11.16 and 12.8 we obtain

$$\oint f d\alpha_\star = \sup \{ \oint u d\alpha : u \in S(\mathfrak{A}) \text{ with } u \leqq f \},$$

$$\int_* f d\alpha = \sup \{ \int_* u d\alpha : u \in S(\mathfrak{A}) \text{ with } u \leqq f \}.$$

Thus the assertion follows from 12.11.

In spite of the above results it is wise to retain the different notations for the two types of integrals, because this will allow to see where the arguments come from. We shall reserve the common notation $\int f d\alpha := \oint f d\alpha = \int_* f d\alpha$ for the particular case that $\alpha : \mathfrak{A} \to [0, \infty]$ is a cmeasure on a σ ring $\mathfrak{A}$ and that $f \in M(\mathfrak{A}) := UM(\mathfrak{A}) = LM(\mathfrak{A})$.

12.13. BIBLIOGRAPHICAL NOTE. The definition of the vertical integral is in essence the usual definition of the integral for functions with values in $[0, \infty]$ in most recent textbooks, except that the latter definition assumes $\alpha : \mathfrak{A} \to [0, \infty]$ to be a cmeasure on a σ ring $\mathfrak{A}$ and $f : X \to [0, \infty]$ to be measurable $\mathfrak{A}$. Also it is effected in two steps, where the first step is for the function class $S(\mathfrak{A})$ of the so-called elementary functions. For a few variants we refer to Bauer [1992], Cohn [1980], and Stroock [1994].

We next comment on the connection with the so-called finitely-additive theory of Riemann integration with respect to a ccontent $\alpha : \mathfrak{A} \to [0, \infty[$ or $[0, \infty]$ on an algebra $\mathfrak{A}$, and its extension due to Dunford [1935]. We refer to Rao-Rao [1983] and in particular to the recent survey of Luxemburg [1991]. These theories are for function classes which arise as the closures of $S(\mathfrak{A})$, and likewise of $UM(\mathfrak{A})$ and/or of $LM(\mathfrak{A})$, under some seminorm or semimetric of the type of an outer integral. We do not follow this procedure, not alone because it need not lead to complete function spaces. The decisive reason is that the present set-up is in perfect accord with the precise requirements of the next chapter. The present comparison theorem 12.11 is contained in Luxemburg [1991] theorem 4.13.

13. The Conventional Integral

The present section develops the conventional integral for functions with values in $\mathbb{R}$ or $\overline{\mathbb{R}}$ on the basis of the two former sections. Apart from this it follows the usual lines from the start. It is for the sake of completeness that we want a short but complete treatment. We note that there will be almost no use of the present section in chapter V.

Measurable Functions

Let X and Y be nonvoid sets and $f : X \to Y$ be a map. Assume that $\mathfrak{A}/\mathfrak{B}$ are pavings in X/Y. Then it is natural to consider for f the properties

$$A \in \mathfrak{A} \Rightarrow f(A) \in \mathfrak{B} \quad \text{and} \quad f^{-1}(B) \in \mathfrak{A} \Leftarrow B \in \mathfrak{B}.$$

It turns out that, opposite to naive opinion, the second relation is the much more profound one. For example, if $\mathfrak{A}/\mathfrak{B}$ are topologies on X/Y then the second relation means that f is continuous, while the first relation means that f is open, which for example is violated as a rule when f is constant. Thus we define f to be **measurable** $\mathfrak{A} \to \mathfrak{B}$ iff it satisfies the second relation above, that is once more iff $B \in \mathfrak{B} \Rightarrow f^{-1}(B) \in \mathfrak{A}$. In order that this notion be non-pathological the pavings $\mathfrak{A}$ and $\mathfrak{B}$ must have a certain richness. For the purpose of measure and integration a minimal requirement appears to be that $\mathfrak{A}$ and $\mathfrak{B}$ be algebras. For example, in case $\mathfrak{B} \neq \{\varnothing\}$ the constant functions $f : X \to Y$ are not all measurable $\mathfrak{A} \to \mathfrak{B}$ unless $X \in \mathfrak{A}$. But this minimal requirement does not suffice, as a look at the notions of section 11 and the subsequent related exercise will show.

13.1. EXAMPLE. On $[0, \infty]$ define $\mathfrak{S}$ to consist of the subsets $[t, \infty]$ and $\mathfrak{T}$ of the subsets $]t, \infty]$ for all real $t > 0$. Let $\mathfrak{A}$ be a lattice with $\varnothing \in \mathfrak{A}$ on X. Then a function $f : X \to [0, \infty]$ is measurable $\mathfrak{A} \to \mathfrak{S}$ iff $f \in \mathrm{UM}(\mathfrak{A})$, and measurable $\mathfrak{A} \to \mathfrak{T}$ iff $f \in \mathrm{LM}(\mathfrak{A})$. We know from 11.1 and 11.3 that these function classes need not be stable under addition and need not be equal even when $\mathfrak{A}$ is an algebra, but that all this is true when $\mathfrak{A}$ is a σ lattice.

13.2. EXERCISE. On $\overline{\mathbb{R}}$ define $\mathfrak{S}$ to consist of the subsets $[t, \infty]$ and $\mathfrak{T}$ of the subsets $]t, \infty]$ for all $t \in \mathbb{R}$. 1) $\mathrm{A}\sigma(\mathfrak{S}) = \mathrm{A}\sigma(\mathfrak{T}) = \mathrm{Bor}(\overline{\mathbb{R}})$. 2) $\mathrm{A}(\mathfrak{S})$ and $\mathrm{A}(\mathfrak{T})$ do not coincide. In fact, for $A \in \mathrm{A}(\mathfrak{S})$ the function $\chi_A|\mathbb{R}$ is right continuous, so that for example the members of $\mathfrak{T}$ are not in $\mathrm{A}(\mathfrak{S})$. Likewise, for $A \in \mathrm{A}(\mathfrak{T})$ the function $\chi_A|\mathbb{R}$ is left continuous, so that for example the members of $\mathfrak{S}$ are not in $\mathrm{A}(\mathfrak{T})$. Hint: Use 3.6.1) combined with 1.17.$\star$).

It is thus clear that certain important points will need σ algebras instead of algebras. We start with some formal properties.

13.3. REMARK. *Assume that $\mathfrak{A}$ and $\mathfrak{B}$ are σ algebras, and that $\mathfrak{B} = A\sigma(\mathfrak{T})$ for a set system $\mathfrak{T} \subset \mathfrak{B}$. Then f is measurable $\mathfrak{A} \to \mathfrak{B} \Leftrightarrow f$ is measurable $\mathfrak{A} \to \mathfrak{T}$.*

Proof of $\Leftarrow$). By 1.11 we have $f^{-1}(\mathfrak{B}) = f^{-1}(A\sigma(\mathfrak{T})) = A\sigma(f^{-1}(\mathfrak{T})) \subset A\sigma(\mathfrak{A}) = \mathfrak{A}$.

13.4. PROPERTIES. 1) *If $f : X \to Y$ is measurable $\mathfrak{A} \to \mathfrak{B}$ and $g : Y \to Z$ is measurable $\mathfrak{B} \to \mathfrak{C}$ then $g \circ f : X \to Z$ is measurable $\mathfrak{A} \to \mathfrak{C}$.* 2) *Assume that $f : X \to T \subset Y$. Then f is measurable $\mathfrak{A} \to \mathfrak{B} \Leftrightarrow f$ is measurable $\mathfrak{A} \to \mathfrak{B} \cap T$.* 3) *Assume that $f : X \to Y$ is measurable $\mathfrak{A} \to \mathfrak{B}$. For nonvoid $S \subset X$ then $f|S : S \to Y$ is measurable $\mathfrak{A} \cap S \to \mathfrak{B}$.* 4) *Assume that $X = \bigcup_{l=1}^{\infty} S_l$ with nonvoid $S_l \in \mathfrak{A}$ $\forall l \in \mathbb{N}$, and that $\mathfrak{A}$ is a σ lattice. If $f|S_l : S_l \to Y$ is measurable $\mathfrak{A} \cap S_l \to \mathfrak{B}$ $\forall l \in \mathbb{N}$ then f is measurable $\mathfrak{A} \to \mathfrak{B}$.*

Proof. 1)2)3) are obvious. 4) For $B \in \mathfrak{B}$ we have $f^{-1}(B) \cap S_l = (f|S_l)^{-1}(B) \in \mathfrak{A} \cap S_l \subset \mathfrak{A}$ since $S_l \in \mathfrak{A}$ $\forall l \in \mathbb{N}$, and hence $f^{-1}(B) \in \mathfrak{A}$ since $\mathfrak{A}^\sigma = \mathfrak{A}$.

The next point is the connection with topology and continuity. Let X be equipped with a paving $\mathfrak{A}$ and Y be a topological space. Then for $f : X \to Y$ one has to distinguish between measurable $\mathfrak{A} \to \mathrm{Bor}(Y)$ and measurable $\mathfrak{A} \to \mathrm{Baire}(Y)$. But in case that $\mathrm{Bor}(Y) = \mathrm{Baire}(Y)$, in particular when Y is semimetrizable, we can in short call this **measurable $\mathfrak{A}$**. Let us note a useful consequence of 13.2.1) and 13.3.

13.5. REMARK. *Let $\mathfrak{A}$ be a σ algebra in X. Then $f : X \to \overline{\mathbb{R}}$ is measurable $\mathfrak{A}$ iff $[f \geqq t] \in \mathfrak{A}$ $\forall t \in \mathbb{R}$, and likewise iff $[f > t] \in \mathfrak{A}$ $\forall t \in \mathbb{R}$.*

Another shorthand notation is as follows. Let X and Y be topological spaces. Then $f : X \to Y$ will be called **Borel measurable** iff it is measurable $\mathrm{Bor}(X) \to \mathrm{Bor}(Y)$, and **Baire measurable** iff it is measurable $\mathrm{Baire}(X) \to \mathrm{Baire}(Y)$.

13.6. EXERCISE. *If $f : X \to Y$ is continuous then it is Borel measurable as well as Baire measurable.*

We combine these remarks for a useful reduction principle.

13.7. PROPOSITION. *Let X be equipped with a σ algebra $\mathfrak{A}$ and Y be a topological space. Then $f : X \to Y$ is measurable $\mathfrak{A} \to \mathrm{Baire}(Y)$*

$\Leftrightarrow$ *for each $\varphi \in C(Y, \mathbb{R})$ the function $\varphi \circ f : X \to \mathbb{R}$ is measurable $\mathfrak{A}$.*

Proof. $\Rightarrow$) is obvious from 13.6. $\Leftarrow$) Let $B \in \mathrm{CCl}(Y)$, that is $B = [\varphi = 0]$ for some $\varphi \in C(Y, \mathbb{R})$. Then $f^{-1}(B) = [\varphi \circ f = 0]$ which is in $\mathfrak{A}$ by assumption. The assertion follows from 13.3.

We turn to the combination of functions, in particular to sequences of functions. For the remainder of the subsection we assume that $\mathfrak{A}$ is a σ algebra in X. We start with the easiest facts.

13.8. REMARK. *Let* $f, g : X \to \overline{\mathbb{R}}$ *be measurable* $\mathfrak{A}$. *Then the sets* $[f < g], [f \leq g], [f = g]$, *and* $[f \neq g]$ *are in* $\mathfrak{A}$.

Proof. It suffices to prove the assertion for $[f < g]$. But $[f < g]$ is the union of the subsets $[f < t] \cap [t < g]$ for all $t \in D$, where $D \subset \mathbb{R}$ is any countable dense subset.

13.9. PROPOSITION. *Let* $f_l : X \to \overline{\mathbb{R}}$ *be measurable* $\mathfrak{A}$ $\forall l \in \mathbb{N}$. 1) *The functions* $\sup_{l \in \mathbb{N}} f_l$ *and* $\inf_{l \in \mathbb{N}} f_l$ *are measurable* $\mathfrak{A}$. *Of course this implies the same fact for finite sequences of functions.* 2) *The functions* $\limsup_{l \to \infty} f_l$ *and* $\liminf_{l \to \infty} f_l$ *are measurable* $\mathfrak{A}$.

Proof. 1) For $t \in \mathbb{R}$ we have

$$[\sup_{l \in \mathbb{N}} f_l > t] = \bigcup_{l=1}^{\infty} [f_l > t] \text{ and } [\inf_{l \in \mathbb{N}} f_l \geq t] = \bigcap_{l=1}^{\infty} [f_l \geq t],$$

so that the assertion follows from 13.5. 2) We have $\limsup_{l \to \infty} f_l = \inf_{n \in \mathbb{N}} \sup_{l \geq n} f_l$ and $\liminf_{l \to \infty} f_l = \sup_{n \in \mathbb{N}} \inf_{l \geq n} f_l$.

13.10. PROPOSITION. *Let* Y *be a topological space. Assume that the* $f_l : X \to Y$ *are measurable* $\mathfrak{A} \to \mathrm{Baire}(Y)$ $\forall l \in \mathbb{N}$, *and that* $f_l \to f : X \to Y$ *pointwise. Then* f *is measurable* $\mathfrak{A} \to \mathrm{Baire}(Y)$.

Proof. In view of 13.7 we can assume that $Y = \mathbb{R}$; but the proof below works for all metrizable Y. Let $B \in \mathrm{CCl}(Y) = \mathrm{Cl}(Y)$. Then

$$A(l, n) := f_l^{-1}([\mathrm{dist}(\cdot, B) \leq 1/n]) \in \mathfrak{A} \quad \forall l, n \in \mathbb{N}, \text{ and hence}$$

$$A := \bigcap_{n=1}^{\infty} \bigcup_{p=1}^{\infty} \bigcap_{l=p}^{\infty} A(l, n) \in \mathfrak{A}.$$

For $x \in X$ now $x \in A$ means that for each $n \in \mathbb{N}$ there exists $p \in \mathbb{N}$ with $x \in A(l, n)$ $\forall l \geq p$, that is with $\mathrm{dist}(f_l(x), B) \leq 1/n$ $\forall l \geq p$. Thus $x \in A$ means that $\mathrm{dist}(f_l(x), B) \to 0$ for $l \to \infty$, that is $\mathrm{dist}(f(x), B) = 0$, and that is $f(x) \in B$ since B is closed. Thus $A = f^{-1}(B)$. The assertion follows.

The next assertion is an important addendum.

13.11. PROPOSITION. *Let* Y *be a Polish space. Assume that the* $f_l : X \to Y$ *are measurable* $\mathfrak{A}$ $\forall l \in \mathbb{N}$. *Then the subset*

$$T := \{x \in X : \text{there exists } \lim_{l \to \infty} f_l(x) =: f(x) \in Y\} \subset X$$

is in $\mathfrak{A}$. *Note that by 13.10 the function* $f : T \to Y$ *is measurable* $\mathfrak{A} \cap T$.

Proof. Let d be a complete metric on Y which furnishes its topology, and let $D \subset Y$ be a countable dense subset. We form

$$M_n(p, q) := \bigcup_{u \in D} f_p^{-1}\big(\nabla(u, 1/n)\big) \cap f_q^{-1}\big(\nabla(u, 1/n)\big) \in \mathfrak{A} \quad \text{for } p, q, n \in \mathbb{N};$$

$$A := \bigcap_{n=1}^{\infty} \bigcup_{r=1}^{\infty} \bigcap_{p=r}^{\infty} \bigcap_{q=r}^{\infty} M_n(p, q) \in \mathfrak{A}.$$

It is a routine verification that

$$\{x \in X : d(f_p(x), f_q(x)) < 1/n\} \subset M_n(p, q)$$
$$\subset \{x \in X : d(f_p(x), f_q(x)) \leqq 2/n\} \quad \text{for } p, q, n \in \mathbb{N}.$$

Thus A consists of those $x \in X$ in which the sequence $(f_l(x))_l$ is Cauchy in d, that is convergent in d. Therefore $T = A \in \mathfrak{A}$ as claimed.

In view of 11.1.3) and 11.2 it is perhaps no surprise that the most complicated task is to handle sums, products,$\cdots$ of scalar-valued measurable functions. For this purpose we introduce a notion which will be central in chapter VII.

Assume that $\mathfrak{S}_1, \cdots, \mathfrak{S}_r$ are pavings in $X_1, \cdots, X_r$. Then we form their **product paving**

$$\mathfrak{S}_1 \times \cdots \times \mathfrak{S}_r := \{S_1 \times \cdots \times S_r : S_l \in \mathfrak{S}_l \ \forall l = 1, \cdots, r\},$$

which is a paving in $X_1 \times \cdots \times X_r$. Furthermore we form $\mathfrak{S}_1 \otimes \cdots \otimes \mathfrak{S}_r := A\sigma(\mathfrak{S}_1 \times \cdots \times \mathfrak{S}_r)$, which is called their **product σ algebra** in case that $\mathfrak{S}_1, \cdots, \mathfrak{S}_r$ are σ algebras themselves. The next assertion is then obvious but will be useful.

13.12. REMARK. Let $\mathfrak{A}$ be a σ algebra in X and $\mathfrak{T}_1, \cdots, \mathfrak{T}_r$ be pavings in $Y_1, \cdots Y_r$. Assume that $f_l : X \to Y_l$ is measurable $\mathfrak{A} \to \mathfrak{T}_l \ \forall l = 1, \cdots, r$. Then $f = (f_1, \cdots, f_r) : X \to Y_1 \times \cdots \times Y_r$ is measurable $\mathfrak{A} \to \mathfrak{T}_1 \otimes \cdots \otimes \mathfrak{T}_r$.

In fact, for $B = B_1 \times \cdots \times B_r \in \mathfrak{T}_1 \times \cdots \times \mathfrak{T}_r$ we have

$$f^{-1}(B) = \bigcap_{l=1}^{r} f_l^{-1}(B_l) \in \mathfrak{A},$$

so that the assertion follows from 13.3. In the sequel we restrict ourselves to products of two factors.

13.13. EXERCISE. Consider pavings $\mathfrak{S}$ in X and $\mathfrak{T}$ in Y. 1) We have $Y \in \mathfrak{T}^\sigma \Rightarrow \mathfrak{S} \times A\sigma(\mathfrak{T}) \subset \mathfrak{S} \otimes \mathfrak{T}$; and of course $X \in \mathfrak{S}^\sigma \Rightarrow A\sigma(\mathfrak{S}) \times \mathfrak{T} \subset \mathfrak{S} \otimes \mathfrak{T}$ as well. Hint: Show that

$\mathfrak{B} := \{B \subset Y : S \times B \in \mathfrak{S} \otimes \mathfrak{T} \; \forall S \in \mathfrak{S}\}$ is a σ algebra in Y.

2) If $X \in \mathfrak{S}^\sigma$ and $Y \in \mathfrak{T}^\sigma$ then $A\sigma(\mathfrak{S}) \otimes A\sigma(\mathfrak{T}) = \mathfrak{S} \otimes \mathfrak{T}$.

13.14. CONSEQUENCE. *For topological spaces X and Y we have*

$$\begin{aligned} \mathrm{Bor}(X) \otimes \mathrm{Bor}(Y) &= \mathrm{Op}(X) \otimes \mathrm{Op}(Y) \\ &\subset A\sigma\big(\mathrm{Op}(X \times Y)\big) = \mathrm{Bor}(X \times Y), \end{aligned}$$

of course for the product topology on $X \times Y$.

We shall see in 13.19 below that $\mathrm{Bor}(X) \otimes \mathrm{Bor}(Y)$ and $\mathrm{Bor}(X \times Y)$ need not be equal. But one has an important partial result.

13.15. REMARK. *For topological spaces X and Y with countable bases we have $\mathrm{Bor}(X) \otimes \mathrm{Bor}(Y) = \mathrm{Bor}(X \times Y)$.*

Proof. Let $\{A_l : l \in \mathbb{N}\}$ and $\{B_l : l \in \mathbb{N}\}$ be respective countable bases. By the definition of the product topology then $\mathrm{Op}(X \times Y) = \{A_p \times B_q : p, q \in \mathbb{N}\}^\sigma \subset \mathrm{Op}(X) \otimes \mathrm{Op}(Y)$ and hence the assertion.

We come to the main consequence in the present context.

13.16. PROPOSITION. *Assume that the function $H : \overline{\mathbb{R}} \times \overline{\mathbb{R}} \to \overline{\mathbb{R}}$ is Borel measurable* (which in particular is true when $\overline{\mathbb{R}} \times \overline{\mathbb{R}}$ is an at most countable union of Borel subsets on which H is continuous). *Then*

$$f, g : X \to \overline{\mathbb{R}} \text{ measurable } \mathfrak{A} \Rightarrow H(f, g) : X \to \overline{\mathbb{R}} \text{ is measurable } \mathfrak{A}.$$

Proof. The assertion in brackets follows from 13.6 and 13.4.4). Now $(f, g) : X \to \overline{\mathbb{R}} \times \overline{\mathbb{R}}$ is measurable $\mathrm{Bor}(\overline{\mathbb{R}}) \otimes \mathrm{Bor}(\overline{\mathbb{R}})$ by 13.12. Thus from 13.15 and 13.4.1) the assertion follows.

13.17. EXAMPLES. The functions $H : H(u, v) = u \dotplus v$ and $= uv \; \forall u, v \in \overline{\mathbb{R}}$ are as required in 13.16, the product with the usual convention $0(\pm\infty) := 0$. Therefore if $f, g : X \to \overline{\mathbb{R}}$ are measurable $\mathfrak{A}$ then the functions $f \dotplus g$ and $fg : X \to \overline{\mathbb{R}}$ are measurable $\mathfrak{A}$ as well.

13.18. SPECIAL CASE. 1) *A function $f : X \to [0, \infty]$ is measurable $\mathfrak{A}$ iff $f \in \mathrm{M}(\mathfrak{A})$.* 2) *A function $f : X \to \overline{\mathbb{R}}$ is measurable $\mathfrak{A}$ iff the functions $f^+, f^- : X \to [0, \infty]$ are measurable $\mathfrak{A}$, that is iff $f^+, f^- \in \mathrm{M}(\mathfrak{A})$.*

Proof. 1) We have $[f \geqq t] = X \in \mathfrak{A}$ for $t \leqq 0$, so that the assertion follows from 13.5. 2) $\Rightarrow$ follows from 13.9.1), and $\Leftarrow$ follows from 13.17 since $f = f^+ \dotplus (-f^-)$. Note that the last implication also has an obvious direct proof.

13.19. EXERCISE. This exercise serves to demonstrate the possible smallness of $\mathrm{Bor}(X) \otimes \mathrm{Bor}(Y)$. We follow Dudley [1989] exercise 4.1.11 and start with a set-theoretical result. 1) Let X be a set which has no injective map into $\mathbb{R}$. Then the diagonal $D := \{(x, x) : x \in X\} \subset X \times X$ is not a member of $\mathfrak{P}(X) \otimes \mathfrak{P}(X)$. Hints: i) Let $\mathfrak{S}$ be a paving in a nonvoid set X. Then

$$A\sigma(\mathfrak{S}) = \bigcup_{\mathfrak{P}} A\sigma(\mathfrak{P}) \quad \text{over the countable pavings } \mathfrak{P} \subset \mathfrak{S}.$$

ii) Let $\mathfrak{S}$ and $\mathfrak{T}$ be pavings in nonvoid sets X and Y. Then

$$A\sigma(\mathfrak{S} \times \mathfrak{T}) \;=\; \bigcup_{\mathfrak{P},\mathfrak{Q}} A\sigma(\mathfrak{P} \times \mathfrak{Q})$$

over the countable pavings $\mathfrak{P} \subset \mathfrak{S}$ and $\mathfrak{Q} \subset \mathfrak{T}$.

iii) Let $\mathfrak{P} = \{P_n : n \in \mathbb{N}\}$ be a countable paving in a nonvoid set X. Then the subsets

$$P(S) := \bigcap_{n\in S} P_n \cap \bigcap_{n\notin S} P_n' \quad \text{for the subsets } S \subset \mathbb{N}$$

form a disjoint cover of X. iv) Let $\mathfrak{P} := \{P_n : n \in \mathbb{N}\}$ and $\mathfrak{Q} := \{Q_n : n \in \mathbb{N}\}$ be countable pavings in nonvoid sets X and Y. Assume the notation of iii). Then each subset $E \in A\sigma(\mathfrak{P} \times \mathfrak{Q})$ is a union of subsets $P(S) \times Q(T)$ with $S, T \in \mathbb{N}$. v) Consider iv) in the special case that $X = Y$ and $E := D \in A\sigma(\mathfrak{P} \times \mathfrak{Q})$. Then each of the $P(S)$ and $Q(T)$ is either void or a singleton. vi) Deduce the assertion from ii) and v).

2) Let E be a nonvoid set. Prove with bare hands that there is no injective map $f : \mathfrak{P}(E) \to E$. Hint: Consider

$$A := \{u \in X : u = f(U) \text{ for some } U \subset X \text{ with } u \notin U\},$$

and $a := f(A)$. Show that both $a \in A$ and $a \notin A$ are impossible.

3) Assume that X carries a Hausdorff topology and has no injective map into $\mathbb{R}$. Then the diagonal $D \subset X \times X$ is not in $\mathrm{Bor}(X) \otimes \mathrm{Bor}(X)$. On the other hand D is closed and hence in $\mathrm{Bor}(X \times X)$.

Integrable Functions and the Integral

In the present subsection we assume that $\mathfrak{A}$ is a σ algebra in X and that $\alpha : \mathfrak{A} \to [0,\infty]$ is a cmeasure. We are thus much more restrictive than in sections 11 and 12. The basic link to these former sections is 13.18. We recall from 12.11 that for $f : X \to [0,\infty]$ measurable $\mathfrak{A}$, we have the common integral

$$\int f d\alpha := \fint f d\alpha = \int_* f d\alpha \in [0,\infty],$$

and its properties obtained in sections 11 and 12. In particular, both 11.11 with 11.1.3) and 12.3 show that the common integral is additive.

We define a function $f : X \to \overline{\mathbb{R}}$ to be **integrable** α iff it is measurable $\mathfrak{A}$ and has $\int f^+ d\alpha < \infty$ and $\int f^- d\alpha < \infty$. Thus a function $f : X \to [0,\infty]$ is integrable α iff it is measurable $\mathfrak{A}$ and has $\int f d\alpha < \infty$. For $f : X \to \overline{\mathbb{R}}$ integrable α we define the **integral** to be $\int f d\alpha := \int f^+ d\alpha - \int f^- d\alpha \in \mathbb{R}$. In case $f : X \to [0,\infty]$ it coincides with the previous one, so that the notation is correct. We also write as usual $\int f d\alpha = \int f(x) d\alpha(x) = \cdots$, in particular when other dependences are involved.

13.20. PROPERTIES. 1) *If $f : X \to \overline{\mathbb{R}}$ is integrable α then so is cf with $c \in \mathbb{R}$, and we have $\int (cf) d\alpha = c \int f d\alpha$. 2) If $f, g : X \to \overline{\mathbb{R}}$ are integrable α*

then so are $f \dotplus g$, and we have $\int (f \dotplus g) d\alpha = \int f d\alpha + \int g d\alpha$.

3) If $f :\to \overline{\mathbb{R}}$ is integrable α then so is $|f|$, and we have $| \int f d\alpha | \leqq \int |f| d\alpha$. Furthermore

$$\int |f| d\alpha \leqq (\sup |f|) \alpha ([f \neq 0]).$$

We isolate the relation on which the proof of 13.20.2) is based.

13.21. EXERCISE. For $u, v, w \in \overline{\mathbb{R}}$ we have $u \dotplus v = w \Rightarrow u^+ + v^+ + w^- = u^- + v^- + w^+$. We also have $\Leftarrow$, except when u and v have opposite values $\pm \infty$.

Proof of 13.20. 1) The case $c = 0$ results from the convention $0(\pm\infty) := 0$, and the case $c > 0$ from the respective previous assertions. The case $c = -1$ is obvious. 2) The function $f \dotplus g =: h$ is measurable $\mathfrak{A}$ by 13.17. By 13.21 we have $f^+ + g^+ + h^- = f^- + g^- + h^+$. It follows that $h^+ \leqq f^+ + g^+$ and $h^- \leqq f^- + g^-$, so that h is integrable α. Then we obtain the final relation, since the integral is known to be additive on $M(\mathfrak{A})$. 3) The first assertion results from $|f| = f^+ + f^-$, and the second one from 12.7.

The next properties of the integral are based on the notion of a null set for α, which in a much wider context had been considered in section 2. The null sets for α are the sets $N \in \mathfrak{A}$ with $\alpha(N) = 0$. We note two obvious facts. 1) Each subset of a null set which is in $\mathfrak{A}$ is a null set as well. 2) Each countable union of null sets is a null set as well.

13.22. EXERCISE. 1) If $N \subset X$ is a null set for α then a subset $A \subset N$ need not be in $\mathfrak{A}$. 2) All subsets of null sets for α are in $\mathfrak{A}$ (and hence are null sets as well) iff the measure α is complete in the sense of section 10.

One says that a property of the points of X holds **almost everywhere with respect to** α, in short **ae** α, iff the subset of those points in which it is violated is contained in a null set for α. We see from the above that some caution is required with this expression.

13.23. PROPERTIES. 1) Assume that $f : X \to \overline{\mathbb{R}}$ is integrable α. Then $[|f| = \infty]$ is a null set for α, that is f is finite ae α. 2) Assume that $f : X \to [0, \infty]$ is measurable $\mathfrak{A}$. Then $\int f d\alpha = 0 \Leftrightarrow [f > 0]$ is a null set for α, that is f is $= 0$ ae α.

Proof. 1) From 11.8.5) we obtain

$$t\alpha([|f| = \infty]) \leqq t\alpha([|f| \geqq t]) \leqq \int |f| d\alpha < \infty \quad \text{for real } t > 0,$$

and hence the assertion for $t \to \infty$. 2$\Rightarrow$) By 11.8.5) the $[f \geqq t] \, \forall t > 0$ are null sets for α, and hence $[f > 0]$ is a null set for α. 2$\Leftarrow$) follows from 12.7.

For the next proofs we introduce a useful new formation. If X is a nonvoid set and $A \subset X$ then we define, besides the characteristic function χ_A of A, the function $\omega_A : X \to [0, \infty]$ to be $\omega_A(x) = \infty$ for $x \in A$ and $\omega_A(x) = 0$ for $x \notin A$. It will be applied as follows.

13.24. EXERCISE. 1) Let $\mathfrak{A}$ be a lattice in X with $\varnothing \in \mathfrak{A}$. For $A \subset X$ then both $\omega_A \in \mathrm{UM}(\mathfrak{A})$ and $\omega_A \in \mathrm{LM}(\mathfrak{A})$ are equivalent to $A \in \mathfrak{A}$. For $\alpha : \mathfrak{A} \to [0, \infty]$ isotone with $\alpha(\varnothing) = 0$ then

$$\int \omega_A d\alpha = \left\{ \begin{array}{ll} \infty & \text{if } \alpha(A) > 0 \\ 0 & \text{if } \alpha(A) = 0 \end{array} \right\}.$$

2) Let $\mathfrak{A}$ be a ring in X and $\alpha : \mathfrak{A} \to [0, \infty]$ be a ccontent. For $A \subset X$ then

$$\int_* \omega_A d\alpha = \left\{ \begin{array}{ll} \infty & \text{if } \alpha_\star(A) > 0 \\ 0 & \text{if } \alpha_\star(A) = 0 \end{array} \right\}.$$

3) We return to the context of the present subsection. For $A \subset X$ then ω_A is measurable $\mathfrak{A}$ iff $A \in \mathfrak{A}$, and ω_A is integrable α iff A is a null set for α. In the latter case $\int \omega_A d\alpha = 0$.

13.25. PROPERTIES. 1) *Let* $f, g : X \to [0, \infty]$ *be measurable* $\mathfrak{A}$. *Then*

$$f \leq g \text{ ae } \alpha \Rightarrow \int f d\alpha \leq \int g d\alpha,$$

$$f = g \text{ ae } \alpha \Rightarrow \int f d\alpha = \int g d\alpha.$$

2) *Let* $f, g : X \to \overline{\mathbb{R}}$ *be integrable* α. *Then*

$$f \leq g \text{ ae } \alpha \Rightarrow \int f d\alpha \leq \int g d\alpha,$$

$$f = g \text{ ae } \alpha \Rightarrow \int f d\alpha = \int g d\alpha.$$

3) *Let* $f, g : X \to \overline{\mathbb{R}}$ *be integrable* α *with* $f \leq g$ *ae* α *and* $\int f d\alpha = \int g d\alpha$. *Then* $f = g$ *ae* α.

4) *Let* $f : X \to \overline{\mathbb{R}}$ *be measurable* $\mathfrak{A}$ *and* $P, Q : X \to \overline{\mathbb{R}}$ *be integrable* α *with* $P \leq f \leq Q$ *ae* α. *Then* f *is integrable* α. *In particular* P *and* Q *can be constants when* α *is finite.*

Proof. 1) We have to prove the first assertion. The subset $N := [f > g]$ is in $\mathfrak{A}$ by 13.8 and hence a null set for α. By definition $f \leq g + \omega_N$. Since the integral is isotone and additive on $\mathrm{M}(\mathfrak{A})$ we obtain

$$\int f d\alpha \leq \int (g + \omega_N) d\alpha = \int g d\alpha + \int \omega_N d\alpha = \int g d\alpha.$$

2) If $f \leq g$ ae α then $f^+ \leq g^+$ and $g^- \leq f^-$ ae α. Therefore $\int f d\alpha \leq \int g d\alpha$ by the definition of the integral and 1). 3) We know that $N := [f > g]$ is a null set for α, and by 13.8 the subset $M := [f < g]$ is in $\mathfrak{A}$. Likewise for $t > 0$ we have $M(t) := [f + t \leq g] \in \mathfrak{A}$. Then $f + t\chi_{M(t)} \leq g$ on N' and thus ae α. This implies that $t\chi_{M(t)} \leq f^- + g^+$ ae α, and hence by 1) that $\chi_{M(t)}$

is integrable α. Then from 13.20.2) and 2) we obtain

$$\int f d\alpha + t\alpha(M(t)) \;=\; \int f d\alpha + \int t\chi_{M(t)} d\alpha$$

$$=\; \int (f + t\chi_{M(t)}) d\alpha \leqq \int g d\alpha = \int f d\alpha,$$

and hence $\alpha(M(t)) = 0$. Now $M(t) \uparrow M$ for $t \downarrow 0$. It follows that $\alpha(M) = 0$, which is the assertion. 4) We have $f^+ \leqq Q^+$ and $f^- \leqq P^-$ ae α. Thus the assertion follows from 1).

We turn to the classical theorems due to Fatou, Beppo Levi, and Lebesgue on the σ continuity properties of the integral.

13.26. THEOREM (Fatou). *Let* $f_n, f : X \to [0, \infty]$ $\forall n \in \mathbb{N}$ *be measurable* $\mathfrak{A}$ *with* $f \leqq \liminf\limits_{n\to\infty} f_n$ *ae* α. *Then* $\int f d\alpha \leqq \liminf\limits_{n\to\infty} \int f_n d\alpha$.

Proof. Let $N \subset X$ be a null set for α such that $f \leqq \liminf\limits_{n\to\infty} f_n$ outside of N. Then the assertion follows from the previous version 12.9 applied to the $f_n + \omega_N$ and to $f + \omega_N$.

13.27. THEOREM (Beppo Levi). *Let* $f_n : X \to \overline{\mathbb{R}}$ $\forall n \in \mathbb{N}$ *be integrable* α *and* $f : X \to \overline{\mathbb{R}}$ *be measurable* $\mathfrak{A}$ *with* $f_n \uparrow f$ *ae* α. *Thus* $\int f_n d\alpha \uparrow$ *some* $c \in]-\infty, \infty]$ *by* 13.25.2). *Then* f *is integrable* α *iff* $c \in \mathbb{R}$. *In this case* $\int f d\alpha = c$.

Proof. Let $N \subset X$ be a null set for α such that $f_n \uparrow f$ outside of N; by 13.23.1) we can also achieve that f_1 is finite-valued outside of N. Then $f_n \dotplus (-f_1) \dotplus \omega_N \uparrow f \dotplus (-f_1) \dotplus \omega_N$, and all these functions are in $M(\mathfrak{A})$. Thus $\int \big(f_n \dotplus (-f_1) \dotplus \omega_N\big) d\alpha \uparrow \int \big(f \dotplus (-f_1) \dotplus \omega_N\big) d\alpha$ by the previous version 12.10. By 13.20.2) the functions $f_n \dotplus (-f_1) \dotplus \omega_N$ are integrable α and $\int \big(f_n \dotplus (-f_1) \dotplus \omega_N\big) d\alpha = \int f_n d\alpha - \int f_1 d\alpha \to c - \int f_1 d\alpha$. It follows that

$$c = \int f_1 d\alpha + \int \big(f \dotplus (-f_1) \dotplus \omega_N\big) d\alpha.$$

Therefore $c \in \mathbb{R}$ iff $f \dotplus (-f_1) \dotplus \omega_N$ is integrable α; in view of $f \dotplus (-f_1) \dotplus \omega_N \dotplus f_1 = f$ ae α it is equivalent that f is integrable α. In this case then $c = \int f_1 d\alpha + \big(\int f d\alpha - \int f_1 d\alpha \big) = \int f d\alpha$.

13.28. THEOREM (Lebesgue). *Let* $f_n, f : X \to \overline{\mathbb{R}}$ $\forall n \in \mathbb{N}$ *be measurable* $\mathfrak{A}$ *with* $f_n \to f$ *ae* α. *Assume that there exists an* $F : X \to [0, \infty]$ *integrable* α *such that* $|f_n| \leqq F$ *ae* α $\forall n \in \mathbb{N}$. *Then all functions* f_n *and* f *are integrable* α, *and we have* $\int f_n d\alpha \to \int f d\alpha$.

Let us first remark that the two properties

$$(|f_n| \leqq F \text{ ae } \alpha) \,\forall n \in \mathbb{N} \quad \text{and} \quad (|f_n| \leqq F \,\forall n \in \mathbb{N}) \text{ ae } \alpha$$

are equivalent (why?). This is an example of a situation where the expression ae α ought to be handled with caution.

Proof. From $-F \leqq f_n, f \leqq F$ ae α we see by 13.25.4) that all functions f_n and f are integrable α. We form the functions $P_n := \inf\limits_{l \geqq n} f_l$ and $Q_n :=$

$\sup_{l \geq n} f_l$ $\forall n \in \mathbb{N}$, which by 13.9.1) are measurable $\mathfrak{A}$, and hence integrable α
since $-F \leq P_n \leq Q_n \leq F$ ae α. Now $P_n \uparrow \liminf_{l \to \infty} f_l$ and $Q_n \downarrow \limsup_{l \to \infty} f_l$, and
hence $P_n \uparrow f$ and $Q_n \downarrow f$ ae α. Thus $\int P_n d\alpha \uparrow \int f d\alpha$ and $\int Q_n d\alpha \downarrow \int f d\alpha$
by the Beppo Levi theorem. At last we have $P_n \leq f_n \leq Q_n$ $\forall n \in \mathbb{N}$. It
follows that $\int f_n d\alpha \to \int f d\alpha$.

The last topic in the present subsection will be the extension to complex-valued functions.

13.29. REMARK. *Assume that* $f : X \to \mathbb{C}$, *and write* $f = P + iQ$ *with*
$P, Q : X \to \mathbb{R}$. *Then* f *is measurable* $\mathfrak{A} \Leftrightarrow P$ *and* Q *are measurable* $\mathfrak{A}$. *In
this case the function* $|f| : X \to [0, \infty[$ *is measurable* $\mathfrak{A}$ *as well.*

Proof. The implication $\Rightarrow$ and the last assertion follow from 13.4.1) and
13.6, since the functions Re, Im, $|\cdot| : \mathbb{C} \to \mathbb{R}$ are continuous. The implication
$\Leftarrow$ follows from 13.12 and 13.15.

We define $f = P + iQ : X \to \mathbb{C}$ to be **integrable** α iff P and Q are
integrable α, and then its **integral** to be $\int f d\alpha := \int P d\alpha + i \int Q d\alpha \in \mathbb{C}$.
The notions and results on the integral have often obvious counterparts for
complex-valued functions. We shall not enter into the details, except when
there is a particular reason. Here is one such case.

13.30. REMARK. *Assume that* $f : X \to \mathbb{C}$ *is measurable* $\mathfrak{A}$. *Then* f *is
integrable* $\alpha \Leftrightarrow |f|$ *is integrable* α. *In this case* $|\int f d\alpha| \leq \int |f| d\alpha$.

Proof. Let $f = P + iQ$. The first assertion follows from $|P|, |Q| \leq |f| \leq
|P| + |Q|$. Assume now that f is integrable α. Fix a complex c with $|c| = 1$
such that $|\int f d\alpha| = c \int f d\alpha = \int (cf) d\alpha$. Then

$$\left| \int f d\alpha \right| = \int \text{Re}(cf) d\alpha \leq \int |cf| d\alpha = \int |f| d\alpha,$$

as claimed.

Integration over Subsets

As before we fix a σ algebra $\mathfrak{A}$ in X and a cmeasure $\alpha : \mathfrak{A} \to [0, \infty]$.

Let $T \subset X$ be a nonvoid subset. In 1.12 we defined the trace $\mathfrak{A} \cap T :=
\{A \cap T : A \in \mathfrak{A}\}$ of $\mathfrak{A}$ on T, which is a σ algebra as well. In case $T \in \mathfrak{A}$
we have $\mathfrak{A} \cap T = \{A \in \mathfrak{A} : A \subset T\}$. It is then obvious that the restriction
$\alpha | \mathfrak{A} \cap T : \mathfrak{A} \cap T \to [0, \infty]$ is a cmeasure on $\mathfrak{A} \cap T$; it will be called the
restriction $\alpha | T$ **of** α **to** T. We recall one more fact from 13.4.3): If
$f : X \to \overline{\mathbb{R}}$ is measurable $\mathfrak{A}$ then its restriction $f | T : T \to \overline{\mathbb{R}}$ is measurable
$\mathfrak{A} \cap T$. We have the basic theorem which follows.

13.31. THEOREM. *Consider a nonvoid subset* $T \in \mathfrak{A}$, *and note that* $\chi_T :
X \to [0, \infty[$ *is measurable* $\mathfrak{A}$. *1) For* $f : X \to [0, \infty]$ *measurable* $\mathfrak{A}$ *we have*

$$\int (f | T) d(\alpha | T) = \int f \chi_T d\alpha =: \int_T f d\alpha.$$

2) *For* $f : X \to \overline{\mathbb{R}}$ *measurable* $\mathfrak{A}$ *we have* $f|T$ *integrable* $\alpha|T \Leftrightarrow f\chi_T$ *is integrable* α. *Then*

$$\int (f|T)d(\alpha|T) = \int f\chi_T d\alpha =: \int_T f d\alpha.$$

Under the equivalent conditions in 2) the function f is defined to be **integrable** α **over** T. In both cases $\int_T f d\alpha$ is called the **integral over** T. Furthermore we define $\int_{\varnothing} f d\alpha := \int f\chi_\varnothing d\alpha = 0$ for all $f : X \to \overline{\mathbb{R}}$ measurable $\mathfrak{A}$.

Proof. 1) We have

$$\int f\chi_T d\alpha = \int\!\!\!\!\!\!\int f\chi_T d\alpha = \int_{0\leftarrow}^{\to\infty} \alpha([f\chi_T \geq t])dt$$

$$= \int_{0\leftarrow}^{\to\infty} (\alpha|T)([f|T \geq t])dt = \int\!\!\!\!\!\!\int (f|T)d(\alpha|T) = \int (f|T)d(\alpha|T).$$

2) $f|T$ integrable $\alpha|T$ means by definition that $\int (f|T)^{\pm}d(\alpha|T)$ are both finite. By 1) now

$$\int (f|T)^{\pm}d(\alpha|T) = \int (f^{\pm}|T)d(\alpha|T) = \int f^{\pm}\chi_T d\alpha = \int (f\chi_T)^{\pm}d\alpha.$$

Thus we obtain all assertions.

The next result describes the integral as a function of its domain.

13.32. THEOREM. *Let* $f : X \to [0, \infty]$ *be measurable* $\mathfrak{A}$. *Define* $\vartheta : \mathfrak{A} \to [0, \infty]$ *to be*

$$\vartheta(A) = \int_A f d\alpha = \int f\chi_A d\alpha \quad \text{for } A \in \mathfrak{A}.$$

Then ϑ *is a cmeasure. Furthermore* $\alpha(A) = 0 \Rightarrow \vartheta(A) = 0$.

Proof. This follows from the fact that the integral is additive on $M(\mathfrak{A})$ and from 13.23.2), and from the Beppo Levi theorem in the version 12.10.

We conclude with a special case. Let $T \subset X$ be a nonvoid subset. For a function $f : T \to \overline{\mathbb{R}}$ we define its **null extension** $f^\natural : X \to \overline{\mathbb{R}}$ to be $f^\natural|T = f$ and $f^\natural|T' = 0$. The next assertion is then an immediate consequence of 13.31.

13.33. REMARK. *Consider a nonvoid* $T \in \mathfrak{A}$. *For* $f : T \to \overline{\mathbb{R}}$ *measurable* $\mathfrak{A} \cap T$ *then* $f^\natural : X \to \overline{\mathbb{R}}$ *is measurable* $\mathfrak{A}$. 1) *If* $f \geq 0$ *and hence* $f^\natural \geq 0$ *then*

$$\int_T f d(\alpha|T) = \int f^\natural d\alpha = \int f^\natural d\alpha.$$

2) f is integrable $\alpha|T \Leftrightarrow f^\natural$ is integrable α. Then

$$\int f d(\alpha|T) = \int f^\natural d\alpha = \int_T f^\natural d\alpha.$$

In clear situations it is common to abbreviate the restriction $\alpha|T$ as α itself. Then 13.31 can be interpreted to say that this is a harmless step. For example, in the next subsection we shall consider the restriction of the Lebesgue measure Λ on $\mathbb{R}$ to a nondegenerate compact interval $T = [a,b] \subset \mathbb{R}$. We shall then write Λ instead of $\Lambda|T$.

Comparison with the Riemann Integral

As announced we fix a compact interval $T = [a,b] \subset \mathbb{R}$ with real $a < b$. Assume that $f : T \to \mathbb{R}$ is a bounded function, and let denote $\alpha := \inf f$ and $\beta := \sup f$. We start to recall the elementary Riemann integral. For each subdivision $\mathfrak{t} : a = t_0 < t_1 < \cdots < t_r = b$ of T we form $\delta(\mathfrak{t}) := \max\{t_l - t_{l-1} : l = 1, \cdots, r\}$ and the subintervals $T_l := [t_{l-1}, t_l]$ $(l = 1, \cdots, r)$. We associate with $\mathfrak{t}$

$$\text{the lower sum } \underline{S}(f, \mathfrak{t}) \quad := \quad \sum_{l=1}^{r} \inf(f|T_l)(t_l - t_{l-1}), \quad \text{and}$$

$$\text{the upper sum } \overline{S}(f, \mathfrak{t}) \quad := \quad \sum_{l=1}^{r} \sup(f|T_l)(t_l - t_{l-1}).$$

One proves that $\underline{S}(f, \mathfrak{s}) \leq \overline{S}(f, \mathfrak{t})$ for all $\mathfrak{s}$ and $\mathfrak{t}$. Therefore

$$\underline{S}(f) := \sup_{\mathfrak{t}} \underline{S}(f, \mathfrak{t}) \leq \inf_{\mathfrak{t}} \overline{S}(f, \mathfrak{t}) =: \overline{S}(f).$$

The function f is defined to be **Riemann integrable** iff $\underline{S}(f) = \overline{S}(f)$. Then the common value $S(f) := \underline{S}(f) = \overline{S}(f)$ is called the **Riemann integral** of f.

We next form the envelopes $P, Q : T \to \mathbb{R}$ of f, defined to be

$$P(x) = \liminf_{z \to x} f(z) := \sup\{t \in \mathbb{R} : f > t \text{ on some neighbourhood of } x\},$$

$$Q(x) = \limsup_{z \to x} f(z) := \inf\{t \in \mathbb{R} : f < t \text{ on some neighbourhood of } x\}.$$

Thus $\alpha \leq P \leq f \leq Q \leq \beta$. It follows from the definition that

$$P \text{ is lower semicontinuous} : [P > t] \subset T \text{ is relative open } \forall t \in \mathbb{R},$$

$$Q \text{ is upper semicontinuous} : [Q < t] \subset T \text{ is relative open } \forall t \in \mathbb{R}.$$

Furthermore we associate with each subdivision $\mathfrak{t}$ of T the functions

$$P_{\mathfrak{t}} : P_{\mathfrak{t}}(x) \quad = \quad \min\{\inf(f|T_l) : l = 1, \cdots, r \text{ with } x \in T_l\},$$

$$Q_{\mathfrak{t}} : Q_{\mathfrak{t}}(x) \quad = \quad \max\{\sup(f|T_l) : l = 1, \cdots, r \text{ with } x \in T_l\} \quad \text{for } x \in T.$$

Thus $\alpha \leq P_{\mathfrak{t}} \leq Q_{\mathfrak{t}} \leq \beta$, and the functions $P_{\mathfrak{t}}$ and $Q_{\mathfrak{t}}$ are constant on the open intervals $\text{int}(T_l)$.

13.34. LEMMA. 1) $P_{\mathfrak{t}} \leqq P$ and $Q \leqq Q_{\mathfrak{t}}$. Therefore $\alpha \leqq P_{\mathfrak{t}} \leqq P \leqq$ $f \leqq Q \leqq Q_{\mathfrak{t}} \leqq \beta$. 2) For each sequence $(\mathfrak{t}(n))_n$ of subdivisions $\mathfrak{t}(n)$ with $\delta(\mathfrak{t}(n)) \to 0$ we have the pointwise convergence $P_{\mathfrak{t}(n)} \to P$ and $Q_{\mathfrak{t}(n)} \to Q$.

Proof. We restrict ourselves to the assertions on P and the $P_{\mathfrak{t}}$. 1) Let $x \in T$. Then there exists $\delta > 0$ such that $V(x, \delta) \cap T$ is contained in the union of all T_l with $x \in T_l$. It follows that $P_{\mathfrak{t}}(x) \leqq \inf (f|V(x, \delta) \cap T) \leqq P(x)$. 2) Fix $x \in T$ and $t \in \mathbb{R}$ with $P(x) > t$, and then $\delta > 0$ with $f|V(x, \delta) \cap T > t$. Now $\delta(\mathfrak{t}(n)) < \delta$ for almost all $n \in \mathbb{N}$; then the subdivision $\mathfrak{t}(n)$ has all its subintervals T_l with $x \in T_l$ contained in $V(x, \delta) \cap T$. It follows that $P_{\mathfrak{t}(n)}(x) \geqq \inf (f|V(x, \delta) \cap T) \geqq t$ for these $n \in \mathbb{N}$. Combined with 1) this implies the assertion.

We turn to the connection with the restriction $\Lambda|T := \Lambda$ of the Lebesgue measure $\Lambda := \lambda^{\sigma}|\mathcal{L} = \lambda_{\bullet}|\mathcal{L}$ on $\mathbb{R}$. We see from the above that P and Q and all $P_{\mathfrak{t}}$ and $Q_{\mathfrak{t}}$ are measurable $\mathrm{Bor}(T) = \mathrm{Bor}(\mathbb{R}) \cap T$, and hence integrable Λ because they are bounded. By definition we have

$$\int P_{\mathfrak{t}} d\Lambda = \sum_{l=1}^{r} \int P_{\mathfrak{t}} \chi_{\mathrm{int}(T_l)} d\Lambda = \sum_{l=1}^{r} \inf(f|T_l)(t_l - t_{l-1}) = \underline{S}(f, \mathfrak{t}).$$

Thus from 13.34 and the Lebesgue theorem 13.28 we obtain $\int P d\Lambda = \underline{S}(f)$.

13.35. PROPOSITION. For each bounded function $f : T \to \mathbb{R}$ the envelopes $P, Q : T \to \mathbb{R}$ are measurable $\mathrm{Bor}(T)$ and integrable Λ, and we have

$$\int P d\Lambda = \underline{S}(f) \quad and \quad \int Q d\Lambda = \overline{S}(f).$$

We are now close to a famous characterization of the Riemann integrable functions.

13.36. REMARK. For $x \in T$ we have $P(x) = Q(x) \Leftrightarrow f$ is continuous in x.

Proof. $\Leftarrow$). For each $\varepsilon > 0$ we have $f(x) - \varepsilon < f < f(x) + \varepsilon$ on some neighbourhood of x and hence $f(x) - \varepsilon \leqq P(x)$ and $Q(x) \leqq f(x) + \varepsilon$. Thus $Q(x) \leqq f(x) \leqq P(x)$ and hence the assertion. $\Rightarrow$) For each $\varepsilon > 0$ we have $f(x) - \varepsilon = P(x) - \varepsilon < f < Q(x) + \varepsilon = f(x) + \varepsilon$ on some neighbourhood of x. Thus f is continuous in x.

13.37. THEOREM. A bounded function $f : T \to \mathbb{R}$ is Riemann integrable iff it is continuous ae Λ. In this case f is measurable $\mathcal{L} \cap T$ and integrable $\Lambda|T =: \Lambda$ with $S(f) = \int f d\Lambda$.

Proof. 1) We have the equivalences

$$f \text{ is Riemann integrable} \quad \Leftrightarrow \quad \int P d\Lambda = \int Q d\Lambda \quad \text{by 13.35}$$

$$\Leftrightarrow \quad P = Q \text{ ae } \Lambda \quad \text{by 13.25.2)3)}$$

$$\Leftrightarrow \quad f \text{ continuous ae } \Lambda \quad \text{by 13.36.}$$

We then have $S(f) = \int Pd\Lambda = \int Qd\Lambda$. 2) In this case we have $P^\natural \leq f^\natural \leq Q^\natural$ and $P^\natural = f^\natural = Q^\natural$ ae Λ on $\mathbb{R}$. For $t \in \mathbb{R}$ therefore $[P^\natural \geq t] \subset [f^\natural \geq t] \subset [Q^\natural \geq t]$, where $[P^\natural \geq t]$ and $[Q^\natural \geq t]$ are in $\mathfrak{L}$ with $\Lambda([P^\natural \geq t]) = \Lambda([Q^\natural \geq t]) < \infty$. Since Λ is complete by 10.14 it follows that $[f^\natural \geq t] \in \mathfrak{L}$ and hence $[f \geq t] \in \mathfrak{L} \cap T$. Thus f is measurable $\mathfrak{L} \cap T$. The proof is complete.

13.38. EXAMPLE. It is a classical result that a Riemann integrable bounded function $f : T \to \mathbb{R}$ need not be measurable $\text{Bor}(T) = \text{Bor}(\mathbb{R}) \cap T$. In order to present an example we assume without proof another classical result: There exists a compact subset $K \subset T$ with $\Lambda(K) = \lambda(K) = 0$ which contains subsets $A \notin \text{Bor}(T)$ (K can be taken as the so-called Cantor set). In fact, the function χ_A is then Riemann integrable, since it is continuous in the points of $T \setminus K$, but it is not measurable $\text{Bor}(T)$.

We conclude the section with the example announced in connection with 9.24. We construct for $\bullet = \sigma\tau$ an inner $\bullet$ premeasure $\varphi : \mathfrak{S} \to [0, \infty[$ on a lattice $\mathfrak{S}$ with $\varnothing \in \mathfrak{S}$ and $\varphi(\varnothing) = 0$ such that $\phi := \varphi_\bullet|\mathfrak{C}(\varphi_\bullet)$ is not outer regular $(\mathfrak{S}\top\mathfrak{S}_\bullet)\bot$ at $\mathfrak{S}_\bullet$ and that $X \in [\phi < \infty]^\sigma$.

13.39. EXERCISE. Let $X = [0, 1]$. We form the function $R : R(x) = 1/x$ for $0 < x \leq 1$ and $R(0) = 0$. Define $\mathfrak{S}$ to consist of all closed subsets $S \subset X$ with $\int_S Rd\Lambda < \infty$, and $\varphi : \mathfrak{S} \to [0, \infty[$ to be $\varphi(S) = \int_S Rd\Lambda$ for $S \in \mathfrak{S}$. 1) $\mathfrak{S}$ is a lattice with $\varnothing \in \mathfrak{S}$, and φ is isotone and modular with $\varphi(\varnothing) = 0$. 2) φ is upward σ continuous and hence inner $\star$ tight. Hint: 7.10.2). 3) φ is τ continuous at $\varnothing$. Hint: 6.34. Therefore φ is an inner $\bullet$ premeasure. Define $\phi := \varphi_\bullet|\mathfrak{C}(\varphi_\bullet)$. 4) $\mathfrak{S} = \mathfrak{S}_\bullet$ and hence $\varphi_\star = \varphi_\bullet$. 5) $\mathfrak{S}\top\mathfrak{S} = \text{Cl}(X) = \text{Comp}(X)$ and hence $(\mathfrak{S}\top\mathfrak{S})\bot = \text{Op}(X)$. Therefore $\text{Bor}(X) \subset \mathfrak{C}(\varphi_\bullet)$. 6) $\phi(A) = \int_A Rd\Lambda$ for all $A \in \text{Bor}(X)$. 7) All $U \in \text{Op}(X)$ with $0 \in U$ have $\phi(U) = \infty$. Therefore ϕ is not outer regular $(\mathfrak{S}\top\mathfrak{S})\bot = \text{Op}(X)$ at $\{0\}$. 8) $X \in [\phi < \infty]^\sigma$. It follows that φ is as required.

CHAPTER V

The Daniell-Stone and Riesz Representation Theorems

The present chapter contains the most important consequences of the extension theories of chapter II. We shall obtain the representation theorems of Daniell-Stone and Frédéric Riesz in the spirit and scope of the extension theories. The Daniell-Stone theorem will be established in versions $\bullet = \star\sigma\tau$ as above, and based on *inner* regularity this time. The Riesz theorem will be a direct specialization of the case $\bullet = \tau$. It will involve all Borel-Radon measures on all Hausdorff topological spaces. We have sketched all this in the introduction. A substantial tool will be the combination of the horizontal and vertical integrals developed in sections 11 and 12.

14. Elementary Integrals on Lattice Cones

After an introduction the present section defines the elementary integrals on lattice cones. These are the functionals which are to be represented. Then several kinds of representations will be introduced.

Introduction

For nonvoid sets X and Y we let as usual Y^X consist of all functions $X \to Y$. On defines a subset $H \subset \overline{\mathbb{R}}^X$ to be a **lattice** iff $f, g \in H \Rightarrow f \vee g$, $f \wedge g \in H$. If $H \subset \mathbb{R}^X$ is a linear subspace then either condition suffices, and one speaks of a **lattice subspace** (or a vector lattice). Justified by success, we define $H \subset \overline{\mathbb{R}}^X$ to be **Stonean** iff

$$f \in H \Rightarrow f \wedge t, (f - t)^+ \in H \quad \text{for all real } t > 0.$$

If $H \subset \mathbb{R}^X$ is a linear subspace then in view of $f = f \wedge t + (f - t)^+$ this means that $f \in H \Rightarrow f \wedge t \in H \; \forall t > 0$, or that $f \in H \Rightarrow f \wedge 1 \in H$.

We start to recall the traditional Daniell-Stone and Riesz representation theorems; see for example Dudley [1989] and Bauer [1992]. We note that a linear functional $I : H \to \mathbb{R}$ on a linear subspace $H \subset \mathbb{R}^X$ is isotone iff it is **positive**, that is $f \geqq 0 \Rightarrow I(f) \geqq 0$ for $f \in H$.

14.1. THEOREM (Traditional Daniell-Stone Theorem). *Let $I : H \to \mathbb{R}$ be a positive linear functional on a Stonean lattice subspace $H \subset \mathbb{R}^X$. Then the following are equivalent.*

i) *There exists a cmeasure* $\alpha : \mathfrak{A} \to [0,\infty]$ *on a* σ *algebra* $\mathfrak{A}$ *in* X *which represents* I, *that is all* $f \in H$ *are integrable* α *with* $I(f) = \int f d\alpha$.

ii) I *is* σ *continuous at* 0, *that is for each sequence* $(f_l)_l$ *in* H *with pointwise* $f_l \downarrow 0$ *one has* $I(f_l) \downarrow 0$.

Note that i)$\Rightarrow$ii) follows from the Beppo Levi theorem 13.27.

The above theorem has a certain usefulness, because it ensures for example that the classical σ continuity properties 13.26 and 13.28 of the integral hold true for I. But in principle its deficiencies are like those of the main extension theorem in traditional abstract measure theory as discussed in the introduction: There is no room for a nonsequential version, and above all there is no room for the aspect of regularity. These are basic points in the traditional Riesz theorem to which we turn next.

For a Hausdorff topological space X we define $\mathrm{CK}(X,\mathbb{R})$ to consist of all continuous functions $f \in \mathrm{C}(X,\mathbb{R})$ such that f vanishes outside of some compact subset of X (which of course can depend on f). Note that $\mathrm{CK}(X,\mathbb{R})$ is a Stonean lattice subspace. Also note that by 13.25.4) the functions $f \in \mathrm{CK}(X,\mathbb{R})$ are integrable with respect to each Borel-Radon measure $\alpha : \mathrm{Bor}(X) \to [0,\infty]$.

14.2. THEOREM (Traditional Riesz Theorem). *Let* X *be a locally compact Hausdorff topological space. There is a one-to-one correspondence between the positive linear functionals* $I : \mathrm{CK}(X,\mathbb{R}) \to \mathbb{R}$ *and the Borel-Radon measures* $\alpha : \mathrm{Bor}(X) \to [0,\infty]$. *The correspondence is*

$$I(f) = \int f d\alpha \quad \text{for all } f \in \mathrm{CK}(X,\mathbb{R}).$$

This fundamental result will be a direct specialization of our Riesz representation theorem in section 16. We add that it is not hard to obtain a direct proof for it from what we have developed so far. This will be done in form of an exercise at the end of the present subsection, also because it offers an occasion to recall some topological facts.

It is obvious that the traditional Daniell-Stone theorem is of no visible use for the proof of the traditional Riesz theorem. This would require a version of the Daniell-Stone theorem which is based on regularity. The extension theories of chapter II evoke the intuitive impression that such versions must not be based on lattice subspaces $H \subset \mathbb{R}^X$ of functions $X \to \mathbb{R}$, but rather on lattice cones, and hence on lattice cones $E \subset [0,\infty[^X$ of functions $X \to [0,\infty[$, and on the appropriate kind of functionals $I : E \to [0,\infty[$. The present chapter will confirm this impression.

There is another reason to pass from lattice subspaces to lattice cones, which comes from the Riesz theorem itself. In present-day analysis one is often forced to exceed the frame of locally compact Hausdorff topological spaces. Since the Borel-Radon measures have been realized as the fundamental class of cmeasures on the class of *all* Hausdorff topological spaces, it is desirable to have the Riesz representation theorem in this comprehensive frame. But then its traditional version breaks down as soon as one leaves the class of locally compact Hausdorff spaces. The reason is that the lattice subspace $CK(X, \mathbb{R})$ becomes too small. For example, it is the null subspace in case that no nonvoid open subset of X is contained in a compact subset; in this context note exercise 14.3 below. Now on each Hausdorff space X there is a wealth of *semicontinuous* real-valued functions which vanish outside of compact subsets, for example the characteristic functions χ_K of the compact subsets $K \subset X$ and their scalar multiples. Therefore it seems natural to base the extension of the Riesz theorem on the upper semicontinuous or the lower semicontinuous functions on X, of course in such a manner that the traditional Riesz theorem is contained in the new result. But these function classes are lattice cones and as a rule not lattice subspaces. Thus we arrive at lattice cones once more. As above it is natural to work with lattice cones $E \subset [0, \infty[^X$ of functions $X \to [0, \infty[$ which vanish outside of compact subsets of X. This forces us to choose the upper semicontinuous functions; see exercise 14.4 below.

Thus we have obtained the frame for the present chapter. We shall see that the aspects of regularity and of $\bullet = \sigma\tau$ continuity will turn up in a natural manner, to an extent that we shall obtain a complete counterpart to the extension theories of chapter II, but this time restricted to the inner situation as remarked above. We shall first consider the cases $\bullet = \sigma\tau$. There is also a Daniell-Stone theorem for $\bullet = \star$, but in this case the complete answer will be different. It will be postponed until section 17.

14.3. EXERCISE. Let X be an infinite-dimensional Hausdorff topological vector space. Then no nonvoid open subset of X is contained in a compact subset.

14.4. EXERCISE. Let X be a Hausdorff topological space such that no nonvoid open subset of X is contained in a compact subset. If $f \in USC(X, \overline{\mathbb{R}})/LSC(X, \overline{\mathbb{R}})$ vanishes outside of some compact subset of X then $f \geqq 0/f \leqq 0$.

14.5. EXERCISE. The aim of this exercise is a direct proof of the traditional Riesz representation theorem 14.2. We assume that X is a locally compact Hausdorff topological space. 1) X is completely regular. Hint: The one-point compactification of X is normal. 2) Let $K \subset X$ be compact nonvoid. Then there exists a function $f \in CK(X, \mathbb{R})$ such that $\chi_K \leqq f \leqq 1$. 3) Define $USCK(X, [-\infty, \infty[)$ to consist of all functions $f \in USC(X, [-\infty, \infty[)$ such that f vanishes outside of some compact subset of X. For each $f \in USCK(X, [-\infty, \infty[)$ then $\{u \in CK(X, \mathbb{R}) : u \geqq f\}$ is nonvoid and has the pointwise infimum f. Hint: Combine 11.21.1) with 2). 4) Let $\alpha : Bor(X) \to [0, \infty]$ be a Borel-Radon measure on X. Define $I : CK(X, \mathbb{R}) \to \mathbb{R}$ to be $I(f) = \int f d\alpha \; \forall f \in CK(X, \mathbb{R})$. Then I is an isotone linear functional. Furthermore

$$\int f d\alpha = \inf\{I(u) : u \in CK(X, \mathbb{R}) \text{ with } u \geqq f\} \quad \forall f \in USCK(X, [0, \infty[).$$

Hint: 11.20. In particular

$$\alpha(K) = \inf\{I(u) : u \in CK(X, \mathbb{R}) \text{ with } u \geq \chi_K\} \quad \forall \text{ compact } K \subset X.$$

Therefore the map $\alpha \mapsto I$ is injective.

5) The Dini Theorem: Let K be a compact topological space. Assume that $M \subset \mathrm{USC}(K, [-\infty, \infty[)$ is nonvoid and downward directed $\downarrow F : K \to [-\infty, \infty[$, so that $F \in \mathrm{USC}(K, [-\infty, \infty[)$ as well. Then $\inf\{\max f : f \in M\} = \max F$.

For the remainder of the exercise we assume that $I : CK(X, \mathbb{R}) \to \mathbb{R}$ is a positive linear functional. We define its extension $I : \mathrm{USCK}(X, [-\infty, \infty[) \to [-\infty, \infty[$ to be

$$I(f) = \inf\{I(u) : u \in CK(X, \mathbb{R}) \text{ with } u \geq f\} \quad \forall f \in \mathrm{USCK}(X, [-\infty, \infty[).$$

6) $\mathrm{USCK}(X, [-\infty, \infty[)$ is a lattice cone. The extended I is isotone and sublinear. 7) Assume that $M \subset \mathrm{USCK}(X, [-\infty, \infty[)$ is nonvoid and downward directed $\downarrow F \in \mathrm{USCK}(X, [-\infty, \infty[)$. Then $\inf\{I(f) : f \in M\} = I(F)$. Hint: Fix $u \in CK(X, \mathbb{R})$ with $u \geq F$ and $P \in M$, and apply 5) to $\{(f - u)^+ : f \in M \text{ with } f \leq P\}$. Then use 2). 8) The extension $I : \mathrm{USCK}(X, [-\infty, \infty[) \to [-\infty, \infty[$ is additive. Hint: Use 7). 9) Define the set function $\varphi : \mathfrak{K} = \mathrm{Comp}(X) \to [0, \infty[$ to be $\varphi(K) = I(\chi_K)$ for $K \in \mathfrak{K}$. Then φ is a Radon premeasure. Hint: Use 9.6. Let $\alpha := \varphi_\bullet|\mathrm{Bor}(X)$ denote its Borel-Radon measure. 10) We have $I(f) = \int f d\alpha$ for all $f \in CK(X, \mathbb{R})$. Hint: Use 11.6 to prove $I(f) = \oint f d\varphi$ for $0 \leq f \in CK(X, \mathbb{R})$.

Lattice Cones

The present subsection starts with a few remarks and examples on lattice cones, and then turns to a fundamental definition. Let X be a nonvoid set. For a subset $M \subset \overline{\mathbb{R}}^X$ we write $M^+ := \{f \in M : f \geq 0\}$. In this connection recall the notations $f^+ := f \vee 0$ and $f^- := (-f) \vee 0$ for $f \in \overline{\mathbb{R}}^X$.

14.6. EXERCISE. If $E \subset [0, \infty[^X$ is a lattice cone then $E - E \subset \mathbb{R}^X$ is a lattice subspace. If in addition E is Stonean then $E - E$ is Stonean as well.

14.7. REMARK. *For a lattice cone $E \subset [0, \infty[^X$ the following are equivalent.*

0) *If $u, v \in E$ with $u \leq v$ then $v - u \in E$.*
1) *There exists a linear subspace $H \subset \mathbb{R}^X$ with $E = H^+$.*
2) *There exists a lattice subspace $H \subset \mathbb{R}^X$ with $E = H^+$.*

In this case there exists a unique lattice subspace $H \subset \mathbb{R}^X$ with $E = H^+$; and this is $H = E - E$.

Proof. If $H \subset \mathbb{R}^X$ is a lattice subspace with $E = H^+$ then $H = H^+ - H^+ = E - E$. i) The implication 2)$\Rightarrow$1) is obvious. ii) To see 1)$\Rightarrow$0) let $u, v \in E$ with $u \leq v$. Then $v - u$ is $\in H$ and ≥ 0, and hence $v - u \in H^+ = E$. iii) To see 0)$\Rightarrow$2) we have to show that $E = (E - E)^+$. But $E \subset (E - E)^+$ is obvious, and $(E - E)^+ \subset E$ is a mere transcription of 0).

The lattice cones $E \subset [0, \infty[^X$ which fulfil the equivalent conditions of 14.7 are called the **primitive** ones. It is of utmost importance that the present chapter is not restricted to primitive lattice cones, which in essence means to lattice subspaces.

14.8. EXAMPLES. Let X be a topological space. We introduce some new notations. 1) $C(X) := C(X, \mathbb{R})$ is a lattice subspace, and $C^+(X) := C(X, [0, \infty[)$ is a primitive lattice cone. $\mathrm{USC}(X) := \mathrm{USC}(X, [-\infty, \infty[)$ and $\mathrm{USC}^+(X) := \mathrm{USC}(X, [0, \infty[)$ are lattice cones which as a rule are not primitive. It is a nontrivial little proof that $\mathrm{USC}(X)$ is stable under addition. All these function classes are Stonean. We recall from 11.19 that $\mathrm{USC}(X, [0, \infty])$ $= \mathrm{UM}(\mathrm{Cl}(X))$ and hence $\mathrm{USC}^+(X) = \mathrm{UM}(\mathrm{Cl}(X)) \cap [0, \infty[^X$. 2) Let X be Hausdorff. $\mathrm{CK}(X) := \mathrm{CK}(X, \mathbb{R})$ is a lattice subspace, and $\mathrm{CK}^+(X) := \mathrm{CK}(X, [0, \infty[)$ is a primitive lattice cone. The obvious $\mathrm{USCK}(X) := \mathrm{USCK}(X, [-\infty, \infty[)$ and $\mathrm{USCK}^+(X) := \mathrm{USCK}(X, [0, \infty[)$ are lattice cones which as a rule are not primitive. All these function classes are Stonean. There is one more class which will be of particular importance. From

$$\mathrm{UM}(\mathrm{Comp}(X)) = \{ f \in [0, \infty]^X : [f \geq t] \text{ compact } \forall 0 < t < \infty \},$$

as defined in section 11, we derive the function class

$$\begin{aligned} \mathrm{UMK}(X) : \quad &= \quad \mathrm{UM}(\mathrm{Comp}(X)) \cap [0, \infty[^X \\ &= \quad \{ f \in [0, \infty[^X : [f \geq t] \text{ compact } \forall 0 < t < \infty \}. \end{aligned}$$

It follows that $\mathrm{USCK}^+(X) \subset \mathrm{UMK}(X) \subset \mathrm{USC}^+(X)$. $\mathrm{UMK}(X)$ is a lattice cone which is Stonean and as a rule not primitive as well. We think of the functions $f \in \mathrm{UMK}(X)$ as to be concentrated on the compact subsets of X.

14.9. EXERCISE. 1) Give an example of a lattice cone $E \subset [0, \infty[^X$ which is not Stonean. 2) Give an example of a cone $E \subset [0, \infty[^X$ which is Stonean but not a lattice. Hint: Let $X = \mathbb{N}$. Define E to consist of the sequences $u = (u_l)_l \in [0, \infty[^{\mathbb{N}}$ with $u_l \to 0$ for $l \to \infty$, such that either $u_l = 0$ for almost all $l \in \mathbb{N}$ or $\sum_{l=1}^{\infty} u_l = \infty$.

14.10. REMARK. *Let $E \subset [0, \infty[^X$ be a Stonean lattice cone. Then $f \wedge t - f \wedge s \in E$ for all $f \in E$ and $0 < s < t$.*

Proof. We have $f \wedge t - f \wedge s = f \wedge t - (f \wedge t) \wedge s = (f \wedge t - s)^+ \in E$.

We come to the most important notion of the subsection. For a lattice cone $E \subset [0, \infty[^X$ we define the two natural pavings

$$\begin{aligned} \mathfrak{t}(E) \quad &:= \quad \{ A \subset X : \chi_A \in E \}, \\ \mathfrak{T}(E) \quad &:= \quad \{ [f \geq t] : f \in E \text{ and } t > 0 \} = \{ [f \geq 1] : f \in E \}. \end{aligned}$$

We list some properties which are all obvious. The third one will soon become relevant when we need the horizontal integral of section 11 for the members of E.

14.11. PROPERTIES. 1) $\mathfrak{t}(E)$ and $\mathfrak{T}(E)$ are lattices in X with $\varnothing \in \mathfrak{t}(E) \subset \mathfrak{T}(E)$. 2) If E is Stonean then $\mathfrak{T}(E) = \{[f = 1] : f \in E \text{ with } f \leqq 1\}$. 3) Let $\mathfrak{S}$ be a lattice in X with $\varnothing \in \mathfrak{S}$. Then $E \subset \mathrm{UM}(\mathfrak{S}) \Leftrightarrow \mathfrak{T}(E) \subset \mathfrak{S}$.

14.12. EXERCISE. 1) Consider $E = \mathrm{CK}^+(\mathbb{R})$ on $X = \mathbb{R}$. Then $\mathfrak{t}(E) = \{\varnothing\}$ and $\mathfrak{T}(E) = \mathrm{Comp}(\mathbb{R})$. Thus $\mathfrak{t}(E)$ and $\mathfrak{T}(E)$ can be far apart. 2) Let X be a topological space. For $E = \mathrm{USC}^+(X)$ then $\mathfrak{t}(E) = \mathfrak{T}(E) = \mathrm{Cl}(X)$. 3) Let X be a Hausdorff topological space. For $E = \mathrm{USCK}^+(X)$ and $E = \mathrm{UMK}(X)$ then $\mathfrak{t}(E) = \mathfrak{T}(E) = \mathrm{Comp}(X)$.

Elementary Integrals

Let $E \subset [0, \infty[^X$ be a lattice cone of functions $X \to [0, \infty[$. We define an **elementary integral** on E to be a functional $I : E \to [0, \infty[$ which is positive-linear, that is additive and fulfils $I(tf) = tI(f)$ for all $f \in E$ and $t \geq 0$, and isotone. These are the natural functionals to be considered on a lattice cone.

14.13. EXERCISE. Let $E \subset [0, \infty[^X$ be a primitive lattice cone and $H := E - E$. Then there is a one-to-one correspondence between the elementary integrals $I : E \to [0, \infty[$ and the isotone linear functionals $L : H \to \mathbb{R}$. The correspondence is $I = L|E$.

Let $I : E \to [0, \infty[$ be an elementary integral. We start to consider the most naive type of integral representations of I. Let $\varphi : \mathfrak{T}(E) \to [0, \infty]$ be an isotone set function with $\varphi(\varnothing) = 0$. We have seen in 14.11.3) that $E \subset \mathrm{UM}(\mathfrak{T}(E))$. Therefore the horizontal integral $\fint f d\varphi$ is defined for all $f \in E$. We define φ to be a **source** of I iff $I(f) = \fint f d\varphi$ for all $f \in E$. Then $I < \infty$ enforces that $\varphi : \mathfrak{T}(E) \to [0, \infty[$.

These sources of I are of course far from those representations which we want to obtain. But on the other hand we shall see that the sources of I have a natural and simple characterization.

To this end we define the **crude outer envelope** $I^\star : [0, \infty]^X \to [0, \infty]$ and the **crude inner envelope** $I_\star : [0, \infty]^X \to [0, \infty]$ of I to be

$$I^\star(f) = \inf\{I(u) : u \in E \text{ with } u \geqq f\} \quad \text{and}$$
$$I_\star(f) = \sup\{I(u) : u \in E \text{ with } u \leqq f\} \quad \text{for } f \in [0, \infty]^X.$$

We list some simple properties.

14.14. PROPERTIES. 1) $I^\star$ and $I_\star$ are isotone. 2) $I_\star \leqq I^\star$ and $I_\star|E = I^\star|E = I$. 3) $I^\star$ is sublinear, and $I_\star$ is superlinear. 4) For $f, g \in [0, \infty]^X$ we have

$$I^\star(f \vee g) + I^\star(f \wedge g) \leqq I^\star(f) + I^\star(g) \quad \text{and}$$
$$I_\star(f \vee g) + I_\star(f \wedge g) \geqq I_\star(f) + I_\star(g).$$

5) For $A \in \mathfrak{T}(E)$ we have $I_\star(\chi_A) \leqq I^\star(\chi_A) < \infty$.

Proof. All properties except the last one are obvious. To see 5) note that $A \in \mathfrak{T}(E)$ means that $A = [f \geq 1]$ for some $f \in E$. Then $\chi_A \leq f$ and hence $I^*(\chi_A) \leq I(f) < \infty$.

The two envelopes I^* and I_* induce the two set functions $\Delta : \mathfrak{T}(E) \to [0, \infty[$ and $\nabla : \mathfrak{T}(E) \to [0, \infty[$, defined to be

$$\Delta(A) = I^*(\chi_A) \quad \text{and} \quad \nabla(A) = I_*(\chi_A) \quad \text{for } A \in \mathfrak{T}(E).$$

Their finiteness follows from 14.14.5). Once more we list some simple properties.

14.15. PROPERTIES. 1) Δ and ∇ are isotone. 2) $\nabla \leq \Delta$ and $\nabla(A) = \Delta(A) = I(\chi_A)$ for $A \in \mathfrak{t}(E)$. 3) Δ is submodular, and ∇ is supermodular.

We come to the representation theorem announced above. It is in essence due to Greco [1982]; see also Denneberg [1994] chapter 13.

14.16. THEOREM. Let $I : E \to [0, \infty[$ be an elementary integral on the Stonean lattice cone $E \subset [0, \infty[^X$. 0) I admits sources iff it has the truncation properties

(0) $I(f \wedge t) \downarrow 0$ for $t \downarrow 0$ and $I(f \wedge t) \uparrow I(f)$ for $t \uparrow \infty$ for all $f \in E$.

1) Assume that I fulfils (0). Then an isotone set function $\varphi : \mathfrak{T}(E) \to [0, \infty[$ is a source of I iff $\nabla \leq \varphi \leq \Delta$. 2) If $\varphi : \mathfrak{T}(E) \to [0, \infty[$ is a source of I and downward σ continuous then $\varphi = \Delta$.

Proof. i) Assume that $\varphi : \mathfrak{T}(E) \to [0, \infty[$ is a source of I, that is

$$I(f) = \fint f d\varphi = \int\limits_{0\leftarrow}^{\to\infty} \varphi([f \geq s]) ds \quad \text{for all } f \in E.$$

Since for $t > 0$ we have $[f \wedge t \geq s] = [f \geq s]$ when $s \leq t$ and $[f \wedge t \geq s] = \varnothing$ when $s > t$, we conclude that

$$I(f \wedge t) = \int\limits_{0\leftarrow}^{t} \varphi([f \geq s]) ds.$$

It follows that I has the truncation properties (0). ii) Assume now that I fulfils (0). We claim that Δ and ∇ are sources of I. Thus all isotone set functions $\varphi : \mathfrak{T}(E) \to [0, \infty[$ with $\nabla \leq \varphi \leq \Delta$ will be sources of I as well. In fact, for $f \in E$ and $0 < s < t$ one verifies that

$$\chi_{[f \geq t]} \leq \frac{1}{t-s}(f \wedge t - f \wedge s) \leq \chi_{[f \geq s]}.$$

Also the middle term is in E by 14.10, and hence we have

$$I\left(\frac{1}{t-s}(f \wedge t - f \wedge s)\right) = \frac{1}{t-s}\Big(I(f \wedge t) - I(f \wedge s)\Big).$$

It follows that

$$\Delta([f \geq t]) = I^\star(\chi_{[f \geq t]}) \leq \frac{1}{t-s}\Big(I(f \wedge t) - I(f \wedge s)\Big)$$
$$\leq I_\star(\chi_{[f \geq s]}) = \nabla([f \geq s]).$$

Thus for $0 < a = t(0) < t(1) < \cdots < t(r) = b < \infty$ we obtain

$$\sum_{l=1}^{r} \big(t(l) - t(l-1)\big)\Delta([f \geq t(l)]) \leq I(f \wedge b) - I(f \wedge a)$$

$$\leq \sum_{l=1}^{r}\big(t(l) - t(l-1)\big)\nabla([f \geq t(l-1)]).$$

The definition of the Riemann integral implies that

$$\int_a^b \Delta([f \geq t])dt \leq I(f \wedge b) - I(f \wedge a) \leq \int_a^b \nabla([f \geq t])dt,$$

and hence $=$ both times. For $a \downarrow 0$ and $b \uparrow \infty$ we obtain

$$I(f) = \int_{0\leftarrow}^{\rightarrow\infty} \Delta([f \geq t])dt = \int_{0\leftarrow}^{\rightarrow\infty} \nabla([f \geq t])dt,$$

as claimed. iii) Next assume that $\varphi : \mathfrak{T}(E) \to [0,\infty[$ is a source of I. Fix $A \in \mathfrak{T}(E)$. For $f \in E$ with $f \leq \chi_A$ we have $[f \geq t] \subset A$ when $t > 0$ and $[f \geq t] = \varnothing$ when $t > 1$, and hence $I(f) \leq \varphi(A)$. Therefore $\nabla(A) = I_\star(\chi_A) \leq \varphi(A)$. Likewise for $f \in E$ with $f \geq \chi_A$ we have $[f \geq t] \supset A$ when $0 < t \leq 1$, and hence $I(f) \geq \varphi(A)$. Therefore $\Delta(A) = I^\star(\chi_A) \geq \varphi(A)$. Thus we have proved $\nabla \leq \varphi \leq \Delta$. iv) It remains to prove 2). Assume that $\varphi : \mathfrak{T}(E) \to [0,\infty[$ is a source of I which is downward σ continuous. We fix $f \in E$ and form

$$P : P(t) = \varphi([f \geq t]) \quad \text{and} \quad Q : Q(t) = \Delta([f \geq t]) \quad \text{for } t > 0.$$

Then P and Q are monotone decreasing with $P \leq Q$. We know from i) that

$$\int_{0\leftarrow}^{t} P(s)ds = \int_{0\leftarrow}^{t} Q(s)ds = I(f \wedge t) \quad \text{for } t > 0.$$

This implies that $P(t-) = Q(t-) \geq Q(t)$ for $t > 0$. Now by assumption the function P is left continuous. Thus $P(t) \geq Q(t)$ and hence $P(t) = Q(t)$ for $t > 0$. It follows that $\varphi = \Delta$. The proof is complete.

14.17. CONSEQUENCE. *Let $I : E \to [0,\infty[$ be an elementary integral on the Stonean lattice cone $E \subset [0,\infty[^X$ which fulfils (0). Then*

$$I_\star(f) = \int f d\nabla_\star \quad \text{for all } f \in [0,\infty]^X.$$

Proof of $\leqq$. From 14.16.1) we see for $u \in E$ with $u \leqq f$ that $I(u) = \int u d\nabla = \int u d\nabla_\star \leqq \int f d\nabla_\star$. It follows that $I_\star(f) \leqq \int f d\nabla_\star$. Proof of $\geqq$. Fix $f \in [0, \infty]^X$. By 11.16 we have

$$\int f d\nabla_\star = \sup\{\int u d\nabla : u \in S(\mathfrak{T}(E)) \text{ with } u \leqq f\}.$$

Now for $u \in S(\mathfrak{T}(E))$, represented after 11.4.iv) in the form

$$u = \sum_{l=1}^{r} t_l \chi_{A(l)} \text{ with } A(1) \supset \cdots \supset A(r) \text{ in } \mathfrak{T}(E) \text{ and } t_1, \cdots, t_r > 0,$$

we have from 11.8.1) and 14.14.3)

$$\int u d\nabla = \sum_{l=1}^{r} t_l \nabla(A(l)) = \sum_{l=1}^{r} t_l I_\star(\chi_{A(l)}) \leqq I_\star\left(\sum_{l=1}^{r} t_l \chi_{A(l)}\right) = I_\star(u).$$

It follows that $\int f d\nabla_\star \leqq I_\star(f)$. The proof is complete.

The last two results will be the basis for the future representation theorems. But the main work is still to be done in the subsequent sections. It will be prepared in the next subsection with the definition of the relevant type of representations.

14.18. EXERCISE. Let $\mathfrak{S}$ be a lattice in X with $\varnothing \in \mathfrak{S}$, and take $E = S(\mathfrak{S})$ as defined after 11.4. 1) E is a Stonean lattice cone. 2) $\mathfrak{t}(E) = \mathfrak{T}(E) = \mathfrak{S}$. 3) Each elementary integral $I : E \to [0, \infty[$ fulfils (0). 4) Let $I : E \to [0, \infty[$ be an elementary integral, and define $\varphi : \varphi(A) = I(\chi_A)$ for $A \in \mathfrak{S}$. Then $\varphi : \mathfrak{S} \to [0, \infty[$ is isotone and modular with $\varphi(\varnothing) = 0$, and is the unique source of I. 5) Let $\varphi : \mathfrak{S} \to [0, \infty[$ be isotone and modular with $\varphi(\varnothing) = 0$, and define $I : I(f) = \int f d\varphi$ for $f \in E = S(\mathfrak{S})$. Then $I : E \to [0, \infty[$ is an elementary integral, and of course $\varphi(A) = I(\chi_A)$ for $A \in \mathfrak{S}$. Hint: 11.11. 6) The two maps defined in 4)5) constitute a one-to-one correspondence between the elementary integrals $I : E = S(\mathfrak{S}) \to [0, \infty[$ and the isotone and modular set functions $\varphi : \mathfrak{S} \to [0, \infty[$ with $\varphi(\varnothing) = 0$.

Representations of Elementary Integrals

We fix a lattice cone $E \subset [0, \infty[^X$ of functions $X \to [0, \infty[$ on a nonvoid set X. Let $I : E \to [0, \infty[$ be an elementary integral. We start with the crude notion of a representation. We define a **representation** of I to be a ccontent $\alpha : \mathfrak{A} \to [0, \infty]$ on some $\mathfrak{A} \subset \mathfrak{P}(X)$ such that

$$E \subset \mathrm{UM}(\mathfrak{A}) \quad \text{and} \quad I(f) = \int f d\alpha = \int_\star f d\alpha \text{ for all } f \in E;$$

the two integrals are equal by 12.11. By 14.11.3) and 11.15 the definition means that

$$\mathfrak{T}(E) \subset \mathfrak{A} \quad \text{and} \quad \alpha|\mathfrak{T}(E) \text{ is a source of } I,$$

in short that α is an extension of some source of I. When $\alpha : \mathfrak{A} \to [0, \infty]$ is a cmeasure on a σ algebra $\mathfrak{A}$ then the definition means that all $f \in E$ are measurable $\mathfrak{A}$ with $I(f) = \int f d\alpha$.

We obtain an existence result which looks perfect, but as we shall see is not.

14.19. PROPOSITION. *If E is Stonean and I fulfils* (0) *then I has at least one representation.*

Proof. From 11.14.2), which combines the basic fact 11.11 with the Hahn-Banach type result 11.14.1), and from 14.15 we see that there exists an isotone and modular set function $\varphi : \mathfrak{T}(E) \to [0, \infty[$ with $\nabla \leqq \varphi \leqq \Delta$. By 14.16.1) φ is a source of I. Then the Smiley-Horn-Tarski theorem 3.4 asserts that φ can be extended to a ccontent.

Thus the result is a counterpart of the Smiley-Horn-Tarski theorem 3.4. With this theorem it shares a disastrous defect: When I is downward σ continuous in the natural sense then it need not have representations which are almost downward σ continuous, that means representations which are cmeasures. It suffices to return to the old counterexample.

14.20. EXAMPLE. Consider the set function $\varphi : \mathfrak{S} \to [0, \infty[$ of example 3.11. By 14.18 there is an elementary integral $I : E \to [0, \infty[$ with (0) on the Stonean lattice cone $E = S(\mathfrak{S})$ such that φ is the unique source of I. Since φ is downward • continuous for • $= \sigma\tau$ it follows from 11.17 that I is downward • continuous in the sense of that theorem. But we know that φ cannot be extended to a cmeasure.

To be sure, the desired existence of a cmeasure representation holds true in case that the Stonean lattice cone E is *primitive*. This is the traditional Daniell-Stone theorem 14.1, combined with 14.6 and 14.13. It is the precise counterpart to the traditional main theorem at the outset of the present text, which said that the then desired existence of a cmeasure extension holds true in case that the initial lattice $\mathfrak{S}$ is a *ring*. Herewith we were back to the frame which we want to surpass. We have to conclude that our basic notion of a representation must be another one, and must be of the same kind as before. At this point the natural definition is as follows.

DEFINITION. Let $I : E \to [0, \infty[$ be an elementary integral. For • $= \star\sigma\tau$ we define a • **representation** of I to be a representation $\alpha : \mathfrak{A} \to [0, \infty]$ of I such that α is an inner • extension of $\alpha | \mathfrak{T}(E)$. We define I to be a • **preintegral** iff it admits • representations.

This time it is redundant to use the word *inner*, because there will be no outer counterpart. Our aim is to characterize those I which are • preintegrals, and then to describe all • representations of I. The first formulation below is not more than a combination of the definitions which are involved.

14.21. PROPOSITION. *Let $I : E \to [0, \infty[$ be an elementary integral and* • $= \star\sigma\tau$. *Then I is a • preintegral iff it has sources which are inner •*

premeasures. In this case the • representations of I coincide with the inner • extensions of those sources of I which are inner • premeasures.

14.22. EXAMPLE. An elementary integral $I : E \rightarrow [0, \infty[$ with (0) on a Stonean lattice cone E can be without • representations for all • $= \star\sigma\tau$, even when I has a unique source and this source can be extended to a cmeasure which is almost downward τ continuous. As an example consider the elementary integral $I : E \rightarrow [0, \infty[$ from 14.20 and its unique source $\varphi : \mathfrak{S} \rightarrow [0, \infty[$. We know from 3.11 that φ is downward τ continuous, and one verifies that it is not inner τ tight. Hence φ is not an inner • premeasure for all • $= \star\sigma\tau$.

Now we have to realize that the cases • $= \sigma\tau$ and • $= \star$ fall far apart. In fact, we see from 14.16.2) that each source of an elementary integral $I : E \rightarrow [0, \infty[$ on a Stonean lattice cone E which is an inner • premeasure for • $= \sigma\tau$ must be $= \Delta$, while this cannot be concluded for • $= \star$. Thus we obtain what follows.

14.23. THEOREM (for • $= \sigma\tau$). *Let $I : E \rightarrow [0, \infty[$ be an elementary integral on a Stonean lattice cone $E \subset [0, \infty[^X$. Then I is a • preintegral iff I fulfils (0) and $\Delta : \mathfrak{T}(E) \rightarrow [0, \infty[$ is an inner • premeasure. In this case the • representations of I are the inner • extensions of Δ. In particular I has the unique maximal • representation $\alpha := \Delta_\bullet | \mathfrak{C}(\Delta_\bullet)$.*

The next section will be devoted to the cases • $= \sigma\tau$. We see that its prime task will be to characterize those elementary integrals $I : E \rightarrow [0, \infty[$ which fulfil (0) and for which $\Delta : \mathfrak{T}(E) \rightarrow [0, \infty[$ is an inner • premeasure. Then section 17 will be devoted to the case • $= \star$ which will be quite different. In fact, we present on the spot an example of an elementary integral $I : E \rightarrow [0, \infty[$ on a Stonean E with two different sources $\neq \Delta$ which are both inner $\star$ premeasures.

14.24. EXERCISE. Let $X = \mathbb{N} \cup (-\mathbb{N})$. We define $E \subset [0, \infty[^X$ to consist of all functions $f : X \rightarrow [0, \infty[$ such that the two sequences $n \mapsto f(n), f(-n) \; \forall n \in \mathbb{N}$ are monotone increasing and have equal finite limits for $n \rightarrow \infty$. 1) E is a Stonean lattice cone. Next define the paving $\mathfrak{T}$ on $\mathbb{N}$ to consist of the subsets $\{n, n + 1, \cdots\} \; \forall n \in \mathbb{N}$. 2) $\mathfrak{t}(E)$ consists of $\varnothing$ and of the subsets $P \cup (-Q) \subset X$ with $P, Q \in \mathfrak{T}$. $\mathfrak{T}(E)$ consists of $\varnothing$ and of the subsets $P, -Q, P \cup (-Q) \subset X$ with $P, Q \in \mathfrak{T}$. Now we define

$$I : E \rightarrow [0, \infty[\text{ to be } I(f) = \lim_{n \rightarrow \pm\infty} f(n) \text{ for } f \in E.$$

3) I is an elementary integral which fulfils (0). 4) We have

$$I^\star(f) = \sup f \text{ and } I_\star(f) = \liminf_{n \rightarrow \pm\infty} f(n) \text{ for } f \in [0, \infty]^X.$$

In particular we see for $A \in \mathfrak{T}(E)$ that

$$\Delta(A) = \left\{ \begin{array}{ll} 1 & \text{if } A \neq \varnothing \\ 0 & \text{if } A = \varnothing \end{array} \right\} \text{ and}$$

$$\nabla(A) = \left\{ \begin{array}{ll} 1 & \text{if } A = P \cup (-Q) \text{ with } P, Q \in \mathfrak{T} \\ 0 & \text{otherwise} \end{array} \right\}.$$

At last we define $\varphi, \psi : \mathfrak{T}(E) \to [0, \infty[$ to be

$$\varphi(A) = \left\{ \begin{array}{ll} 1 & \text{if } A \cap \mathbb{N} \neq \varnothing \\ 0 & \text{if } A \subset -\mathbb{N} \end{array} \right\} \text{ and } \psi(A) = \left\{ \begin{array}{ll} 1 & \text{if } A \cap (-\mathbb{N}) \neq \varnothing \\ 0 & \text{if } A \subset \mathbb{N} \end{array} \right\}.$$

5) φ and ψ are isotone with $\nabla \leq \varphi, \psi \leq \Delta$ and hence sources of I. Furthermore φ, ψ and ∇, Δ are all different. 6) φ and ψ are inner $\star$ premeasures.

We conclude the section with an exercise which centers around a natural problem: To characterize those inner $\bullet$ premeasures $\varphi : \mathfrak{S} \to [0, \infty[$ with $\varphi(\varnothing) = 0$ which occur in 14.23. We shall see that all of them can occur.

14.25. EXERCISE (for $\bullet = \sigma\tau$). 1) Let $\varphi : \mathfrak{S} \to [0, \infty[$ be an inner $\bullet$ premeasure on a lattice $\mathfrak{S}$ in X with $\varnothing \in \mathfrak{S}$ and $\varphi(\varnothing) = 0$. Then there exists an elementary integral $I : E \to [0, \infty[$ on a Stonean lattice cone $E \subset [0, \infty[^X$ which is a $\bullet$ preintegral and satisfies $\mathfrak{T}(E) = \mathfrak{S}$ and $\Delta = \varphi$ (and has in fact the unique source φ). Hint: 14.18.

In the remainder of the exercise we assume that $E \subset [0, \infty[^X$ is a *primitive* Stonean lattice cone. We shall see that the situation will then be different. 2) For $f \in E$ and $s > 0$ we have

$$[f > s] \subset \big(\mathfrak{T}(E) \top \mathfrak{T}(E)\big) \bot.$$

Thus for each $A \in \mathfrak{T}(E)$ there exist $U \in \big(\mathfrak{T}(E) \top \mathfrak{T}(E)\big) \bot$ and $B \in \mathfrak{T}(E)$ with $A \subset U \subset B$. 3) Let $I : E \to [0, \infty[$ be an elementary integral which is a $\bullet$ preintegral, and let as above $\Delta : \mathfrak{T}(E) \to [0, \infty[$ and $\alpha := \Delta_\bullet | \mathfrak{C}(\Delta_\bullet)$. Then α is outer regular $\big(\mathfrak{T}(E) \top (\mathfrak{T}(E))_\bullet\big) \bot$ at $(\mathfrak{T}(E))_\bullet$. We recall from 13.39 that there exist inner $\bullet$ premeasures $\varphi : \mathfrak{S} \to [0, \infty[$ with $\varphi(\varnothing) = 0$ such that $\phi := \varphi_\bullet | \mathfrak{C}(\varphi_\bullet)$ is not outer regular $(\mathfrak{S} \top \mathfrak{S}_\bullet) \bot$ at $\mathfrak{S}_\bullet$.

15. The Continuous Daniell-Stone Theorem

Preparations on Lattice Cones

Let $E \subset [0, \infty[^X$ be a lattice cone of functions $X \to [0, \infty[$ on the nonvoid set X. We define for $\bullet = \star\sigma\tau$ the function classes

$$E_\bullet := \big\{ \inf_{f \in M} f : M \subset E \text{ nonvoid of type } \bullet \big\} \subset [0, \infty[^X.$$

Since

$$\inf_{f \in M} f = \inf_{f \in N} f \quad \text{for} \quad N := \{f_1 \wedge \cdots \wedge f_n : f_1, \cdots, f_n \in M \text{ and } n \in \mathbb{N}\},$$

we can restrict ourselves to the nonvoid subsets $M \subset E$ of type • which are downward directed. Thus $E_\star = E$ and

$$E_\sigma = \left\{ \lim_{l\to\infty} f_l : (f_l)_l \text{ in } E \text{ antitone} \right\},$$

$$E_\tau = \left\{ F \in [0,\infty[^X : F = \inf_{f\in E, f\geq F} f \right\}.$$

We list the relevant properties.

15.1. EXERCISE. $(E_\bullet)_\bullet = E_\bullet$. Furthermore if $M \subset E_\bullet$ is nonvoid of type • with $M \downarrow F$ then there exists $N \subset E$ nonvoid of type • with $N \downarrow F$, and such that each $u \in N$ is $\geq$ some $f \in M$. This is an obvious counterpart of 6.6 and proved in the same manner.

15.2. PROPERTIES. 1) $E_\bullet$ is a lattice cone, and $E = E_\star \subset E_\sigma \subset E_\tau$. 2) If E is Stonean then $E_\bullet$ is Stonean. 3) $\mathfrak{T}(E_\bullet) = (\mathfrak{T}(E))_\bullet$ and $\mathfrak{t}(E_\bullet) \supset (\mathfrak{t}(E))_\bullet$. 4) If • $= \sigma\tau$ and E is Stonean then $\mathfrak{T}(E_\bullet) = \mathfrak{t}(E_\bullet)$. Therefore

$$(\mathfrak{T}(E))_\bullet = \mathfrak{T}(E_\bullet) = \mathfrak{t}(E_\bullet) \supset (\mathfrak{t}(E))_\bullet.$$

Proof. 1)2) are obvious. 3) On the one hand we have

$$\left[\inf_{f\in M} f \geq 1 \right] = \bigcap_{f\in M} [f \geq 1] \quad \text{for nonvoid } M \subset E.$$

This proves $\mathfrak{T}(E_\bullet) = (\mathfrak{T}(E))_\bullet$. On the other hand we have

for $A \subset X$ and nonvoid $\mathfrak{M} \subset \mathfrak{P}(X) : A = \bigcap_{M\in\mathfrak{M}} M \Rightarrow \chi_A = \inf_{m\in\mathfrak{M}} \chi_M.$

This proves $\mathfrak{t}(E_\bullet) \supset (\mathfrak{t}(E))_\bullet$. 4) requires the lemma which follows. Its proof consists of immediate verifications.

15.3. LEMMA. For real $0 \leq x \leq 1$ and $0 \leq t < 1$ define

$$x_{<t>} := \frac{1}{1-t}(x-t)^+.$$

Then 1) $0 \leq x_{<t>} \leq x \leq 1$. 2) The function $(x,t) \mapsto x_{<t>}$ is continuous. 3) For fixed $0 \leq t < 1$ the function $x \mapsto x_{<t>}$ is monotone increasing with

$$x_{<t>} = 0 \text{ for } 0 \leq x \leq t \quad \text{and} \quad x_{<t>} = 1 \text{ for } x = 1.$$

4) For fixed $0 \leq x \leq 1$ the function $t \mapsto x_{<t>}$ is monotone decreasing, and for $t \uparrow 1$ we have

$$x_{<t>} \downarrow 0 \text{ when } 0 \leq x < 1, \quad \text{and} \quad x_{<t>} \downarrow 1 \text{ when } x = 1.$$

Proof of 15.2.4). i) We first prove $\mathfrak{T}(E) \subset \mathfrak{t}(E_\sigma)$. Let $A \in \mathfrak{T}(E)$, that is $A = [f = 1]$ for some $f \in E$ with $f \leq 1$. For $0 \leq t < 1$ now $f_{<t>} \in E$, and for $t \uparrow 1$ we have $f_{<t>} \downarrow \chi_A$ by 15.3.4). Therefore $\chi_A \in E_\sigma$, that is $A \in \mathfrak{t}(E_\sigma)$. ii) By 15.2.1)2) $E_\bullet$ is a Stonean lattice cone. Thus from i) and 15.1 we obtain $\mathfrak{T}(E_\bullet) \subset \mathfrak{t}((E_\bullet)_\sigma) = \mathfrak{t}(E_\bullet)$ and hence the assertion.

Preparations on Elementary Integrals

We start to formulate the natural notions of continuity, after the model of 11.17. Assume that $E \subset [0, \infty]^X$ is a lattice and that $I : E \to [0, \infty]$ is isotone, and let $\bullet = \sigma\tau$. We define I to be **downward $\bullet$ continuous** iff $\inf\{I(f) : f \in M\} = I(F)$ for all $M \subset E$ nonvoid of type $\bullet$ and downward directed in the pointwise order with $M \downarrow F \in E$; and **almost downward $\bullet$ continuous** iff this holds true whenever $I(f) < \infty \; \forall f \in M$. Furthermore if $0 \in E$ and $I(0) = 0$ then we define I to be $\bullet$ **continuous at 0** iff $\inf\{I(f) : f \in M\} = 0$ for all $M \subset E$ nonvoid of type $\bullet$ and downward directed $\downarrow 0$; and **almost $\bullet$ continuous at 0** iff this holds true whenever $I(f) < \infty \; \forall f \in M$.

Now assume that $E \subset [0, \infty[^X$ is a lattice cone and that $I : E \to [0, \infty[$ is an elementary integral.

15.4. REMARK. *Let $\bullet = \sigma\tau$. 1) I is $\bullet$ continuous at $0 \Rightarrow I^\star|E_\bullet$ is $\bullet$ continuous at 0 as well. 2) I is downward $\bullet$ continuous $\Rightarrow I^\star|E_\bullet$ is an elementary integral and downward $\bullet$ continuous as well.*

Proof. 1) is obvious from 15.1 applied to $F = 0$. 2) We first prove that $I^\star|E_\bullet$ is downward $\bullet$ continuous. Fix $F \in E_\bullet$ and $M \subset E_\bullet$ nonvoid of type $\bullet$ with $M \downarrow F$, and let $N \subset E$ be as in 15.1. For fixed $v \in E$ with $v \geqq F$ the subset $\{u \vee v : u \in N\} \subset E$ is nonvoid of type $\bullet$ and $\downarrow F \vee v = v$. By assumption it follows that

$$\inf_{f \in M} I^\star(f) \leqq \inf_{u \in N} I(u) \leqq \inf_{u \in N} I(u \vee v) = I(v).$$

Thus the infimum on the left is $\leqq I^\star(F)$ and hence $= I^\star(F)$. This proves that $I^\star|E_\bullet$ is downward $\bullet$ continuous. Then an immediate consequence is that $I^\star|E_\bullet$ is additive and hence an elementary integral.

15.5. PROPOSITION. *Assume that E is Stonean and that I fulfils (0). Let $\bullet = \sigma\tau$. 1) I is $\bullet$ continuous at $0 \Rightarrow \Delta$ is $\bullet$ continuous at $\varnothing$. 2) The following are equivalent.*

i) *I is downward $\bullet$ continuous.*
ii) *Δ is modular and downward $\bullet$ continuous.*
iii) *Δ is downward $\bullet$ continuous.*

Proof. 1) By 15.4.1) and by definition the set function $A \mapsto I^\star(\chi_A)$ on $\mathfrak{t}(E_\bullet)$ is $\bullet$ continuous at $\varnothing$. But we know from 15.2.4) that $\mathfrak{t}(E_\bullet) = (\mathfrak{T}(E))_\bullet \supset \mathfrak{T}(E)$. 2) The implication i)$\Rightarrow$ii) follows from 15.4.2) and 15.2.4) as before. ii)$\Rightarrow$iii) is obvious. iii)$\Rightarrow$i) By 14.16.1) Δ is a source of I. Therefore the assertion is contained in 11.17.

At this point we recall from 14.20 that the equivalent properties in 15.5.2) do not enforce that I has at least one cmeasure representation.

The New Envelopes for Elementary Integrals

After this we proceed as in the inner extension theory of chapter II. Let $I : E \to [0, \infty[$ be an elementary integral on the lattice cone $E \subset [0, \infty[^X$.

We define besides $I_\star$ the **inner • envelopes** $I_\bullet : [0,\infty]^X \to [0,\infty]$ for $\bullet = \sigma\tau$ to be

$$I_\bullet(f) \;=\; \sup\Big\{ \inf_{u\in M} I(u) : M \subset E \text{ nonvoid of type } \bullet$$

$$\text{with } M \downarrow \text{ some function } \leq f\Big\} \quad \text{for } f \in [0,\infty]^X.$$

Furthermore we define for fixed $v \in E$ the **satellite inner • envelopes** $I_\bullet^v : [0,\infty]^X \to [0,\infty[$ to be

$$I_\bullet^v(f) \;=\; \sup\Big\{ \inf_{u\in M} I(u) : M \subset E \text{ nonvoid of type } \bullet \text{ with }$$

$$u \leq v \;\forall u \in M \text{ and } M \downarrow \text{ some function } \leq f\Big\} \quad \text{for } f \in [0,\infty]^X.$$

Both formulas include the case $\bullet = \star$. The properties of these functionals correspond to those of the former envelopes.

15.6. PROPERTIES. 1) $I_\bullet$ and $I_\bullet^v$ are isotone. 2) $I_\bullet$ is superlinear. 3) $I_\star \leq I_\sigma \leq I_\tau$ and $I_\star^v \leq I_\sigma^v \leq I_\tau^v$. 4) For $\bullet = \sigma\tau$ we have

$$I_\bullet|E = I \;\;\Leftrightarrow\;\; I \text{ is downward } \bullet \text{ continuous,}$$

$$I_\bullet(0) = 0 \;\;\Leftrightarrow\;\; I \text{ is } \bullet \text{ continuous at } 0.$$

5) $I_\star^v \leq I(v) < \infty$. 6) We have

$$I_\bullet(f) = \sup\{I_\bullet^v(f) : v \in E\} \quad \text{for } f \in [0,\infty]^X.$$

7) Assume that $I = I_\bullet|E$. Then $I_\bullet(f) = I_\bullet^v(f)$ for $f \leq v \in E$. 8) We have $I = I_\bullet|E$ iff $I^\star = I_\bullet$ on $E_\bullet$. 9) If $I = I_\bullet|E$ and $E_\bullet = E$ then $I_\bullet = I_\star$.

Proof. 1)2)3)5) are obvious, and both parts of 4) are routine verifications from the definitions. 6) follows the model of 6.29.4), and 7) follows the model of 6.29.5). 8) One proves as in 6.10 and 6.11 that i) $I^\star \leq I_\bullet$ on $E_\bullet$, and that ii) $I = I_\bullet|E \Rightarrow I^\star \geq I_\bullet$. Then the assertion follows. 9) is an obvious consequence of 8).

The next step are the important two propositions which follow.

15.7. PROPOSITION. Assume that E is Stonean and that I fulfils (0). Let $\bullet = \sigma\tau$, and assume that Δ is downward $\bullet$ continuous. Then

$$I^\star(f) = I_\bullet(f) = \fint f d\Delta_\bullet \quad \text{for } f \in E_\bullet.$$

Furthermore $I^\star|E_\bullet = I_\bullet|E_\bullet$ is downward $\bullet$ continuous.

15.8. PROPOSITION. Assume that E is Stonean and that I fulfils (0). Let $\bullet = \sigma\tau$, and assume that Δ is an inner $\bullet$ premeasure with $\alpha := \Delta_\bullet|\mathfrak{C}(\Delta_\bullet)$. Then

$$I_\bullet(f) = \fint f d\Delta_\bullet = \int_* f d\alpha \quad \text{for all } f \in [0,\infty]^X.$$

Proof of 15.7. 0) We know from 14.16.1) that $I(u) = \fint u d\Delta$ for all $u \in E$. i) We have $\Delta_\bullet|\mathfrak{T}(E) = \Delta$ by 6.5. In particular $\Delta_\bullet(\varnothing) = 0$, so that the horizontal integral $\fint f d\Delta_\bullet$ is defined for all $f \in [0,\infty]^X$. Next $\Delta_\bullet|(\mathfrak{T}(E))_\bullet$ is downward $\bullet$ continuous by 6.5.iii). Hence by 11.17 the horizontal integral

$f \mapsto \int f d\Delta_\bullet$ is almost downward • continuous on $\mathrm{UM}((\mathfrak{T}(E))_\bullet)$. ii) We have $\mathfrak{T}(E_\bullet) = (\mathfrak{T}(E))_\bullet$ by 15.2.3 and hence $E_\bullet \subset \mathrm{UM}((\mathfrak{T}(E))_\bullet)$ by 14.11.3. Since each function $f \in E_\bullet$ is $\leqq$ some member of E it follows from (0) that $\int f d\Delta_\bullet < \infty$. Thus i) implies that $f \mapsto \int f d\Delta_\bullet$ is downward • continuous on $E_\bullet$. iii) On the other hand we know from 15.5 that I is downward • continuous, and hence from 15.4.2) then $I^\star | E_\bullet$ is downward • continuous. Now $I = I_\bullet | E$ by 15.6.4) and hence $I^\star = I_\bullet$ on $E_\bullet$ by 15.6.8). iv) We have seen that the two functionals $I^\star | E_\bullet = I_\bullet | E_\bullet$ and $f \mapsto \int f d\Delta_\bullet$ on $E_\bullet$ are equal on E and are both downward • continuous, and therefore must be equal on $E_\bullet$. This completes the proof of 15.7. v) In order to prepare the proof of 15.8 we add the remark that $I_\bullet(f) \leqq \int f d\Delta_\bullet$ for all $f \in [0,\infty]^X$. In fact, fix $f \in [0,\infty]^X$ and let $M \subset E$ be nonvoid of type • such that $M \downarrow F \leqq f$. Then $F \in E_\bullet$, and from the above we obtain

$$\inf_{u \in M} I(u) = \inf_{u \in M} I_\bullet(u) = I_\bullet(F) = \int F d\Delta_\bullet \leqq \int f d\Delta_\bullet.$$

The assertion follows.

Proof of 15.8. i) We first note that $\Delta_\bullet = \alpha_\star$. In fact, the two set functions are equal on $\mathfrak{C}(\Delta_\bullet)$, and because of $(\mathfrak{T}(E))_\bullet \subset \mathfrak{C}(\Delta_\bullet)$ are both inner regular $\mathfrak{C}(\Delta_\bullet)$. ii) From i) and 12.12 we conclude that

$$\int f d\Delta_\bullet = \int_* f d\alpha \quad \text{for all } f \in [0,\infty]^X.$$

iii) In view of part v) of the last proof it remains to show that $\int f d\Delta_\bullet \leqq I_\bullet(f)$ for $f \in [0,\infty]^X$. Because of ii) and 15.7 this is a consequence of

$$\int_* f d\alpha = \sup\{\int_* u d\alpha : u \in E_\bullet \text{ with } u \leqq f\} \quad \text{for } f \in [0,\infty]^X.$$

iv) In order to prove the last assertion we recall from 12.1 that

$$\int_* f d\alpha = \sup\left\{\sum_{l=1}^r t_l \alpha(A(l)) : A(1), \cdots, A(r) \in \mathfrak{C}(\Delta_\bullet)\right.$$

$$\text{and } t_1, \cdots, t_r > 0 \text{ with } f \geqq \sum_{l=1}^r t_l \chi_{A(l)}\Bigg\}.$$

For each choice of the entities in the brackets we have a function

$$u := \sum_{l=1}^r t_l \chi_{A(l)} \in \mathrm{S}(\mathfrak{C}(\Delta_\bullet)), \text{ and } \int_* u d\alpha = \sum_{l=1}^r t_l \alpha(A(l)) \text{ by 12.5 .}$$

But since α is inner regular $(\mathfrak{T}(E))_\bullet$ we can restrict ourselves to choices $A(1), \cdots, A(r) \in (\mathfrak{T}(E))_\bullet$, which by 15.2.4) is $= \mathfrak{T}(E_\bullet) = \mathfrak{t}(E_\bullet)$. It follows that $\chi_{A(l)} \in E_\bullet$ $(l = 1, \cdots, r)$ and hence $u \in E_\bullet$. This completes the proof of 15.8.

The Main Theorem

We come to the main theorem of the present context. It furnishes the characterization of the • preintegrals for • = $\sigma\tau$ as announced after 14.23. The theorem is structured like the main extension theorems of chapter II.

15.9. THEOREM (Continuous Daniell-Stone Theorem • = $\sigma\tau$). *Let* $I :$ $E \to [0, \infty[$ *be an elementary integral on a Stonean lattice cone* $E \subset [0, \infty[^X$. *Then the following are equivalent.*

1) *There exist* • *representations of* I, *that is* I *is a* • *preintegral.*

1') I *fulfils* (0); *and* $\Delta : \mathfrak{T}(E) \to [0, \infty[$ *is an inner* • *premeasure.*

2) $I(v) = I(u) + I_\bullet(v - u)$ *for all* $u \leqq v$ *in* E.

3) $I_\bullet|E = I$; *and* $I(v) \leqq I(u) + I_\bullet(v - u)$ *for all* $u \leqq v$ *in* E.

3') $I_\bullet(0) = 0$; *and* $I(v) \leqq I(u) + I_\bullet^v(v - u)$ *for all* $u \leqq v$ *in* E.

In this case I *has the unique maximal* • *representation* $\alpha := \Delta_\bullet|\mathfrak{C}(\Delta_\bullet)$. *Furthermore we have*

$$I^*(f) = I_\bullet(f) \quad = \quad \int f d\alpha \qquad \text{for all } f \in E_\bullet,$$

$$I_\bullet(f) \quad = \quad \int_* f d\alpha \qquad \text{for all } f \in [0, \infty]^X.$$

The equivalence 1)⇔1') has been obtained in 14.23. Also 1) implies the first final assertion by 14.23, and 1') implies the second final assertion by 15.7 and 15.8.

The implications 1')⇒2)⇒3)⇒3') can be settled in a few words. 1')⇒2) Let $\alpha := \Delta_\bullet|\mathfrak{C}(\Delta_\bullet)$ be the maximal inner • extension of Δ. For $u \leqq v$ in E we combine 12.3 with $E \subset \mathrm{UM}(\mathfrak{T}(E)) \subset \mathrm{UM}(\mathfrak{C}(\Delta_\bullet))$ to obtain

$$\int_* v d\alpha = \int_* (u + (v - u)) d\alpha = \int_* u d\alpha + \int_* (v - u) d\alpha.$$

By 15.7 and 15.8 this means that $I(v) = I(u) + I_\bullet(v - u)$. 2)⇒3) is obvious. 3)⇒3') follows with the aid of 15.6.7).

It remains to prove the implication 3')⇒1'). Thus we shall assume 3') all the time. By 15.6.4) the first condition $I_\bullet(0) = 0$ means that I is • continuous at 0. The proof will be divided into five steps.

i) I fulfils (0).

ii) Δ is • continuous at $\varnothing$.

iii) Assume that $B \in \mathfrak{T}(E)$ and $B = [v = 1]$ for some $v \in E$ with $v \leqq 1$. In the notation of 15.3 we have $v_{<t>} \in E$ for $0 \leqq t < 1$, and 15.3.4) implies that $v_{<t>} \downarrow \chi_B$ for $t \uparrow 1$. We claim that $I(v_{<t>}) \downarrow I^*(\chi_B) =: \Delta(B)$ for $t \uparrow 1$.

iv) Δ is modular.

v) Δ is inner • tight.

It is clear that this will complete the proof of the implication 3')⇒1') and hence of the theorem.

Proof of i). For $f \in E$ and $t > 0$ we have $f \wedge t, (f - t)^+ \in E$ with $f = f \wedge t + (f - t)^+$ and hence $I(f) = I(f \wedge t) + I((f - t)^+)$. We use that I

is $\bullet$ continuous at 0. For $t \downarrow 0$ we have $f \wedge t \downarrow 0$ and hence $I(f \wedge t) \downarrow 0$, and for $t \uparrow \infty$ we have $(f - t)^+ \downarrow 0$ and hence $I(f \wedge t) \uparrow I(f)$.

Proof of ii). This follows from 15.5.1).

Proof of iii). iii.1) We fix $u \in E$ with $u \geqq \chi_B$. To be shown is

$$\lim_{t \uparrow 1} I(v_{<t>}) \leqq I(u).$$

In view of $v \geqq \chi_B$ we can assume that $u \leqq v \leqq 1$. Furthermore fix $\varepsilon > 0$. iii.2) By the second condition in 3') there exists an $M \subset E$ nonvoid of type $\bullet$ with $M \downarrow$ some $F \leqq v - u$ such that

$$h \leqq v \text{ and } I(h) \geqq I(v) - I(u) - \varepsilon \text{ for all } h \in M.$$

iii.3) The subset $\{h \wedge v_{<t>} : h \in M \text{ and } 0 \leqq t < 1\} \subset E$ is nonvoid of type $\bullet$ and downward directed $\downarrow F \wedge \chi_B$. Here we have $F \wedge \chi_B = 0$ because $0 \leqq F \wedge \chi_B \leqq (v - u) \wedge \chi_B$ and $v = u = 1$ on B. Since I is $\bullet$ continuous at 0 it follows that

$$\inf \{I(h \wedge v_{<t>}) : h \in M \text{ and } 0 \leqq t < 1\} = 0.$$

iii.4) For $h \in M$ and $0 \leqq t < 1$ we have

$$I(v) - I(u) - \varepsilon + I(v_{<t>}) \leqq I(h) + I(v_{<t>})$$
$$= I(h \vee v_{<t>}) + I(h \wedge v_{<t>}) \leqq I(v) + I(h \wedge v_{<t>}),$$

where we have used $h \vee v_{<t>} \leqq v$ because of 15.3.1). It follows that

$$I(v_{<t>}) \leqq I(u) + \varepsilon + I(h \wedge v_{<t>}).$$

From iii.3) we obtain the assertion.

Proof of iv). Let $A, B \in \mathfrak{T}(E)$, and fix $u, v \in E$ with $u, v \leqq 1$ such that $A = [u = 1]$ and $B = [v = 1]$. Then $A \cup B = [u \vee v = 1]$ and $A \cap B = [u \wedge v = 1]$. Now for $0 \leqq t < 1$

$$(u \vee v)_{<t>} = u_{<t>} \vee v_{<t>} \text{ and } (u \wedge v)_{<t>} = u_{<t>} \wedge v_{<t>};$$

this is clear since the function $x \mapsto x_{<t>}$ is monotone increasing by 15.3.3). Thus we have

$$I((u \vee v)_{<t>}) + I((u \wedge v)_{<t>}) = I(u_{<t>}) + I(v_{<t>}).$$

For $t \uparrow 1$ we conclude from iii) that $\Delta(A \cup B) + \Delta(A \cap B) = \Delta(A) + \Delta(B)$.

Proof of v). We fix $A \subset B$ in $\mathfrak{T}(E)$ and $\varepsilon > 0$. v.1) Let $u, v \in E$ with $u, v \leqq 1$ such that $A = [u = 1]$ and $B = [v = 1]$. We can achieve that

$$I(u) \leqq I^\star(\chi_A) + \varepsilon = \Delta(A) + \varepsilon \text{ and } I(v) \leqq I^\star(\chi_B) + \varepsilon = \Delta(B) + \varepsilon,$$

and then that $u \leqq v$. v.2) By the second condition in 3') there exists an $M \subset E$ nonvoid of type $\bullet$ with $M \downarrow$ some $F \leqq v - u$ such that

$$h \leqq v \text{ and } I(h) \geqq I(v) - I(u) - \varepsilon \geqq \Delta(B) - \Delta(A) - 2\varepsilon \text{ for all } h \in M.$$

v.3) For $h \in M$ and $0 \leqq t < 1$ we have

$$h + v_{<t>} \leqq \left(\frac{h}{\varepsilon} \wedge v\right)_{<t>} + \frac{\varepsilon}{t} v + v.$$

In fact, this holds true

$$\text{for } v \leqq t \qquad\qquad \text{since } v_{<t>} = 0 \text{ and } h \leqq v,$$

$$\text{for } v \geqq t \text{ and } h \leqq \varepsilon \qquad \text{since } v_{<t>} \leqq v \text{ and } h \leqq \varepsilon \leqq \frac{\varepsilon}{t} v,$$

$$\text{for } v \geqq t \text{ and } h \geqq \varepsilon \qquad \text{since } \frac{h}{\varepsilon} \wedge v = v \text{ and } h \leqq v.$$

It follows that

$$I(h) + I(v_{<t>}) \leqq I\Big(\big(\frac{h}{\varepsilon} \wedge v\big)_{<t>}\Big) + \frac{\varepsilon}{t} I(v) + I(v).$$

For $t \uparrow 1$ we conclude from iii) that

$$\begin{aligned}
I(h) + \Delta(B) &\leqq \Delta\Big(\big[\frac{h}{\varepsilon} \wedge v = 1\big]\Big) + \varepsilon I(v) + I(v) \\
&= \Delta\big([h \geqq \varepsilon] \cap B\big) + \varepsilon I(v) + I(v).
\end{aligned}$$

We combine this with v.1) and v.2) to obtain

$$\Delta(B) - \Delta(A) \leqq \Delta([h \geqq \varepsilon] \cap B) + \varepsilon\big(\Delta(B) + \varepsilon + 3\big) \quad \text{for all } h \in M.$$

v.4) The paving $\{[h \geqq \varepsilon] \cap B : h \in M\} \subset \mathfrak{T}(E)$ is of type $\bullet$ and downward directed $\downarrow [F \geqq \varepsilon] \cap B \subset [v - u \geqq \varepsilon] \cap B \subset B \setminus A$, where we have used that $v = u = 1$ on A. Thus by definition

$$\inf\big\{\Delta([h \geqq \varepsilon] \cap B) : h \in M\big\} \leqq \Delta_\bullet^B(B \setminus A).$$

v.5) From v.3) and v.4) we obtain

$$\Delta(B) - \Delta(A) \leqq \Delta_\bullet^B(B \setminus A) + \varepsilon\big(\Delta(B) + \varepsilon + 3\big) \quad \text{for all } \varepsilon > 0,$$

and hence the assertion. This completes the proof of the theorem.

We define for $\bullet = \sigma\tau$ an elementary integral $I : E \to [0, \infty[$ on a Stonean lattice cone E to be $\bullet$ **tight** iff it fulfils

$$I(v) \leqq I(u) + I_\bullet^v(v - u) \quad \text{for all } u \leqq v \text{ in } E,$$

as it appears in condition 3') above. Besides these notions one also defines I to be $\star$ **tight** iff it fulfils

$$I(v) \leqq I(u) + I_\star(v - u) \quad \text{for all } u \leqq v \text{ in } E;$$

here of course $I_\star$ and $I_\star^v$ amount to the same. We shall see that this notion is sometimes convenient but has no basic importance. As in chapter II it is obvious that

$$I \text{ is } \star \text{ tight} \Rightarrow I \text{ is } \sigma \text{ tight} \Rightarrow I \text{ is } \tau \text{ tight}.$$

We show on the spot that both converses $\Leftarrow$ are false. The counterexample for the left converse will be an I which is a σ preintegral and a $\star$ preintegral. Therefore the above theorem becomes false at least for $\bullet = \sigma$ when one formulates condition 3') with $\star$ tight instead of $\bullet$ tight. The modified 3') remains a sufficient condition for 1)1') but ceases to be an equivalent one.

15.10. EXAMPLE. Let $\mathfrak{S}$ be a lattice with $\varnothing \in \mathfrak{S}$, and $\varphi : \mathfrak{S} \to [0, \infty[$ be an isotone and modular set function with $\varphi(\varnothing) = 0$ which is downward τ continuous. From 14.18 we obtain an elementary integral $I : E = S(\mathfrak{S}) \to [0, \infty[$ such that φ is the unique source of I. Then 14.16 implies $\varphi = \Delta$. Furthermore $I_\bullet | E = I$ by 15.5.2) and 15.6.4). Therefore 15.9 says for $\bullet = \sigma\tau$ that the following are equivalent. 1) I is a $\bullet$ preintegral. 1') φ is an inner $\bullet$ premeasure. 3') I is $\bullet$ tight. Thus 6.32 furnishes an example where I is τ tight but not σ tight.

15.11. EXERCISE. We turn to the counterexample for the left converse. Define $E \subset [0, \infty[^X$ on $X =]0, 1[$ to consist of the bounded *lower* semicontinuous functions $X \to [0, \infty[$. 1) E is a Stonean lattice cone which is not primitive. 2) $\mathfrak{t}(E) = \mathrm{Op}(X)$ and $\mathfrak{T}(E) = (\mathrm{Op}(X))_\sigma$. Hint for the second inclusion $\supset$: For $A = \bigcap\limits_{l=1}^{\infty} A(l)$ with a sequence $(A(l))_l$ in $\mathrm{Op}(X)$ consider the function $f := \sum\limits_{l=1}^{\infty} 1/2^l \chi_{A(l)}$. For the sequel note that $\mathfrak{T}(E) \supset \mathrm{Comp}(X)$. Now define $I : E \to [0, \infty[$ to be

$$I(f) \quad = \quad \int f d\Lambda = \textstyle\int\!\!\!\!-\, f d\Lambda$$

$$\text{for } f \in E \subset \mathrm{UM}(\mathfrak{T}(E)) \subset \mathrm{UM}(\mathrm{Bor}(X)) \subset \mathrm{UM}(\mathfrak{L} \cap X),$$

with $\Lambda | X =: \Lambda$ as usual. 3) I is an elementary integral. 4) Λ is a σ representation and a $\star$ representation of I. Thus I is a σ preintegral and a $\star$ preintegral. 5) I is not $\star$ tight. Hint: Take $u = \chi_A$ for an open dense subset $A \subset X$ with $\Lambda(A) < 1$ as in the proof of 5.14.3), and $v = 1$.

Next we want to incorporate the traditional Daniell-Stone theorem 14.1. It is contained in the special case of 15.9 that the Stonean lattice cone E is primitive. In this case all elementary integrals $I : E \to [0, \infty[$ are $\star$ tight and hence $\bullet$ tight for $\bullet = \star\sigma\tau$. Thus the same happens as in the conventional $\bullet$ extension theories when one specializes from lattices $\mathfrak{S}$ with $\varnothing \in \mathfrak{S}$ to rings. The result is as follows.

15.12. SPECIAL CASE (for $\bullet = \sigma\tau$). Let $I : E \to [0, \infty[$ be an elementary integral on a primitive Stonean lattice cone $E \subset [0, \infty[^X$. Then the following are equivalent.
1) There exist $\bullet$ representations of I, that is I is a $\bullet$ preintegral.
1') I fulfils (0); and $\Delta : \mathfrak{T}(E) \to [0, \infty[$ is an inner $\bullet$ premeasure.
3') $I_\bullet(0) = 0$.

We specialize once more to the case $\bullet = \sigma$.

15.13. SPECIAL CASE. Let $I : E \to [0, \infty[$ be an elementary integral on a primitive Stonean lattice cone $E \subset [0, \infty[^X$. Then the following are equivalent.
0) There exists a cmeasure representation of I.
1) There exist σ representations of I, that is I is a σ preintegral.

1') I fulfils (0); and $\Delta : \mathfrak{T}(E) \to [0, \infty[$ is an inner σ premeasure.

3') $I_\sigma(0) = 0$.

In fact, 1)1')3') are equivalent by 15.10. It remains to incorporate condition 0). On the one hand 1) implies that I has the cmeasure representation $\alpha := \Delta_\sigma | \mathfrak{C}(\Delta_\sigma)$, so that 0) is fulfilled. On the other hand 0)⇒3') follows from the Beppo Levi case $\bullet = \sigma$ of 11.17.

The last equivalence 0)⇔3') is the traditional Daniell-Stone theorem 14.1. We see that the specialization 15.13 contains much more: It furnishes the cmeasure representation $\alpha := \Delta_\sigma | \mathfrak{C}(\Delta_\sigma)$ of I which is explicit in a sense, and is inner regular $(\mathfrak{T}(E))_\sigma = \mathfrak{T}(E_\sigma) = \mathfrak{t}(E_\sigma)$.

15.14. BIBLIOGRAPHICAL NOTE. The roots of the traditional Daniell-Stone theorem 14.1 are Daniell [1917-1918] and Stone [1948-1949]. However, it required additional work to obtain the theorem in its present form. For example, Glicksberg [1952] considered a certain topological special case of 14.1 to be a new result which needed an ab-ovo proof. The above theorem 14.1 is in the textbooks since the sixties. See for example Royden [1963], Zaanen [1967], Meyer [1966] theorem II.24, and Bauer [1968] Satz 39.4. The last-mentioned version included an assertion of regularity, which at once led to important consequences like Korollar 39.5. The nonsequential counterpart of the traditional Daniell-Stone theorem 14.1 has been treated in terms of the so-called abstract Bourbaki integral. For recent versions we refer to Leinert [1995] chapter 14 and König [1992a] appendix. It is of course in $\bullet = \tau$ of the present 15.12.

The work in the present context started, as far as the author is aware, in Pollard-Topsøe [1975] and Topsøe [1976]. The first paper was still restricted to primitive Stonean lattice cones E and obtained the equivalence 15.12.1)⇔3'). The second paper then obtained the implication 15.9.3')⇒1), but with 3') fortified to $\star$ tightness instead of $\bullet$ tightness which was still unknown. We know that the converse of this modified implication is false, so that it is not an equivalence result. Thus the situation was the same as described in the bibliographical annex to chapter II on the extension theories. Both papers also contained variants where the role of $\mathfrak{T}(E)$ passed over to other lattices. The essentials are contained in 15.15 below. The work of Anger-Portenier [1992a] presents related refinements and is more comprehensive in certain aspects, but it is complicated and also based on $\star$ tightness from the start. The present version 15.9 is due to the author [1995]. Its proof owes much to Topsøe [1976].

An Extended Situation

We conclude the section with a final proposition, in order to cover an extended situation which has been considered in the literature as mentioned above. Let $I : E \to [0, \infty[$ be a $\bullet$ preintegral on a Stonean lattice cone $E \subset [0, \infty[^X$. The aim is that the role of $\mathfrak{T}(E)$ passes over to some other lattice $\Re$ with $\varnothing \in \Re$, which of course must be related to $\mathfrak{T}(E)$. But we shall not require that $\Re \subset \mathfrak{T}(E)$.

15.15. PROPOSITION (for $\bullet = \sigma\tau$). *Let $I : E \to [0, \infty[$ be a $\bullet$ preintegral on a Stonean lattice cone $E \subset [0, \infty[^X$, and $\alpha := \Delta_\bullet | \mathfrak{C}(\Delta_\bullet)$. Assume that $\Re$ is a paving in X such that $\Re \subset (\mathfrak{T}(E))_\bullet \subset \Re \top \Re$. Then the following are equivalent.*

0) *For each pair $f \in E$ and $\varepsilon > 0$ there exists $K \in \Re$ such that*

$$I(v) \leq I(u) + \varepsilon \text{ for all } u, v \in E \text{ with } u \leq v \leq f \text{ and } u|K = v|K.$$

1) *$\Delta_\bullet$ is inner regular $\Re$.*
1') *$\Delta_\bullet$ is inner regular $\Re$ at $\mathfrak{T}(E)$.*
2) *All $\bullet$ representations of I are inner regular $\Re$.*
2') *There exists a $\bullet$ representation of I which is inner regular $\Re$.*

In this case, and if $\Re$ is a lattice with $\varnothing \in \Re$, then α is an inner $\bullet$ extension of $\alpha | \Re = \Delta_\bullet | \Re$. Furthermore $(\alpha | \Re)_\bullet = \Delta_\bullet$ and $\Re \top \Re_\bullet \subset \mathfrak{C}(\Delta_\bullet)$.

Condition 0) says that in a sense the elementary integral I is concentrated on $\Re$.

In view of 14.23 the implications 1)$\Rightarrow$2)$\Rightarrow$2')$\Rightarrow$1') are clear. Thus for the first part we have to prove that 0)$\Rightarrow$1) and 1')$\Rightarrow$0).

Proof of 0)$\Rightarrow$1). By 6.3.4) we have to show that $\Delta_\bullet$ is inner regular $\Re$ at $(\mathfrak{T}(E))_\bullet$. We recall from 15.2.4) that $(\mathfrak{T}(E))_\bullet = \mathfrak{T}(E_\bullet) = \mathrm{t}(E_\bullet)$. Fix $A \in (\mathfrak{T}(E))_\bullet$ and $\varepsilon > 0$. Thus $\chi_A \in E_\bullet$, so that there exists $f \in E$ with $\chi_A \leq f \leq 1$. From the assumption 0) we obtain $K \in \Re$ such that all $u \in E$ with $u \leq f$ and $u|K = f|K$ fulfil $I(f) \leq I(u) + \varepsilon$. If now $v \in E$ with $v \geq \chi_K$ then $f \wedge v \in E$ fulfils $f \wedge v = f$ on $\Re$, and hence $I(f) \leq I(f \wedge v) + \varepsilon$ or $I(f \vee v) \leq I(v) + \varepsilon$. By assumption $K \in (\mathfrak{T}(E))_\bullet$ and hence $\chi_K \in E_\bullet$. Thus on the one hand $I(f \vee v) = \int (f \vee v) d\alpha \geq \alpha(A \cup K)$. On the other hand 15.7 with 11.8.1) implies that

$$\inf\{I(v) : v \in E \text{ with } v \geq \chi_K\} = I^*(\chi_K) = \fint \chi_K d\Delta_\bullet = \Delta_\bullet(K) = \alpha(K).$$

It follows that $\alpha(A \cup K) \leq \alpha(K) + \varepsilon$ or $\alpha(A) \leq \alpha(A \cap K) + \varepsilon$. Since $A \cap K \in \Re$ in view of $(\mathfrak{T}(E))_\bullet \subset \Re \top \Re$ the assertion follows.

Proof of 1')$\Rightarrow$0). Fix $f \in E$ and $\varepsilon > 0$, and then $0 < a < b < \infty$ such that

$$\int\limits_{0\leftarrow}^{a} \alpha([f \geq t]) dt \leq \varepsilon/4 \quad \text{and} \quad \int\limits_{b}^{\to\infty} \alpha([f \geq t]) dt \leq \varepsilon/4.$$

From the assumption 1') applied to $[f \geq a] \in \mathfrak{T}(E)$ we obtain $K \in \Re$ with $K \subset [f \geq a]$ such that $\alpha([f \geq a]) \leq \alpha(K) + \varepsilon/2(b - a)$. Now let $u, v \in E$ with $u \leq v \leq f$ and $u|K = v|K$. Then $[v \geq t] \subset [f \geq t]$ for $t > 0$ and hence

$$I(v) = \fint v d\alpha \leq \int\limits_{a}^{b} \alpha([v \geq t]) dt + \varepsilon/2.$$

For $t \geqq a$ furthermore $[v \geqq t] \subset [u \geqq t] \cup ([f \geqq a] \setminus K)$ and hence

$$\alpha([v \geqq t]) \leqq \alpha([u \geqq t]) + \varepsilon/2(b-a).$$

It follows that $I(v) \leqq I(u) + \varepsilon$.

The remainder of the proof is routine. We see from $\mathfrak{K}_{\bullet} \subset (\mathfrak{T}(E))_{\bullet}$ and 2) that α is an inner $\bullet$ extension of $\alpha|\mathfrak{K}$. Thus $\alpha|\mathfrak{K}$ is an inner $\bullet$ premeasure. From 6.18 it follows that $\alpha = (\alpha|\mathfrak{K})_{\bullet}|\mathfrak{C}(\Delta_{\bullet})$. Therefore $(\alpha|\mathfrak{K})_{\bullet} = \Delta_{\bullet}$ on $(\mathfrak{T}(E))_{\bullet}$ and hence on $\mathfrak{P}(X)$. Then $\mathfrak{K}\top\mathfrak{K}_{\bullet} \subset \mathfrak{C}((\alpha|\mathfrak{K})_{\bullet}) = \mathfrak{C}(\Delta_{\bullet})$ follows from 6.31. The proof is complete.

16. The Riesz Theorem

In the present section we obtain the extended Riesz representation theorem as a specialization of the continuous Daniell-Stone theorem 15.9 in the τ version.

Preliminaries

We assume X to be a Hausdorff topological space and write as before $\mathfrak{K} :=$ Comp(X). We recall from 14.8 the Stonean lattice cones

$$\mathrm{CK}^+(X) \subset \mathrm{USCK}^+(X) \subset \mathrm{UMK}(X) \subset \mathrm{USC}^+(X),$$

with the definition

$$\mathrm{UMK}(X) := \mathrm{UM}(\mathfrak{K}) \cap [0,\infty[^X = \{f \in [0,\infty[^X : [f \geqq t] \in \mathfrak{K} \text{ for all } t > 0\}.$$

The members of $\mathrm{UMK}(X)$ are viewed as concentrated on the compact subsets of X.

In the present context we have to consider Stonean lattice cones $E \subset [0,\infty[^X$ which are related to compactness. Thus on the one hand we assume that $\mathfrak{T}(E) \subset \mathfrak{K}$, which means that $E \subset \mathrm{UMK}(X)$. On the other hand we have to impose a certain richness condition on E. We define E to be **rich** iff it satisfies the equivalent conditions which follow.

16.1. REMARK. *For a Stonean lattice cone $E \subset \mathrm{UMK}(X)$ the following are equivalent.* 1) $(\mathfrak{T}(E))_\tau = \mathfrak{K}$. 2) $\chi_K \in E_\tau$ *for all* $K \in \mathfrak{K}$. 3) $\mathrm{USCK}^+(X) \subset E_\tau$.

16.2. LEMMA. $S(\mathfrak{K}) \subset \mathrm{USCK}^+(X)$. *Furthermore*

$$f = \inf_{u \in S(\mathfrak{K}), u \geqq f} u \quad \text{for all } f \in \mathrm{USCK}^+(X).$$

Proof of 16.2. Assume that $f \in \mathrm{USCK}^+(X)$ vanishes outside of the nonvoid $K \in \mathfrak{K}$. Fix $a \in X$ and $\varepsilon > 0$. Then

$$u := \big(\max f - f(a)\big)\chi_{[f \geqq f(a)+\varepsilon]} + \big(f(a) + \varepsilon\big)\chi_K$$

is in $S(\mathfrak{K})$, and it satisfies $u \geqq f$ and $u(a) \leqq f(a) + \varepsilon$.

Proof of 16.1. 1)$\Leftrightarrow$2) We have $(\mathfrak{T}(E))_\tau = \mathfrak{T}(E_\tau) = \mathfrak{t}(E_\tau)$ by 15.2.4). 2)$\Rightarrow$3) follows from 16.2, and 3)$\Rightarrow$2) is obvious.

16.3. REMARK. $CK^+(X)$ is rich $\Leftrightarrow X$ is locally compact.

Proof. $\Rightarrow$) Let $K \in \mathfrak{K}$. Then $\chi_K \leq f$ for some $f \in CK^+(X)$. Thus K is contained in the open subset $[f > 0]$, which in turn is contained in some compact subset of X. In particular X is locally compact. $\Leftarrow$) is known from 14.5.3).

We need one more fact. The main assertion is a consequence of the classical Dini theorem in 14.5.5).

16.4. REMARK. For an elementary integral $I : E \to [0, \infty[$ on a Stonean lattice cone $E \subset UMK(X)$ the following are equivalent.

1) I is τ continuous at 0.
2) I fulfils (0).
3) $I(f \wedge t) \to 0$ for $t \downarrow 0$ for all $f \in E$.

These conditions are satisfied when E is rich and $\subset USCK^+(X)$, but need not be satisfied when E is rich $\subset UMK(X)$.

Proof. i) We have seen 1)$\Rightarrow$2) as the first part in the proof of 15.9. 2)$\Rightarrow$3) is obvious. Thus we have to prove 3)$\Rightarrow$1). Let $M \subset E$ be nonvoid with $M \downarrow 0$. To be shown is $\inf\{I(u) : u \in M\} = 0$. Fix $f \in M$ and $\varepsilon > 0$. We claim that

$$\inf\{I(u) : u \in M \text{ with } u \leq f\} \leq I(f \wedge \varepsilon),$$

which implies the assertion. Consider the compact subset $K := [f \geq \varepsilon]$. The claim is obvious when $K = \varnothing$, since then $f \wedge \varepsilon = f$. In case $K \neq \varnothing$ the Dini theorem asserts that

$$\inf\{\sup(u|K) : u \in M \text{ with } u \leq f\} = 0.$$

Now for $u \in M$ with $u \leq f$ we have

$$u \leq 1/\varepsilon \sup(u|K)f + f \wedge \varepsilon,$$
$$I(u) \leq 1/\varepsilon \sup(u|K)I(f) + I(f \wedge \varepsilon),$$

and hence the claim. ii) Assume that E is rich and $\subset USCK^+(X)$. Let $f \in E$. Then f vanishes outside of some $K \in \mathfrak{K}$. Since E is rich we have $\chi_K \leq F$ for some $F \in E$. For $t > 0$ therefore $f \wedge t \leq t\chi_K \leq tF$ and hence $I(f \wedge t) \leq tI(F)$. Thus 3) is fulfilled. iii) The last assertion follows from the counterexample below.

16.5. EXAMPLE. On $X = [0, \infty[$ define $E \subset [0, \infty[^X$ to consist of the continuous functions $f : X \to [0, \infty[$ such that the finite limit $I(f) := \lim_{x \to \infty} (xf(x))$ exists. One verifies that E is a rich Stonean lattice cone $\subset UMK(X)$ and that $I : E \to [0, \infty[$ is an elementary integral. However, we have $I(f) = I(f \wedge t)$ for all $f \in E$ and $t > 0$, so that condition 16.4.3) is violated.

The Main Theorem

16.6. THEOREM (Riesz Representation Theorem). *Let* $I : E \to [0, \infty[$ *be an elementary integral on a rich Stonean lattice cone* $E \subset \mathrm{UMK}(X)$. *Then the following are equivalent.*

0) *There exists a Borel-Radon measure* $\alpha : \mathrm{Bor}(X) \to [0, \infty]$ *which is a representation of* I.

1) *There exist* τ *representations of* I, *that is* I *is a* τ *preintegral.*

2) $I(v) = I(u) + I_\tau(v - u)$ *for all* $u \leqq v$ *in* E.

3') $I(f \wedge t) \to 0$ *for* $t \downarrow 0$ *for all* $f \in E$ (this is automatic when $E \subset \mathrm{USCK}^+(X)$); *and* I *is* τ *tight.*

In this case I *has the unique Borel-Radon measure representation* $\alpha := \Delta_\tau|\mathrm{Bor}(X)$ *with* $\mathrm{Bor}(X) \subset \mathfrak{C}(\Delta_\tau)$, *which therefore is a* τ *representation of* I. *Furthermore we have*

$$I^*(f) = I_\tau(f) = \int f d\alpha \quad \text{for all } f \in E_\tau,$$

$$I_\tau(f) = \int_* f d\alpha \quad \text{for all } f \in [0, \infty]^X.$$

Proof. We recall that

$$\mathfrak{T}(E) \subset (\mathfrak{T}(E))_\tau = \mathfrak{K} \subset \mathrm{Cl}(X) \subset \mathrm{Bor}(X).$$

i) Let $\alpha : \mathrm{Bor}(X) \to [0, \infty]$ be a Borel-Radon measure which is a representation of I. Thus on the one hand $\alpha|\mathfrak{T}(E)$ is a source of I, and 14.16.2) implies that $\alpha|\mathfrak{T}(E) = \Delta$. On the other hand

α is inner regular $\mathfrak{K} = (\mathfrak{T}(E))_\tau$,

$\alpha|\mathfrak{K} = \alpha|(\mathfrak{T}(E))_\tau < \infty$ is downward τ continuous by 9.4.

Thus α is an inner τ extension of the source Δ of I and hence a τ representation of I. Also α is a restriction of $\Delta_\tau|\mathfrak{C}(\Delta_\tau)$. Thus we have the implication 0)$\Rightarrow$1) and the first of the two final assertions.

ii) The conditions 1)2)3') are equivalent by 15.9 and 16.4.

iii) Assume 1). Then 14.23 says that $\Delta : \mathfrak{T}(E) \to [0, \infty[$ is an inner τ premeasure, and that I has the unique maximal τ representation $\Delta_\tau|\mathfrak{C}(\Delta_\tau)$. From

$$\mathrm{Cl}(X) \subset \mathfrak{K} \top \mathfrak{K} \subset \mathfrak{T}(E) \top \mathfrak{K} = \mathfrak{T}(E) \top (\mathfrak{T}(E))_\tau \subset \mathfrak{C}(\Delta_\tau)$$

we see that $\mathrm{Bor}(X) \subset \mathfrak{C}(\Delta_\tau)$. Furthermore $\Delta_\tau|\mathfrak{K} = \Delta_\tau|(\mathfrak{T}(E))_\tau < \infty$, and Δ_τ is inner regular $\mathfrak{K} = (\mathfrak{T}(E))_\tau$. Thus $\alpha := \Delta_\tau|\mathrm{Bor}(X)$ is a Borel-Radon measure. Of course α is a representation of I since it is a restriction of $\Delta_\tau|\mathfrak{C}(\Delta_\tau)$. It follows that 1)$\Rightarrow$0).

iv) Assume the equivalent conditions 0)1)2)3'). For $\phi := \Delta_\tau|\mathfrak{C}(\Delta_\tau)$ then $\phi_\star = \Delta_\tau$, since this holds true on $\mathfrak{C}(\Delta_\tau)$, and both sides are inner regular $\mathfrak{C}(\Delta_\tau)$. For $\alpha := \Delta_\tau|\mathrm{Bor}(X)$ likewise $\alpha_\star = \Delta_\tau$. Therefore $\alpha_\star = \phi_\star$. Thus 12.12 implies that

$$\int_* f d\alpha = \oint f d\alpha_\star = \oint f d\phi_\star = \int_* f d\phi \quad \text{for all } f \in [0, \infty]^X.$$

From 15.7 and 15.8 we obtain the second final assertion. The proof is complete.

The theorem attains a more familiar form when we look at the particular case that $E \subset \mathrm{USCK}^+(X)$. First note that for each Borel-Radon measure α : $\mathrm{Bor}(X) \to [0, \infty]$ and all $f \in \mathrm{USCK}^+(X)$ we have $\int f d\alpha < \infty$. Now consider a rich Stonean lattice cone $E \subset \mathrm{USCK}^+(X)$; examples are $E = \mathrm{USCK}^+(X)$ and $E = S(\mathfrak{K})$. Then each Borel-Radon measure α : $\mathrm{Bor}(X) \to [0, \infty]$ defines an elementary integral $I : E \to [0, \infty[$ via $I(f) = \int f d\alpha$ for $f \in E$. Of course α is a representation of I. Now we invoke 16.6. The implication 0)$\Rightarrow$3') shows that I is τ tight. The final uniqueness assertion shows that the map $\alpha \mapsto I$ is injective. But above all it follows from 3')$\Rightarrow$0) that each elementary integral $I : E \to [0, \infty[$ which is τ tight can be obtained in this manner. Thus we have the result which follows.

16.7. THEOREM (Riesz Representation Theorem). *Assume that* $E \subset \mathrm{USCK}^+(X)$ *is a rich Stonean lattice cone. There is a one-to-one correspondence between the elementary integrals* $I : E \to [0, \infty[$ *which are* τ *tight and the Borel-Radon measures* $\alpha : \mathrm{Bor}(X) \to [0, \infty]$. *The correspondence is* $I(f) = \int f d\alpha$ *for all* $f \in E$.

We emphasize that this result involves *all* Borel-Radon measures on X, and in particular is not restricted to the locally finite ones.

16.8. REMARK. 1) The question whether 16.6.3') and 16.7 hold true with $\star$ tight instead of τ tight must remain open (in one direction). We know that the answer is no in the continuous Daniell-Stone theorem 15.9.3'), at least for $\bullet = \sigma$. But the former example 15.11 does not settle the present question. 2) However, in the particular case that $E_\tau = E$ we can write 16.6.3') and 16.7 with $\star$ tight instead of τ tight. This follows from 15.6.9) combined with the implication 16.6.3')$\Rightarrow$2).

We turn to the special case that E is primitive. Of course this means that one is not far from continuous functions.

16.9. SPECIAL CASE. *Let* $I : E \to [0, \infty[$ *be an elementary integral on a primitive rich Stonean lattice cone* $E \subset \mathrm{UMK}(X)$. *Then the following are equivalent.*

0) *There exists a Borel-Radon measure* $\alpha : \mathrm{Bor}(X) \to [0, \infty]$ *which is a representation of* I.

1) *There exist* τ *representations of* I, *that is* I *is a* τ *preintegral.*

3') $I(f \wedge t) \to 0$ *for* $t \downarrow 0$ *for all* $f \in E$ (*this is automatic when* $E \subset \mathrm{USCK}^+(X)$).

In this case I *has the unique Borel-Radon measure representation* $\alpha := \Delta_\tau | \mathrm{Bor}(X)$ *with* $\mathrm{Bor}(X) \subset \mathfrak{C}(\Delta_\tau)$, *and this is a* τ *representation of* I.

16.10. SPECIAL CASE. *Assume that* $E \subset \mathrm{USCK}^+(X)$ *is a primitive rich Stonean lattice cone. There is a one-to-one correspondence between the elementary integrals* $I : E \to [0, \infty[$ *and the Borel-Radon measures* $\alpha :$ $\mathrm{Bor}(X) \to [0, \infty]$. *The correspondence is* $I(f) = \int f \, d\alpha$ *for all* $f \in E$.

If X is locally compact then $E := \mathrm{CK}^+(X)$ is as required in view of 16.3. This furnishes the traditional Riesz representation theorem 14.2. Thus 14.2 is indeed a direct special case of the new Riesz representation theorem 16.7.

An Extended Situation

After these main results we conclude the section with a final theorem, in order to cover an extended situation which has been considered in the literature. In the sequel we consider a Stonean lattice cone $E \subset \mathrm{USC}^+(X)$, which means that $\mathfrak{T}(E) \subset \mathrm{Cl}(X)$. We do not require that $E \subset \mathrm{UMK}(X)$, that is $\mathfrak{T}(E) \subset \mathfrak{K}$. Nevertheless we ask the question when an elementary integral $I : E \to [0, \infty[$ has a representation which is a Borel-Radon measure. It is clear that then I itself must be concentrated on the compact subsets of X in some sense or other. We shall see that an appropriate condition is the specialization of 15.15.0) above, that is

(comp) For each pair $f \in E$ and $\varepsilon > 0$ there exists $K \in \mathfrak{K}$ such that

$$I(v) \leqq I(u) + \varepsilon \text{ for all } u, v \in E \text{ with } u \leqq v \leqq f \text{ and } u|K = v|K.$$

Furthermore we extend the former *richness* condition on E to mean that $\mathfrak{K} \subset (\mathfrak{T}(E))_\tau$. This amounts to a condition of separation. Then we obtain the result which follows.

16.11. THEOREM. *Let* $I : E \to [0, \infty[$ *be an elementary integral on a rich Stonean lattice cone* $E \subset \mathrm{USC}^+(X)$. *Then the following are equivalent.*

0) *There exists a Borel-Radon measure* $\alpha : \mathrm{Bor}(X) \to [0, \infty]$ *which is a representation of* I.

1) I *is a* τ *preintegral and fulfils* (comp).

3') I *is* τ *tight and fulfils* (comp).

In this case I *has the unique Borel-Radon measure representation* $\alpha :=$ $\Delta_\tau|\mathrm{Bor}(X)$ *with* $\mathrm{Bor}(X) \subset \mathfrak{C}(\Delta_\tau)$, *and this is a* τ *representation of* I.

Proof. 0)$\Rightarrow$1) We start as in the proof of 16.6. Let $\alpha : \mathrm{Bor}(X) \to [0, \infty]$ be a Borel-Radon measure which is a representation of I. Then on the one hand $\alpha|\mathfrak{T}(E)$ is a source of I and hence $= \Delta$. On the other hand

α is inner regular $\mathfrak{K}$ and hence inner regular $(\mathfrak{T}(E))_\tau$,

$\alpha|(\mathfrak{T}(E))_\tau < \infty$ is downward τ continuous by 9.4.

Thus α is an inner τ extension of $\alpha|\mathfrak{T}(E) = \Delta$ and hence a τ representation of I, so that I is a τ preintegral. Since α is inner regular $\mathfrak{K}$ it follows from 15.15.2')$\Rightarrow$0) that I fulfils (comp). Also α is a restriction of $\Delta_\tau|\mathfrak{C}(\Delta_\tau)$. Thus we have 0)$\Rightarrow$1) and the final assertion. 1)$\Rightarrow$0) We obtain from 15.15.0)$\Rightarrow$2) that $\Delta_\tau|\mathfrak{C}(\Delta_\tau)$ is inner regular $\mathfrak{K}$. Furthermore $\mathrm{Cl}(X) \subset \mathfrak{K}\top\mathfrak{K} \subset \mathfrak{C}(\Delta_\tau)$ and hence $\mathrm{Bor}(X) \subset \mathfrak{C}(\Delta_\tau)$. Thus $\alpha := \Delta_\tau|\mathrm{Bor}(X)$ is a Borel-Radon measure representation of I.

1)⇒3') is in 15.9. 3')⇒1) In view of 15.9.3')⇒1) it remains to prove that $I_\tau(0) = 0$. To see this let $M \subset E$ be nonvoid $\downarrow 0$. To be shown is $\inf\{I(f) : f \in M\} = 0$. Fix $f \in M$ and $\varepsilon > 0$, and choose $K \in \mathfrak{K}$ as in (comp). Since $K \in \mathfrak{K} \subset (\mathfrak{T}(E))_\tau = \mathfrak{T}(E_\tau) = \mathfrak{t}(E_\tau)$ or $\chi_K \in E_\tau$ by 15.2.4) there exists $h \in E$ with $\chi_K \leq h$. Now for each $\delta > 0$ the Dini theorem 14.5.5) furnishes a function $v \in M$ with $v|K \leq \delta$, and we can achieve that $v \leq f$. From (comp) applied to $v \wedge (\delta h) \leq v$ we obtain

$$I(v) \leq I(v \wedge (\delta h)) + \varepsilon \leq I(\delta h) + \varepsilon = \delta I(h) + \varepsilon.$$

From this the assertion follows. The proof is complete.

Once more we specialize to the case that E is primitive. Note that in this case condition (comp) admits an obvious simplification.

16.12. SPECIAL CASE. *Let $I : E \to [0, \infty[$ be an elementary integral on a primitive rich Stonean lattice cone $E \subset \mathrm{USC}^+(X)$. Then the following are equivalent.*

0) *There exists a Borel-Radon measure $\alpha : \mathrm{Bor}(X) \to [0, \infty]$ which is a representation of I.*

1) *I is a τ preintegral and fulfils* (comp).

3') *I fulfils* (comp).

In this case I has the unique Borel-Radon measure representation $\alpha := \Delta_\tau|\mathrm{Bor}(X)$ with $\mathrm{Bor}(X) \subset \mathfrak{C}(\Delta_\tau)$, and this is a τ representation of I.

If X is completely regular then $E = \mathrm{C}^+(X)$ and $E = \mathrm{CB}^+(X)$, defined to consist of the bounded functions in $\mathrm{C}^+(X)$, are examples of primitive rich Stonean lattice cones $\subset \mathrm{USC}^+(X)$.

16.13. BIBLIOGRAPHICAL NOTE. The traditional Riesz representation theorem 14.2 for locally compact Hausdorff topological spaces X is in the recent textbooks. See for example Cohn [1980] theorem 7.2.8 and Bauer [1992] Satz 29.1 with 29.3. For the historical development which led to this theorem we refer to the survey article of Batt [1973]. A more comprehensive version is due to Bauer [1956]; see for example Anger-Portenier [1992a] theorem 14.14.5. This version is contained in the present 16.9.

The next step culminated in the work of Pollard-Topsøe [1975]. These authors applied their Daniell-Stone type theorem to obtain a Riesz type theorem on arbitrary Hausdorff topological spaces X, but restricted to $\mathrm{C}^+(X)$ since the paper assumed *primitive* Stonean lattice cones from the start. Their result is in essence equal to 16.12. An excellent ab-ovo presentation is in Berg-Christensen-Ressel [1984] theorem 2.2.2; see also Topsøe [1983]. For the particular cases of 16.12 noted after its formulation we refer to Bourbaki [1969] proposition 5.5 and also to Behrends [1987] chapter V, based on ideas of Erik Thomas.

The subsequent article of Topsøe [1976] started to abandon the primitive Stonean lattice cones. Its basic intention was to pave the road to incorporate $\mathrm{USC}^+(X)$, but there was no explicit treatment of the Riesz theorem itself.

In Anger-Portenier [1992a] there are results of the type 3')⇒0) in the present theorems 16.6 and 16.11. We also refer to the companion article Anger-Portenier [1992b] which is less complicated, in particular to theorem 4.2 combined with 5.5. The entire work presupposes conditions like $\star$ tightness from the start (the authors adhere to the Bourbaki viewpoint to place positive-linear functionals above measures).

The present version 16.6 is due to the author [1995], while 16.11 appears here for the first time. As far as the author can see there were no such equivalence theorems in the literature before.

17. The Non-continuous Daniell-Stone Theorem

We return to the abstract situation. The present section obtains the counterpart $\bullet = \star$ of the so-called continuous Daniell-Stone theorem 15.9 which was for $\bullet = \sigma\tau$. We recall that 15.9 was structured like the basic extension theorems of chapter II, nearest for certain to the conventional inner main theorem 6.31. Since these former theorems were uniform in $\bullet = \star\sigma\tau$ one could expect that the same holds true in the present context. That this is not so can be no surprise after the end of section 14. But the complete answer will be a surprise, and that it has no resemblance to the continuous cases $\bullet = \sigma\tau$. Then in the next chapter it will come as another surprise that the basic existence theorems on the so-called transplantation of premeasures will be of the same structure like the main result of the present section.

Introduction

The present section assumes an elementary integral $I : E \to [0, \infty[$ on a Stonean lattice cone $E \subset [0, \infty[^X$ in a nonvoid set X. We recall the final subsection of section 14. Our aim is to characterize those I which are $\star$ preintegrals, and then to describe all $\star$ representations of I. The $\star$ representations are those representations $\alpha : \mathfrak{A} \to [0, \infty]$ of I which are inner regular $\mathfrak{T}(E)$. The connection with the sources of I has been formulated in 14.21.

We recall the essential differences between the cases $\bullet = \sigma\tau$ and $\bullet = \star$ which previous examples have disclosed.

1) In the cases $\bullet = \sigma\tau$ no source other than $\Delta : \mathfrak{T}(E) \to [0, \infty[$ can be an inner $\bullet$ premeasure. However, example 14.25 shows that I can have several sources different from Δ and ∇ which are inner $\star$ premeasures. In example 14.25 the extremal sources Δ and ∇ themselves are not modular and hence not inner $\star$ premeasures.

2) In the cases $\bullet = \sigma\tau$ the continuous Daniell-Stone theorem 15.9 characterizes those I which are $\bullet$ preintegrals. The natural counterpart would be that

$$I \text{ is } \star \text{ preintegral} \iff I \text{ is } \star \text{ tight} .$$

However, example 15.11 shows that the implication $\Rightarrow$ is false. We shall see in 17.11 that the implication $\Leftarrow$ is true whenever I fulfils (0). This result is due to Topsøe [1976]. Its value is that $\star$ tightness can be easily verified in important situations. But $\star$ tightness fails to be a fundamental condition in case $\bullet = \star$, as it failed to be in the cases $\bullet = \sigma\tau$.

The Maximality Lemma

Let us turn to positive results. The first step toward the main theorem is the lemma which follows.

17.1. LEMMA. *Assume that* $\varphi, \psi : \mathfrak{T}(E) \to [0, \infty[$ *are sources of* I *with* $\varphi \leqq \psi$ *such that* φ *is an inner* $\star$ *premeasure and* ψ *is superadditive. Then* $\varphi = \psi$.

We isolate part of the assertion for later reference.

17.2. LEMMA. *Let* $\mathfrak{S}$ *be a lattice with* $\varnothing \in \mathfrak{S}$, *and* $\varphi, \psi : \mathfrak{S} \to [0, \infty[$ *with* $\varphi(\varnothing) = \psi(\varnothing) = 0$ *and* $\varphi \leqq \psi$. *Assume that* φ *is an inner* $\star$ *premeasure and that* ψ *is isotone and superadditive. If* $A \in \mathfrak{S}$ *is such that there exists* $B \supset A$ *with* $\varphi_\star(B) = \psi_\star(B) < \infty$ *then* $\varphi(A) = \psi(A)$.

Proof of 17.2. Fix $A \in \mathfrak{S}$. 1) We claim that

$$\varphi(S) - \varphi(A) \leqq \psi(S) - \psi(A) \quad \text{for all } S \in \mathfrak{S} \text{ with } S \supset A.$$

In fact, we have

$$\varphi(S) - \varphi(A) \leqq \varphi_\star(S \setminus A) = \sup\{\varphi(K) : K \in \mathfrak{S} \text{ with } K \subset S \setminus A\},$$

and here $\varphi(K) \leqq \psi(K) \leqq \psi(K \cup A) - \psi(A) \leqq \psi(S) - \psi(A)$. 2) From 1) it follows that

$$\varphi_\star(B) - \varphi(A) \leqq \psi_\star(B) - \psi(A) \quad \text{for all } B \supset A.$$

If we choose $B \supset A$ as assumed then $\varphi(A) \geqq \psi(A)$ and hence $\varphi(A) = \psi(A)$.

Proof of 17.1. 1) Fix $f \in E$. From 14.16 we have

$$I(f) = \int\limits_{0\leftarrow}^{\to\infty} \nabla([f \geqq t])dt = \int\limits_{0\leftarrow}^{\to\infty} \Delta([f \geqq t])dt.$$

Therefore $\nabla([f \geqq t]) = \Delta([f \geqq t])$ for all $t > 0$ except on some countable subset. 2) Now let $A \in \mathfrak{T}(E)$, that is $A = [f \geqq 1]$ for some $f \in E$. From 1) we obtain $0 < t < 1$ such that $B := [f \geqq t] \in \mathfrak{T}(E)$ satisfies $\nabla(B) = \Delta(B)$. Then $B \supset A$ and $\varphi(B) = \psi(B)$. From 17.2 we obtain the assertion.

We add a first application of the maximality lemma. We define I to be **separative** iff for each choice of $A, B \in \mathfrak{T}(E)$ with $A \cap B = \varnothing$ and $\varepsilon > 0$ there exist $u, v \in E$ with $u \geqq \chi_A$ and $v \geqq \chi_B$ such that $I(u \wedge v) \leqq \varepsilon$.

17.3. REMARK. *I is separative iff* Δ *is superadditive, that is iff* Δ *is additive.*

Proof. i) Assume that I is separative, and let $A, B \in \mathfrak{T}(E)$ with $A \cap B = \varnothing$. Fix $\varepsilon > 0$ and first choose $w \in E$ with $w \geqq \chi_{A \cup B}$ and $I(w) \leqq \Delta(A \cup B) + \varepsilon$. Then choose $u, v \in E$ as in the definition above. We can assume that $u, v \leqq w$. It follows that

$$\Delta(A) + \Delta(B) \leqq I(u) + I(v) = I(u \vee v) + I(u \wedge v)$$
$$\leqq I(w) + I(u \wedge v) \leqq \Delta(A \cup B) + 2\varepsilon,$$

and hence the assertion. ii) To see the converse fix $A, B \in \mathfrak{T}(E)$ with $A \cap B = \varnothing$ and $\varepsilon > 0$. Then let $u, v \in E$ with $u \geqq \chi_A$ and $v \geqq \chi_B$ such that $I(u) \leqq \Delta(A) + \varepsilon$ and $I(v) \leqq \Delta(B) + \varepsilon$. It follows that

$$\Delta(A \cup B) + I(u \wedge v) \leqq I(u \vee v) + I(u \wedge v) = I(u) + I(v)$$
$$\leqq \Delta(A) + \Delta(B) + 2\varepsilon \leqq \Delta(A \cup B) + 2\varepsilon,$$

and hence $I(u \wedge v) \leqq 2\varepsilon$.

The combination of 17.1 and 17.3 produces a certain substitute for the uniqueness result 14.16.2) which was intended for $\bullet = \sigma\tau$.

17.4. PROPOSITION. *Assume that I is separative. If $\varphi : \mathfrak{T}(E) \to [0, \infty[$ is a source of I which is an inner $\star$ premeasure then $\varphi = \Delta$.*

Subtight Sources

We come to the central notion of the section. We prepare the definition with the next remark.

17.5. REMARK. *Let $\varphi : \mathfrak{T}(E) \to [0, \infty[$ be a source of I which is super-modular. Then*

$$I(v) - I(u) \geqq \int (v - u) d\varphi_\star \quad \text{for all } u \leqq v \text{ in } E.$$

Proof. By 6.3.5) the set function $\varphi_\star : \mathfrak{P}(X) \to [0, \infty]$ is supermodular as well. Consider $u \leqq v$ in E. From 11.11 we obtain

$$\int (v - u) d\varphi_\star + \int u \, d\varphi_\star \leqq \int v \, d\varphi_\star.$$

Now $\int u \, d\varphi_\star = \int u \, d\varphi = I(u)$, and the same for v. The assertion follows.

After this we define a source $\varphi : \mathfrak{T}(E) \to [0, \infty[$ of I to be **subtight** iff it satisfies the opposite relation to the above, that is

$$I(v) - I(u) \leqq \int (v - u) d\varphi_\star \quad \text{for all } u \leqq v \text{ in } E.$$

We define $\Sigma(I)$ to consist of all sources $\varphi : \mathfrak{T}(E) \to [0, \infty[$ of I which are supermodular and subtight. We shall see that $\Sigma(I)$ can be void.

17.6. REMARK. *If $\Sigma(I)$ is nonvoid then it is upward inductive in the argumentwise order.*

Proof. Let $T \subset \Sigma(I)$ be nonvoid and totally ordered. Then $\vartheta := \sup_{\varphi \in T} \varphi$ is a well-defined isotone set function with $\nabla \leqq \vartheta \leqq \Delta$. Thus ϑ is a source of I. i) To see that ϑ is supermodular fix $A, B \in \mathfrak{T}(E)$. If $\alpha, \beta \in T$ then one of them $\varphi \in T$ fulfils $\alpha, \beta \leqq \varphi$. Therefore

$$\alpha(A) + \beta(B) \leqq \varphi(A) + \varphi(B) \leqq \varphi(A \cup B) + \varphi(A \cap B)$$
$$\leqq \vartheta(A \cup B) + \vartheta(A \cap B).$$

It follows that $\vartheta(A) + \vartheta(B) \leqq \vartheta(A \cup B) + \vartheta(A \cap B)$. ii) To see that ϑ is subtight consider $u \leqq v$ in E. Then $I(v) - I(u)$ is $\leqq \int (v - u)d\varphi_\star$ for all $\varphi \in T$ and hence $\leqq \int (v - u)d\vartheta_\star$. It follows that $\vartheta \in \Sigma(I)$.

The importance of the subtight sources results from the theorem which follows.

17.7. THEOREM. *For a source* $\varphi : \mathfrak{T}(E) \to [0, \infty[$ *of I the following are equivalent.* 1) φ *is an inner $\star$ premeasure.* 2) φ *is a maximal member of* $\Sigma(I)$.

Proof of 1)⇒2). i) We first prove that φ is subtight and hence in $\Sigma(I)$. Let $\alpha := \varphi_\star | \mathfrak{C}(\varphi_\star)$ be its maximal inner $\star$ extension. Then $\alpha_\star = \varphi_\star$, because this is true on $\mathfrak{C}(\varphi_\star)$ and both sides are inner regular $\mathfrak{C}(\varphi_\star)$. By 12.12 we have

$$\int_\star f d\alpha = \int f d\alpha_\star = \int f d\varphi_\star \quad \text{for all } f \in [0, \infty]^X,$$

in particular $\int_\star f d\alpha = I(f)$ for $f \in E$. Consider now $u \leqq v$ in E. By 12.3 then

$$\int_\star v d\alpha = \int_\star (v - u)d\alpha + \int_\star u d\alpha \text{ and hence } I(v) - I(u) = \int (v - u)d\varphi_\star.$$

ii) φ is a maximal member of $\Sigma(I)$. In fact, we see from 17.1 that φ is even equal to each source $\psi : \mathfrak{T}(E) \to [0, \infty[$ of I with $\varphi \leqq \psi$ which is supermodular.

The proof of the implication 2)⇒1) is more involved. We isolate the basic step in a lemma. We recall the class $\sqsubset \mathfrak{T}(E)$ of the subsets of X which are upward enclosable $\mathfrak{T}(E)$. Thus if $\varphi : \mathfrak{T}(E) \to [0, \infty[$ is isotone then $\varphi_\star < \infty$ (and even $\varphi^\star < \infty$) on $\sqsubset \mathfrak{T}(E)$.

17.8. LEMMA. *Assume* $\varphi \in \Sigma(I)$. *Fix a subset T in* $\sqsubset \mathfrak{T}(E)$, *and define* $\psi : \mathfrak{T}(E) \to [0, \infty[$ *to be*

$$\psi(A) = \varphi_\star(A \cup T) + \varphi_\star(A \cap T) - \varphi_\star(T) \quad \text{for } A \in \mathfrak{T}(E).$$

Then $\psi \in \Sigma(I)$ *and* $\varphi \leqq \psi$.

Proof of 17.8. i) ψ is isotone with $\psi(\varnothing) = 0$. ii) $\psi \geqq \varphi \geqq \nabla$ since $\varphi_\star$ is supermodular by 6.3.5). iii) ψ is supermodular, once more since $\varphi_\star$ is supermodular. This is a routine verification. iv) $I(v) - I(u) \leq \int (v-u)d\psi_\star$ for all $u \leq v$ in E. This follows from $\varphi_\star \leq \psi_\star$. v) We have

$$I(f) = \int f d\psi \quad \text{for all } f \in E \text{ with } f \leq 1 \text{ and } [f = 1] \supset T.$$

In fact, by the definition of ψ we have

$$\int f d\psi = \int_{0\leftarrow}^{\rightarrow\infty} \psi([f \geq t])dt = \int_{0\leftarrow}^{1} \psi([f \geq t])dt$$

$$= \int_{0\leftarrow}^{1} \Big(\varphi_\star([f \geq t] \cup T) + \varphi_\star([f \geq t] \cap T) - \varphi_\star(T) \Big) dt;$$

in view of $[f \geq t] \supset [f \geq 1] = [f = 1] \supset T$ for $0 < t \leq 1$ this is

$$= \int_{0\leftarrow}^{1} \Big(\varphi_\star([f \geq t]) + \varphi_\star(T) - \varphi_\star(T) \Big) dt = \int_{0\leftarrow}^{1} \varphi_\star([f \geq t])dt$$

$$= \int_{0\leftarrow}^{1} \varphi([f \geq t])dt = \int_{0\leftarrow}^{\rightarrow\infty} \varphi([f \geq t])dt = \int f d\varphi = I(f).$$

vi) We can now prove that $I(f) = \int f d\psi$ for all $f \in E$. This will complete the present proof. Fix $f \in E$. For $0 < t < \infty$ then $u := 1/t(f \wedge t) \in E$ with $u \leq 1$. Now by assumption there exists $B \in \mathfrak{T}(E)$ with $B \supset T$. Let $B = [g = 1]$ for some $g \in E$ with $g \leq 1$. We put $v := g \vee u \in E$. Then $u \leq v \leq 1$ and $[v = 1] \supset [g = 1] = B \supset T$, and hence $I(v) = \int v d\psi$ by v). From iv) and 11.11 we obtain

$$\int u d\psi + I(v) - I(u) \leq \int u d\psi_\star + \int (v-u)d\psi_\star \leq \int v d\psi_\star = I(v),$$

and hence $I(u) \geq \int u d\psi$. By the definition of u it follows that

$$I(f \wedge t) = tI(u) \geq \int (tu)d\psi = \int (f \wedge t)d\psi$$

$$= \int_{0\leftarrow}^{\rightarrow\infty} \psi([f \wedge t \geq s])ds = \int_{0\leftarrow}^{t} \psi([f \geq s])ds.$$

For $t \uparrow \infty$ we obtain $I(f) \geq \int f d\psi$. On the other hand $I(f) = \int f d\varphi \leq \int f d\psi$ by ii). The proof is complete.

Proof of 17.7.2)$\Rightarrow$1). Assume that $\varphi : \mathfrak{T}(E) \to [0, \infty[$ is a maximal member of $\Sigma(I)$. Then first of all φ is an isotone and supermodular set function with $\varphi(\varnothing) = 0$. From 17.8 we obtain

$$\varphi_\star(A \cup T) + \varphi_\star(A \cap T) = \varphi(A) + \varphi_\star(T)$$
$$\text{for all } A \in \mathfrak{T}(E) \text{ and } T \in \sqsubset \mathfrak{T}(E).$$

If we take $A \subset B$ in $\mathfrak{T}(E)$ and $T := B \setminus A$ then it follows that $\varphi(B) = \varphi(A) + \varphi_\star(B \setminus A)$. Thus φ is inner $\star$ tight and hence an inner $\star$ cpremeasure. The proof is complete.

The Main Theorem

We combine the last result with the former 14.21 which relates the $\star$ representations of I to those sources of I which are inner $\star$ premeasures. The main theorem which follows has indeed no resemblance to the continuous Daniell-Stone theorem 15.9.

17.9. THEOREM (Non-continuous Daniell-Stone Theorem). *Let $I : E \to [0, \infty[$ be an elementary integral on a Stonean lattice cone $E \subset [0, \infty[^X$. Then the following are equivalent.*

1) *There exist $\star$ representations of I, that is I is a $\star$ preintegral.*

2) *The set $\Sigma(I)$ of the supermodular and subtight sources of I is nonvoid.*

In this case $\Sigma(I)$ is upward inductive in the argumentwise order. The sources of I which are inner $\star$ premeasures are the maximal members of $\Sigma(I)$.

We add two important assertions which express the particular roles of the two extremal sources Δ and ∇.

17.10. PROPOSITION. *Assume that I fulfils (0). Then the following are equivalent.*

    1) *I is a $\star$ preintegral and separative.*

    2) *Δ is the unique source of I which is an inner $\star$ premeasure.*

    3) *Δ is an inner $\star$ premeasure.*

Proof. 1)$\Rightarrow$2) follows from 17.4. 2)$\Rightarrow$3) is obvious. 3)$\Rightarrow$1) The assumption implies that Δ is modular. Hence I is separative by 17.3.

The second assertion considers the condition that I be $\star$ tight.

17.11. PROPOSITION. *Assume that I fulfils (0). Then*

$$I \text{ is } \star \text{ tight} \iff \nabla \text{ is a member of } \Sigma(I).$$

Thus if I is $\star$ tight then I is a $\star$ preintegral. We know that the converse is false.

Proof. ∇ is a source of I, and it is supermodular by 14.15.3). Furthermore 14.17 shows that I is $\star$ tight iff

$$I(v) - I(u) \leqq \int (v - u) d\nabla_\star \quad \text{for all } u \leqq v \text{ in } E,$$

that is iff ∇ is a subtight source of I.

We insert another example which contributes to both these assertions. It constructs an I such that a unique one $\varphi : \mathfrak{T}(E) \to [0, \infty[$ of its sources is an inner $\star$ premeasure. On the one hand this φ will be different from the two extremal sources Δ and ∇. On the other hand I will not be $\star$ tight (whereas the example I in 14.24 is $\star$ tight). The present example is quite complicated.

17.12. EXERCISE. Let $X = \mathbb{N} \cup \{\infty\}$, and let $E \subset [0, \infty[^X$ consist of all functions $f : X \to [0, \infty[$ such that $f|\mathbb{N}$ is monotone increasing and bounded with $f(\infty) \leqq \lim\limits_{n \to \infty} f(n)$. 1) E is a Stonean lattice cone. Next define the paving $\mathfrak{T}$ on $\mathbb{N}$ to consist of the subsets $\{n, n+1, \cdots\}$ $\forall n \in \mathbb{N}$. 2) $\mathfrak{t}(E)$ consists of $\varnothing$ and of the subsets $T, T \cup \{\infty\}$ with $T \in \mathfrak{T}$. $\mathfrak{T}(E)$ consists of $\mathfrak{t}(E)$ and $\{\infty\}$. Now define

$$I : E \to [0, \infty[\text{ to be } I(f) = f(\infty) + \lim_{n \to \infty} f(n) \quad \text{for } f \in E.$$

3) I is an elementary integral which fulfils (0). 4) For $f \in [0, \infty]^X$ we have

$$I^\star(f) = f(\infty) + \sup f \quad \text{and} \quad I_\star(f) = f(\infty) \wedge \left(\liminf_{n \to \infty} f(n)\right) + \liminf_{n \to \infty} f(n).$$

5) I is not $\star$ tight. Hint: Take $u, v \in E$ with $u(\infty) < v(\infty)$ and $u(n) = v(n)$ for $n \in \mathbb{N}$. 6) Compute the set functions $\Delta, \nabla : \mathfrak{T}(E) \to [0, \infty[$. 7) I has a unique source $\varphi : \mathfrak{T}(E) \to [0, \infty[$ which is an inner $\star$ premeasure. φ is $\neq \Delta, \nabla$.

17.13. EXERCISE. Let $I : E \to [0, \infty[$ be an elementary integral on a Stonean lattice cone $E \subset [0, \infty[^X$. Assume that $\alpha : \mathfrak{A} \to [0, \infty]$ is a $\star$ representation of I. Then

$$I_\star(f) = \int_\star f d\alpha \text{ for all } f \in [0, \infty]^X \Leftrightarrow \alpha|\mathfrak{T}(E) = \nabla.$$

Hint: Use 12.12 and 14.17.

As before we specialize to the case that E is primitive. Then each elementary integral $I : E \to [0, \infty[$ is $\star$ tight.

17.14. SPECIAL CASE. Let $E \subset [0, \infty[^X$ be a primitive Stonean lattice cone. Then each elementary integral $I : E \to [0, \infty[$ with (0) is a $\star$ preintegral.

We add a consequence which looks quite narrow but is of considerable interest.

17.15. CONSEQUENCE. Let $E \subset [0, \infty[^X$ be a primitive Stonean lattice cone such that

i) E consists of bounded functions and contains the positive constants.
ii) All elementary integrals $I : E \to [0, \infty[$ are separative.

Then all elementary integrals $I : E \to [0, \infty[$ fulfil (0). There is a one-to-one correspondence between the elementary integrals $I : E \to [0, \infty[$ and the inner $\star$ premeasures $\varphi : \mathfrak{T}(E) \to [0, \infty[$ with $\varphi(\varnothing) = 0$. The correspondence is

$I \mapsto \varphi :=$ *the unique source* Δ *of* I *which is an inner* $\star$ *premeasure;*

$\varphi \mapsto I : I(f) = \int f d\varphi$ *for* $f \in E$.

Proof. 1) It is an obvious consequence of i) that each elementary integral $I : E \to [0, \infty[$ fulfils (0), and hence is a $\star$ preintegral by 17.14. Thus by ii) and 17.10 each I has its Δ as the unique source which is an inner $\star$ premeasure. 2) On the other hand i) implies that $X \in \mathfrak{T}(E)$. Furthermore for each isotone set function $\varphi : \mathfrak{T}(E) \to [0, \infty[$ with $\varphi(\varnothing) = 0$ we have

$$\int f d\varphi \leqq (\sup f)\varphi(X) < \infty \quad \text{for all } f \in E.$$

If φ is modular then the functional $I : I(f) = \int f d\varphi \; \forall f \in E$ thus defined is an elementary integral by 11.11. 3) It follows that the two maps $I \mapsto \varphi$ and $\varphi \mapsto I$ as above are well-defined, and that their two compositions are the respective identities. This is the assertion.

As an example let X be a topological space, and define as above $E = \mathrm{CB}^+(X)$ to consist of the bounded functions in $\mathrm{C}^+(X)$. Thus $E \subset [0, \infty[^X$ is a primitive Stonean lattice cone with 17.15.i). One verifies that

1) $\mathfrak{T}(E) = \mathrm{CCl}(X)$.

2) For each pair $A, B \in \mathrm{CCl}(X)$ with $A \cap B = \varnothing$ and each $\varepsilon > 0$ there exist functions $u, v \in E$ with $u \geqq \chi_A$ and $v \geqq \chi_B$ such that $u \wedge v \leqq \varepsilon$. In fact, if $u, v \in E$ with $u, v \leqq 1$ such that $A = [u = 1]$ and $B = [v = 1]$ then

$$u_n := \Big(\frac{1-v}{(1-u)+(1-v)}\Big)^n \quad \text{and} \quad v_n := \Big(\frac{1-u}{(1-u)+(1-v)}\Big)^n,$$

with sufficiently large $n \in \mathbb{N}$, are as required.

It follows that E fulfils 17.15.ii). We thus obtain the famous representation theorem due to Alexandroff [1940-43].

17.16. THEOREM (Alexandroff Representation Theorem). *Let* X *be a topological space. There is a one-to-one correspondence between the elementary integrals* $I : \mathrm{CB}^+(X) \to [0, \infty[$ *and the inner* $\star$ *premeasures* $\varphi : \mathrm{CCl}(X) \to [0, \infty[$ *with* $\varphi(\varnothing) = 0$. *The correspondence is* $I(f) = \int f d\varphi$ *for all* $f \in \mathrm{CB}^+(X)$.

17.17. BIBLIOGRAPHICAL NOTE. The main result of Topsøe [1976] is equivalent to the final positive assertion of 17.11, combined with the uniqueness assertion 17.10 in case I is separative. The special case that E is primitive was in Pollard-Topsøe [1975], and also the notion of a separative I. Both papers contained variants as described in 15.14 and afterwards. For the Alexandroff representation theorem we refer to Varadarajan [1965] and Knowles [1967]. In Anger-Portenier [1992a] there are fortified results in the frame of $\star$ tightness. However, the present main theorem 17.9 is new, as far as the author is aware, and appears here for the first time.

CHAPTER VI

Transplantation of Contents and Measures

The present chapter is another application of the extension theories of chapter II. It develops a certain method for the formation of contents and measures from initial ones on the basis of regularity. The point is that one has to switch from the initial lattice of regularity to another prescribed one. Thus the procedure is more than an extension, it seems that transplantation is a better name.

We start in section 18 with the non-continuous case $\bullet = \star$. We consider the full inner situation in the sense of chapter II, this time for inner $\star$ premeasures with *finite* values. In view of the upside-down transform method it would be equivalent to consider the full outer situation. The main theorem 18.9 contains some well-known previous results which could be named the Henry-Lembcke-Bachman-Sultan-Lipecki-Adamski theorem. As announced in section 17 it is in close formal resemblance to the former main theorem 17.9.

Then section 19 considers the case $\bullet = \sigma$, via reduction to the former $\bullet = \star$ (in case $\bullet = \tau$ certain basic facts are not true). Here we restrict ourselves to the conventional inner situation, in order not to overload the chapter. We conclude with the application to existence and uniqueness theorems on the extension of Baire measures to Borel measures in topological spaces, as announced before 8.11.

18. Transplantation of Contents

Introduction and Preparations

We start with a short description of what will be done. Let X be a nonvoid set.

Assume that $\varphi : \mathfrak{S} \to \mathbb{R}$ is an inner $\star$ premeasure on a lattice $\mathfrak{S}$ in X, and $\phi := \varphi_\star | \mathfrak{C}(\varphi_\star, +)$ its maximal inner $\star$ extension. We fix another lattice $\mathfrak{T}$ in X and ask for the inner $\star$ premeasures $\psi : \mathfrak{T} \to \mathbb{R}$ on $\mathfrak{T}$ such that

$(\star)$ $\qquad \psi_\star | \mathfrak{C}(\psi_\star, +)$ is an extension of $\phi = \varphi_\star | \mathfrak{C}(\varphi_\star, +)$.

These ψ can be viewed as the transplants of φ onto $\mathfrak{T}$. Of course without further assumptions there will be neither existence nor uniqueness. We do not assume from the start that $\mathfrak{S} \subset \mathfrak{T}$, when the transplants ψ were extensions of φ to $\mathfrak{T}$. But an essential overall assumption will be that the members of $\mathfrak{T}$ be upward and downward enclosable $\mathfrak{S}$, that is $\mathfrak{T} \subset (\mathfrak{S} \sqsubset \mathfrak{S})$. Then we shall prove what follows.

1) An isotone and supermodular set function $\psi : \mathfrak{T} \to \mathbb{R}$ satisfies $(\star)$ iff $\psi_\star|\mathfrak{S} = \varphi$. This follows from the important earlier result 6.15 with addendum 6.17, which is the inner transcription of 4.20 with 4.22. It will be obtained at the end of the present subsection.

2) After this we form the collection $\star(\varphi, \mathfrak{T})$ of all isotone and supermodular set functions $\psi : \mathfrak{T} \to \mathbb{R}$ with $(\star)$, that is with $\psi_\star|\mathfrak{S} = \varphi$. It turns out that $\star(\varphi, \mathfrak{T})$ is upward inductive in the argumentwise order. The main fact is that a member $\psi : \mathfrak{T} \to \mathbb{R}$ of $\star(\varphi, \mathfrak{T})$ is maximal in this order iff it is an inner $\star$ premeasure. Thus we obtain a fundamental existence theorem: There are inner $\star$ premeasures $\psi : \mathfrak{T} \to \mathbb{R}$ with $(\star)$ iff $\star(\varphi, \mathfrak{T})$ is nonvoid. This result has several important specializations. The most obvious one is the case $\mathfrak{S} \subset \mathfrak{T}$, where $\star(\varphi, \mathfrak{T})$ is nonvoid because it contains $\psi := \varphi_\star|\mathfrak{T}$.

3) At last we obtain a uniqueness result, which however requires the assumption that $\mathfrak{S} \subset \mathfrak{T}$. First note that all $\psi \in \star(\varphi, \mathfrak{T})$ fulfil $\psi \leqq \phi^\star|\mathfrak{T}$. Then the result says that there is a unique inner $\star$ premeasure $\psi : \mathfrak{T} \to \mathbb{R}$ with $(\star)$ iff the set function $\phi^\star|\mathfrak{T}$ (which is finite and isotone) is supermodular, and in this case we have $\psi = \phi^\star|\mathfrak{T}$. One direction is an immediate consequence of the previous facts, while the other direction requires extensive further preparations.

We conclude the introduction with the remark that the earlier presentations were all restricted to the conventional situation and to the particular case that $\varphi : \mathfrak{S} \to [0, \infty[$ is a ccontent on a ring (but sometimes the value ∞ was admitted).

After this we turn to some preparations. The exercise below will be a useful tool.

18.1. EXERCISE. Let $\varphi : \mathfrak{S} \to \overline{\mathbb{R}}$ and $\psi : \mathfrak{T} \to \overline{\mathbb{R}}$ be isotone set functions on lattices $\mathfrak{S}$ and $\mathfrak{T}$. Then the following are equivalent. 0) For each pair $S \in \mathfrak{S}$ and $T \in \mathfrak{T}$ with $T \subset S$ one has $\psi(T) \leqq \varphi(S)$. 1) $\psi \leqq \varphi^\star|\mathfrak{T}$. 2) $\psi_\star|\mathfrak{S} \leqq \varphi$.

We pass to the above result 1).

18.2. PROPOSITION. *Let* $\varphi : \mathfrak{S} \to \mathbb{R}$ *and* $\psi : \mathfrak{T} \to \mathbb{R}$ *be isotone and supermodular on lattices* $\mathfrak{S}$ *and* $\mathfrak{T}$. *Assume that* $\mathfrak{T} \subset (\mathfrak{S} \sqsubset \mathfrak{S})$. *Then* $\psi_\star|\mathfrak{S} = \varphi$ *implies that* $\psi_\star|\mathfrak{C}(\psi_\star, +)$ *is an extension of* $\varphi_\star|\mathfrak{C}(\varphi_\star, +)$ *(note that the converse is true when* φ *is an inner* $\star$ *premeasure).*

Proof. 1) $\varphi_\star$ and $\psi_\star$ are isotone and supermodular $+$ by 6.3.5). On $\mathfrak{S}$ we have by assumption $\psi_\star = \varphi = \varphi_\star$, and this is finite. This implies that $\psi_\star \geqq \varphi_\star$ since $\varphi_\star$ is inner regular $\mathfrak{S}$ by 6.3.4). 2) Now the assertion follows from 6.17 applied to $\mathfrak{P} = \mathfrak{Q} = \mathfrak{H} := \mathfrak{S}$ and to $\phi := \psi_\star$ and $\theta := \varphi_\star$. In fact, the assumptions of 6.15 are fulfilled since $\phi = \psi_\star$ is inner regular $\mathfrak{T}$ by 6.3.4) and hence inner regular $\mathfrak{S} \sqsubset \mathfrak{S}$ by assumption.

The Existence Theorem

We assume that $\varphi : \mathfrak{S} \to \mathbb{R}$ is an isotone and supermodular set function on a lattice $\mathfrak{S}$. We fix another lattice $\mathfrak{T}$ such that $\mathfrak{T} \subset (\mathfrak{S} \sqsubset \mathfrak{S})$. As above we define $\star(\varphi, \mathfrak{T})$ to consist of all isotone and supermodular set functions $\psi : \mathfrak{T} \to \mathbb{R}$ such that $\psi_\star|\mathfrak{S} = \varphi$.

18.3. PROPERTIES. *Let $\psi \in \star(\varphi, \mathfrak{T})$. 1) $\psi_\star|\mathfrak{C}(\psi_\star, +)$ is an extension of $\varphi_\star|\mathfrak{C}(\varphi_\star, +)$. 2) We have $\psi \leqq \phi^\star|\mathfrak{T}$ with $\phi := \varphi_\star|\mathfrak{C}(\varphi_\star, +)$. 3) $\psi_\star \geqq \varphi_\star$.*

Proof. 1) is 18.2. 2) follows from 1) and 18.1 applied to ϕ and ψ. 3) follows from $\psi_\star|\mathfrak{S} = \varphi$ since $\varphi_\star$ is inner regular $\mathfrak{S}$.

18.4. PROPOSITION. *If $\star(\varphi, \mathfrak{T})$ is nonvoid then it is upward inductive in the argumentwise order.*

Proof. Assume that $E \subset \star(\varphi, \mathfrak{T})$ is nonvoid and totally ordered. We form $\varepsilon := \sup\limits_{\psi \in E} \psi$. Thus $\varepsilon : \mathfrak{T} \to] - \infty, \infty]$ is isotone. 1) For $\psi \in E$ we have $\psi_\star|\mathfrak{S} = \varphi$ and hence $\psi \leqq \varphi^\star|\mathfrak{T}$ by 18.1. It follows that $\varepsilon \leqq \varphi^\star|\mathfrak{T}$ and hence $\varepsilon_\star|\mathfrak{S} \leqq \varphi$ by 18.1. On the other hand $\psi \leqq \varepsilon$ for $\psi \in E$ and hence $\varphi = \psi_\star|\mathfrak{S} \leqq \varepsilon_\star|\mathfrak{S}$. Therefore $\varepsilon_\star|\mathfrak{S} = \varphi$. 2) We have $\varphi^\star|\mathfrak{T} < \infty$ since $\mathfrak{T}$ is upward enclosable $\mathfrak{S}$. Thus from 1) we obtain $\varepsilon : \mathfrak{T} \to \mathbb{R}$. 3) To see that ε is supermodular fix $A, B \in \mathfrak{T}$. If $\alpha, \beta \in E$ then one of them $\psi \in E$ fulfils $\alpha, \beta \leqq \psi$. Therefore

$$\begin{aligned} \alpha(A) + \beta(B) \leqq \psi(A) + \psi(B) &\leqq \psi(A \cup B) + \psi(A \cap B) \\ &\leqq \varepsilon(A \cup B) + \varepsilon(A \cap B). \end{aligned}$$

It follows that $\varepsilon(A) + \varepsilon(B) \leqq \varepsilon(A \cup B) + \varepsilon(A \cap B)$. The proof is complete.

18.5. PROPOSITION. *Assume that $\psi \in \star(\varphi, \mathfrak{T})$ is an inner $\star$ premeasure. Then ψ is a maximal member of $\star(\varphi, \mathfrak{T})$.*

The proof is based on the lemma below, which is a close relative to 17.2. The difference is that 17.2 has a weaker assumption but is limited to the conventional situation.

18.6. LEMMA. *Let $\mathfrak{T}$ be a lattice. Assume that $\psi : \mathfrak{T} \to [-\infty, \infty[$ is an inner $\star$ premeasure, and that $\vartheta : \mathfrak{T} \to \overline{\mathbb{R}}$ is isotone and supermodular $+$ with $\psi \leqq \vartheta$. If $T \in \mathfrak{T}$ is such that there exist $A, B \subset X$ with $A \subset T \subset B$ and*

$$a := \psi_\star(A) = \vartheta_\star(A) \in \mathbb{R} \text{ and } b := \psi_\star(B) = \vartheta_\star(B) \in \mathbb{R},$$

then $\psi(T) = \vartheta(T)$.

Proof of 18.6. Fix $\varepsilon > 0$ and then

$$P \in \mathfrak{T} \text{ with } P \subset A \text{ and } a - \varepsilon \leqq \psi(P) \leqq \vartheta(P) \leqq a,$$
$$Q \in \mathfrak{T} \text{ with } Q \subset B \text{ and } b - \varepsilon \leqq \psi(Q) \leqq \vartheta(Q) \leqq b.$$

We can assume that $T \subset Q \subset B$. Thus $P \subset A \subset T \subset Q \subset B$, with $P \subset T \subset B$ in $\mathfrak{T}$. Since ψ is an inner $\star$ premeasure and $\vartheta_\star$ is supermodular

$\overset{.}{+}$ by 6.3.5) we have

$$
\begin{aligned}
a + b - 2\varepsilon &= (a - \varepsilon) + (b - \varepsilon) \leqq \psi(P) + \psi(Q) = \psi(T) + \psi_\star(P|T'|Q) \\
&\leqq \vartheta(T) + \vartheta_\star(P|T'|Q) \leqq \vartheta(P) + \vartheta(Q) \leqq a + b,
\end{aligned}
$$

where all values are finite. It follows that $0 \leqq \vartheta(T) - \psi(T) \leqq 2\varepsilon$ for all $\varepsilon > 0$ and hence $\psi(T) = \vartheta(T)$.

Proof of 18.5. Let $\vartheta \in \star(\varphi, \mathfrak{T})$ with $\psi \leqq \vartheta$. Fix $T \in \mathfrak{T}$. By assumption there exist $A, B \in \mathfrak{S}$ with $A \subset T \subset B$. Therefore

$$
\psi_\star(A) = \vartheta_\star(A) = \varphi(A) \in \mathbb{R} \text{ and } \psi_\star(B) = \vartheta_\star(B) = \varphi(B) \in \mathbb{R}.
$$

Thus we obtain $\psi(T) = \vartheta(T)$ from 18.6.

18.7. PROPOSITION. *Assume that φ is an inner $\star$ premeasure. Then each maximal $\psi \in \star(\varphi, \mathfrak{T})$ is an inner $\star$ premeasure.*

Once more we isolate the basic step in a lemma.

18.8. LEMMA. *Assume that φ is an inner $\star$ premeasure. Fix $\psi \in \star(\varphi, \mathfrak{T})$ and $E \in \mathfrak{S} \sqsubset \mathfrak{S}$ and define*

$$
\vartheta : \vartheta(T) = \psi_\star(T \cup E) + \psi_\star(T \cap E) - \psi_\star(E) \quad \text{for } T \in \mathfrak{T}.
$$

Note that all arguments are in $\mathfrak{S} \sqsubset \mathfrak{S}$ and hence all values are finite, so that $\vartheta : \mathfrak{T} \to \mathbb{R}$. *We claim that $\vartheta \in \star(\varphi, \mathfrak{T})$ and $\vartheta \geqq \psi$.*

Proof of 18.8. i) $\psi_\star$ is supermodular $\overset{.}{+}$ by 6.3.5). ii) From i) we obtain $\vartheta \geqq \psi$ and hence $\vartheta_\star \geqq \psi_\star$. iii) From i) we conclude that ϑ is supermodular. Of course ϑ is isotone. iv) For $U \subset E$ we have $\vartheta_\star(U) \leqq \psi_\star(U)$ and hence $\vartheta_\star(U) = \psi_\star(U)$. To see this we can assume that $\vartheta_\star(U) > -\infty$. Let $T \in \mathfrak{T}$ with $T \subset U$. Then $T \subset E$ and hence $\vartheta(T) = \psi_\star(T) \leqq \psi_\star(U)$. It follows that $\vartheta_\star(U) \leqq \psi_\star(U)$. v) For $V \supset E$ we have $\vartheta_\star(V) \leqq \psi_\star(V)$ and hence $\vartheta_\star(V) = \psi_\star(V)$. To see this we can assume that $\vartheta_\star(V) > -\infty$. Let $T \in \mathfrak{T}$ with $T \subset V$. Then $T \cup E \subset V$ and hence $\vartheta(T) \leqq \psi_\star(T \cup E) \leqq \psi_\star(V)$. It follows that $\vartheta_\star(V) \leqq \psi_\star(V)$. vi) We can now prove that $\vartheta_\star|\mathfrak{S} = \varphi$. This will complete the present proof. By assumption φ is an inner $\star$ premeasure, so that $\varphi_\star|\mathfrak{C}(\varphi_\star, \overset{.}{+})$ is a content on the algebra $\mathfrak{C}(\varphi_\star, \overset{.}{+}) \supset \mathfrak{S}$. By 18.3.1) then $\psi_\star|\mathfrak{C}(\psi_\star, \overset{.}{+})$ is an extension of $\varphi_\star|\mathfrak{C}(\varphi_\star, \overset{.}{+})$, that is $\psi_\star = \varphi_\star$ on $\mathfrak{C}(\varphi_\star, \overset{.}{+}) \subset \mathfrak{C}(\psi_\star, \overset{.}{+})$. Now fix $A \in \mathfrak{S}$ and then $U, V \in \mathfrak{S}$ with $U \subset A, E \subset V$. We obtain

$$
\begin{aligned}
\psi_\star(U) + \psi_\star(V) &= \psi_\star(A) + \psi_\star(U|A'|V) \\
&\leqq \vartheta_\star(A) + \vartheta_\star(U|A'|V) \quad \text{termwise by ii)} \\
&\leqq \vartheta_\star(U) + \vartheta_\star(V) \quad \text{by 6.3.5)} \\
&= \psi_\star(U) + \psi_\star(V) \quad \text{by iv)v).}
\end{aligned}
$$

Thus all terms are finite. It follows that $\vartheta_\star(A) = \psi_\star(A) = \varphi(A)$.

Proof of 18.7. Fix $U \subset T \subset V$ in $\mathfrak{T}$ and put $E := U|T'|V = U \cup (V \cap T') \in \mathfrak{S} \sqsubset \mathfrak{S}$. Then $T \cup E = V$ and $T \cap E = U$. Now for the set function $\vartheta \in \star(\varphi, \mathfrak{T})$

formed in 18.8 we have $\vartheta = \psi$. Therefore

$$\psi_\star(T \cup E) + \psi_\star(T \cap E) = \psi_\star(E) + \psi(T),$$
$$\psi(U) + \psi(V) = \psi(T) + \psi_\star(U|T'|V).$$

From 6.22.4)$\Rightarrow$1) we obtain the assertion.

The three propositions proved so far furnish result 2) of the introduction.

18.9. THEOREM. *Assume that* $\varphi : \mathfrak{S} \to \mathbb{R}$ *is an inner* $\star$ *premeasure on a lattice* $\mathfrak{S}$. *Let* $\mathfrak{T}$ *be a lattice with* $\mathfrak{T} \subset (\mathfrak{S} \sqsubset \mathfrak{S})$. *If* $\star(\varphi, \mathfrak{T})$ *is nonvoid then it is upward inductive. The maximal members of* $\star(\varphi, \mathfrak{T})$ *are the inner* $\star$ *premeasures* $\psi : \mathfrak{T} \to \mathbb{R}$ *with* $\psi_\star|\mathfrak{S} = \varphi$.

Thus we obtain a fundamental existence theorem, which will be formulated once more in explicit terms.

18.10. THEOREM. *Assume that* $\varphi : \mathfrak{S} \to \mathbb{R}$ *is an inner* $\star$ *premeasure on a lattice* $\mathfrak{S}$. *Let* $\mathfrak{T}$ *be a lattice with* $\mathfrak{T} \subset (\mathfrak{S} \sqsubset \mathfrak{S})$. *Then*

> *for each* $\vartheta : \mathfrak{T} \to \mathbb{R}$ *isotone and supermodular with* $\vartheta_\star|\mathfrak{S} = \varphi$
>
> *there exists* $\psi : \mathfrak{T} \to \mathbb{R}$ *inner* $\star$ *premeasure with* $\psi_\star|\mathfrak{S} = \varphi$

such that $\psi \geqq \vartheta$.

18.11. EXERCISE. The above theorem becomes false when one weakens the assumption $\mathfrak{T} \subset (\mathfrak{S} \sqsubset \mathfrak{S})$ to $\mathfrak{T} \subset (\sqsubset \mathfrak{S})$ or to $\mathfrak{T} \subset (\sqsupset \mathfrak{S})$, even if $\mathfrak{S} \subset \mathfrak{T}$. Hint for $\mathfrak{T} \subset (\sqsupset \mathfrak{S})$: Fix a nonvoid proper subset $A \subset X$. Let $\mathfrak{S}$ consist of $\varnothing$ and A, and define $\varphi : \mathfrak{S} \to \mathbb{R}$ to be $\varphi(\varnothing) = 0$ and $\varphi(A) = 1$. Then let $\mathfrak{T}$ consist of $\mathfrak{S}$ and X, and define $\vartheta : \mathfrak{T} \to \mathbb{R}$ to extend φ with $\vartheta(X) > 1$.

Specializations of the Existence Theorem

A remarkable specialization is $\mathfrak{S} = \{\varnothing, X\}$. Then each real $c \geqq 0$ defines the inner $\star$ premeasure $\varphi : \mathfrak{S} \to [0, \infty[$ with $\varphi(\varnothing) = 0$ and $\varphi(X) = c$. Thus 18.10 reads as follows.

18.12. THEOREM. *Let* $\mathfrak{T}$ *be a lattice with* $\varnothing \in \mathfrak{T}$. *Then for each bounded isotone and supermodular* $\vartheta : \mathfrak{T} \to [0, \infty[$ *with* $\vartheta(\varnothing) = 0$ *there exists an inner* $\star$ *premeasure* $\psi : \mathfrak{T} \to [0, \infty[$ *with* $\psi(\varnothing) = 0$ *such that* $\psi \geqq \vartheta$ *and* $\sup \psi = \sup \vartheta$.

This result is a close relative to the so-called core theorem in cooperative game theory.

18.13. EXERCISE. The above theorem becomes false when one removes the assumption that $\vartheta : \mathfrak{T} \to [0, \infty[$ be bounded. Hint for a counterexample: Let $\mathfrak{T}$ consist of the finite subsets of an uncountable set X. Define $\vartheta : \mathfrak{T} \to [0, \infty[$ to be $\vartheta(T) = (\#(T))^2$ for $T \in \mathfrak{T}$.

18.14. EXERCISE. Compare 18.12 with the Hahn-Banach type result 11.14.2).

We turn to the specialization $\mathfrak{S} \subset \mathfrak{T}$ which has been mentioned in the introduction.

18.15. THEOREM. *Assume that* $\varphi : \mathfrak{S} \to \mathbb{R}$ *is an inner* $\star$ *premeasure on a lattice* $\mathfrak{S}$. *Let* $\mathfrak{T}$ *be a lattice with* $\mathfrak{S} \subset \mathfrak{T} \subset (\mathfrak{S} \sqsubset \mathfrak{S})$. *Then there exists an inner* $\star$ *premeasure* $\psi : \mathfrak{T} \to \mathbb{R}$ *such that* $\psi | \mathfrak{S} = \varphi$.

In fact, the set function $\vartheta := \varphi_\star | \mathfrak{T}$ is as required in 18.10.

18.16. CONSEQUENCE. *Assume that* $\varphi : \mathfrak{S} \to \mathbb{R}$ *is an inner* $\star$ *premeasure on a lattice* $\mathfrak{S}$. *Then* φ *can be extended to a* $\dotplus$ *content* $\beta : \mathfrak{P}(X) \to \overline{\mathbb{R}}$ *with* $\beta(X) = \sup \varphi$.

Proof. From 18.15 applied to $\mathfrak{T} := \mathfrak{S} \sqsubset \mathfrak{S}$ we obtain an inner $\star$ premeasure $\psi : \mathfrak{T} \to \mathbb{R}$ such that $\psi | \mathfrak{S} = \varphi$. By 18.3.1) $\psi_\star | \mathfrak{C}(\psi_\star, +)$ is an extension of $\varphi_\star | \mathfrak{C}(\varphi_\star, +)$. Now from 6.12 applied to $\phi := \psi_\star$ we obtain $\mathfrak{C}(\psi_\star, +) = \mathfrak{P}(X)$. In fact, for $A \subset X$ and $P, Q \in \mathfrak{T}$ we have $P|A|Q, P|A'|Q \in \mathfrak{T}$ as well, and $\phi | \mathfrak{T} = \psi_\star | \mathfrak{T} = \psi$ is modular. Therefore $\beta := \psi_\star | \mathfrak{C}(\psi_\star, +)$ is as required.

We come to an important extension of the specialization $\mathfrak{S} \subset \mathfrak{T}$. It is based on the so-called Marczewski condition.

18.17. REMARK. *Let* $\varphi : \mathfrak{S} \to \overline{\mathbb{R}}$ *be an isotone set function on a lattice* $\mathfrak{S}$, *and* $\mathfrak{T}$ *be a lattice.* 1) $(\varphi_\star | \mathfrak{T})_\star | \mathfrak{S} \leqq (\varphi^\star | \mathfrak{T})_\star | \mathfrak{S} \leqq \varphi$, *with equalities in case* $\mathfrak{S} \subset \mathfrak{T}$. *The relation* $(\varphi_\star | \mathfrak{T})_\star | \mathfrak{S} = \varphi$ *is called the* **Marczewski condition**. *It means that* $\varphi_\star$ *is inner regular* $\mathfrak{T}$ *at* $\mathfrak{S}$. 2) *If the Marczewski condition is fulfilled then* $(\varphi_\star | \mathfrak{T})_\star = \varphi_\star$. *This means that* $\varphi_\star$ *is inner regular* $\mathfrak{T}$.

Proof. 1) is obvious. 2) We obtain $(\varphi_\star | \mathfrak{T})_\star \geqq \varphi_\star$ since $\varphi_\star$ is inner regular $\mathfrak{S}$. The converse $\leqq$ is obvious.

18.18. THEOREM. *Assume that* $\varphi : \mathfrak{S} \to \mathbb{R}$ *is an inner* $\star$ *premeasure on a lattice* $\mathfrak{S}$. *Let* $\mathfrak{T}$ *be a lattice with* $\mathfrak{T} \subset (\mathfrak{S} \sqsubset \mathfrak{S})$. *Assume that the Marczewski condition is fulfilled. Then there exists an inner* $\star$ *premeasure* $\psi : \mathfrak{T} \to \mathbb{R}$ *such that* $\psi_\star | \mathfrak{S} = \varphi$.

This follows from 18.10 applied to $\vartheta := \varphi_\star | \mathfrak{T}$.

18.19. REMARK. *Assume that* $\varphi : \mathfrak{S} \to \mathbb{R}$ *is an inner* $\star$ *premeasure on a lattice* $\mathfrak{S}$. *Let* $\mathfrak{T}$ *be a lattice with* $\mathfrak{T} \subset (\mathfrak{S} \sqsubset \mathfrak{S})$. *Consider the condition*

(ex) *there exists an inner* $\star$ *premeasure* $\psi : \mathfrak{T} \to \mathbb{R}$ *with* $\psi_\star | \mathfrak{S} = \varphi$.

Then we have the implications

$$\text{Marczewski condition} \ \Rightarrow \ (\text{ex}) \ \Rightarrow \ (\varphi^\star | \mathfrak{T})_\star | \mathfrak{S} = \varphi.$$

The converse implications $\Leftarrow$ *are both false, even when* $\mathfrak{S}$ *is an algebra.*

Proof. The left implication $\Rightarrow$ is 18.18. To prove the right implication $\Rightarrow$ note that $\psi \leqq \varphi^\star | \mathfrak{T}$ by 18.1 and hence $\varphi = \psi_\star | \mathfrak{S} \leqq (\varphi^\star | \mathfrak{T})_\star | \mathfrak{S}$. The converse $\geqq$ is in 18.17.1). Counterexamples for the two converse implications $\Leftarrow$ will be in 18.20 and 18.21 below.

18.20. EXAMPLE. Let $\mathfrak{S}$ consist of $\varnothing$ and X, and define $\varphi : \mathfrak{S} \to \mathbb{R}$ to be $\varphi(\varnothing) = 0$ and $\varphi(X) = 1$. Then fix a nonvoid proper subset $A \subset X$. Let $\mathfrak{T}$ consist of $\varnothing$ and A, and define $\psi : \mathfrak{T} \to \mathbb{R}$ to be $\psi(\varnothing) = 0$ and $\psi(A) = 1$. Of course φ and ψ are inner $\star$ premeasures, and $\mathfrak{T} \subset (\mathfrak{S} \sqsubset \mathfrak{S})$. One verifies that $\psi_\star|\mathfrak{S} = \varphi$. On the other hand $\varphi_\star|\mathfrak{T} = 0$, so that the Marczewski condition is violated.

18.21. EXAMPLE. Let $X :=]0, 1]$. 1) We define $\mathfrak{S}$ to consist of $\varnothing$ and of the finite unions of the subsets

$$I_n^l :=]l - 1/2^n, l/2^n] \quad \text{for integer } n \geq 0 \text{ and } l = 1, \cdots, 2^n.$$

Thus $\mathfrak{S}$ is an algebra, and each nonvoid $S \in \mathfrak{S}$ has the form

$$S = \bigcup_{l \in M} I_n^l \quad \text{for some } n \geq 0 \text{ and nonvoid } M \subset \{1, \cdots, 2^n\}.$$

Let $\varphi := \Lambda|\mathfrak{S}$, which is a ccontent and hence an inner $\star$ premeasure. 2) Fix dense subsets $E_n^l \subset I_n^l$ such that the E_n^l for all $n \geq 0$ and $l = 1, \cdots, 2^n$ are pairwise disjoint. For example one can take $E_n^l := (c_n + \mathbb{Q}) \cap I_n^l$ with $c_n \in \mathbb{R} \; \forall n \geq 0$ which are linearly independent over $\mathbb{Q}$. Define $\mathfrak{T}$ to consist of $\varnothing$ and of the finite unions of the sets E_n^l for $n \geq 0$ and $l = 1, \cdots, 2^n$. Thus $\mathfrak{T}$ is a lattice (and even a ring). 3) The condition $(\varphi^\star|\mathfrak{T})_\star|\mathfrak{S} \geq \varphi$ is fulfilled. In fact, write the nonvoid $S \in \mathfrak{S}$ in the form $S = \bigcup_{l \in M} I_n^l$ as above, and let $T := \bigcup_{l \in M} E_n^l \in \mathfrak{T}$. Then $T \subset S$, and each $A \in \mathfrak{S}$ with $T \subset A$ fulfils $S \subset A$. Thus $\varphi^\star(T) = \varphi(S)$ and hence the assertion. 4) We do not have (ex). To see this let $\psi : \mathfrak{T} \to \mathbb{R}$ be an inner $\star$ premeasure with $\psi_\star|\mathfrak{S} = \varphi$. Fix a nonvoid $T \in \mathfrak{T}$. Let $p \geq 0$ be such that the sets $E_n^l \subset T$ have all $n \leq p$, and let $q > p$. Then $T \cap I_q^k \; \forall k = 1, \cdots, 2^q$ contains no nonvoid member of $\mathfrak{T}$, so that $\psi_\star(T \cap I_q^k) = 0$. Now $T \cap I_q^k \in \mathfrak{C}(\psi_\star, +)$ since $\mathfrak{C}(\varphi_\star, +) \subset \mathfrak{C}(\psi_\star, +)$ by 18.3.1). Thus

$$T = \bigcup_{k=1}^{2^q} T \cap I_q^k \quad \text{implies that} \quad \psi(T) = \psi_\star(T) = \sum_{k=1}^{2^q} \psi_\star(T \cap I_q^k) = 0.$$

Therefore $\psi = 0$ and hence $\varphi = 0$. This is the desired contradiction.

We conclude with a concrete particular case of 18.18.

18.22. CONSEQUENCE. Let X be a Hausdorff topological space. Assume that $\varphi : \mathfrak{S} \to [0, \infty[$ is an inner $\star$ premeasure on a lattice $\mathfrak{S}$ with $\varnothing \in \mathfrak{S}$ and $\varphi(\varnothing) = 0$. Also assume that $\mathrm{Comp}(X)$ is upward enclosable $\mathfrak{S}$. If $\varphi_\star$ is inner regular $\mathrm{Comp}(X)$ at $\mathfrak{S}$ then $\alpha := \varphi_\star|\mathfrak{C}(\varphi_\star)$ can be extended to a Radon measure (in the sense of section 9).

18.23. BIBLIOGRAPHICAL NOTE. The main theorem 18.9 has been obtained in König [1992b] in a more special situation, but with the same ideas. This work owes much to Adamski [1987].

We turn to the subsequent specializations. For 18.12 and its context we refer to Adamski [1987] section 2, also for the counterexample 18.13, and to Kindler [1987]. In the mainstream the basic specialization 18.15 was the independent result of Lembcke [1970] and Bachman-Sultan [1980]. Its extension 18.18 has been obtained in Lipecki [1987] and Adamski [1987]. The topological result 18.22 is due to Henry [1969]. At last the nontrivial example 18.21 is from König [1992b] example 2.11.

The Theorem of Łoś-Marczewski

The present subsection has the aim to prove a basic extension theorem due to Łoś-Marczewski [1949], transferred to the frame of ovals. The Łoś-Marczewski result furnishes explicit formulas for certain extensions of a simple-step type. It corresponds to the simple-step extension procedure used in the traditional proof of the Hahn-Banach theorem. We shall need the extended Łoś-Marczewski theorem for the uniqueness theorem of the next subsection.

We start with three lemmata. The first two ones require the rules 4.2 and 6.4 for semimodular set functions which are based on separation.

18.24. LEMMA. *Let* $\alpha : \mathfrak{A} \to \overline{\mathbb{R}}$ *be an isotone set function on an oval* $\mathfrak{A}$. *Fix subsets* $E, F \subset X$ *with* $E \cap F = \varnothing$. *Then*

$$\alpha \text{ content } \dot{+} \quad \Rightarrow \quad S \mapsto \alpha^\star\big((S \cap E) \cup F\big) \text{ is a content } \dot{+} \text{ on } \mathfrak{A};$$
$$\alpha \text{ content } + \quad \Rightarrow \quad S \mapsto \alpha_\star\big((S \cap E) \cup F\big) \text{ is a content } + \text{ on } \mathfrak{A}.$$

Proof. Fix $P, Q \in \mathfrak{A}$ and put $A := (P \cap E) \cup F$ and $B := (Q \cap E) \cup F$. Then $A \cup B = \big((P \cup Q) \cap E\big) \cup F$ and $A \cap B = \big((P \cap Q) \cap E\big) \cup F$. i) For the first assertion we have to prove that $\alpha^\star(A \cup B) \dot{+} \alpha^\star(A \cap B) = \alpha^\star(A) \dot{+} \alpha^\star(B)$. From 4.1.5) we know that $\leqq$. In order to deduce $\geqq$ from 4.2 we have to show that A, B are separated $\mathfrak{A}$. Thus let $M \in \mathfrak{A}$ with $A \cap B = (P \cap Q \cap E) \cup F \subset M$. We form

$$S := M \cup (P \cap Q') = P \cup M | Q | M \in \mathfrak{A},$$
$$T := M \cup (Q \cap P') = Q \cup M | P | M \in \mathfrak{A}.$$

It is obvious that $S \cap T = M$. Furthermore

$$A = (P \cap E) \cup F \subset (P \cap Q \cap E) \cup (P \cap Q') \cup F \subset M \cup (P \cap Q') = S,$$

and likewise $B \subset T$. The assertion follows. ii) For the second assertion we use 6.4 and have to show that A, B are coseparated $\mathfrak{A}$. Thus let $M \in \mathfrak{A}$ with $M \subset A \cup B = \big((P \cup Q) \cap E\big) \cup F$. We form

$$S := M \cap (P \cup Q') = M | Q | M \cap P \in \mathfrak{A},$$
$$T := M \cap (Q \cup P') = M | P | M \cap Q \in \mathfrak{A}.$$

It is obvious that $M = S \cup T$. Furthermore

$$S = M \cap (P \cup Q') \subset \big(A \cup (Q \cap P')\big) \cap (P \cup Q') \subset A,$$

and likewise $T \subset B$. The assertion follows.

18.25. LEMMA. *Let* $\alpha : \mathfrak{A} \to \overline{\mathbb{R}}$ *be a content* $\dotplus$ *on an oval* $\mathfrak{A}$. *Fix* $E \subset X$. *For* $P, Q, U, V \in \mathfrak{A}$ *with* 1) $P \subset Q$ *and* $U \subset V$ *and* 2) $Q \cap V \subset P \cup U$ *then*

$$\alpha^\star(U|E|Q) \dotplus \alpha^\star(V|E|P) = \alpha^\star(V|E|Q) \dotplus \alpha^\star(U|E|P).$$

Proof. From 1) we see that

$$(U|E|Q) \cup (V|E|P) = V|E|Q \quad \text{and} \quad (U|E|Q) \cap (V|E|P) = U|E|P.$$

Thus from 4.1.5) we know that $\geqq$. In order to deduce $\leqq$ from 4.2 we have to show that $U|E|Q$ and $V|E|P$ are separated $\mathfrak{A}$. Thus let $M \in \mathfrak{A}$ with $U|E|P \subset M$. We form

$$S := M \cup (Q \cap P') = Q \cup M|P|M \in \mathfrak{A},$$
$$T := M \cup (V \cap U') = V \cup M|U|M \in \mathfrak{A}.$$

Then on the one hand

$$U|E|Q \subset (U|E|P) \cup (Q \cap P') \subset M \cup (Q \cap P') = S,$$

and likewise $V|E|P \subset T$. On the other hand we obtain from 2)

$$S \cap T = M \cup (Q \cap P' \cap V \cap U') = M \cup (Q \cap V \cap (P \cup U)') = M.$$

The assertion follows.

The third lemma below is a basic step.

18.26. LEMMA. *Let* $\alpha : \mathfrak{A} \to \mathbb{R}$ *be a finite content on an oval* $\mathfrak{A}$. 0) *The set functions* $\alpha^\star$ *and* $\alpha_\star$ *are finite on* $\mathfrak{A} \sqsubset \mathfrak{A}$. 1) *For* $E \subset X$ *and* $A, B \in \mathfrak{A}$ *we have*

$$\alpha^\star(A|E|B) + \alpha_\star(A|E'|B) = \alpha(A) + \alpha(B).$$

Proof. 0) For $M \in \mathfrak{A} \sqsubset \mathfrak{A}$ there are by definition $A, B \in \mathfrak{A}$ with $A \subset M \subset B$. Therefore

$$-\infty < \alpha(A) \leqq \alpha_\star(M) \leqq \alpha^\star(M) \leqq \alpha(B) < \infty.$$

1) Fix $E \subset X$ and $A, B \in \mathfrak{A}$, and put $S := A|E|B$ and $T := A|E'|B = B|E|A$. Then $S \cup T = A \cup B =: V$ and $S \cap T = A \cap B =: U$ are in $\mathfrak{A}$. It remains to prove that $\alpha^\star(S) + \alpha_\star(T) = \alpha(U) + \alpha(V)$, since the right side is $= \alpha(A) + \alpha(B)$. i) Fix $M \in \mathfrak{A}$ with $U \subset S \subset M \subset V$ and form $N := U \cup (V \cap M') = V|M|U \in \mathfrak{A}$. It is obvious that $M \cup N = V$ and $M \cap N = U$. Also $U \subset N \subset T$. Therefore $\alpha(U) + \alpha(V) = \alpha(M) + \alpha(N)$ is $\leqq \alpha(M) + \alpha_\star(T)$ and hence $\leqq \alpha^\star(S) + \alpha_\star(T)$. ii) Fix $N \in \mathfrak{A}$ with $U \subset N \subset T \subset V$ and form $M := U \cup (V \cap N') = V|N|U \in \mathfrak{A}$. It is obvious that $M \cup N = V$ and $M \cap N = U$. Also $S \subset M \subset V$. Therefore $\alpha(U) + \alpha(V) = \alpha(M) + \alpha(N)$ is $\geqq \alpha^\star(S) + \alpha(N)$ and hence $\geqq \alpha^\star(S) + \alpha_\star(T)$. The proof is complete.

We head for the main results. For a lattice $\mathfrak{A}$ in X and a subset $E \subset X$ we define

$$\mathfrak{A}(E) := \{M|E|N : M, N \in \mathfrak{A}\},$$
$$\mathfrak{A}[E] := \{M|E|N : M \subset N \text{ in } \mathfrak{A}\} = \{M \cup (N \cap E) : M, N \in \mathfrak{A}\}.$$

We list the relevant properties.

18.27. PROPERTIES. 1) $\mathfrak{A} \subset \mathfrak{A}[E] \subset \mathfrak{A}(E) \subset (\mathfrak{A} \sqsubset \mathfrak{A})$. 2) $E \in \mathfrak{A}[E] \Leftrightarrow$ $E \in \mathfrak{A}(E) \Leftrightarrow E \in \mathfrak{A} \sqsubset \mathfrak{A}$. 3) $\mathfrak{A}(E)$ and $\mathfrak{A}[E]$ are lattices. 4) $\mathfrak{A}$ oval $\Rightarrow \mathfrak{A}(E)$ is an oval. $\mathfrak{A}$ ring $\Rightarrow \mathfrak{A}(E)$ is a ring. 5) $\mathfrak{A}(E) = \mathfrak{A}(E')$.

Proof. 1)2)3)5) are clear. 4) The ring case is clear as well. So assume that $\mathfrak{A}$ is an oval. Fix $P, Q, A \in \mathfrak{A}(E)$. Thus $P = P_1|E|P_2$, $Q = Q_1|E|Q_2$, and $A = A_1|E|A_2$ with $P_1, P_2, Q_1, Q_2, A_1, A_2 \in \mathfrak{A}$. Then on E' we have $P = P_1, Q = Q_1, A = A_1$ and hence $A' = A_1'$, and therefore $P|A|Q = P_1|A_1|Q_1$. Likewise $P|A|Q = P_2|A_2|Q_2$ on E. It follows that

$$P|A|Q = (P_1|A_1|Q_1)|E|(P_2|A_2|Q_2) \in \mathfrak{A}(E).$$

18.28. REMARK. Let $\alpha : \mathfrak{A} \to \overline{\mathbb{R}}$ be an isotone set function on an oval $\mathfrak{A}$. Fix subsets $E, F \subset X$ with $E \cap F = \varnothing$. Then

$$\alpha \text{ content } \dotplus \;\; \Rightarrow \;\; S \mapsto \alpha^\star\big((S \cap E) \cup F\big) \text{ is a content } \dotplus \text{ on } \mathfrak{A}(E);$$
$$\alpha \text{ content } \underset{.}{+} \;\; \Rightarrow \;\; S \mapsto \alpha_\star\big((S \cap E) \cup F\big) \text{ is a content } \underset{.}{+} \text{ on } \mathfrak{A}(E).$$

This is an immediate consequence of 18.24.

18.29. PROPOSITION. Let $\alpha : \mathfrak{A} \to \mathbb{R}$ be a finite content on an oval $\mathfrak{A}$. Fix $E \subset X$ and $U \in \mathfrak{A}$. Define $\varphi, \psi : \mathfrak{A}(E) \to \mathbb{R}$ to be

$$\varphi(S) = \alpha^\star(U|E|S) + \alpha_\star(U|E'|S) - \alpha(U),$$
$$\psi(S) = \alpha^\star(U|E'|S) + \alpha_\star(U|E|S) - \alpha(U);$$

note that all terms are finite by 18.27.1) and 18.26.0).

1) φ and ψ are contents on $\mathfrak{A}(E)$.
2) $\varphi|\mathfrak{A} = \psi|\mathfrak{A} = \alpha$.
3) $\varphi(S) = \alpha^\star(S)$ and $\psi(S) = \alpha_\star(S)$ for all $S \in \mathfrak{A}[E]$ with $U \subset S$.

Proof. 1) follows from 18.28, and 2) follows from 18.26.1). It remains to prove 3). Fix $S \in \mathfrak{A}[E]$ with $U \subset S$. Thus $S = P|E|Q = P \cup (Q \cap E)$ with $P \subset Q$ in $\mathfrak{A}$. We can assume that $U \subset P \subset Q$. i) From 18.26.1) we obtain

$$\varphi(S) = \alpha^\star(U|E|Q) + \alpha_\star(U|E'|P) - \alpha(U)$$
$$= \alpha^\star(U|E|Q) + \alpha(P) - \alpha^\star(U|E|P)$$
$$= \alpha^\star(U|E|Q) + \alpha^\star(P|E|P) - \alpha^\star(U|E|P).$$

Thus from 18.25 applied to P, Q, U and $V := P$ it follows that $\varphi(S) = \alpha^\star(P|E|Q) = \alpha^\star(S)$. ii) Define $T := P|E'|Q = P \cup (Q \cap E')$. Thus $T \in \mathfrak{A}[E']$ with $U \subset T$. Now the first assertion in 3) which has been proved in i) can be applied to E', U and T. We obtain

$$\alpha^\star(U|E'|T) + \alpha_\star(U|E|T) - \alpha(U) - \alpha^\star(T) = 0,$$
$$\alpha^\star(U|E'|Q) + \alpha_\star(U|E|P) - \alpha(U) - \alpha^\star(P|E'|Q) = 0.$$

On the other hand

$$\psi(S) - \alpha_\star(S) = \alpha^\star(U|E'|P) + \alpha_\star(U|E|Q) - \alpha(U) - \alpha_\star(P|E|Q).$$

Addition of the last two equations and three applications of 18.26.1) furnish $\psi(S) - \alpha_\star(S) = 0$. The proof is complete.

18.30. THEOREM. *Let* $\alpha : \mathfrak{A} \to \mathbb{R}$ *be a finite content on an oval* $\mathfrak{A}$. *Fix* $E \subset X$. *Then* $\alpha^\star$ *and* $\alpha_\star$ *are finite and modular on* $\mathfrak{A}[E]$.

Proof. The finiteness is known from 18.26.0). Fix $S, T \in \mathfrak{A}[E]$ and then $U \in \mathfrak{A}$ such that $U \subset S, T$. Then from 18.29 applied to E and U we obtain $\alpha^\star(S \cup T) + \alpha^\star(S \cap T) = \alpha^\star(S) + \alpha^\star(T)$ and the same for $\alpha_\star$. This is the assertion.

18.31. CONSEQUENCE. *Let* $\alpha : \mathfrak{A} \to \overline{\mathbb{R}}$ *be a content* $\dot{+}$ *on an oval* $\mathfrak{A}$, *and let* $\mathfrak{S} \subset \mathfrak{A}$ *be a lattice on which* α *is finite. Fix* $E \subset X$. *Then* $\alpha^\star$ *and* $\alpha_\star$ *are finite and modular on* $\mathfrak{S}[E]$.

Proof. $\mathfrak{E} := \{A \in \mathfrak{A} : \alpha(A) \in \mathbb{R}\} \subset \mathfrak{A}$ is an oval, and $\varepsilon := \alpha|\mathfrak{E}$ is a finite content on $\mathfrak{E}$. One verifies that $\varepsilon^\star \geqq \alpha^\star$ and $\varepsilon_\star \leqq \alpha_\star$, and that $\varepsilon^\star = \alpha^\star$ and $\varepsilon_\star = \alpha_\star$ on $\mathfrak{E} \sqsubset \mathfrak{E}$. By 18.30 therefore $\alpha^\star$ and $\alpha_\star$ are finite and modular on $\mathfrak{E}[E]$. The assertion follows.

The Uniqueness Theorem

One direction of the uniqueness theorem is a simple consequence of the first two subsections.

18.32. PROPOSITION. *Assume that* $\varphi : \mathfrak{S} \to \mathbb{R}$ *is an inner* $\star$ *premeasure on a lattice* $\mathfrak{S}$ *with* $\phi := \varphi_\star|\mathfrak{C}(\varphi_\star, +)$. *Let* $\mathfrak{T}$ *be a lattice with* $\mathfrak{T} \subset (\mathfrak{S} \sqsubset \mathfrak{S})$ *such that* $\star(\varphi, \mathfrak{T})$ *is nonvoid. If* $\phi^\star|\mathfrak{T}$ *is supermodular then it is the unique inner* $\star$ *premeasure* $\psi : \mathfrak{T} \to \mathbb{R}$ *with* $\psi_\star|\mathfrak{S} = \varphi$.

18.33. ADDENDUM. *In the conventional situation* $\varnothing \in \mathfrak{S}$ *with* $\varphi(\varnothing) = 0$ *and* $\varnothing \in \mathfrak{T}$ *it suffices to assume that* $\phi^\star|\mathfrak{T}$ *is superadditive.*

Proof of 18.32 and 18.33. For $\psi := \phi^\star|\mathfrak{T}$ we see from 18.1 that $\psi_\star|\mathfrak{C}(\varphi_\star, +) \leqq \phi$ and hence $\psi_\star|\mathfrak{S} \leqq \varphi$. By 18.9 there are inner $\star$ premeasures $\vartheta : \mathfrak{T} \to \mathbb{R}$ with $\vartheta_\star|\mathfrak{S} = \varphi$. By 18.3.2) each of them fulfils $\vartheta \leqq \psi$. Therefore $\varphi = \vartheta_\star|\mathfrak{S} \leqq \psi_\star|\mathfrak{S} \leqq \varphi$ and hence $\vartheta_\star|\mathfrak{S} = \psi_\star|\mathfrak{S} = \varphi$. It follows that $\vartheta = \psi$, in the full situation from 18.6 and in the conventional situation from 17.2.

The other direction requires the restriction $\mathfrak{S} \subset \mathfrak{T}$. We do not know whether it can be avoided.

18.34. PROPOSITION. *Assume that* $\varphi : \mathfrak{S} \to \mathbb{R}$ *is an inner* $\star$ *premeasure on a lattice* $\mathfrak{S}$ *with* $\phi := \varphi_\star|\mathfrak{C}(\varphi_\star, +)$. *Let* $\mathfrak{T}$ *be a lattice with* $\mathfrak{S} \subset \mathfrak{T} \subset (\mathfrak{S} \sqsubset \mathfrak{S})$. *If there is a unique inner* $\star$ *premeasure* $\psi : \mathfrak{T} \to \mathbb{R}$ *with* $\psi|\mathfrak{S} = \varphi$ *then* $\psi = \phi^\star|\mathfrak{T}$.

Proof. Fix $E \in \mathfrak{T}$. We use the last subsection in that we conclude from 18.31 that $\phi^\star|\mathfrak{S}[E]$ is supermodular. By 18.32 thus $\eta := \phi^\star|\mathfrak{S}[E]$ is an inner $\star$ premeasure $\eta : \mathfrak{S}[E] \to \mathbb{R}$ with $\eta|\mathfrak{S} = \varphi$ (and in fact the unique one). By 18.27.2) we have $E \in \mathfrak{S}[E]$ and then $\eta(E) = \phi^\star(E)$. Now $\mathfrak{S}[E] \subset \mathfrak{T}$. By 18.15 there exists an inner $\star$ premeasure $\vartheta : \mathfrak{T} \to \mathbb{R}$ such that $\vartheta|\mathfrak{S}[E] = \eta$. Thus $\vartheta|\mathfrak{S} = \eta|\mathfrak{S} = \varphi$ and $\vartheta(E) = \eta(E) = \phi^\star(E)$. Now the uniqueness assumption enforces that $\vartheta = \psi$. Therefore $\psi(E) = \phi^\star(E)$. This is the assertion.

18.35. THEOREM. *Assume that $\varphi : \mathfrak{S} \to \mathbb{R}$ is an inner $\star$ premeasure on a lattice $\mathfrak{S}$ with $\phi := \varphi_\star | \mathfrak{C}(\varphi_\star, +)$. Let $\mathfrak{T}$ be a lattice with $\mathfrak{S} \subset \mathfrak{T} \subset (\mathfrak{S} \sqsubset \mathfrak{S})$. Then there is a unique inner $\star$ premeasure $\psi : \mathfrak{T} \to \mathbb{R}$ with $\psi | \mathfrak{S} = \varphi$ iff $\phi^\star | \mathfrak{T}$ is supermodular. In this case $\psi = \phi^\star | \mathfrak{T}$.*

18.36. ADDENDUM. *In the conventional situation $\varnothing \in \mathfrak{S}$ with $\varphi(\varnothing) = 0$ and hence $\varnothing \in \mathfrak{T}$ it is equivalent that $\phi^\star | \mathfrak{T}$ is superadditive.*

18.37. BIBLIOGRAPHICAL NOTE. The last result 18.36 is due to Tarash-chanskii [1989]. For other uniqueness assertions we refer to Lipecki [1983] [1988][1990].

In the present section we have stressed the formal resemblance to section 17 in the existence results. There is no such connection with respect to uniqueness. In fact, one can see from 17.12 that 18.35 has no counterpart in section 17.

We conclude with a simple example for non-uniqueness.

18.38. EXAMPLE. Let $\mathfrak{S}$ consist of $\varnothing$ and X, and define $\varphi : \mathfrak{S} \to \mathbb{R}$ to be $\varphi(\varnothing) = 0$ and $\varphi(X) = 1$. Then fix nonvoid subsets $P, Q \subset X$ with $P \cup Q = X$ and $P \cap Q = \varnothing$. Let $\mathfrak{T}$ consist of $\mathfrak{S}$ and P, Q. Then the inner $\star$ premeasures $\psi : \mathfrak{T} \to \mathbb{R}$ with $\psi | \mathfrak{S} = \varphi$ are the extensions of φ with

$$\psi(P) = t \text{ and } \psi(Q) = 1 - t \quad \text{for some } 0 \leqq t \leqq 1.$$

On the other hand $\phi := \varphi_\star | \mathfrak{C}(\varphi_\star) = \varphi$ has $\phi^\star(P) = \phi^\star(Q) = 1$.

19. Transplantation of Measures

The present section considers the case $\bullet = \sigma$. We restrict ourselves to the conventional inner situation since the full inner situation would be much less simple. But there are some preparations which can be done for the full inner situation with the same effort.

Preparations

We start with the application of 6.15 with 6.17 as in the last section.

19.1. PROPOSITION. *Let $\varphi : \mathfrak{S} \to \mathbb{R}$ and $\psi : \mathfrak{T} \to \mathbb{R}$ be isotone and supermodular on lattices $\mathfrak{S}$ and $\mathfrak{T}$ with $\varphi_\sigma | \mathfrak{S} = \varphi$. Assume that $\mathfrak{T} \subset (\mathfrak{S} \sqsubset \mathfrak{S})$. Then $\psi_\sigma | \mathfrak{S} = \varphi$ implies that $\psi_\sigma | \mathfrak{C}(\psi_\sigma, +)$ is an extension of $\varphi_\sigma | \mathfrak{C}(\varphi_\sigma, +)$ (note that the converse is true when φ is an inner σ premeasure).*

Proof. 1) φ_σ and ψ_σ are isotone and supermodular + by 6.3.5). On $\mathfrak{S}$ we have by assumption $\psi_\sigma = \varphi = \varphi_\sigma$, and this is finite. Hence $\psi_\sigma = \varphi_\sigma$ on $\mathfrak{S}_\sigma$ by 6.7, and $\psi_\sigma \geqq \varphi_\sigma$ since φ_σ is inner regular $\mathfrak{S}_\sigma$ by 6.3.4). 2) Now the assertion follows from 6.17 applied to $\mathfrak{P} = \mathfrak{Q} := \mathfrak{S}$ and $\mathfrak{H} := \mathfrak{S}_\sigma$ and to $\phi := \psi_\sigma$ and $\theta := \varphi_\sigma$, combined with $\mathfrak{T}_\sigma \subset (\mathfrak{S}_\sigma \sqsubset \mathfrak{S})$ and 6.7.

We need one more lemma, which is devoted to an obvious comparison.

19.2. LEMMA. *Let $\mathfrak{S}$ be a lattice. 1) Assume that $\varphi : \mathfrak{S} \to [-\infty, \infty[$ is an inner σ premeasure. Then $\xi := \varphi_\sigma | \mathfrak{S}_\sigma$ is an inner $\star$ premeasure, and downward σ continuous and hence an inner σ premeasure as well. Furthermore $\xi_\sigma = \xi_\star = \varphi_\sigma$. 2) Assume that $\xi : \mathfrak{S}_\sigma \to [-\infty, \infty[$ is an inner σ premeasure. Then $\varphi := \xi | \mathfrak{S}$ is an inner σ premeasure as well. Furthermore $\xi_\sigma = \xi_\star = \varphi_\sigma$, and hence in particular $\xi = \varphi_\sigma | \mathfrak{S}_\sigma$.*

Proof of 1). $\xi : \mathfrak{S}_\sigma \to [-\infty, \infty[$ is isotone and supermodular, and an extension of φ and hence $\not\equiv -\infty$. We have $\xi_\star = \varphi_\sigma$ since φ_σ is inner regular $\mathfrak{S}_\sigma$. Furthermore $\xi_\star | \mathfrak{C}(\xi_\star, +) = \varphi_\sigma | \mathfrak{C}(\varphi_\sigma, +)$ is an extension of ξ in the crude sense. Thus 6.22 implies that ξ is an inner $\star$ premeasure. Next ξ is downward σ continuous by 6.5.iii), and hence an inner σ premeasure as a consequence of 6.22. At last $\xi_\sigma = \xi_\star$ follows from 6.5.iv).

Proof of 2). It is clear that $\varphi \not\equiv -\infty$. i) By 6.5.iv) we have $\xi_\sigma = \xi_\star$. ii) $\alpha := \xi_\sigma | \mathfrak{C}(\xi_\sigma, +)$ is a content $+$ on an algebra which is an extension of ξ and hence of φ. In particular $\mathfrak{S}_\sigma \subset \mathfrak{C}(\xi_\sigma, +)$. Furthermore $\alpha | \mathfrak{S}_\sigma = \xi$ is downward σ continuous, and α is inner regular $\mathfrak{S}_\sigma$. Therefore α is an inner σ extension of φ. Thus φ is an inner σ premeasure. iii) Now 6.18 implies that α is a restriction of $\varphi_\sigma | \mathfrak{C}(\varphi_\sigma, +)$. Thus $\xi_\sigma = \varphi_\sigma$ on $\mathfrak{C}(\xi_\sigma, +)$ and in particular on $\mathfrak{S}_\sigma$. Therefore $\xi_\sigma = \varphi_\sigma$ since both sides are inner regular $\mathfrak{S}_\sigma$. The proof is complete.

The Existence Theorem

Let $\mathfrak{S}$ be a lattice with $\varnothing \in \mathfrak{S}$, and $\varphi : \mathfrak{S} \to [0, \infty[$ be an isotone and supermodular set function with $\varphi(\varnothing) = 0$ and $\varphi_\sigma | \mathfrak{S} = \varphi$. We fix another lattice $\mathfrak{T}$ such that $\varnothing \in \mathfrak{T} \subset (\sqsubset \mathfrak{S})$. As before we define $\sigma(\varphi, \mathfrak{T})$ to consist of all isotone and supermodular set functions $\psi : \mathfrak{T} \to [0, \infty[$ such that $\psi_\sigma | \mathfrak{S} = \varphi$.

19.3. PROPERTIES. *Let $\psi \in \sigma(\varphi, \mathfrak{T})$. 0) $\psi(\varnothing) = \psi_\sigma(\varnothing) = 0$. 1) $\psi_\sigma | \mathfrak{C}(\psi_\sigma)$ is an extension of $\varphi_\sigma | \mathfrak{C}(\varphi_\sigma)$. 2) We have $\psi_\sigma | \mathfrak{T}_\sigma \leq \phi^\star | \mathfrak{T}_\sigma$ with $\phi := \varphi_\sigma | \mathfrak{C}(\varphi_\sigma)$. 3) $\psi_\sigma | \mathfrak{S}_\sigma = \varphi_\sigma | \mathfrak{S}_\sigma$. Therefore $\psi_\sigma \geq \varphi_\sigma$.*

Proof. 0) follows from $0 \leq \psi(\varnothing) \leq \psi_\sigma(\varnothing) = \varphi(\varnothing) = 0$. 1) is contained in 19.1. 2) We have $(\psi_\sigma | \mathfrak{T}_\sigma)_\star = \psi_\sigma$ since ψ_σ is inner regular $\mathfrak{T}_\sigma$. Thus 1) implies that $(\psi_\sigma | \mathfrak{T}_\sigma)_\star | \mathfrak{C}(\varphi_\sigma) = \psi_\sigma | \mathfrak{C}(\varphi_\sigma) = \phi$. Then the assertion follows from 18.1. 3) $\psi_\sigma | \mathfrak{S}_\sigma = \varphi_\sigma | \mathfrak{S}_\sigma$ follows from $\psi_\sigma | \mathfrak{S} = \varphi = \varphi_\sigma | \mathfrak{S}$ and 6.7. Then $\psi_\sigma \geq \varphi_\sigma$ since φ_σ is inner regular $\mathfrak{S}_\sigma$.

We now obtain the counterparts to 18.9 and 18.10, except that we are restricted to the conventional situation.

19.4. THEOREM. *Assume that $\varphi : \mathfrak{S} \to [0, \infty[$ is an inner σ premeasure on a lattice $\mathfrak{S}$ with $\varnothing \in \mathfrak{S}$ and $\varphi(\varnothing) = 0$, and $\phi := \varphi_\sigma | \mathfrak{C}(\varphi_\sigma)$. Let $\mathfrak{T}$ be a lattice with $\varnothing \in \mathfrak{T} \subset (\sqsubset \mathfrak{S})$, and assume that $\phi^\star | \mathfrak{T}$ is σ continuous at $\varnothing$. If $\sigma(\varphi, \mathfrak{T})$ is nonvoid then it is upward inductive in the argumentwise order. The maximal members of $\sigma(\varphi, \mathfrak{T})$ are the inner σ premeasures $\psi : \mathfrak{T} \to [0, \infty[$ with $\psi_\sigma | \mathfrak{S} = \varphi$.*

19.5. THEOREM. *Assume that $\varphi : \mathfrak{S} \to [0, \infty[$ is an inner σ premeasure on a lattice $\mathfrak{S}$ with $\varnothing \in \mathfrak{S}$ and $\varphi(\varnothing) = 0$, and $\phi := \varphi_\sigma | \mathfrak{C}(\varphi_\sigma)$. Let $\mathfrak{T}$ be a lattice with $\varnothing \in \mathfrak{T} \subset (\sqsubset \mathfrak{S})$, and assume that $\phi^\star | \mathfrak{T}$ is σ continuous at $\varnothing$. Then*

> for each $\vartheta : \mathfrak{T} \to [0, \infty[$ *isotone and supermodular with* $\vartheta_\sigma | \mathfrak{S} = \varphi$
>
> there exists $\psi : \mathfrak{T} \to [0, \infty[$ *inner σ premeasure with* $\psi_\sigma | \mathfrak{S} = \varphi$

such that $\psi \geqq \vartheta$.

For the proof of 19.4 and 19.5 we first isolate the three main points. Then the assertions will be reduced to the former 18.9 and 18.10.

0) $\xi := \varphi_\sigma | \mathfrak{S}_\sigma$ is an extension of φ, and an inner $\star$ premeasure by 19.2.1). Furthermore $\xi_\star = \varphi_\sigma$ and hence $\phi = \xi_\star | \mathfrak{C}(\xi_\star)$.

1) Assume that $\vartheta \in \sigma(\varphi, \mathfrak{T})$ and form $\theta := \vartheta_\sigma | \mathfrak{T}_\sigma$. Then $\theta_\star = \vartheta_\sigma$ since ϑ_σ is inner regular $\mathfrak{T}_\sigma$. We have $\theta(\varnothing) = 0$ by 19.3.0), and $\theta_\star | \mathfrak{S}_\sigma = \vartheta_\sigma | \mathfrak{S}_\sigma = \varphi_\sigma | \mathfrak{S}_\sigma = \xi$ and hence $\theta \in \star(\xi, \mathfrak{T}_\sigma)$ by 19.3.3).

2) Let $\eta \in \star(\xi, \mathfrak{T}_\sigma)$ be an inner $\star$ premeasure. Then $\psi := \eta | \mathfrak{T}$ is an inner σ premeasure with $\psi_\sigma = \eta_\star$ and hence $\psi \in \sigma(\varphi, \mathfrak{T})$. In fact, we have $\eta \leqq \phi^\star | \mathfrak{T}_\sigma$ by 18.3.2). Thus η is σ continuous at $\varnothing$ as a consequence of 8.12, and hence an inner σ premeasure by 6.31. Then the assertion follows from 19.2.2).

Proof of 19.5. Let $\vartheta \in \sigma(\varphi, \mathfrak{T})$ and then $\theta \in \star(\xi, \mathfrak{T}_\sigma)$ as in 1) above. By 18.10 and 0) there exists an inner $\star$ premeasure $\eta \in \star(\xi, \mathfrak{T}_\sigma)$ with $\eta \geqq \theta$. Then 2) says that $\psi := \eta | \mathfrak{T}$ is an inner σ premeasure in $\sigma(\varphi, \mathfrak{T})$. On $\mathfrak{T}$ we have $\psi = \eta \geqq \theta = \vartheta_\sigma \geqq \vartheta$. This is the assertion.

Proof of 19.4. i) Fix $\vartheta \in \sigma(\varphi, \mathfrak{T})$. If ϑ is a maximal member of $\sigma(\varphi, \mathfrak{T})$ then 19.5 shows that ϑ is an inner σ premeasure. Now assume that ϑ is an inner σ premeasure, and that $\vartheta \leqq \psi$ for some $\psi \in \sigma(\varphi, \mathfrak{T})$. We have to prove that $\vartheta = \psi$. By 1) above $\theta := \vartheta_\sigma | \mathfrak{T}_\sigma$ and $\eta := \psi_\sigma | \mathfrak{T}_\sigma$ are in $\star(\xi, \mathfrak{T}_\sigma)$ with $\theta \leqq \eta$, and θ is an inner $\star$ premeasure by 19.2.1). From 18.9 it follows that $\theta = \eta$. On $\mathfrak{T}$ therefore $\vartheta = \vartheta_\sigma = \theta = \eta = \psi_\sigma \geqq \psi$ and hence $\vartheta = \psi$. ii) It remains to prove that $\sigma(\varphi, \mathfrak{T})$ is upward inductive in the argumentwise order. Assume that $E \subset \sigma(\varphi, \mathfrak{T})$ is nonvoid and totally ordered. By 1) above then $\{\vartheta_\sigma | \mathfrak{T}_\sigma : \vartheta \in E\}$ is $\subset \star(\xi, \mathfrak{T}_\sigma)$, and of course nonvoid and totally ordered as well. From 18.9 we obtain an inner $\star$ premeasure $\eta \in \star(\xi, \mathfrak{T}_\sigma)$ such that $\vartheta_\sigma | \mathfrak{T}_\sigma \leqq \eta$ for all $\vartheta \in E$. By 2) above $\psi := \eta | \mathfrak{T}$ is an inner σ premeasure in $\sigma(\varphi, \mathfrak{T})$. On $\mathfrak{T}$ now $\vartheta \leqq \vartheta_\sigma \leqq \eta = \psi$ for all $\vartheta \in E$. The proof is complete.

Specializations of the Existence Theorem

The specializations in the last section started with $\mathfrak{S} = \{\varnothing, X\}$. It is simple to see that in this case, due to its inherent discreteness, the new result 19.5 is contained in the old one 18.10.

The counterparts of the subsequent former specializations require a nontrivial lemma.

19.6. LEMMA. *Let* $\varphi : \mathfrak{S} \to [-\infty, \infty[$ *be an isotone and supermodular set function on a lattice* $\mathfrak{S}$. *Then* $(\varphi_\star | \sqsubset \mathfrak{S})_\sigma = \varphi_\sigma$.

Proof. We have $\geqq$ since $\varphi_\star | \sqsubset \mathfrak{S}$ is an extension of φ. Thus to be shown is $\leqq$. We fix $A \subset X$ and can assume that $(\varphi_\star | \sqsubset \mathfrak{S})_\sigma(A) > -\infty$. We fix a sequence $(D_l)_l$ in $\sqsubset \mathfrak{S}$ with $D_l \downarrow$ some $D \subset A$ and $\lim_{l\to\infty} \varphi_\star(D_l) =: c > -\infty$, which implies that all $\varphi_\star(D_l)$ are finite. To be shown is $c \leqq \varphi_\sigma(A)$. Let $\varepsilon > 0$ and choose $S_l \in \mathfrak{S}$ with $S_l \subset D_l$ such that $\varphi(S_l) \geqq \varphi_\star(D_l) - \varepsilon/2^l$. Then form $T_l := S_1 \cap \cdots \cap S_l \in \mathfrak{S}$, so that $T_l \subset S_l \subset D_l$ and hence $T_l \downarrow$ some $T \subset D \subset A$. i) We claim that

$$\varphi(T_l) \geqq \varphi_\star(D_l) - \varepsilon(1 - 1/2^l) \quad \text{for } l \geqq 1.$$

This is clear for $l = 1$. For the induction step $1 \leqq l \Rightarrow l + 1$ note that $T_l \cap S_{l+1} = T_{l+1}$ and $T_l \cup S_{l+1} \subset D_l \cup D_{l+1} = D_l$. By assumption and by the induction hypothesis therefore

$$\begin{aligned}
\varphi(T_l) + \varphi(S_{l+1}) &\leqq \varphi(T_l \cup S_{l+1}) + \varphi(T_l \cap S_{l+1}) \\
&\leqq \varphi_\star(D_l) + \varphi(T_{l+1}) \\
&\leqq \varphi(T_l) + \varepsilon(1 - 1/2^l) + \varphi(T_{l+1}), \\
\varphi(S_{l+1}) &\leqq \varepsilon(1 - 1/2^l) + \varphi(T_{l+1}),
\end{aligned}$$

which implies the assertion. ii) From i) we obtain $c := \lim_{l\to\infty} \varphi_\star(D_l) \leqq \lim_{l\to\infty} \varphi(T_l) + \varepsilon \leqq \varphi_\sigma(A) + \varepsilon$ for all $\varepsilon > 0$. The assertion follows.

The next result is the counterpart to the above 18.18 which was based on the Marczewski condition.

19.7. THEOREM. *Assume that* $\varphi : \mathfrak{S} \to [0, \infty[$ *is an inner* σ *premeasure on a lattice* $\mathfrak{S}$ *with* $\varnothing \in \mathfrak{S}$ *and* $\varphi(\varnothing) = 0$, *and* $\phi := \varphi_\sigma | \mathfrak{C}(\varphi_\sigma)$. *Let* $\mathfrak{T}$ *be a lattice with* $\varnothing \in \mathfrak{T} \subset (\sqsubset \mathfrak{S})$, *and let* $\phi^\star | \mathfrak{T}$ *be* σ *continuous at* $\varnothing$. *Assume that* $(\varphi_\star | \mathfrak{T})_\sigma | \mathfrak{S} \geqq \varphi$, *which implies that* $(\varphi_\star | \mathfrak{T})_\sigma | \mathfrak{S} = \varphi$. *Then there exists an inner* σ *premeasure* $\psi : \mathfrak{T} \to [0, \infty[$ *such that* $\psi_\sigma | \mathfrak{S} = \varphi$.

Proof. For the set function $\vartheta := \varphi_\star | \mathfrak{T}$ we see from 19.6 that

$$\vartheta_\sigma = (\varphi_\star | \mathfrak{T})_\sigma \leqq (\varphi_\star | \sqsubset \mathfrak{S})_\sigma = \varphi_\sigma \quad \text{and hence} \quad \vartheta_\sigma | \mathfrak{S} \leqq \varphi.$$

Thus $\vartheta_\sigma | \mathfrak{S} \geqq \varphi$ implies in fact that $\vartheta_\sigma | \mathfrak{S} = \varphi$. Therefore ϑ is as required in 19.5.

An obvious special case is the subsequent counterpart to the above 18.15.

19.8. THEOREM. *Assume that* $\varphi : \mathfrak{S} \to [0, \infty[$ *is an inner* σ *premeasure on a lattice* $\mathfrak{S}$ *with* $\varnothing \in \mathfrak{S}$ *and* $\varphi(\varnothing) = 0$, *and* $\phi := \varphi_\sigma | \mathfrak{C}(\varphi_\sigma)$. *Let* $\mathfrak{T}$ *be a lattice with* $\mathfrak{S} \subset \mathfrak{T} \subset (\sqsubset \mathfrak{S})$, *and let* $\phi^\star | \mathfrak{T}$ *be* σ *continuous at* $\varnothing$. *Then there exists an inner* σ *premeasure* $\psi : \mathfrak{T} \to [0, \infty[$ *such that* $\psi | \mathfrak{S} = \varphi$.

19.9. BIBLIOGRAPHICAL NOTE. The above 19.7 is in Adamski [1987] theorem 3.4(b), but with the sharper old Marczewski condition instead of the new one. A close relative of 19.8 is in Adamski [1984a] theorem 3.3(a).

The Uniqueness Theorem

This time we have to assume that $\mathfrak{S} \subset \mathfrak{T}$ from the start. We also need another nontrivial lemma. It is true for $\bullet = \star \sigma \tau$.

19.10. LEMMA. *Assume that $\varphi : \mathfrak{S} \to [0, \infty[$ is an inner $\bullet$ premeasure on a lattice $\mathfrak{S}$ with $\varnothing \in \mathfrak{S}$ and $\varphi(\varnothing) = 0$, and $\phi := \varphi_\bullet | \mathfrak{C}(\varphi_\bullet)$. Let $\mathfrak{T}$ be a lattice with $\mathfrak{S} \subset \mathfrak{T} \subset (\sqsubset \mathfrak{S})$. If $\phi^\star | \mathfrak{T}$ is supermodular then it is inner $\bullet$ tight.*

Proof. Let $\psi := \phi^\star | \mathfrak{T}$. By assumption $\psi : \mathfrak{T} \to [0, \infty[$ is isotone and supermodular and an extension of φ. 1) We have

$$\varphi(P) \leqq \phi^\star(A) + \varphi_\bullet(P \setminus A) \quad \text{for } P \in \mathfrak{S} \text{ and } A \subset P.$$

To see this let $H \in \mathfrak{C}(\varphi_\bullet)$ with $A \subset H \subset P$. Then

$$\varphi(P) = \phi(P) = \phi(H) + \phi(P \setminus H) \quad = \quad \phi(H) + \varphi_\bullet(P \setminus H)$$
$$\leqq \quad \phi(H) + \varphi_\bullet(P \setminus A).$$

This implies the assertion. 2) Let $A \subset B$ in $\mathfrak{T}$. We fix $\varepsilon > 0$ and $P \in \mathfrak{S}$ with $A \subset B \subset P$. From 1) we obtain

$$\psi(B) - \psi(A) = \psi(B) - \phi^\star(A) \leqq \psi(B) - \varphi(P) + \varphi_\bullet(P \setminus A).$$

Note that $\varphi_\bullet(P \setminus A) = \varphi_\bullet^P(P \setminus A)$ by 6.29.5). Thus there is a paving $\mathfrak{M} \subset \mathfrak{S}$ of type $\bullet$ with $\mathfrak{M} \downarrow \subset P \setminus A$ and $S \subset P \; \forall S \in \mathfrak{M}$ such that $\inf\limits_{S \in \mathfrak{M}} \varphi(S) \geqq \varphi_\bullet(P \setminus A) - \varepsilon$. For $S \in \mathfrak{M}$ therefore

$$\psi(B) - \psi(A) \leqq \psi(B) - \varphi(P) + \varphi(S) + \varepsilon = \psi(B) - \psi(P) + \psi(S) + \varepsilon.$$

Since ψ is supermodular and $B \cup S \subset P$ it follows that $\psi(B) - \psi(A) \leqq \psi(B \cap S) + \varepsilon$. Now $\{B \cap S : S \in \mathfrak{M}\} \subset \mathfrak{T}$ is a paving of type $\bullet$ with $\downarrow \subset B \cap (P \setminus A) = B \setminus A$ and all members $\subset B \cap P = B$. It follows that $\psi(B) - \psi(A) \leqq \psi_\bullet^B(B \setminus A) + \varepsilon$ and hence the assertion.

We come to the uniqueness theorem. The simultaneous appearance of properties 2)3)4) is a little imperfection which we could not avoid.

19.11. THEOREM. *Assume that $\varphi : \mathfrak{S} \to [0, \infty[$ is an inner σ premeasure on a lattice $\mathfrak{S}$ with $\varnothing \in \mathfrak{S}$ and $\varphi(\varnothing) = 0$, and $\phi := \varphi_\sigma | \mathfrak{C}(\varphi_\sigma)$. Let $\mathfrak{T}$ be a lattice with $\mathfrak{S} \subset \mathfrak{T} \subset (\sqsubset \mathfrak{S})$, and assume that $\phi^\star | \mathfrak{T}$ is σ continuous at $\varnothing$. Then the following are equivalent.*
1) *There is a unique inner σ premeasure $\psi : \mathfrak{T} \to [0, \infty[$ with $\psi | \mathfrak{S} = \varphi$ (note that the existence is clear from 19.8).*
2) *$\phi^\star | \mathfrak{T}$ is supermodular.*
3) *$\phi^\star | \mathfrak{T}_\sigma$ is supermodular.*
4) *$\phi^\star | \mathfrak{T}_\sigma$ is superadditive.*
In this case the unique inner σ premeasure $\psi : \mathfrak{T} \to [0, \infty[$ with $\psi | \mathfrak{S} = \varphi$ is $\psi = \phi^\star | \mathfrak{T}$.

Proof. We start with the implication 1) $\Rightarrow \psi = \phi^\star | \mathfrak{T}$. Note that it contains the implication 1) $\Rightarrow$ 0) $\phi^\star | \mathfrak{T}$ is an inner σ premeasure .

i) Fix $E \in \mathfrak{T}$. We see from the definition and from 18.27.1)2)3) that $\mathfrak{S}[E]$ is a lattice with $\mathfrak{S} \subset \mathfrak{S}[E] \subset \mathfrak{T}$ and $E \in \mathfrak{S}[E]$. ii) We conclude from 18.31 that $\phi^\star|\mathfrak{S}[E]$ is finite and supermodular. Then 19.10 implies that $\phi^\star|\mathfrak{S}[E]$ is inner σ tight. Furthermore $\phi^\star|\mathfrak{S}[E]$ is σ continuous at $\varnothing$ by assumption. It follows that $\eta := \phi^\star|\mathfrak{S}[E]$ is an inner σ premeasure. Let us put $\nu := \eta_\sigma|\mathfrak{C}(\eta_\sigma)$. iii) We have $\eta|\mathfrak{S} = \phi^\star|\mathfrak{S} = \phi|\mathfrak{S} = \varphi$. Thus $\eta_\sigma|\mathfrak{S} = \varphi$ so that $\eta \in \sigma(\varphi, \mathfrak{S}[E])$. From 19.3.1) we see that ν is an extension of ϕ. This implies that $0 \leq \nu^\star \leq \phi^\star$. Therefore $\nu^\star|\mathfrak{T}$ is σ continuous at $\varnothing$. iv) Now 19.8 can be applied to $\eta : \mathfrak{S}[E] \to [0, \infty[$ with ν and to $\mathfrak{T}$. It follows that there exists an inner σ premeasure $\vartheta : \mathfrak{T} \to [0, \infty[$ such that $\vartheta|\mathfrak{S}[E] = \eta$. In particular $\vartheta|\mathfrak{S} = \eta|\mathfrak{S} = \varphi$. Thus assumption 1) implies that $\vartheta = \psi$. Therefore $\psi(E) = \vartheta(E) = \eta(E) = \phi^\star(E)$. This is the assertion.

The remainder of the proof consists of the two chains of implications 0)$\Rightarrow$3)$\Rightarrow$4)$\Rightarrow$1) and 0)$\Rightarrow$3)$\Rightarrow$2)$\Rightarrow$0). Here 3)$\Rightarrow$4) and 3)$\Rightarrow$2) are obvious, while 2)$\Rightarrow$0) is an immediate consequence of 19.10. Thus it remains to prove 0)$\Rightarrow$3) and 4)$\Rightarrow$1).

Proof of 0)$\Rightarrow$3). Since $\psi := \phi^\star|\mathfrak{T}$ is an inner σ premeasure it is clear that $\psi_\sigma|\mathfrak{T}_\sigma$ is modular. Thus it suffices to prove that $\psi_\sigma|\mathfrak{T}_\sigma = \phi^\star|\mathfrak{T}_\sigma$. Now on the one hand $\psi \in \sigma(\varphi, \mathfrak{T})$, and hence $\psi_\sigma|\mathfrak{T}_\sigma \leq \phi^\star|\mathfrak{T}_\sigma$ by 19.3.2). On the other hand let $P \in \mathfrak{T}_\sigma$ and $(P_l)_l$ in $\mathfrak{T}$ with $P_l \downarrow P$. Then $\phi^\star(P) \leq \phi^\star(P_l) = \psi(P_l) = \psi_\sigma(P_l)$ implies that $\phi^\star(P) \leq \psi_\sigma(P)$.

Proof of 4)$\Rightarrow$1). Let $\psi : \mathfrak{T} \to [0, \infty[$ be an inner σ premeasure with $\psi|\mathfrak{S} = \varphi$. By 19.2.1) then $\eta := \psi_\sigma|\mathfrak{T}_\sigma$ is an inner $\star$ premeasure, and $\eta \leq \phi^\star|\mathfrak{T}_\sigma$ by 19.3.2). Now 17.2 can be applied to these two set functions because of the hypothesis 4). For each $A \in \mathfrak{T}_\sigma$ there exists $B \in \mathfrak{S}$ with $A \subset B$, and we have

$$\eta_\star(B) = \eta(B) = \psi_\sigma(B) = \psi(B) \quad = \quad \varphi(B),$$
$$(\phi^\star|\mathfrak{T}_\sigma)_\star(B) = \phi^\star(B) = \phi(B) \quad = \quad \varphi(B).$$

It follows that $\eta = \phi^\star|\mathfrak{T}_\sigma$, and hence in particular $\psi = \phi^\star|\mathfrak{T}$. The proof is complete.

Extension of Baire Measures to Borel Measures

Let X be a topological space. The present subsection returns to the problem to extend a Baire measure $\alpha : \text{Baire}(X) \to [0, \infty]$ to Borel measures $\beta : \text{Bor}(X) \to [0, \infty]$. The previous treatment in section 8 was based on the inner τ theory. This time the problem will be treated as an application of the transplantation procedure of the present section. Thus as before we shall obtain, in the spirit of the present text and in consequence of its main theorems, not the sheer existence or uniqueness of Borel extensions, but rather the existence or uniqueness of Borel extensions with certain natural regularity properties. In the present context we have of course inner regularity, as in the earlier treatment.

Let $\alpha : \text{Baire}(X) \to [0, \infty]$ be a cmeasure. We define as before $\mathfrak{S} := \{A \in \text{CCl}(X) : \alpha(A) < \infty\}$ and $\varphi := \alpha|\mathfrak{S}$. The basic properties of these formations are collected in 8.2.1) and 8.5. Thus $\mathfrak{S}$ is a lattice with $\varnothing \in \mathfrak{S}$ and $\mathfrak{S}_\sigma = \mathfrak{S}$, and $\varphi : \mathfrak{S} \to [0, \infty[$ is an inner σ premeasure. Furthermore α is inner regular $\mathfrak{S}$ at $[\alpha < \infty]^\sigma$, and is inner regular $\mathfrak{S}$ iff it is semifinite above. We form the cmeasure $\phi := \varphi_\sigma | \mathfrak{C}(\varphi_\sigma) = \varphi_\star | \mathfrak{C}(\varphi_\star)$. It follows that $\phi = \alpha$ on $[\alpha < \infty]^\sigma$, and $\phi = \alpha$ on $\mathrm{Baire}(X)$ iff α is semifinite above.

19.12. REMARK. *For* $\phi := \varphi_\sigma | \mathfrak{C}(\varphi_\sigma)$ *we have* $\phi^\star \leqq \alpha^\star$, *and* $\phi^\star = \alpha^\star$ *on* $\sqsubset \mathfrak{S}^\sigma$.

Proof. We use some of the properties listed in 8.2.1). With the notation $\mathfrak{C} := \mathfrak{C}(\varphi_\sigma) = \mathfrak{C}(\varphi^\sigma) \supset \mathrm{Baire}(X)$ as in iii) we have

$$
\begin{aligned}
\phi^\star &= (\varphi_\sigma | \mathfrak{C})^\star \leqq (\varphi_\sigma | \mathrm{Baire}(X))^\star && \text{since we pass to a restriction} \\
&\leqq \alpha^\star \leqq (\varphi^\sigma | \mathrm{Baire}(X))^\star && \text{by v)} \\
&= \varphi^\sigma = (\varphi^\sigma | \mathfrak{C} \cap (\sqsubset \mathfrak{S}^\sigma))^\star && \text{since } \varphi^\sigma \text{ is outer regular } \mathfrak{S}^\sigma \\
&= (\varphi_\sigma | \mathfrak{C} \cap (\sqsubset \mathfrak{S}^\sigma))^\star && \text{by iii)} \\
&= (\varphi_\sigma | \mathfrak{C})^\star = \phi^\star && \text{on } \sqsubset \mathfrak{S}^\sigma.
\end{aligned}
$$

This implies both assertions.

Next we look at the Borel side. The remark below is a recollection of essentials from 6.31 with 6.18.

19.13. REMARK. Let $\mathfrak{T}$ be a lattice in X with $\mathfrak{T} \subset \mathrm{Cl}(X) \subset (\mathfrak{T} \sqsubset \mathfrak{T})$ (which implies $\varnothing \in \mathfrak{T}$). Then there is a one-to-one correspondence between the inner σ premeasures $\psi : \mathfrak{T} \to [0, \infty[$ with $\psi(\varnothing) = 0$ and the cmeasures $\beta : \mathrm{Bor}(X) \to [0, \infty]$ with $\beta | \mathfrak{T} < \infty$ which are inner regular $\mathfrak{T}_\sigma$. The correspondence is

$$\psi \mapsto \beta := \psi_\sigma | \mathrm{Bor}(X) \quad \text{and} \quad \beta \mapsto \psi := \beta | \mathfrak{T};$$

also recall that $\mathrm{Bor}(X) \subset \mathfrak{C}(\psi_\sigma)$.

We turn to the transition from the Baire side to the Borel side. Let as above $\alpha : \mathrm{Baire}(X) \to [0, \infty]$ be a cmeasure with $\mathfrak{S}$ and $\varphi := \alpha | \mathfrak{S}$. We define $\mathfrak{T} := \mathrm{Cl}(X) \cap (\sqsubset \mathfrak{S})$. Thus $\mathfrak{T}$ is a lattice in X with $\mathfrak{T} \subset \mathrm{Cl}(X) \subset (\mathfrak{T} \top \mathfrak{T})$ and $\mathfrak{T}_\sigma = \mathfrak{T}$ (and even with $\mathfrak{T}_\tau = \mathfrak{T}$). For these data then 19.13 furnishes a one-to-one correspondence between the inner σ premeasures $\psi : \mathfrak{T} \to [0, \infty[$ with $\psi | \mathfrak{S} = \varphi$ and the cmeasures $\beta : \mathrm{Bor}(X) \to [0, \infty]$ with $\beta = \alpha$ on $\mathfrak{S}$ which are inner regular $\mathfrak{T}$. We use this correspondence in order to obtain the desired extension theorem.

On the one hand we translate known properties of ψ into properties of β. From 19.3.1) we know that $\psi_\sigma | \mathfrak{C}(\psi_\sigma)$ is an extension of $\varphi_\sigma | \mathfrak{C}(\varphi_\sigma)$. On $\mathrm{Baire}(X)$ this means that $\beta = \varphi_\sigma = \varphi_\star$ by v) in 8.2.1). Therefore $\beta | \mathrm{Baire}(X)$ is inner regular $\mathfrak{S}$. It follows that $\beta | \mathrm{Baire}(X) = \alpha$ iff α is inner regular $\mathfrak{S}$, that is iff α is semifinite above. In all cases we have $\beta = \alpha$ on $[\alpha < \infty]^\sigma \subset \mathrm{Baire}(X)$ by 8.5.i).

On the other hand the existence theorem 19.8 and the uniqueness theorem 19.11 translate into existence and uniqueness results for the cmeasures

$\beta : \text{Bor}(X) \to [0, \infty]$. For an adequate formulation we note from 19.12 that $\phi^\star | \mathfrak{T} = \alpha^\star | \mathfrak{T}$. Thus we have proved what follows.

19.14. THEOREM. *Let* $\alpha : \text{Baire}(X) \to [0, \infty]$ *be a cmeasure with* $\mathfrak{S} := \{A \in \text{CCl}(X) : \alpha(A) < \infty\}$ *and* $\varphi := \alpha | \mathfrak{S}$. *Define* $\mathfrak{T} := \text{Cl}(X) \cap (\sqsubset \mathfrak{S})$.
0) *If* $\beta : \text{Bor}(X) \to [0, \infty]$ *is a cmeasure with*

  0.i) $\beta = \alpha$ *on* $\mathfrak{S}$,

  0.ii) β *is inner regular* $\mathfrak{T}$,

then it also fulfils

  0.iii) $\beta = \alpha$ *on* $[\alpha < \infty]^\sigma \subset \text{Baire}(X)$,

  0.iv) $\beta | \text{Baire}(X) = \alpha$ *iff* α *is semifinite above,*

  0.v) $\beta | \text{Baire}(X)$ *is inner regular* $\mathfrak{S}$.

1) *Assume that* $\alpha^\star | \mathfrak{T}$ *is* σ *continuous at* $\varnothing$. *Then*

1.i) *there exist cmeasures* $\beta : \text{Bor}(X) \to [0, \infty]$ *as in* 0).

1.ii) *There exists a unique cmeasure* $\beta : \text{Bor}(X) \to [0, \infty]$ *as in* 0) *iff* $\alpha^\star | \mathfrak{T}$ *is superadditive. In this case* $\beta = \alpha^\star$ *on* $\mathfrak{T}$.

Before we proceed we want to compare the present result with the former extension theorem 8.11. We fix a cmeasure $\alpha : \text{Baire}(X) \to [0, \infty]$ with $\mathfrak{S}$ and $\varphi := \alpha | \mathfrak{S}$ as in both theorems, and also assume α to be inner regular $\mathfrak{S}$ from the start as in 8.11. Then 8.11 imposes the two essential further assumptions

  1) X is completely regular, and

  2) φ is τ continuous at $\varnothing$.

From assumption 1) and 8.1.5) we see that $\mathfrak{T} = \mathfrak{S}_\tau$, which means that the central sublattices of $\text{Bor}(X)$ in the two theorems are equal. Then assumption 2) combined with 8.12 implies that $\alpha^\star | \mathfrak{T}$ is τ continuous at $\varnothing$. Thus with simple means we obtain much more than the initial assumption in the new 19.14.1), which implies its existence assertion. In adequate relation the existence result in 8.11.1) is much sharper. It is the explicit assertion that $\varphi : \mathfrak{S} \to [0, \infty[$ is an inner τ premeasure with $\mathfrak{C}(\varphi_\tau) \supset \text{Bor}(X)$, and that $\beta := \varphi_\tau | \text{Bor}(X)$ is an extension as required. Now let us turn to the uniqueness assertions. The explicit existence result in 8.11.1) implies that $\beta | \mathfrak{T} = \varphi_\tau | \mathfrak{S}_\tau$ is downward τ continuous. It follows that $\beta = \alpha^\star$ on $\mathfrak{T} = \mathfrak{S}_\tau$, and hence that $\alpha^\star | \mathfrak{T}$ is modular. Therefore we are in the case of uniqueness in 19.14.1). Thus in the situation of 8.11 the new assertion of uniqueness is equal to the old one.

19.15. EXERCISE. The last theorem has a remarkable consequence which does not involve regularity: *Let* $\alpha : \text{Baire}(X) \to [0, \infty]$ *be a cmeasure with* $\mathfrak{S}$ *and* $\mathfrak{T}$ *as above. Assume that* $\alpha^\star | \mathfrak{T}$ *is* σ *continuous at* $\varnothing$. *Then there exist cmeasures* $\beta : \text{Bor}(X) \to [0, \infty]$ *which extend* α. Hint: Combine the existence assertion in 19.14.1) with 2.13.2).

In the remainder of the subsection we consider properties of the topological space X which ensure that all Baire measures on X fulfil the conditions which occur in 19.14.1). This will lead to earlier forms of extension theorems.

The two definitions which follow are for lattices $\mathfrak{S}$ and $\mathfrak{T}$ with $\varnothing \in \mathfrak{S}, \mathfrak{T}$ in a nonvoid set X. For $\bullet = \sigma\tau$ one defines on the one hand $\mathfrak{T}$ to be $\bullet$ **dominated** $\mathfrak{S}$ iff for each paving $\mathfrak{N} \subset \mathfrak{T}$ of type $\bullet$ with $\mathfrak{N} \downarrow \varnothing$ there exists a paving $\mathfrak{M} \subset \mathfrak{S}$ of type $\bullet$ with $\mathfrak{M} \downarrow \varnothing$ such that $\mathfrak{M} \subset (\sqsupset \mathfrak{N})$. On the other hand one defines $\mathfrak{T}$ to be **separated** $\mathfrak{S}$ iff each pair of subsets $A, B \in \mathfrak{T}$ with $A \cap B = \varnothing$ is separated $\mathfrak{S}$ in the sense of 4.2, that is there are $S, T \in \mathfrak{S}$ with $A \subset S$ and $B \subset T$ such that $S \cap T = \varnothing$.

19.16. REMARK. 1) $\mathfrak{S}_\bullet$ *is* $\bullet$ *dominated* $\mathfrak{S}$ *whenever* $\mathfrak{S}$ *is a lattice with* $\varnothing \in \mathfrak{S}$. This is a special case of 18.5. 2) $\mathfrak{T}$ *is* σ *dominated* $\mathfrak{S} \Leftrightarrow$ *for each sequence* $(T_l)_l$ *in* $\mathfrak{T}$ *with* $T_l \downarrow \varnothing$ *there is a sequence* $(S_l)_{l \geq k}$ *in* $\mathfrak{S}$ *for some* $k \geq 1$ *such that* $S_l \downarrow \varnothing$ *and* $T_l \subset S_l$ *for* $l \geq k$.

Proof of 19.16.2). $\Rightarrow$) By assumption there exists a countable paving $\mathfrak{M} \subset \mathfrak{S}$ with $\mathfrak{M} \downarrow \varnothing$ such that each $M \in \mathfrak{M}$ contains some T_l. Let $(M_k)_k$ be a sequence in $\mathfrak{M}$ with $M_k \downarrow \varnothing$. We can assume that $M_k \supset T_{l(k)}$ with $1 \leq l(1) < \cdots < l(k) < \cdots$. Now define $S_l := M_k$ for $l(k) \leq l < l(k+1)$ $\forall k \geq 1$. Then the sequence $(S_l)_{l \geq l(1)}$ is as required. $\Leftarrow$) Let $\mathfrak{N} \subset \mathfrak{T}$ be a countable paving with $\mathfrak{N} \downarrow \varnothing$. Fix a sequence $(T_l)_l$ in $\mathfrak{N}$ with $T_l \downarrow \varnothing$, and let $(S_l)_{l \geq k}$ in $\mathfrak{S}$ be as in the assumption. Then the paving $\mathfrak{M} := \{S_l : l \geq k\} \subset \mathfrak{S}$ is as required.

19.17. REMARK. *Let* X *be a topological space.* 1) *Assume that* $\mathrm{Cl}(X)$ *is* σ *dominated* $\mathrm{Baire}(X)$. *For each cmeasure* $\alpha : \mathrm{Baire}(X) \to [0, \infty]$ *then* $\alpha^\star | \mathfrak{T}$ *is* σ *continuous at* $\varnothing$. 2) *Assume that* $\mathrm{Cl}(X)$ *is separated* $\mathrm{Baire}(X)$. *For each cmeasure* $\alpha : \mathrm{Baire}(X) \to [0, \infty]$ *then* $\alpha^\star | \mathfrak{T}$ *is subadditive.* Of course $\mathfrak{T} \subset \mathrm{Cl}(X)$ is as defined in 19.14.

Proof. 1) Let $(T_l)_l$ be a sequence in $\mathfrak{T}$ with $T_l \downarrow \varnothing$. Fix a sequence $(S_l)_{l \geq k}$ in $\mathrm{Baire}(X)$ such that $S_l \downarrow \varnothing$ and $T_l \subset S_l$ for $l \geq k$. By the definition of $\mathfrak{T}$ we can assume that $\alpha(S_l) < \infty$ for $l \geq k$. Thus $\alpha(S_l) \downarrow 0$. From $0 \leq \alpha^\star(T_l) \leq \alpha(S_l)$ it follows that $\alpha^\star(T_l) \downarrow 0$. 2) is an immediate consequence of 4.3 applied to $\alpha : \mathrm{Baire}(X) \to [0, \infty]$.

19.18. THEOREM. *If* $\mathrm{Cl}(X)$ *is* σ *dominated* $\mathrm{Baire}(X)$ *then for each cmeasure* $\alpha : \mathrm{Baire}(X) \to [0, \infty]$ *there exists a cmeasure* $\beta : \mathrm{Bor}(X) \to [0, \infty]$ *such that*

19.14.0.i) $\beta = \alpha$ *on* $\mathfrak{S} := \{A \in \mathrm{CCl}(X) : \alpha(A) < \infty\}$,

19.14.0.ii) β *is inner regular* $\mathfrak{T} := \mathrm{Cl}(X) \cap (\sqsubset \mathfrak{S})$.

If in addition $\mathrm{Cl}(X)$ *is separated* $\mathrm{Baire}(X)$ *then* β *is unique, and* $\beta = \alpha^\star$ *on* $\mathfrak{T}$.

There are well-known sufficient properties of the topological space X which are in direct terms of $\mathrm{Op}(X)$ and $\mathrm{Cl}(X)$. Let us define X to be σ **paracompact** iff $\mathrm{Cl}(X)$ is σ dominated $\mathrm{Op}(X)$. For a normal Hausdorff space this is equivalent to countable paracompactness in the usual sense; see Engelking [1989] corollary 5.2.2. Recall that X is defined to be **normal** iff $\mathrm{Cl}(X)$ is separated $\mathrm{Op}(X)$. From these definitions and from 8.1.6) it follows that

i) if X is σ paracompact and normal then $\mathrm{Cl}(X)$ is σ dominated $\mathrm{COp}(X)$ and $\mathrm{CCl}(X)$, and hence in particular σ dominated $\mathrm{Baire}(X)$;

ii) if X is normal then $\mathrm{Cl}(X)$ is separated $\mathrm{COp}(X)$ and $\mathrm{CCl}(X)$, and hence in particular separated $\mathrm{Baire}(X)$.

Therefore 19.18 implies the final result which follows.

19.19. CONSEQUENCE. *Assume that X is σ paracompact and normal. Then for each cmeasure $\alpha : \mathrm{Baire}(X) \to [0, \infty]$ there exists a unique cmeasure $\beta : \mathrm{Bor}(X) \to [0, \infty]$ such that*

19.14.0.i) $\beta = \alpha$ *on* $\mathfrak{S} := \{A \in \mathrm{CCl}(X) : \alpha(A) < \infty\}$,

19.14.0.ii) β *is inner regular* $\mathfrak{T} := \mathrm{Cl}(X) \cap (\sqsubset \mathfrak{S})$.

We have $\beta = \alpha^\star$ *on* $\mathfrak{T}$.

19.20. BIBLIOGRAPHICAL NOTE. 19.18 is in Adamski [1984a] theorem 3.14, and likewise the respective special case of 19.15. The author has not seen the complete 19.14 in the literature.

The ancestor of all extension theorems of the present kind is the famous result due to Mařík [1957]. It is under the same assumptions as 19.19, but supposes the cmeasure $\alpha : \mathrm{Baire}(X) \to [0, \infty]$ to be outer regular $\mathrm{COp}(X)$. It obtains a unique Borel measure extension $\beta : \mathrm{Bor}(X) \to [0, \infty]$ of α which is characterized in the spirit of the two-step extension method as described in the bibliographical annex to chapter II. The same applies to the related result in Sapounakis-Sion [1987] theorem 7.1.

CHAPTER VII

Products of Contents and Measures

The present chapter develops the product formation for contents and measures in the spirit of chapter II. We also use the extended integration procedures of chapter IV. These means will be adequate for a comprehensive treatment. The central part is the second section which in particular contains the Radon product measure of Radon measures. We restrict ourselves to the case of two factors.

20. The Traditional Product Formations

The first subsection uses the horizontal integral of section 11 to define a product formation on which all subsequent ones will be based. The remainder of the section will be devoted to the traditional product theory, in the sense that the factors are ccontents and cmeasures. In the latter case it is well-known that there need not be a unique product measure except under countable finiteness of the factors. However, we shall obtain natural uniqueness assertions in terms of regularity.

The Basic Product Formation

Let X and Y be nonvoid sets. For a subset $E \subset X \times Y$ we define the **vertical sections**

$$E(x) := \{y \in Y : (x, y) \in E\} \quad \text{for } x \in X,$$

and the **vertical projection**

$$\Pr(E) := \{y \in Y : (x, y) \in E \text{ for some } x \in X\} = \bigcup_{x \in X} E(x) \subset Y.$$

Of course one likewise forms the **horizontal sections** $E[y] \subset X$ for $y \in Y$ and the **horizonal projection** $\Pr[E] \subset X$. We list some properties which are all obvious.

20.1. PROPERTIES. 1) *For $A \subset X$ and $B \subset Y$ we have*

$$(A \times B)(x) = \left\{ \begin{array}{ll} B & \text{when } x \in A \\ \varnothing & \text{when } x \in A' \end{array} \right\},$$

$$\Pr(A \times B) = \left\{ \begin{array}{ll} B & \text{when } A \neq \varnothing \\ \varnothing & \text{when } A = \varnothing \end{array} \right\}.$$

2) *For $E \subset X \times Y$ and $x \in X$ we have $E'(x) = (E(x))'$. 3) Let $(E_t)_{t \in I}$ be a family of subsets of $X \times Y$. For all $x \in X$ then*

$$\left(\bigcup_{t \in I} E_t \right)(x) = \bigcup_{t \in I} E_t(x) \quad and \quad \left(\bigcap_{t \in I} E_t \right)(x) = \bigcap_{t \in I} E_t(x).$$

In section 13 we defined for pavings $\mathfrak{S}$ in X and $\mathfrak{T}$ in Y the **product paving**

$$\mathfrak{S} \times \mathfrak{T} := \{ S \times T : S \in \mathfrak{S} \text{ and } T \in \mathfrak{T} \} \quad \text{in } X \times Y.$$

We list some properties.

20.2. PROPERTIES. 1) *If $\mathfrak{S}$ and $\mathfrak{T}$ are*

$$\begin{aligned} lattices \text{ then } & \mathrm{L}(\mathfrak{S} \times \mathfrak{T}) = (\mathfrak{S} \times \mathfrak{T})^{\star}, \\ rings \text{ then } & \mathrm{R}(\mathfrak{S} \times \mathfrak{T}) = (\mathfrak{S} \times \mathfrak{T})^{\star}, \\ algebras \text{ then } & \mathrm{A}(\mathfrak{S} \times \mathfrak{T}) = (\mathfrak{S} \times \mathfrak{T})^{\star}. \end{aligned}$$

2) *If $\mathfrak{S}$ is a ring and $\mathfrak{T}$ is a lattice then each $E \in (\mathfrak{S} \times \mathfrak{T})^{\star}$ can be represented in the form*

$$E = \bigcup_{l=1}^{r} A_l \times B_l \text{ with } A_1, \cdots, A_r \in \mathfrak{S} \text{ pairwise disjoint and } B_1, \cdots, B_r \in \mathfrak{T}.$$

Proof. 1) The lattice case follows from 1.2.10). To see the ring case one verifies for $A, S \subset X$ and $B, T \subset Y$ that

$$(A \times B) \cap (S \times T)' = (A \times (B \cap T')) \cup ((A \cap S') \times B).$$

The algebra case is then clear. 2) has an obvious proof via 3.5.

20.3. REMARK. *Let $\mathfrak{S}$ and $\mathfrak{T}$ be lattices with $\varnothing$. For $E \in (\mathfrak{S} \times \mathfrak{T})^{\star}$ then*

1) $E(x) \in \mathfrak{T}$ *for all $x \in X$.*

2) *Assume that $\psi : \mathfrak{T} \to [0, \infty]$ is isotone with $\psi(\varnothing) = 0$. Then the function $\psi(E(\cdot)) : X \to [0, \infty]$ has a finite value set and is in $\mathrm{UM}(\mathfrak{S}) \cap \mathrm{LM}(\mathfrak{S})$.*

Proof. Let $E = \bigcup_{l=1}^{r} A_l \times B_l$ with $A_1, \cdots, A_r \in \mathfrak{S}$ and $B_1, \cdots, B_r \in \mathfrak{T}$. For $x \in X$ one verifies that

$$E(x) = \{ y \in Y : (x, y) \in E = \bigcup_{l=1}^{r} A_l \times B_l \} = \bigcup_{l : x \in A_l} B_l.$$

It follows that $E(x) \in \mathfrak{T}$, and the first assertion in 2). The proof of the second assertion in 2) can be restricted to $\mathrm{UM}(\mathfrak{S})$. Fix $t > 0$. For $x \in X$ then

$$\psi(E(x)) = \psi\left(\bigcup_{l : x \in A_l} B_l \right) \geqq t$$

$$\Leftrightarrow \quad \exists \text{ nonvoid } D \subset \{ 1, \cdots, r \} \text{ such that } x \in \bigcap_{l \in D} A_l \text{ and } \psi\left(\bigcup_{l \in D} B_l \right) \geqq t.$$

Therefore

$$[\psi(E(\cdot)) \geqq t] = \bigcup_D \bigcap_{l \in D} A_l$$

over the nonvoid $D \subset \{1, \cdots, r\}$ with $\psi(\bigcup_{l \in D} B_l) \geqq t$.

It follows that $[\psi(E(\cdot)) \geqq t] \in \mathfrak{S}$ and hence the assertion.

We come to the basic product formation. The definition makes sense in virtue of 20.3.

20.4. PROPOSITION. *Let $\mathfrak{S}$ and $\mathfrak{T}$ be lattices with $\varnothing$, and*

$$\varphi : \mathfrak{S} \to [0, \infty] \text{ be isotone with } \varphi(\varnothing) = 0,$$
$$\psi : \mathfrak{T} \to [0, \infty] \text{ be isotone with } \psi(\varnothing) = 0.$$

We define the set function $\varphi \times \psi : (\mathfrak{S} \times \mathfrak{T})^\star \to [0, \infty]$ to be

$$(\varphi \times \psi)(E) = \int \psi(E(\cdot)) d\varphi \quad \text{for } E \in (\mathfrak{S} \times \mathfrak{T})^\star.$$

Then $\varphi \times \psi =: \vartheta$ has the properties

1) ϑ *is isotone.*

2) $\vartheta(A \times B) = \varphi(A)\psi(B)$ *for $A \in \mathfrak{S}$ and $B \in \mathfrak{T}$, with the usual convention* $0\infty := 0$. *In particular $\vartheta(\varnothing) = 0$. Furthermore $\vartheta < \infty$ when $\varphi, \psi < \infty$.*

3) *If φ and ψ are modular then ϑ is modular as well.*

Proof. 1) is obvious. 2) Let $E := A \times B$. From 20.1.1) we have

$$\psi(E(x)) = \left\{ \begin{array}{ll} \psi(B) & \text{when } x \in A \\ 0 & \text{when } x \in A' \end{array} \right\}.$$

Therefore

$$\vartheta(E) = \int_{0\leftarrow}^{\to\infty} \varphi\Big([\psi(E(\cdot)) \geqq t]\Big) dt$$

$$\text{with } [\psi(E(\cdot)) \geqq t] = \left\{ \begin{array}{ll} A & \text{when } \psi(B) \geqq t \\ \varnothing & \text{when } \psi(B) < t \end{array} \right\}.$$

It follows that $\vartheta(E) = \varphi(A)\psi(B)$ in all cases.

3) Let $E, F \in (\mathfrak{S} \times \mathfrak{T})^\star$. i) We first deduce from 20.3.2) that $\psi(E(\cdot)) + \psi(F(\cdot))$ is in $\mathrm{UM}(\mathfrak{S})$. In fact, let $W \subset [0, \infty]$ be a finite set which contains the value sets of $\psi(E(\cdot))$ and $\psi(F(\cdot))$. For $t > 0$ then

$$[\psi(E(\cdot)) + \psi(F(\cdot)) \geqq t] = \{x \in X : \psi(E(x)) + \psi(F(x)) \geqq t\}$$

$$= \bigcup_{u,v \in W \text{ with } u+v \geqq t} \{x \in X : \psi(E(x)) \geqq u\} \cap \{x \in X : \psi(F(x)) \geqq v\}$$

$$= \bigcup_{u,v \in W \text{ with } u+v \geqq t} [\psi(E(\cdot)) \geqq u] \cap [\psi(F(\cdot)) \geqq v].$$

This union is in $\mathfrak{S}$ since each time at least one of the numbers u and v is > 0. ii) Since ψ is modular we see from 20.1.3) that

$$\psi\big((E\cup F)(\cdot)\big) + \psi\big((E\cap F)(\cdot)\big) \;=\; \psi\big(E(\cdot)\cup F(\cdot)\big) + \psi\big(E(\cdot)\cap F(\cdot)\big)$$
$$= \; \psi\big(E(\cdot)\big) + \psi\big(F(\cdot)\big).$$

Since φ is modular the assertion follows from 11.11 and i).

20.5. PROPOSITION. *Let $\mathfrak{S}$ and $\mathfrak{T}$ be lattices with $\varnothing$, and*

$$\varphi : \mathfrak{S} \to [0,\infty] \text{ *be isotone and modular with* } \varphi(\varnothing) = 0,$$
$$\psi : \mathfrak{T} \to [0,\infty] \text{ *be isotone and modular with* } \psi(\varnothing) = 0.$$

Then there exists a unique set function $\vartheta : (\mathfrak{S} \times \mathfrak{T})^\star \to [0,\infty]$ which is isotone and modular and fulfils

$$\vartheta(A \times B) = \varphi(A)\psi(B) \quad \text{*for all* } A \in \mathfrak{S} \text{ *and* } B \in \mathfrak{T}.$$

This is $\vartheta = \varphi \times \psi$.

Proof. Assume that $\vartheta : (\mathfrak{S} \times \mathfrak{T})^\star \to [0,\infty]$ is as formulated above. To be shown is that ϑ equals $\varphi \times \psi$. Fix

$$E = \bigcup_{l=1}^{r} A_l \times B_l \quad \text{with } A_1, \cdots, A_r \in \mathfrak{S} \text{ and } B_1, \cdots, B_r \in \mathfrak{T}.$$

i) Assume that $\varphi(A_l)\psi(B_l) = \infty$ for some $l = 1, \cdots, r$. Then $\vartheta(A_l \times B_l) = \infty = (\varphi \times \psi)(A_l \times B_l)$ and hence $\vartheta(E) = \infty = (\varphi \times \psi)(E)$. ii) Assume now that $\varphi(A_l)\psi(B_l) < \infty$ and hence $\vartheta(A_l \times B_l) = \varphi(A_l)\psi(B_l) = (\varphi \times \psi)(A_l \times B_l) < \infty$ for all $l = 1, \cdots, r$. Then $\vartheta(E), (\varphi \times \psi)(E) < \infty$. From 2.5.1) applied to the restrictions of ϑ and $\varphi \times \psi$ to the lattice $\{M \in (\mathfrak{S} \times \mathfrak{T})^\star : M \subset E\}$ we obtain $\vartheta(E) = (\varphi \times \psi)(E)$.

20.6. CONSEQUENCE. *The set function $\varphi \times \psi : (\mathfrak{S} \times \mathfrak{T})^\star \to [0,\infty]$ has the symmetrized representation*

$$(\varphi \times \psi)(E) = \int\!\!\!\!\!\!-\,\varphi\big(E[\cdot]\big)\,d\psi \quad \text{*for* } E \in (\mathfrak{S} \times \mathfrak{T})^\star.$$

We deduce from 11.15 that the basic product formation is compatible with restrictions.

20.7. REMARK. *Let $\mathfrak{S} \subset \mathfrak{A}$ and $\mathfrak{T} \subset \mathfrak{B}$ be lattices with $\varnothing$, and*

$$\alpha : \mathfrak{A} \to [0,\infty] \text{ *be isotone with* } \alpha(\varnothing) = 0,$$
$$\beta : \mathfrak{B} \to [0,\infty] \text{ *be isotone with* } \beta(\varnothing) = 0.$$

Then the restrictions $\varphi := \alpha|\mathfrak{S}$ and $\psi := \beta|\mathfrak{T}$ fulfil

$$\varphi \times \psi = (\alpha \times \beta)|(\mathfrak{S} \times \mathfrak{T})^\star.$$

Proof. Let $E \in (\mathfrak{S} \times \mathfrak{T})^\star \subset (\mathfrak{A} \times \mathfrak{B})^\star$. For $x \in X$ then $E(x) \in \mathfrak{T} \subset \mathfrak{B}$, and the function $\beta\big(E(\cdot)\big) = \psi\big(E(\cdot)\big)$ is in UM($\mathfrak{S}$) $\subset$ UM($\mathfrak{A}$). From 11.15 we obtain

$$(\alpha \times \beta)(E) = \int\!\!\!\!\!\!-\,\beta\big(E(\cdot)\big)\,d\alpha = \int\!\!\!\!\!\!-\,\psi\big(E(\cdot)\big)\,d\alpha = \int\!\!\!\!\!\!-\,\psi\big(E(\cdot)\big)\,d\varphi = (\varphi \times \psi)(E).$$

We conclude with a first look at the particular case which will be met in the remainder of the present section.

20.8. PROPOSITION. *Assume that*

$$\varphi : \mathfrak{S} \to [0, \infty] \ \text{is a ccontent on a ring } \mathfrak{S} \text{ in } X,$$
$$\psi : \mathfrak{T} \to [0, \infty] \ \text{is a ccontent on a ring } \mathfrak{T} \text{ in } Y,$$

so that $\varphi \times \psi$ *is a ccontent on the ring* $(\mathfrak{S} \times \mathfrak{T})^\star$. *Then*

$$(\varphi \times \psi)(E) = \int_* \psi(E(\cdot)) \, d\varphi \quad \text{for } E \in (\mathfrak{S} \times \mathfrak{T})^\star.$$

If φ *and* ψ *are upward* σ *continuous then* $\varphi \times \psi$ *is upward* σ *continuous as well.*

Proof. The first assertion follows from 12.11, and the second one from 12.10. In fact, let $(E_l)_l$ be a sequence in $(\mathfrak{S} \times \mathfrak{T})^\star$ with $E_l \uparrow E \in (\mathfrak{S} \times \mathfrak{T})^\star$. By 20.1.3) then $E_l(x) \uparrow E(x)$ for $x \in X$. Since ψ is upward σ continuous we have $\psi(E_l(\cdot)) \uparrow \psi(E(\cdot))$ pointwise on X. Since φ is upward σ continuous we obtain $(\varphi \times \psi)(E_l) \uparrow (\varphi \times \psi)(E)$ from 12.10.

The Traditional Product Situation

For the remainder of the section we assume that

$$\alpha : \mathfrak{A} \to [0, \infty] \text{ is a ccontent on an algebra } \mathfrak{A} \text{ in } X,$$
$$\beta : \mathfrak{B} \to [0, \infty] \text{ is a ccontent on an algebra } \mathfrak{B} \text{ in } Y.$$

From the above subsection we obtain the **product content**

$$\Delta := \alpha \times \beta \quad \text{on the algebra } \mathfrak{P} := (\mathfrak{A} \times \mathfrak{B})^\star \text{ in } X \times Y.$$

We also consider the finite restrictions of α and β to the rings $\mathfrak{a} := [\alpha < \infty]$ and $\mathfrak{b} := [\beta < \infty]$. They produce the finite **narrow product content**

$$\delta := (\alpha|\mathfrak{a}) \times (\beta|\mathfrak{b}) \quad \text{on the ring } \mathfrak{p} := (\mathfrak{a} \times \mathfrak{b})^\star \text{ in } X \times Y.$$

We have $\delta = \Delta|\mathfrak{p}$ by 20.7. If α and β are upward σ continuous then Δ and δ are upward σ continuous as well by 20.8.

Since $\mathfrak{P}$ and $\mathfrak{p}$ are rings it is clear that

$$\Delta \text{ is outer } \star \text{ tight and hence outer } \sigma \text{ tight,}$$
$$\delta \text{ is outer } \star \text{ tight and hence outer } \sigma \text{ tight,}$$
$$\delta \text{ is inner } \star \text{ tight and hence inner } \sigma \text{ tight,}$$

while the cases $\bullet = \tau$ are not realistic. Our main concerns are the outer σ extensions of Δ and hence the envelope Δ^σ, and the inner σ extensions of δ and hence the envelope δ_σ. We prepare the next subsection with some relevant facts.

20.9. LEMMA. *Let* $E \in \mathfrak{P}$ *with* $\Delta(E) < \infty$. *Then there exist* $K \in \mathfrak{p}$ *and null sets* $M \in [\alpha = 0]$ *and* $N \in [\beta = 0]$ *such that*

$$K \subset E \subset K \cup (X \times N) \cup (M \times Y).$$

Therefore we have

$$K \subset E \subset K \cup F \quad \text{for some } F \in \mathfrak{P} \text{ with } \Delta(F) = 0.$$

Note that this implies $\Delta(E) = \Delta(K) = \delta(K)$.

Proof. Let $E \in \mathfrak{P}$ be represented as in 20.2.2). Thus

$$\Delta(E) = \sum_{l=1}^{r} \alpha(A_l)\beta(B_l) < \infty.$$

It is obvious that the subsets

$$K := \bigcup_{l:\alpha(A_l),\beta(B_l)<\infty} A_l \times B_l, \quad M := \bigcup_{l:\beta(B_l)=\infty} A_l \text{ and } N := \bigcup_{l:\alpha(A_l)=\infty} B_l$$

are as required.

20.10. LEMMA. $\star$) $\delta_\star|\mathfrak{P} \leqq \Delta$. *If α and β are semifinite above then* $\delta_\star|\mathfrak{P} = \Delta$. σ) *Assume that α and β are upward σ continuous. Then* $\delta_\sigma|\mathfrak{P} \leqq \Delta$. *If α and β are semifinite above then* $\delta_\sigma|\mathfrak{P} = \Delta$.

Proof of $\star$). i) We have $\delta_\star = (\Delta|\mathfrak{p})_\star \leqq \Delta_\star$ and hence $\delta_\star|\mathfrak{P} \leqq \Delta$. ii) Let $E \in \mathfrak{P}$ with $\Delta(E) < \infty$. Then 20.9 shows that $\Delta(E) = \Delta(K) = \delta(K) \leqq \delta_\star(E)$. iii) Assume that α and β are semifinite above. It remains to prove for $E \in \mathfrak{P}$ with $\Delta(E) = \infty$ that $\delta_\star(E) = \infty$. There exist $A \in \mathfrak{A}$ and $B \in \mathfrak{B}$ with $A \times B \subset E$ and $\alpha(A)\beta(B) = \infty$. In all cases there are increasing sequences

$(A_l)_l$ in $\mathfrak{A}$ with $A_l \subset A$ such that $0 < \alpha(A_l) < \infty$ and $\alpha(A_l) \uparrow \alpha(A)$,
$(B_l)_l$ in $\mathfrak{B}$ with $B_l \subset B$ such that $0 < \beta(B_l) < \infty$ and $\beta(B_l) \uparrow \beta(B)$.

Therefore $(A_l \times B_l)_l$ is an increasing sequence in $\mathfrak{p}$ with $A_l \times B_l \subset A \times B \subset E$ such that $\delta(A_l \times B_l) = \alpha(A_l)\beta(B_l) \uparrow \infty$. The assertion follows.

Proof of σ). i) We fix $P \in \mathfrak{P}$. Let $(E_l)_l$ be a sequence in $\mathfrak{p}$ with $E_l \downarrow E \subset P$. Then $(E_l \cup (E_1 \cap P))_l$ is a sequence in $\mathfrak{p}$ with $E_l \cup (E_1 \cap P) \downarrow E \cup (E_1 \cap P) = E_1 \cap P \in \mathfrak{p}$. Since δ is downward σ continuous it follows that

$$\delta(E_l) \leqq \delta(E_l \cup (E_1 \cap P)) \downarrow \delta(E_1 \cap P) \leqq \Delta(P).$$

Therefore $\delta_\sigma(P) \leqq \Delta(P)$. ii) Assume that α and β are semifinite above. Then $\star$) and i) combine to $\Delta = \delta_\star|\mathfrak{P} \leqq \delta_\sigma|\mathfrak{P} \leqq \Delta$. Therefore $\delta_\sigma|\mathfrak{P} = \Delta$.

Product Measures

In the present subsection we assume that the ccontents $\alpha : \mathfrak{A} \to [0, \infty]$ and $\beta : \mathfrak{B} \to [0, \infty]$ are upward σ continuous. From section 13 we recall the notation $\mathfrak{A} \otimes \mathfrak{B} := A\sigma(\mathfrak{A} \times \mathfrak{B})$. We define a cmeasure $\nu : \mathfrak{A} \otimes \mathfrak{B} \to [0, \infty]$ to be a **product measure** of α and β iff

$$\nu(A \times B) = \alpha(A)\beta(B) \text{ for all } A \in \mathfrak{A} \text{ and } B \in \mathfrak{B}, \text{ that is iff } \nu|\mathfrak{P} = \Delta,$$

and a **narrow product measure** of α and β iff

$$\nu(A \times B) = \alpha(A)\beta(B) \text{ for all } A \in \mathfrak{a} \text{ and } B \in \mathfrak{b}, \text{ that is iff } \nu|\mathfrak{p} = \delta,$$

where the two equivalences follow from 20.2.2) or 20.5. We shall obtain such cmeasures from the simplest specializations of the extension theories of chapter II.

We know that the product content $\Delta : \mathfrak{P} \to [0, \infty]$ is an outer σ premeasure. Of course $\mathfrak{A} \otimes \mathfrak{B} \subset \mathfrak{C}(\Delta^\sigma)$. It follows that $\Delta^\sigma | \mathfrak{A} \otimes \mathfrak{B}$ is a product measure of α and β, and the unique one which is an outer σ extension of Δ. Now a product measure $\nu : \mathfrak{A} \otimes \mathfrak{B} \to [0, \infty]$ of α and β is an outer σ extension of Δ iff it is outer regular $\mathfrak{P}^\sigma$, by the basic definition of section 4. The cases $\delta : \mathfrak{p} \to [0, \infty[$ are alike, except that one has to note that $\mathfrak{P} \subset \mathfrak{p} \top \mathfrak{p}$. Combined with 20.10.$\sigma$) we obtain the theorem which follows.

20.11. THEOREM. *Assume that α and β are upward σ continuous.*

1) $\Delta^\sigma | \mathfrak{A} \otimes \mathfrak{B}$ *is the unique product measure of α and β which is outer regular* $\mathfrak{P}^\sigma$.

2) $\delta^\sigma | \mathfrak{A} \otimes \mathfrak{B}$ *is the unique narrow product measure of α and β which is outer regular* $\mathfrak{p}^\sigma$.

3) $\delta_\sigma | \mathfrak{A} \otimes \mathfrak{B}$ *is the unique narrow product measure of α and β which is inner regular* $\mathfrak{p}_\sigma$. *If α and β are semifinite above then it is a product measure of α and β.*

We continue with the classical uniqueness theorem.

20.12. THEOREM. *The narrow product measures of α and β are all equal on those subsets $E \in \mathfrak{A} \otimes \mathfrak{B}$ which are upward enclosable $\mathfrak{p}^\sigma$, that is upward enclosable $\mathfrak{a}^\sigma \times \mathfrak{b}^\sigma$. In particular if*

$$X \in \mathfrak{a}^\sigma \text{ and } Y \in \mathfrak{b}^\sigma,$$

then there is a unique narrow product measure of α and β.

We present two proofs. The first proof is within the present extension theories. The second proof is close to the usual one via the classical uniqueness theorem 3.1.σ).

First proof. Let $\nu : \mathfrak{A} \otimes \mathfrak{B} \to [0, \infty]$ be a narrow product measure of α and β. From 7.1.σ) applied to δ and ν we obtain $\delta_\sigma \leqq \nu \leqq \delta^\sigma$ on $\mathfrak{A} \otimes \mathfrak{B}$. Then 7.5 applied to δ furnishes $\delta_\sigma(E) = \nu(E) = \delta^\sigma(E)$ for all $E \in \mathfrak{A} \otimes \mathfrak{B}$ upward enclosable $\mathfrak{p}^\sigma$.

Second proof. In view of 3.1.σ) all narrow product measures of α and β coincide on $\mathrm{R}\sigma(\mathfrak{p}) \subset \mathfrak{A} \otimes \mathfrak{B}$. From 1.19.$\sigma$) and $\mathfrak{p} \subset \mathfrak{P} \subset \mathfrak{p} \top \mathfrak{p}$ we see that $\mathrm{A}\sigma(\mathfrak{p}) = \{E \in \mathfrak{A} \otimes \mathfrak{B} : E \text{ or } E' \text{ in } \sqsubset \mathfrak{p}^\sigma\}$, and hence from 1.17.$\sigma$) that $\mathrm{R}\sigma(\mathfrak{p}) = \{E \in \mathfrak{A} \otimes \mathfrak{B} : E \text{ in } \sqsubset \mathfrak{p}^\sigma\}$. The assertion follows.

The first example below will show that α and β can have more than one product measure, even when one of them is finite and the other one is semifinite above. The second example will show that under the same circumstances a narrow product measure need not be a product measure of α and β.

20.13. EXAMPLE. Let $X = Y$ be the unit interval $I = [0, 1]$ and $\mathfrak{A} = \mathfrak{B} = \mathrm{Bor}(I)$. We know from 13.15 that $\mathfrak{A} \otimes \mathfrak{B} = \mathrm{Bor}(I \times I)$. Let $\alpha : \mathrm{Bor}(I) \to [0, \infty[$ be the Borel-Lebesgue measure and $\beta : \mathrm{Bor}(I) \to [0, \infty]$ defined to be

$$\beta(B) = \sum_{x \in B} G(x) \quad \text{for } B \in \mathrm{Bor}(I),$$

where $G : I \to [0, \infty[$ is some prescribed function. By 20.11.1)3) both $\Delta^\sigma | \mathrm{Bor}(I \times I)$ and $\delta_\sigma | \mathrm{Bor}(I \times I)$ are product measures of α and β. For the present purpose we assume that $G \geqq$ some $\varepsilon > 0$. We claim that then $\Delta^\sigma(D) = \infty$ and $\delta_\sigma(D) = 0$ for the diagonal

$$D := \{(x, x) : x \in I\} \in \mathrm{Cl}(I \times I) \subset \mathrm{Bor}(I \times I).$$

Therefore the two product measures are different.

Proof. i) The subsets $E \in \mathfrak{p}$ are of the form

$$E = \bigcup_{l=1}^{r} A_l \times \{b_l\} \quad \text{with } A_1, \cdots, A_r \in \mathrm{Bor}(I) \text{ and } b_1, \cdots, b_r \in I.$$

Therefore $\mathfrak{p}_\sigma = \mathfrak{p}$, and hence $\delta_\sigma = \delta_\star$ by 6.5.iv). In case $E \subset D$ we have in the above representation $A_l \subset \{b_l\}$ $(l = 1, \cdots, r)$, and hence $\delta(E) = 0$. It follows that $\delta_\sigma(D) = \delta_\star(D) = 0$. ii) Let $E \in \mathfrak{P}$ with $\Delta(E) < \infty$. We use 20.9 to obtain $E \subset K \cup (I \times N) \cup (M \times I)$ with

$K \in \mathfrak{p}$ and hence from i) $K \subset I \times F$ with finite $F \subset I$,

$M \in [\alpha = 0]$, and $N \in [\beta = 0]$ and hence $N = \varnothing$.

Thus $E \subset (I \times F) \cup (M \times I)$. iii) Assume now that $\Delta^\sigma(D) < \infty$. Then there exists a sequence $(E_l)_l$ in $\mathfrak{P}$ with all $\Delta(E_l) < \infty$ such that $E_l \uparrow \supset D$. Thus ii) implies that $D \subset (I \times N) \cup (M \times I)$ with countable $N \subset I$ and $M \in [\alpha = 0]$, and hence $I \subset M \cup N$. This is a contradiction which proves that $\Delta^\sigma(D) = \infty$.

20.14. EXAMPLE. We know that $\nu := \delta^\sigma | \mathfrak{A} \otimes \mathfrak{B}$ is a narrow product measure of α and β. We claim that in the above example it is not a product measure. In fact, if $A \in \mathfrak{A}$ is nonvoid and $B \in \mathfrak{B}$ is not upward enclosable $\mathfrak{b}^\sigma$ then $A \times B$ is not upward enclosable $\mathfrak{p}^\sigma$ and hence $\nu(A \times B) = \delta^\sigma(A \times B) = \infty$. Therefore in case $\alpha(A) = 0$ we do not have $\nu(A \times B) = \alpha(A)\beta(B)$.

The last point in the subsection will be to extend the sectional representation from the product contents to the product measures. We shall need the next result, whose second part will be deduced from the transporter theorem 1.16.σ).

20.15. PROPOSITION. Assume that $\mathfrak{A}$ and $\mathfrak{B}$ are σ algebras. 0) If $E \in \mathfrak{A} \otimes \mathfrak{B}$ then $E(x) \in \mathfrak{B}$ for all $x \in X$. i) Assume that $E \in \mathfrak{A} \otimes \mathfrak{B}$ has $\mathrm{Pr}(E) \subset Y$ upward enclosable $\mathfrak{b}^\sigma$. Then the function $\beta(E(\cdot)) : X \to [0, \infty]$ is measurable $\mathfrak{A}$.

Proof of 0). Fix $x \in X$ and consider the map $h : Y \to X \times Y$ defined to be $y \mapsto (x, y)$. For $E \subset X \times Y$ then $\overset{-1}{h}(E) = E(x)$. Thus to be shown is that

h is measurable $\mathfrak{B} \to \mathfrak{A} \otimes \mathfrak{B}$. Now $\overset{-1}{h}(\mathfrak{A} \times \mathfrak{B}) = \mathfrak{B}$ by 20.1.1). Therefore 1.11 implies that

$$\overset{-1}{h}(\mathfrak{A} \otimes \mathfrak{B}) = \overset{-1}{h}\big(\mathrm{A}\sigma(\mathfrak{A} \times \mathfrak{B})\big) = \mathrm{A}\sigma\big(\overset{-1}{h}(\mathfrak{A} \times \mathfrak{B})\big) = \mathrm{A}\sigma(\mathfrak{B}) = \mathfrak{B}.$$

The assertion follows.

Proof of i). Define $\mathfrak{N}$ to consist of all subsets $E \in \mathfrak{A} \otimes \mathfrak{B}$ such that the function $\beta(E(\cdot)) : X \to [0, \infty]$ is $< \infty$ and measurable $\mathfrak{A}$. 1) We have $\mathfrak{A} \times \mathfrak{b} \subset \mathfrak{N}$ by 20.1.1). Furthermore $\mathfrak{N}$ fulfils $\setminus$ by 20.1.2) and 13.16, and $\downarrow \sigma$ by 20.1.3) and 13.10. Thus 1.16.σ) applied to $\mathfrak{M} := \mathfrak{A} \times \mathfrak{b}$ and combined with $\mathfrak{A} \times \mathfrak{B} \subset \mathfrak{M}\top$ furnishes $\mathfrak{A} \otimes \mathfrak{B} \subset (\mathfrak{A} \times \mathfrak{b})\top\mathfrak{N}$, that is

$$E \cap (A \times B) \in \mathfrak{N} \quad \text{for } E \in \mathfrak{A} \otimes \mathfrak{B} \text{ and } A \times B \in \mathfrak{A} \times \mathfrak{b}.$$

2) Assume now that $E \in \mathfrak{A} \otimes \mathfrak{B}$ has $\mathrm{Pr}(E) \subset Y$ upward enclosable $\mathfrak{b}^\sigma$. Let $(B_l)_l$ be a sequence in $\mathfrak{b}$ with $B_l \uparrow \supset \mathrm{Pr}(E)$. Then $E_l := E \cap (X \times B_l) \uparrow E$, and hence $\beta(E_l(\cdot)) \uparrow \beta(E(\cdot))$ from 0). We know from 1) that the E_l are in $\mathfrak{N}$. It follows that the function $\beta(E(\cdot))$ is measurable $\mathfrak{A}$.

20.16. EXAMPLE. The size restriction in 20.15.i) cannot be dispensed with. In fact, in example 20.13 with $E := D \in \mathrm{Bor}(I \times I) = \mathfrak{A} \otimes \mathfrak{B}$ one has $D(x) = \{x\}$ and hence $\beta(D(x)) = G(x)$ for $x \in I$. Thus $\beta(D(\cdot)) = G$, which of course need not be measurable $\mathfrak{A} = \mathrm{Bor}(I)$.

The above proposition is the basis for the result which follows. Its proof is a series of obvious verifications.

20.17. THEOREM. *Assume that α and β are cmeasures on σ algebras $\mathfrak{A}$ and $\mathfrak{B}$. Define $\delta : \mathfrak{A} \otimes \mathfrak{B} \to [0, \infty]$ to be*

$$\delta(E) = \left\{ \begin{array}{ll} \int \beta(E(\cdot))\,d\alpha & \text{if } \mathrm{Pr}(E) \subset Y \text{ is upward enclosable } \mathfrak{b}^\sigma \\ \infty & \text{if not} \end{array} \right\}.$$

Then δ is a cmeasure which fulfils

$$\delta(A \times B) = \alpha(A)\beta(B) \quad \text{for all } A \in \mathfrak{A} \text{ and } B \in \mathfrak{b}^\sigma,$$

and hence is a narrow product measure of α and β.

We combine the last theorem with the uniqueness result 20.12.

20.18. CONSEQUENCE. *Each narrow product measure $\nu : \mathfrak{A} \otimes \mathfrak{B} \to [0, \infty]$ fulfils*

$$\nu(E) = \int \beta(E(\cdot))\,d\alpha \quad \text{for all } E \in \mathfrak{A} \otimes \mathfrak{B} \text{ upward enclosable } \mathfrak{p}^\sigma,$$

that is for all $E \in \mathfrak{A} \otimes \mathfrak{B}$ upward enclosable $\mathfrak{a}^\sigma \times \mathfrak{b}^\sigma$.

20.19. BIBLIOGRAPHICAL NOTE. The material of the first subsection should be known, but we cannot name a complete reference. The traditional product theory is in most textbooks restricted to the so-called σ finite case that

$$X \in \mathfrak{a}^\sigma \text{ and } Y \in \mathfrak{b}^\sigma.$$

Exceptions are Weir [1974] and Behrends [1987]. The author has not seen characterizations as in 20.11 in the literature. The most important aspect is the use of inner regularity in terms of the envelope δ_σ. This is the route which will be pursued in the next section.

21. The Product Formations Based on Inner Regularity

The present section obtains the main theorems of the chapter. We develop the product formation on the basis of the conventional inner extension theory of chapter II. This includes the case $\bullet = \tau$ and hence can be specialized to Radon measures.

Further Properties of the Basic Product Formation

Let X and Y be nonvoid sets. We fix lattices $\mathfrak{S}$ in X and $\mathfrak{T}$ in Y with $\varnothing$, and the notation $\mathfrak{R} := (\mathfrak{S} \times \mathfrak{T})^\star$. Like $\mathfrak{S}_\bullet$ and $\mathfrak{T}_\bullet$ we shall often meet $\mathfrak{R}_\bullet = ((\mathfrak{S} \times \mathfrak{T})^\star)_\bullet$, an expression which does not look simple. Therefore we start with a few remarks on this expression, with formal relations and with natural and important examples.

21.1. EXERCISE. *We have* $\mathfrak{S}_\bullet \times \mathfrak{T}_\bullet \subset \mathfrak{R}_\bullet$ *and*

$$(\mathfrak{S}\top\mathfrak{S}_\bullet) \times (\mathfrak{T}\top\mathfrak{T}_\bullet) \subset \mathfrak{R}\top\mathfrak{R}_\bullet.$$

21.2. REMARK. *Assume that*

> $\mathfrak{U}$ *is a lattice in* X *with* $\varnothing, X \in \mathfrak{U}$,
> $\mathfrak{V}$ *is a lattice in* Y *with* $\varnothing, Y \in \mathfrak{V}$.

Then $((\mathfrak{U} \times \mathfrak{V})^\bullet)\bot = (((\mathfrak{U}\bot) \times (\mathfrak{V}\bot))^\star)_\bullet.$

Proof. We put $\mathfrak{M} := \mathfrak{U}\bot$ and $\mathfrak{N} := \mathfrak{V}\bot$. From

$$(A \times B)' = (A' \times Y) \cup (X \times B') \quad \text{for } A \subset X \text{ and } B \subset Y$$

we see that $(\mathfrak{U} \times \mathfrak{V})\bot \subset (\mathfrak{M} \times \mathfrak{N})^\star$ or $\mathfrak{U} \times \mathfrak{V} \subset ((\mathfrak{M} \times \mathfrak{N})^\star)\bot$, and hence $(\mathfrak{U} \times \mathfrak{V})^\star \subset ((\mathfrak{M} \times \mathfrak{N})^\star)\bot$ since the second member is a lattice. Thus $(\mathfrak{U} \times \mathfrak{V})^\star = ((\mathfrak{M} \times \mathfrak{N})^\star)\bot$ since the situation is symmetric. From 1.5.2) we obtain

$$(\mathfrak{U} \times \mathfrak{V})^\bullet = (((\mathfrak{M} \times \mathfrak{N})^\star)\bot)^\bullet = (((\mathfrak{M} \times \mathfrak{N})^\star)_\bullet)\bot.$$

This is the assertion.

21.3. REMARK. *Let* X *and* Y *be topological spaces.* 1) *Then*

$$\begin{aligned}\mathrm{Op}(X \times Y) &= (\mathrm{Op}(X) \times \mathrm{Op}(Y))^\tau,\\ \mathrm{Cl}(X \times Y) &= ((\mathrm{Cl}(X) \times \mathrm{Cl}(Y))^\star)_\tau.\end{aligned}$$

2) *Assume that* X *and* Y *are Hausdorff. Then*

$$\mathrm{Comp}(X \times Y) = ((\mathrm{Comp}(X) \times \mathrm{Comp}(Y))^\star)_\tau.$$

Proof. 1) The first relation is the definition of the product topology. The second relation then follows from 21.2. 2) A subset $E \subset X \times Y$ is compact iff it is closed and contained in the product $A \times B$ of some compact subsets $A \subset X$ and $B \subset Y$.

We turn to the main results of the subsection. We assume that

$$\varphi : \mathfrak{S} \to [0, \infty[\text{ is isotone with } \varphi(\varnothing) = 0,$$
$$\psi : \mathfrak{T} \to [0, \infty[\text{ is isotone with } \psi(\varnothing) = 0,$$

and let $\vartheta = \varphi \times \psi : \mathfrak{R} \to [0, \infty[$ be the basic product formation defined in 20.4.

21.4. PROPOSITION. *Assume that φ and ψ are downward • continuous. Then ϑ is downward • continuous as well.*

Proof. Let $\mathfrak{M} \subset \mathfrak{R}$ be a paving of type • with $\mathfrak{M} \downarrow E \in \mathfrak{R}$. For $x \in X$ we see from 20.3.1) and 20.1.3) that $\{M(x) : M \in \mathfrak{M}\} \subset \mathfrak{T}$ is a paving of type • with $\downarrow E(x) \in \mathfrak{T}$. Since ψ is downward • continuous we have

$$\inf\{\psi(M(x)) : M \in \mathfrak{M}\} = \psi(E(x)).$$

By 20.3.2) therefore $\{\psi(M(\cdot)) : M \in \mathfrak{M}\} \subset \mathrm{UM}(\mathfrak{S})$ is nonvoid of type • and downward directed in the pointwise order with $\downarrow \psi(E(\cdot)) \in \mathrm{UM}(\mathfrak{S})$. Furthermore $\vartheta(M) = \int \psi(M(\cdot)) d\varphi < \infty$ for $M \in \mathfrak{M}$. Since φ is downward • continuous it follows from 11.17 that

$$\inf\{\int \psi(M(\cdot)) d\varphi : M \in \mathfrak{M}\} = \int \psi(E(\cdot)) d\varphi,$$

which is the assertion.

21.5. EXERCISE. *Assume that φ and ψ are • continuous at $\varnothing$. Then ϑ is • continuous at $\varnothing$ as well.* Hint: Use 11.22 instead of 11.17.

21.6. LEMMA. *Assume that φ and ψ are downward • continuous. For each $E \in \mathfrak{R}_\bullet$ then*

i) $E(x) \in \mathfrak{T}_\bullet$ *and* $\psi_\bullet(E(x)) < \infty$ *for all* $x \in X$;
ii) *the function* $\psi_\bullet(E(\cdot)) : X \to [0, \infty[$ *is in* $\mathrm{UM}(\mathfrak{S}_\bullet)$;
iii) $\vartheta_\bullet(E) = \int \psi_\bullet(E(\cdot)) d\varphi_\bullet < \infty$.

We can assume that • $= \sigma\tau$. We recall from 6.5 that $\varphi_\bullet|\mathfrak{S}_\bullet$ and $\psi_\bullet|\mathfrak{T}_\bullet$ are finite extensions of φ and ψ and are downward • continuous. By 21.4 likewise $\vartheta_\bullet|\mathfrak{R}_\bullet$ is a finite extension of ϑ and is downward • continuous. We also recall the restriction formula 20.7. We note that

$$\varphi_\bullet \times \psi_\bullet : (\mathfrak{P}(X) \times \mathfrak{P}(Y))^\star \to [0, \infty],$$

while the expression $\int \psi_\bullet(E(\cdot)) d\varphi_\bullet \in [0, \infty]$ is defined for all $E \subset X \times Y$.

Proof. Define $\mathfrak{K}$ to consist of all $E \in \mathfrak{R}_\bullet$ which fulfil i)ii)iii). Thus $\mathfrak{R} \subset \mathfrak{K}$ from 20.3. It remains to prove that $\mathfrak{K}$ satisfies $\downarrow$ •. Let $\mathfrak{M} \subset \mathfrak{K}$ be a paving of type • with $\mathfrak{M} \downarrow E$. Then $E \in \mathfrak{R}_\bullet$. i) For $x \in X$ we see from 20.1.3) that $\{M(x) : M \in \mathfrak{M}\} \subset \mathfrak{T}_\bullet$ is a paving of type $\bullet$ with $\downarrow E(x)$. Therefore $E(x) \in \mathfrak{T}_\bullet$. It is obvious that $\psi_\bullet(E(x)) < \infty$. ii) We have

$$\inf\{\psi_\bullet(M(x)) M \in \mathfrak{M}\} = \psi_\bullet(E(x)).$$

For $t > 0$ therefore

$$\bigcap_{M \in \mathfrak{M}} [\psi_\bullet(M(\cdot)) \geqq t] = [\psi_\bullet(E(\cdot)) \geqq t],$$

which is $\in \mathfrak{S}_\bullet$. It follows that $\psi_\bullet(E(\cdot)) : X \to [0, \infty[$ is in $\mathrm{UM}(\mathfrak{S}_\bullet)$. iii) $\{\psi_\bullet(M(\cdot)) : M \in \mathfrak{M}\} \subset \mathrm{UM}(\mathfrak{S}_\bullet)$ is nonvoid of type $\bullet$ and downward directed in the pointwise order with $\downarrow \psi_\bullet(E(\cdot)) \in \mathrm{UM}(\mathfrak{S}_\bullet)$. Thus 11.17 combined with 11.15 implies that

$$\inf\{\int \psi_\bullet(M(\cdot)) d\varphi_\bullet : M \in \mathfrak{M}\} = \int \psi_\bullet(E(\cdot)) d\varphi_\bullet.$$

On the other hand we know that $\inf\{\vartheta_\bullet(M) : M \in \mathfrak{M}\} = \vartheta_\bullet(E)$. The assertion follows.

21.7. EXERCISE. Assume that φ and ψ are downward $\bullet$ continuous. Deduce from 21.6 that $\vartheta_\bullet(A \times B) = \varphi_\bullet(A)\psi_\bullet(B)$ for all $A \subset X$ and $B \subset Y$.

We conclude with one more lemma.

21.8. LEMMA. *Assume that*

$\alpha : \mathfrak{A} \to [0, \infty]$ *is a ccontent on a ring* $\mathfrak{A} \supset \mathfrak{S}$ *and inner regular* $\mathfrak{S}$,

$\beta : \mathfrak{B} \to [0, \infty]$ *is a ccontent on a ring* $\mathfrak{B} \supset \mathfrak{T}$ *and inner regular* $\mathfrak{T}$.

Then the ccontent $\alpha \times \beta : (\mathfrak{A} \times \mathfrak{B})^* \to [0, \infty]$ *is inner regular* $(\mathfrak{S} \times \mathfrak{T})^* = \mathfrak{R}$.

Proof. Fix $E \in (\mathfrak{A} \times \mathfrak{B})^*$. By 20.2.2) then

$$E = \bigcup_{l=1}^{r} A_l \times B_l \text{ with } A_1, \cdots, A_r \in \mathfrak{A} \text{ pairwise disjoint and } B_1, \cdots, B_r \in \mathfrak{B},$$

so that $(\alpha \times \beta)(E) = \sum_{l=1}^{r} \alpha(A_l)\beta(B_l)$. To be shown is

$$(\alpha \times \beta)(E) = \sup\{(\alpha \times \beta)(M) : M \in \mathfrak{R} \text{ with } M \subset E\}.$$

i) Assume that $(\alpha \times \beta)(E) = \infty$. Then $\alpha(A_l)\beta(B_l) = \infty$ for some $l = 1, \cdots, r$, and thus $\alpha(A_l), \beta(B_l) > 0$ by the usual convention. In this case the assertion is obvious. ii) Assume that $(\alpha \times \beta)(E) < \infty$ and hence $\alpha(A_l)\beta(B_l) < \infty$ for all $l = 1, \cdots, r$. Fix $\varepsilon > 0$. Then take

$$S_l := \varnothing \text{ and } T_l := \varnothing \text{ in case } \alpha(A_l)\beta(B_l) = 0;$$

and in case $\alpha(A_l)\beta(B_l) > 0$ and hence $0 < \alpha(A_l), \beta(B_l) < \infty$ take

$$S_l \in \mathfrak{S} \text{ with } S_l \subset A_l \text{ and } \alpha(S_l) \geqq (1 - \varepsilon)\alpha(A_l),$$
$$T_l \in \mathfrak{T} \text{ with } T_l \subset B_l \text{ and } \beta(T_l) \geqq (1 - \varepsilon)\beta(B_l).$$

Then we have

$$M := \bigcup_{l=1}^{r} S_l \times T_l \in \mathfrak{R} \text{ with } M \subset E \text{ and } (\alpha \times \beta)(M) = \sum_{l=1}^{r} \alpha(S_l)\beta(T_l),$$

since $S_1, \cdots, S_r$ are pairwise disjoint, and hence

$$(\alpha \times \beta)(M) \geqq (1 - \varepsilon)^2 (\alpha \times \beta)(E).$$

The assertion follows.

The Main Theorem

The theorem below fulfils what can be expected.

21.9. THEOREM. *Assume that*

$$\varphi : \mathfrak{S} \to [0, \infty[\text{ is an inner } \bullet \text{ premeasure with } \varphi(\varnothing) = 0,$$
$$\psi : \mathfrak{T} \to [0, \infty[\text{ is an inner } \bullet \text{ premeasure with } \psi(\varnothing) = 0.$$

Then $\vartheta = \varphi \times \psi : \mathfrak{R} \to [0, \infty[$ *is an inner* $\bullet$ *premeasure. Moreover if*

$$\Phi := \varphi_\bullet | \mathfrak{C}(\varphi_\bullet) \quad and \quad \Psi := \psi_\bullet | \mathfrak{C}(\psi_\bullet),$$

then $\theta := \vartheta_\bullet | \mathfrak{C}(\vartheta_\bullet)$ *is an extension of* $\Phi \times \Psi$.

21.10. REMARK. Assume that φ and ψ are both inner σ and inner τ premeasures. Hence $\vartheta = \varphi \times \psi$ is both an inner σ and an inner τ premeasure as well. By 6.24 and 6.25 then $\vartheta_\sigma | \mathfrak{C}(\vartheta_\sigma)$ is a restriction of $\vartheta_\tau | \mathfrak{C}(\vartheta_\tau)$. However, we shall see that the increase

in the domains from $\mathfrak{C}(\vartheta_\sigma)$ to $\mathfrak{C}(\vartheta_\tau)$, and
in the lattices of inner regularity from $\mathfrak{R}_\sigma$ to $\mathfrak{R}_\tau$

can have fundamental implications.

Proof of 21.9. 1) We consider the product formation $\eta := (\varphi_\bullet | \mathfrak{S}_\bullet) \times (\psi_\bullet | \mathfrak{T}_\bullet) < \infty$ on the lattice $\mathfrak{H} := (\mathfrak{S}_\bullet \times \mathfrak{T}_\bullet)^\star$. From 21.1 we have $\mathfrak{R} \subset \mathfrak{H} \subset \mathfrak{R}_\bullet$ and hence $\mathfrak{H}_\bullet = \mathfrak{R}_\bullet$. From 21.4 we know that η is downward $\bullet$ continuous. Furthermore $\vartheta = \eta | \mathfrak{R}$ by 20.7 and $\vartheta_\bullet | \mathfrak{H} = \eta$ by 21.6. We claim that $\vartheta_\bullet = \eta_\bullet$. The relation $\leqq$ is obvious since ϑ is a restriction of η. In order to prove $\geqq$ we fix $E \subset X \times Y$. Let $\mathfrak{M} \subset \mathfrak{H}$ be a paving of type $\bullet$ with $\mathfrak{M} \downarrow$ some $D \subset E$. Then $D \in \mathfrak{H}_\bullet = \mathfrak{R}_\bullet$. It follows that

$$\inf\{\eta(M) : M \in \mathfrak{M}\} = \inf\{\vartheta_\bullet(M) : M \in \mathfrak{M}\} = \vartheta_\bullet(D) \leqq \vartheta_\bullet(E),$$

since $\vartheta_\bullet | \mathfrak{R}_\bullet$ is known to be downward $\bullet$ continuous. Therefore $\eta_\bullet(E) \leqq \vartheta_\bullet(E)$ as claimed.

2) From $\mathfrak{S}_\bullet \subset \mathfrak{C}(\varphi_\bullet)$ and $\mathfrak{T}_\bullet \subset \mathfrak{C}(\psi_\bullet)$ we have $\mathfrak{H} \subset (\mathfrak{C}(\varphi_\bullet) \times \mathfrak{C}(\psi_\bullet))^\star$. From 20.7 also $\eta = (\Phi \times \Psi) | \mathfrak{H}$. Now 21.8 implies that $\Phi \times \Psi$ is inner regular $\mathfrak{H}$. Therefore $\Phi \times \Psi$ is an inner $\star$ extension of η.

3) We next prove that ϑ is an inner $\bullet$ premeasure. By the conventional inner main theorem 6.31 we have to show that

$$\vartheta(Q) \leqq \vartheta(P) + \vartheta_\bullet(Q \setminus P) \quad \text{for all } P \subset Q \text{ in } \mathfrak{R}.$$

Now we see from 2) and 1) that

$$\begin{aligned}
\vartheta(Q) - \vartheta(P) &= (\Phi \times \Psi)(Q) - (\Phi \times \Psi)(P) = (\Phi \times \Psi)(Q \setminus P) \\
&= \sup\{\eta(H) : H \in \mathfrak{H} \text{ with } H \subset Q \setminus P\} \\
&= \sup\{\vartheta_\bullet(H) : H \in \mathfrak{H} \text{ with } H \subset Q \setminus P\} \leqq \vartheta_\bullet(Q \setminus P).
\end{aligned}$$

This is the assertion.

4) We know from 2) that $\Phi \times \Psi$ is an inner $\star$ extension of η. Thus η is an inner $\star$ premeasure, and $\Phi \times \Psi$ is a restriction of $\eta_\star | \mathfrak{C}(\eta_\star)$. Since η is downward $\bullet$ continuous it follows from 6.24 and 6.25 that $\eta_\star | \mathfrak{C}(\eta_\star)$ is a restriction of $\eta_\bullet | \mathfrak{C}(\eta_\bullet)$. At last we know from 1) that $\eta_\bullet = \vartheta_\bullet$. Therefore $\Phi \times \Psi$ is a restriction of $\vartheta_\bullet | \mathfrak{C}(\vartheta_\bullet) = \theta$. This finishes the proof of the theorem.

We add a few words of comparison with the traditional product formations in section 20. The proofs alone make clear that the former procedures are much more superficial than the present one. Nevertheless the traditional product formations are not superfluous, because the new procedure contains part iii) of the former main result 20.11 as a primitive specialization, but not parts i) and ii). In the other direction the previous inner results can be of some use in case $\bullet = \sigma$, but are of no use at all in case $\bullet = \tau$. However, this case can be spectacular, as we shall see next.

We turn to the specialization to Radon measures. Let X and Y be Hausdorff topological spaces. We assume that

$$\alpha : \mathrm{Bor}(X) \to [0, \infty] \text{ is a Borel-Radon measure on } X,$$
$$\beta : \mathrm{Bor}(Y) \to [0, \infty] \text{ is a Borel-Radon measure on } Y,$$

and form the restrictions

$$\varphi := \alpha | \mathfrak{S} \quad \text{to } \mathfrak{S} := \mathrm{Comp}(X),$$
$$\psi := \beta | \mathfrak{T} \quad \text{to } \mathfrak{T} := \mathrm{Comp}(Y).$$

Thus φ and ψ are Radon premeasures in the sense of 9.1, that is inner $\bullet$ premeasures for all $\bullet = \star\sigma\tau$. We write as usual

$$\Phi := \varphi_\bullet | \mathfrak{C}(\varphi_\bullet), \quad \text{so that } \alpha = \Phi | \mathrm{Bor}(X),$$
$$\Psi := \psi_\bullet | \mathfrak{C}(\psi_\bullet), \quad \text{so that } \beta = \Psi | \mathrm{Bor}(Y).$$

Then 21.9 says that $\vartheta = \varphi \times \psi : \mathfrak{R} \to [0, \infty[$ is an inner $\bullet$ premeasure for $\bullet = \star\sigma\tau$ as well. Moreover it says that

$$(\mathrm{Bor}(X) \times \mathrm{Bor}(Y))^\star \subset (\mathfrak{C}(\varphi_\bullet) \times \mathfrak{C}(\psi_\bullet))^\star \subset \mathfrak{C}(\vartheta_\bullet),$$

and that $\theta := \vartheta_\bullet | \mathfrak{C}(\vartheta_\bullet)$ satisfies

$$\theta | (\mathrm{Bor}(X) \times \mathrm{Bor}(Y))^\star = \alpha \times \beta.$$

In both cases $\bullet = \sigma\tau$ therefore $\theta | \mathrm{Bor}(X) \otimes \mathrm{Bor}(Y)$ is a product measure of α and β. But $\theta | \mathrm{Bor}(X) \otimes \mathrm{Bor}(Y)$ need not be a Radon measure on $X \times Y$ in the sense of section 9, for the simple reason that its domain $\mathrm{Bor}(X) \otimes \mathrm{Bor}(Y)$ need not contain the compact subsets of $X \times Y$, as we have seen in 13.19. It would be unnatural to expect much more from $\mathfrak{C}(\vartheta_\sigma)$, but we do not plan to deepen this point.

However, in case $\bullet = \tau$ the picture is different. From 21.3.2) we obtain

$$\mathfrak{R}_\tau = \mathrm{Comp}(X \times Y),$$

$$\mathrm{Cl}(X \times Y) \subset \mathfrak{R}\top\mathfrak{R}_\tau \subset \mathfrak{C}(\vartheta_\tau) \quad \text{and hence} \quad \mathrm{Bor}(X \times Y) \subset \mathfrak{C}(\vartheta_\tau).$$

Since ϑ_τ is inner regular $\mathfrak{R}_\tau = \mathrm{Comp}(X \times Y)$ it follows that $\theta := \vartheta_\tau | \mathfrak{C}(\vartheta_\tau)$ is a Radon measure. Hence $\nu := \theta | \mathrm{Bor}(X \times Y) = \vartheta_\tau | \mathrm{Bor}(X \times Y)$ is a Borel-Radon measure on $X \times Y$. Of course its restriction $\nu | \mathrm{Bor}(X) \otimes \mathrm{Bor}(Y)$ is a product measure of α and β, that is

$$\nu(A \times B) = \alpha(A)\beta(B) \quad \text{for all } A \in \mathrm{Bor}(X) \text{ and } B \in \mathrm{Bor}(Y).$$

Our final result looks as follows.

21.11. THEOREM. *Assume that $\alpha : \mathrm{Bor}(X) \to [0, \infty]$ and $\beta : \mathrm{Bor}(Y) \to [0, \infty]$ are Borel-Radon measures on Hausdorff topological spaces X and Y. Then there exists a unique Borel-Radon measure $\nu : \mathrm{Bor}(X \times Y) \to [0, \infty]$ on $X \times Y$ such that*

$$\nu(A \times B) = \alpha(A)\beta(B) \quad \text{for all } A \in \mathrm{Comp}(X) \text{ and } B \in \mathrm{Comp}(Y).$$

We have

$$\nu = (\varphi \times \psi)_\tau | \mathfrak{C}\big((\varphi \times \psi)_\tau\big)$$
$$\text{with } \varphi := \alpha | \mathrm{Comp}(X) \text{ and } \psi := \beta | \mathrm{Comp}(Y).$$

The restriction $\nu | \mathrm{Bor}(X) \otimes \mathrm{Bor}(Y)$ is a product measure of α and β.

Proof. It remains to prove the uniqueness assertion. Assume that $\nu : \mathrm{Bor}(X \times Y) \to [0, \infty]$ is a Borel-Radon measure on $X \times Y$ as in the theorem. Thus in the former notations

$$\nu(A \times B) = \vartheta(A \times B) \quad \text{for all } A \in \mathfrak{S} \text{ and } B \in \mathfrak{T}.$$

Then $\nu(E) = \vartheta(E)$ for all $E \in \mathfrak{R}$ from the uniqueness assertion in 20.5. Therefore $\nu(E) = \vartheta_\tau(E) = \theta(E)$ for all $E \in \mathfrak{R}_\tau = \mathrm{Comp}(X \times Y)$ and hence for all $E \in \mathrm{Bor}(X \times Y)$. This completes the proof.

The Sectional Representation

The present subsection is under the assumption that

$$\varphi : \mathfrak{S} \to [0, \infty[\text{ is an inner } \bullet \text{ premeasure with } \varphi(\varnothing) = 0,$$
$$\psi : \mathfrak{T} \to [0, \infty[\text{ is an inner } \bullet \text{ premeasure with } \psi(\varnothing) = 0,$$

where $\bullet = \sigma\tau$. We know from 21.9 that $\vartheta = \varphi \times \psi : \mathfrak{R} \to [0, \infty[$ is an inner $\bullet$ premeasure as well. If as before

$$\Phi := \varphi_\bullet | \mathfrak{C}(\varphi_\bullet) \quad \text{and} \quad \Psi := \psi_\bullet | \mathfrak{C}(\psi_\bullet),$$

then $\theta := \vartheta_\bullet | \mathfrak{C}(\vartheta_\bullet)$ is an extension of $\Phi \times \Psi$. Our aim is to extend the sectional representation of the traditional product measure $\theta | \mathfrak{C}(\varphi_\bullet) \otimes \mathfrak{C}(\psi_\bullet)$, as obtained in 20.18, to the whole of θ, as far as this is possible. We shall see that the situation is more involved than before. The final results for the future will be 21.12, 21.16 with 21.13, and 21.19.

For the sequel we have the choice between a downward and an upward procedure. Both times the central result 21.15 and its consequences would be the same. We prefer the upward procedure, because in it the important 21.12 will not be burdened with finiteness conditions. The upward procedure will be based on the formations $\mathfrak{U} := (\mathfrak{S}\top\mathfrak{S}_\bullet)\bot$ and $\mathfrak{V} := (\mathfrak{T}\top\mathfrak{T}_\bullet)\bot$. We see from 21.2 and 21.1 that

$$((\mathfrak{U} \times \mathfrak{V})^\bullet)\bot = (((\mathfrak{S}\top\mathfrak{S}_\bullet) \times (\mathfrak{T}\top\mathfrak{T}_\bullet))^\star)_\bullet,$$
$$\mathfrak{R}_\bullet \subset ((\mathfrak{U} \times \mathfrak{V})^\bullet)\bot \subset \mathfrak{R}\top\mathfrak{R}_\bullet.$$

This has to be combined with the old fact $\mathfrak{R}\top\mathfrak{R}_\bullet \subset \mathfrak{C}(\vartheta_\bullet)$ from 6.31. It follows that $(\mathfrak{U} \times \mathfrak{V})^\bullet \subset \mathfrak{C}(\vartheta_\bullet)$.

21.12. THEOREM. *Let $\mathfrak{P} \subset \mathfrak{U}$ and $\mathfrak{Q} \subset \mathfrak{V}$ be lattices with $\cup\bullet$. For each $E \in (\mathfrak{P} \times \mathfrak{Q})^\bullet$ then*

0) $E(x) \in \mathfrak{Q}$ *for all* $x \in X$;

i) *the function* $\Psi(E(\cdot)) : X \to [0,\infty]$ *is in* $\mathrm{LM}(\mathfrak{P})$;

ii) $\theta(E) = \int\!\!\!-\, \Psi(E(\cdot))d\Phi$.

Proof. Define $\mathfrak{K}$ to consist of all $E \in (\mathfrak{R}\top\mathfrak{R}_\bullet)\bot$ which fulfil 0)i)ii). Then $(\mathfrak{P} \times \mathfrak{Q})^\star \subset \mathfrak{K}$. In fact, 0)i) result from 20.3, and ii) results from 21.9 since $(\mathfrak{P} \times \mathfrak{Q})^\star \subset (\mathfrak{C}(\varphi_\bullet) \times \mathfrak{C}(\psi_\bullet))^\star$ where $\theta = \Phi \times \Psi$. Therefore it remains to prove that $\mathfrak{K}$ satisfies $\uparrow \bullet$. Let $\mathfrak{M} \subset \mathfrak{K}$ be a paving of type $\bullet$ with $\mathfrak{M} \uparrow E$. Then $E \in (\mathfrak{R}\top\mathfrak{R}_\bullet)\bot$. 0) For $x \in X$ we see from 20.1.3) that $\{M(x) : M \in \mathfrak{M}\} \subset \mathfrak{Q}$ is a paving of type $\bullet$ with $\uparrow E(x)$. Hence $E(x) \in \mathfrak{Q}$. i) The restriction $\Psi|(\mathfrak{T}\top\mathfrak{T}_\bullet)\bot$ is upward $\bullet$ continuous, in case $\bullet = \tau$ by 6.28. Thus we have

$$\sup\{\Psi(M(x)) : M \in \mathfrak{M}\} = \Psi(E(x)) \quad \text{for } x \in X.$$

For $t > 0$ therefore

$$\bigcup_{m \in \mathfrak{M}} [\Psi(M(\cdot)) > t] = [\Psi(E(\cdot)) > t],$$

which is in $\mathfrak{P}$. It follows that $\Psi(E(\cdot)) : X \to [0,\infty]$ is in $\mathrm{LM}(\mathfrak{P})$. ii) $\{\Psi(M(\cdot)) : M \in \mathfrak{M}\} \subset \mathrm{LM}(\mathfrak{P})$ is nonvoid of type $\bullet$ and upward directed in the pointwise order with $\uparrow \Psi(E(\cdot)) \in \mathrm{LM}(\mathfrak{P})$. Furthermore $\Phi|(\mathfrak{S}\top\mathfrak{S}_\bullet)\bot$ is upward $\bullet$ continuous. Thus 11.18 combined with 11.15 implies that

$$\sup\{\int\!\!\!-\,\Psi(M(\cdot))d\Phi : M \in \mathfrak{M}\} = \int\!\!\!-\,\Psi(E(\cdot))d\Phi.$$

On the other hand $\theta|(\mathfrak{R}\top\mathfrak{R}_\bullet)\bot$ is upward $\bullet$ continuous as well. Hence $\sup\{\theta(M) : M \in \mathfrak{M}\} = \theta(E)$. The assertion follows.

Next we combine this with 1.11 and 1.16.σ).

21.13. REMARK. *Let $\mathfrak{P}$ in X and $\mathfrak{Q}$ in Y be lattices. For each $E \in \mathrm{A}\sigma((\mathfrak{P} \times \mathfrak{Q})^\bullet)$ then $E(x) \in \mathrm{A}\sigma(\mathfrak{Q}^\bullet)$ for all $x \in X$.*

Proof. Fix $x \in X$ and consider the map $h : Y \to X \times Y$ defined to be $y \mapsto (x, y)$. For $E \subset X \times Y$ then $\overset{-1}{h}(E) = E(x)$. Thus to be shown is that h is measurable $A\sigma(\mathfrak{Q}^\bullet) \to A\sigma((\mathfrak{P} \times \mathfrak{Q})^\bullet)$. From 1.11 and 20.1.1) we obtain

$$\overset{-1}{h}\left(A\sigma((\mathfrak{P} \times \mathfrak{Q})^\bullet)\right) = A\sigma\left(\overset{-1}{h}((\mathfrak{P} \times \mathfrak{Q})^\bullet)\right) = A\sigma\left((\overset{-1}{h}(\mathfrak{P} \times \mathfrak{Q}))^\bullet\right)$$
$$\subset A\sigma((\mathfrak{Q} \cup \{\varnothing\})^\bullet) = A\sigma(\mathfrak{Q}^\bullet).$$

This is the assertion.

21.14. PROPOSITION. *Assume that* $\Phi(X) < \infty$ *and* $\Psi(Y) < \infty$. *Let* $\mathfrak{P} \subset \mathfrak{U}$ *and* $\mathfrak{Q} \subset \mathfrak{V}$ *be lattices with* $\cup\bullet$. *For each* $E \in A\sigma((\mathfrak{P} \times \mathfrak{Q})^\bullet)$ *then*

i) *the function* $\Psi(E(\cdot)) : X \to [0, \infty[$ *is measurable* $A\sigma(\mathfrak{P})$;

ii) $\theta(E) = \int \Psi(E(\cdot)) d\Phi < \infty$.

Proof. We can assume that $X \in \mathfrak{P}$ and $Y \in \mathfrak{Q}$. 1) Define $\mathfrak{N}$ to consist of all $E \in A\sigma((\mathfrak{P} \times \mathfrak{Q})^\bullet)$ such that

i) *the function* $\Psi(E(\cdot)) : X \to [0, \infty[$ *is measurable* $A\sigma(\mathfrak{P})$;

ii) $\theta(E) = \int \Psi(E(\cdot)) d\Phi$.

Then $\mathfrak{N}$ satisfies $\backslash$ and $\downarrow \sigma$. This is clear from standard facts in section 13 combined with 20.1.2)3). 2) Let $\mathfrak{M} := (\mathfrak{P} \times \mathfrak{Q})^\bullet$. Then $\mathfrak{M} \subset \mathfrak{N}$ from 21.12. Furthermore $\mathfrak{M}\top = \mathfrak{M}$ since $\mathfrak{M}$ is a lattice and $X \times Y \in \mathfrak{M}$. Thus the transporter theorem 1.16.σ) furnishes

$$A\sigma((\mathfrak{P} \times \mathfrak{Q})^\bullet) = A\sigma(\mathfrak{M}\top) \subset \mathfrak{M}\top\mathfrak{N} \subset \mathfrak{N}.$$

This is the assertion.

The above result 21.14 admits an essential extension. We need a little interlude which is based on the former lemma 9.21.

We fix subsets $P \in \mathfrak{C}(\varphi_\bullet)$ with $\Phi(P) < \infty$ and $Q \in \mathfrak{C}(\psi_\bullet)$ with $\Psi(Q) < \infty$, and define

$$\Phi_P : \mathfrak{C}(\varphi_\bullet) \to [0, \infty[\quad \text{to be} \quad \Phi_P(A) = \Phi(A \cap P) \text{ for } A \in \mathfrak{C}(\varphi_\bullet),$$
$$\Psi_Q : \mathfrak{C}(\psi_\bullet) \to [0, \infty[\quad \text{to be} \quad \Psi_Q(B) = \Psi(B \cap Q) \text{ for } B \in \mathfrak{C}(\psi_\bullet).$$

From 9.21 we see that

$$\varphi_P := \Phi_P | \mathfrak{S} \text{ is an inner } \bullet \text{ premeasure,}$$
$$\text{and } \Phi_P \text{ is an inner } \bullet \text{ extension of } \varphi_P,$$
$$\psi_Q := \Psi_Q | \mathfrak{T} \text{ is an inner } \bullet \text{ premeasure,}$$
$$\text{and } \Psi_Q \text{ is an inner } \bullet \text{ extension of } \psi_Q.$$

We conclude from 21.9 that $\vartheta_{PQ} = \varphi_P \times \psi_Q : \mathfrak{R} \to [0, \infty[$ is an inner $\bullet$ premeasure as well. On the other side we consider the subset $P \times Q \in \mathfrak{C}(\varphi_\bullet) \times \mathfrak{C}(\psi_\bullet) \subset \mathfrak{C}(\vartheta_\bullet)$. Once more from 9.21 we see that $\theta(\cdot \cap (P \times Q)) | \mathfrak{R}$ is an inner $\bullet$ premeasure, and $\theta(\cdot \cap (P \times Q))$ is an inner $\bullet$ extension of $\theta(\cdot \cap (P \times Q)) | \mathfrak{R}$. We first claim that

$$\vartheta_{PQ}(E) = \theta(E \cap (P \times Q)) \quad \text{for all } E \in \mathfrak{R}.$$

In fact, in virtue of 2.5.1) it suffices to show this for $E = S \times T$ with $S \in \mathfrak{S}$ and $T \in \mathfrak{T}$. But for these subsets we have

$$
\begin{aligned}
\vartheta_{PQ}(E) &= \vartheta_{PQ}(S \times T) = (\varphi_P \times \psi_Q)(S \times T) = \varphi_P(S)\psi_Q(T) \\
&= \Phi_P(S)\Psi_Q(T) = \Phi(S \cap P)\Psi(T \cap Q) \\
&= \theta\big((S \cap P) \times (T \cap Q)\big) = \theta\big(E \cap (P \times Q)\big).
\end{aligned}
$$

Now let $\theta_{PQ} := (\vartheta_{PQ})_\bullet | \mathfrak{C}((\vartheta_{PQ})_\bullet)$ be the maximal inner $\bullet$ extension of $\vartheta_{PQ} = \theta\big(\cdot \cap (P \times Q)\big)|\mathfrak{R}$. Then it follows that $\mathfrak{C}(\vartheta_\bullet) \subset \mathfrak{C}((\vartheta_{PQ})_\bullet)$ and

$$
\theta_{PQ}(E) = \theta\big(E \cap (P \times Q)\big) \quad \text{for all } E \in \mathfrak{C}(\vartheta_\bullet).
$$

This is the intermediate result which we need.

21.15. THEOREM. *Let* $\mathfrak{P} \subset \mathfrak{U}$ *and* $\mathfrak{Q} \subset \mathfrak{V}$ *be lattices with* $\cup\bullet$. *For each* $E \in \mathrm{A}\sigma\big((\mathfrak{P} \times \mathfrak{Q})^\bullet\big)$, *and for* $Q \in \mathfrak{C}(\psi_\bullet)$ *with* $\Psi(Q) < \infty$, *then*

i) *the function* $\Psi\big(E(\cdot) \cap Q\big) : X \to [0, \infty[$ *is measurable* $\mathrm{A}\sigma(\mathfrak{P})$;

ii) $\theta\big(E \cap (X \times Q)\big) = \int \Psi\big(E(\cdot) \cap Q\big)d\Phi$.

Proof. Besides $Q \in \mathfrak{C}(\psi_\bullet)$ with $\Psi(Q) < \infty$ we fix $P \in \mathfrak{C}(\varphi_\bullet)$ with $\Phi(P) < \infty$ until the last step of the proof, and retain the former notations. The assertions will be obtained upon application of 21.14 to φ_P and ψ_Q with the respective ϑ_{PQ} and θ_{PQ}. Note that the inner $\bullet$ extensions Φ_P and Ψ_Q of φ_P and ψ_Q are not claimed to be their maximal ones. Let $E \in \mathrm{A}\sigma\big((\mathfrak{P} \times \mathfrak{Q})^\bullet\big)$. We obtain assertion i) since $\Psi_Q\big(E(x)\big) = \Psi\big(E(x) \cap Q\big)$ for $x \in X$. Next 21.14.ii) reads

$$
\theta_{PQ}(E) = \int \Psi_Q\big(E(\cdot)\big)d\Phi_P < \infty.
$$

The first member is $= \theta\big(E \cap (P \times Q)\big)$ in virtue of our intermediate result. Let us rewrite the second member. For the cmeasures Φ and Φ_P on $\mathfrak{C}(\varphi_\bullet)$ and for an $f : X \to [0, \infty]$ measurable $\mathfrak{C}(\varphi_\bullet)$ we have

$$
\begin{aligned}
\int f d\Phi_P &= \oint f d\Phi_P = \int_{0\leftarrow}^{\to\infty} \Phi_P([f \geqq t])dt = \int_{0\leftarrow}^{\to\infty} \Phi([f \geqq t] \cap P)dt \\
&= \int_{0\leftarrow}^{\to\infty} \Phi([\chi_P f \geqq t])dt = \oint \chi_P f d\Phi = \int \chi_P f d\Phi.
\end{aligned}
$$

We use this relation for $f := \Psi_Q\big(E(\cdot)\big) = \Psi\big(E(\cdot) \cap Q\big)$. Then we obtain

$$
\theta\big(E \cap (P \times Q)\big) = \int \chi_P \Psi\big(E(\cdot) \cap Q\big)d\Phi.
$$

At last we form the supremum over all $P \in \mathfrak{C}(\varphi_\bullet)$ with $\Phi(P) < \infty$. For the first member we note that θ is inner regular $\mathfrak{R}_\bullet$, and for the second member we invoke 12.2.3). Then assertion ii) follows.

21.16. CONSEQUENCE. *Let* $\mathfrak{P} \subset \mathfrak{U}$ *and* $\mathfrak{Q} \subset \mathfrak{V}$ *be lattices with* $\cup \bullet$. *Assume that* $E \in \mathrm{A}\sigma\big((\mathfrak{P} \times \mathfrak{Q})^{\bullet}\big)$ *has* $\mathrm{Pr}(E) \subset Y$ *upward enclosable* $[\Psi < \infty]^{\sigma}$. *Then*

i) *the function* $\Psi\big(E(\cdot)\big) : X \to [0, \infty]$ *is measurable* $\mathrm{A}\sigma(\mathfrak{P})$;

ii) $\theta(E) = \int \Psi\big(E(\cdot)\big) d\Phi$.

Proof. This is an obvious consequence of 21.15 combined with standard facts in section 13.

The most important of the above results will be specialized to Radon measures at the end of the subsection. Before this we keep the present main road and pass to the final step.

21.17. THEOREM. *Assume that* $E \in \mathfrak{C}(\vartheta_{\bullet})$. *For fixed* $Q \in \mathfrak{C}(\psi_{\bullet})$ *with* $\Psi(Q) < \infty$ *then*

0) $E(x) \cap Q \in \mathfrak{C}(\psi_{\bullet})$ *for all* $x \in X$ *except on some* $N(Q) \in \mathfrak{C}(\varphi_{\bullet})$ *with* $\Phi\big(N(Q)\big) = 0$;

i) *the function* $\psi_{\bullet}\big(E(\cdot) \cap Q\big) : X \to [0, \infty[$ *is measurable* $\mathfrak{C}(\varphi_{\bullet})$;

ii) $\theta\big(E \cap (X \times Q)\big) = \int \psi_{\bullet}\big(E(\cdot) \cap Q\big) d\Phi$.

The null sets $N(\cdot) \in \mathfrak{C}(\varphi_{\bullet})$ in 0) cannot be dispensed with, as the next example will show. It is for $Q := Y$ with $\Psi(Y) < \infty$.

21.18. EXAMPLE. We assume on the one hand a nonvoid subset $N \in \mathfrak{C}(\varphi_{\bullet})$ with $\Phi(N) = 0$, and on the other hand a subset $T \subset Y$ which is not in $\mathfrak{C}(\psi_{\bullet})$. From 21.9 we have

$$N \times Y \in \mathfrak{C}(\varphi_{\bullet}) \times \mathfrak{C}(\psi_{\bullet}) \subset \mathfrak{C}(\vartheta_{\bullet}) \quad \text{with} \quad \theta(N \times Y) = \Phi(N)\Psi(Y) = 0.$$

Thus 10.16 implies that $E := N \times T$ is in $\mathfrak{C}(\vartheta_{\bullet})$. But for $x \in N$ we have $E(x) = T \notin \mathfrak{C}(\psi_{\bullet})$.

Proof of 21.17. For the first steps we fix $Q \in \mathfrak{C}(\psi_{\bullet})$ with $\Psi(Q) < \infty$ and $P \in \mathfrak{C}(\varphi_{\bullet})$ with $\Phi(P) < \infty$. 1) We have $E \cap (P \times Q), E' \cap (P \times Q) \in \mathfrak{C}(\vartheta_{\bullet})$ with values $\theta(\cdot) < \infty$. Since θ is inner regular $\mathfrak{R}_{\bullet}$ there exist sequences

$(A_l)_l$ in $\mathfrak{R}_{\bullet}$ with $A_l \uparrow A \subset E \cap (P \times Q)$ and $\theta(A) = \theta\big(E \cap (P \times Q)\big)$,

$(B_l)_l$ in $\mathfrak{R}_{\bullet}$ with $B_l \uparrow B \subset E' \cap (P \times Q)$ and $\theta(B) = \theta\big(E' \cap (P \times Q)\big)$.

Of course $A, B \in \mathfrak{C}(\vartheta_{\bullet})$. Note that $A, B \subset P \times Q$ implies that $A(x), B(x) \subset Q$ for $x \in X$ and $A(x) = B(x) = \varnothing$ for $x \in P'$. 2) We recall $\mathfrak{U}$ and $\mathfrak{V}$, and note that $A, B \in \mathrm{A}\sigma\big((\mathfrak{U} \times \mathfrak{V})^{\bullet}\big)$. Thus from 21.13 and 21.15 applied to $\mathfrak{P} := \mathfrak{U}$ and $\mathfrak{Q} := \mathfrak{V}$ and to A and B we obtain

0) $A(x), B(x) \in \mathrm{A}\sigma(\mathfrak{V}) \subset \mathfrak{C}(\psi_{\bullet})$ *for all* $x \in X$;

i) *the functions* $\Psi\big(A(\cdot)\big), \Psi\big(B(\cdot)\big) : X \to [0, \infty]$ *are measurable* $\mathrm{A}\sigma(\mathfrak{U}) \subset \mathfrak{C}(\varphi_{\bullet})$;

ii) $\theta(A) = \int \Psi\big(A(\cdot)\big) d\Phi$ *and* $\theta(B) = \int \Psi\big(B(\cdot)\big) d\Phi$.

3) We pass from B to $D := (P \times Q) \setminus B \in \mathfrak{C}(\vartheta_\bullet)$. We have on the one hand $A \subset E \cap (P \times Q) \subset D \subset P \times Q$ with

$$\theta(A) = \theta\big(E \cap (P \times Q)\big) = \theta(P \times Q) - \theta\big(E' \cap (P \times Q)\big)$$
$$= \theta(P \times Q) - \theta(B) = \theta(D).$$

On the other hand $D(x) = \varnothing$ for $x \in P'$, while for $x \in P$ we have $D(x) = Q \setminus B(x) \in \mathfrak{C}(\psi_\bullet)$ and hence $\Psi\big(D(x)\big) = \Psi(Q) - \Psi\big(B(x)\big)$. Therefore the function $\Psi\big(D(\cdot)\big) : X \to [0, \infty[$ is measurable $\mathfrak{C}(\varphi_\bullet)$. We have

$$\theta(D) = \theta(P \times Q) - \theta(B) = \int_P \big(\Psi(Q) - \Psi(B(\cdot))\big) d\Phi = \int \Psi\big(D(\cdot)\big) d\Phi.$$

4) We come to the decisive point. We have $A(x) \subset E(x) \cap Q \subset D(x)$ for $x \in P$, and $\int_P \Psi\big(A(\cdot)\big) d\Phi = \theta(A) = \theta(D) = \int_P \Psi\big(D(\cdot)\big) d\Phi$. Thus if we define

$$N(P, Q) := \{x \in P : \Psi\big(A(x)\big) < \Psi\big(D(x)\big)\} \in \mathfrak{C}(\varphi_\bullet),$$

then 13.25.3) furnishes $\Phi\big(N(P, Q)\big) = 0$. For $x \in P \setminus N(P, Q)$ then $\Psi\big(A(x)\big) = \Psi\big(D(x)\big)$, and 10.16 applied to ψ furnishes $E(x) \cap Q \in \mathfrak{C}(\psi_\bullet)$.

For the next steps we keep $Q \in \mathfrak{C}(\psi_\bullet)$ with $\Psi(Q) < \infty$. 5) In order to prove assertion 0) we form

$$N(Q) := \{x \in X : E(x) \cap Q \notin \mathfrak{C}(\psi_\bullet)\} \subset X.$$

For each $P \in \mathfrak{C}(\varphi_\bullet)$ with $\Phi(P) < \infty$ then 4) implies that $P \cap N(Q) \subset N(P, Q)$. Thus from 10.16 applied to φ we obtain $P \cap N(Q) \in \mathfrak{C}(\varphi_\bullet)$ with $\Phi\big(P \cap N(Q)\big) = 0$. This holds true in particular for all $P \in \mathfrak{S}$. Thus 6.21 implies $N(Q) \in \mathfrak{C}(\varphi_\bullet)$, and we have $\Phi\big(N(Q)\big) = 0$ since Φ is inner regular $\mathfrak{S}_\bullet$. This proves 0). 6) In order to prove assertion i) we form

$$L(Q, t) := [\psi_\bullet\big(E(\cdot) \cap Q\big) \geq t] \subset X \quad \text{for } t > 0.$$

For fixed $P \in \mathfrak{C}(\varphi_\bullet)$ with $\Phi(P) < \infty$ we have on the one hand $L(Q, t) \cap N(P, Q) \in \mathfrak{C}(\varphi_\bullet)$ from 4) and 10.16 applied to φ. On the other hand we see for $x \in P \setminus N(P, Q)$ that $\psi_\bullet\big(E(x) \cap Q\big) = \Psi\big(A(x)\big)$ and hence $L(Q, t) \cap (P \setminus N(P, Q)) = [\Psi(A(\cdot)) \geq t] \cap (P \setminus N(P, Q)) \in \mathfrak{C}(\varphi_\bullet)$. Therefore $L(Q, t) \cap P \in \mathfrak{C}(\varphi_\bullet)$. As before we conclude from 6.21 that $L(Q, t) \in \mathfrak{C}(\varphi_\bullet)$. This proves i). 7) In order to prove assertion ii) we once more fix $P \in \mathfrak{C}(\varphi_\bullet)$ with $\Phi(P) < \infty$. From the representation formula for $\theta(A)$ in 2) and from 4) we obtain

$$\theta\big(E \cap (P \times Q)\big) = \theta(A) = \int \Psi\big(A(\cdot)\big) d\Phi = \int \chi_P \psi_\bullet\big(E(\cdot) \cap Q\big) d\Phi.$$

At last we form the supremum over all $P \in \mathfrak{C}(\varphi_\bullet)$ with $\Phi(P) < \infty$. As at the end of the proof of 21.15 then assertion ii) follows.

21.19. CONSEQUENCE. *Assume that* $E \in \mathfrak{C}(\vartheta_\bullet)$ *has* $\mathrm{Pr}(E) \subset Y$ *upward enclosable* $[\Psi < \infty]^\sigma$. *Then*

0) $E(x) \in \mathfrak{C}(\psi_\bullet)$ *for all* $x \in X$ *except on some* $N \in \mathfrak{C}(\varphi_\bullet)$ *with* $\Phi(N) = 0$;

i) *the function* $\psi_\bullet(E(\cdot)) : X \to [0,\infty]$ *is measurable* $\mathfrak{C}(\varphi_\bullet)$;

ii) $\theta(E) = \int \psi_\bullet(E(\cdot))d\Phi$.

Proof. This is an immediate consequence of 21.17 combined once more with 10.16 applied to φ and with standard facts in section 13.

As announced above we conclude with the specialization to Radon measures. We assume that $\alpha : \text{Bor}(X) \to [0,\infty]$ and $\beta : \text{Bor}(Y) \to [0,\infty]$ are Borel-Radon measures on Hausdorff topological spaces X and Y, and that $\nu : \text{Bor}(X \times Y) \to [0,\infty]$ is the unique Borel-Radon product measure of α and β obtained in 21.11. As before we form the restrictions $\varphi := \alpha|\mathfrak{S}$ to $\mathfrak{S} := \text{Comp}(X)$ and $\psi := \beta|\mathfrak{T}$ to $\mathfrak{T} := \text{Comp}(Y)$, so that $\alpha = \Phi|\text{Bor}(X)$ and $\beta = \Psi|\text{Bor}(Y)$. The present context allows to take $\mathfrak{P} := \text{Op}(X) \subset \mathfrak{U}$ and $\mathfrak{Q} := \text{Op}(Y) \subset \mathfrak{V}$. From 21.3.1) then $(\mathfrak{P} \times \mathfrak{Q})^\tau = \text{Op}(X \times Y)$ and hence $A\sigma((\mathfrak{P} \times \mathfrak{Q})^\tau) = \text{Bor}(X \times Y)$. Therefore the above results 21.12 and 21.16 with 21.13 specialize as follows. We do not specialize the final 21.19, because there is no particular reason to do so.

21.20. THEOREM. *For each* $E \in \text{Op}(X \times Y)$ *we have*

   0) $E(x) \in \text{Op}(Y)$ *for all* $x \in X$;

   i) *the function* $\beta(E(\cdot)) : X \to [0,\infty]$ *is in* $\text{LSC}(X,[0,\infty])$;

   ii) $\nu(E) = \int \beta(E(\cdot))d\alpha$.

21.21. THEOREM. *For each* $E \in \text{Bor}(X \times Y)$ *we have*

   0) $E(x) \in \text{Bor}(Y)$ *for all* $x \in X$.

If $\text{Pr}(E) \subset Y$ *is upward enclosable* $[\beta < \infty]^\sigma$ *then*

   i) *the function* $\beta(E(\cdot)) : X \to [0,\infty[$ *is Borel measurable;*

   ii) $\nu(E) = \int \beta(E(\cdot))d\alpha$.

We add a drastic example which shows that the size restriction in 21.21, and hence in 21.16 and 21.19, cannot be dispensed with.

21.22. EXAMPLE. Let $X = [0,1]$ with the usual topology, and let $\alpha = \Lambda|\text{Bor}(X)$ be the Borel-Lebesgue measure. Let $Y = [0,1]$ with the discrete topology, and let $\beta : \text{Bor}(Y) = \mathfrak{P}(Y) \to [0,\infty]$ be the counting measure. Both α and β are Borel-Radon measures. The diagonal $D \subset X \times Y$ is closed in the product topology and hence in $\text{Bor}(X \times Y)$. We have on the one hand $D(x) = \{x\} \in \text{Comp}(Y)$ and hence $\beta(D(x)) = 1$ for $x \in X$. Therefore $\int \beta(D(\cdot))d\alpha = 1$. On the other hand, if $Q \in \text{Bor}(Y)$ has $\beta(Q) < \infty$ and hence is finite, then $\beta(D(\cdot) \cap Q) = \chi_Q$ and hence $\nu(D \cap (X \times Q)) = \int \beta(D(\cdot) \cap Q)d\alpha = 0$ from 21.21. Since ν is inner regular $\text{Comp}(X \times Y)$ and since each compact subset of $X \times Y$ is contained in $X \times Q$ for some finite $Q \subset Y$, it follows that $\nu(D) = 0$. Thus $\nu(D) = \int \beta(D(\cdot))d\alpha$ is not true.

21.23. BIBLIOGRAPHICAL NOTE. It is one of the serious failures of traditional measure theory that the abstract theory had no natural method of product formation which in case of two Borel-Radon measures, even on locally compact Hausdorff topological spaces, furnishes an outcome of the

same sort. The site where the need for this was most vital, and where an
appropiate answer then arose, was the context of the Haar measure on locally compact topological groups; see for example Hewitt-Ross [1963] with
its historical notes. The answer was the construction of the Borel-Radon
product measure on locally compact Hausdorff topological spaces via the
traditional Riesz representation theorem 14.2. It became a show-piece in
the development à la Bourbaki, and perhaps an essential motivation for it.
For presentations in recent textbooks we refer to Cohn [1980] chapter 7.6
and Floret [1981] section 13.

For the construction of the Borel-Radon product measure on locally
compact Hausdorff spaces also direct methods were developed, for example
in Johnson [1966] and Kelley-Srinivasan [1988]. We also quote the work
of Bledsoe-Morse [1955] to which we shall come back below. There is a
comprehensive comparison theorem in Godfrey-Sion [1969].

After the Borel-Radon measures had been defined on arbitrary Hausdorff
topological spaces, the theorem on the existence and uniqueness of the Borel-
Radon product measure was extended to this context; see Bourbaki [1969]
section 2.5. In Henry [1969] it has been obtained as an application of the
basic transplantation theorem which is the essence of the present 18.22. For
the present notion of a Borel-Radon measure we refer to the presentation in
Berg-Christensen-Ressel [1984] chapter 2.1. It is even for so-called Radon
bimeasures. Results similar to 21.11 are in Sapounakis-Sion [1987] section
8, as before in the spirit of the two-step extension method of this paper.

After the result of Henry [1969] quoted above one can expect that the
abstract transplantation theorems as treated in chapter VI lead to abstract
versions of the Borel-Radon product measure. This has in fact been achieved
in Adamski [1984a] theorem 3.12, based on the close relative of the present
19.8 quoted at that place. However, it appears that the transplantation
techniques are not an adequate direct method for the present purpose.

As far as the author is aware, there is but one other place in the literature where the Borel-Radon product measure could have been deduced
from abstract measure theory. This is the paper of Bledsoe-Morse [1955]
quoted above. However, this work has different and less simple basic ideas
and is quite technical. For easier descriptions we refer to the final part of
its introduction, to Godfrey-Sion [1969] and Bledsoe-Wilks [1972] section 2,
and to the illustrative exercise 4.4.14 in Dudley [1989].

After all the present main theorem 21.9 unmasks the formation of the
Borel-Radon product measure as a special case of an extensive and simple
principle in abstract measure theory. It appears here for the first time.

22. The Fubini-Tonelli and Fubini Theorems

The present section obtains the Fubini-Tonelli and Fubini theorems
which pertain to the product formations treated above. It is natural that
they come in several versions. The substance of the Fubini-Tonelli theorems

is contained in the above results on sectional representations. What remains is to pass from measurable subsets of the product set to measurable functions with values in $[0, \infty]$, both times in the respective sense. This is the work of a standard theorem on monotone approximation which will be supplied in the first subsection. In view of its importance the first subsection will be more extensive than needed in the sequel.

Monotone Approximation of Functions

Let X be a nonvoid set. We start with the result which will be needed in the next subsection.

22.1. PROPOSITION. *Let $\mathfrak{S}$ be a lattice in X with $\varnothing \in \mathfrak{S}$. For each $f \in \mathrm{UM}(\mathfrak{S}) \cup \mathrm{LM}(\mathfrak{S})$ there exists a sequence $(f_n)_n$ in $\mathrm{S}(\mathfrak{S})$ such that*

i) $f_n \uparrow f$ *pointwise;*

ii) $f_n \leqq f \wedge 2^n \leqq f_n + 1/2^n$.

Note that ii) *implies for $0 < c < \infty$ on $[f \leqq c]$ the uniform estimation $0 \leqq f - f_n \leqq 1/2^n$ when $2^n \geqq c$.*

22.2. ADDENDUM. *The functions f_n can be chosen to be*

$$f_n \quad := \quad (1/2^n) \sum_{l=1}^{2^{2n}} \chi_{[f \geqq l/2^n]} \quad \text{in case } f \in \mathrm{UM}(\mathfrak{S}),$$

$$f_n \quad := \quad (1/2^n) \sum_{l=1}^{2^{2n}} \chi_{[f > l/2^n]} \quad \text{in case } f \in \mathrm{LM}(\mathfrak{S}).$$

Proof of 22.1 and 22.2. We see from 11.4 that the functions f_n defined in 22.2 are in $\mathrm{S}(\mathfrak{S})$. In the sequel we restrict ourselves to the case $\geqq$. It will be obvious that the case $>$ has the same proof. 1) From 11.6 applied to $[a, b] := [0, 2^n]$ and to the subdivision points $t(l) := l/2^n$ for $l = 0, 1, \cdots, 2^{2n}$ we obtain

$$f_n \quad \leqq \quad f \wedge 2^n \leqq (1/2^n) \sum_{l=1}^{2^{2n}} \chi_{[f \geqq (l-1)/2^n]}$$

$$= \quad (1/2^n) \sum_{l=1}^{2^{2n}} \left(\chi_{[f \geqq l/2^n]} + \chi_{[l/2^n > f \geqq (l-1)/2^n]} \right)$$

$$= \quad f_n + (1/2^n) \sum_{l=1}^{2^{2n}} \chi_{[l/2^n > f \geqq (l-1)/2^n]} \leqq f_n + 1/2^n.$$

2) We have

$$[f \geqq l/2^n] = [f \geqq 2l/2^{n+1}] \subset [f \geqq (2l-1)/2^{n+1}] \quad \text{for } l = 0, 1, \cdots, 2^{2n},$$

and therefore

$$f_n \leqq (1/2^{n+1}) \sum_{l=1}^{2^{2n}} \left(\chi_{[f \geqq 2l/2^{n+1}]} + \chi_{[f \geqq (2l-1)/2^{n+1}]} \right)$$

$$= (1/2^{n+1}) \sum_{l=1}^{2^{2(n+1)}} \chi_{[f \geqq l/2^{n+1}]} = f_{n+1}.$$

3) From 1)2) it is clear that $f_n(x) \uparrow f(x)$ at all points $x \in X$ where $f(x) < \infty$. But in case $f(x) = \infty$ we have $f_n(x) = 2^n$ from the definition. This completes the proof.

We know from 11.1 and 11.3 that certain simple manipulations with the function classes $\mathrm{UM}(\mathfrak{S})$ and $\mathrm{LM}(\mathfrak{S})$ require severe limitations. This explains the restrictive assumption in the consequence which follows.

22.3. PROPOSITION. *Let* $\mathfrak{A}$ *be a* σ *algebra in* X. *For each* $f : X \to \overline{\mathbb{R}}$ *measurable* $\mathfrak{A}$ *there exists a sequence* $(f_n)_n$ *of functions* $f_n : X \to \mathbb{R}$ *measurable* $\mathfrak{A}$ *with finite value sets such that*

i) $0 \leqq f_n \uparrow f$ *pointwise on* $[f \geqq 0]$, *and* $0 \geqq f_n \downarrow f$ *pointwise on* $[f \leqq 0]$. *Therefore* $f_n \to f$ *pointwise and* $|f_n| \leqq |f|$.

ii) *For* $0 < c < \infty$ *we have on* $[|f| \leqq c]$ *the uniform estimation* $|f - f_n| \leqq 1/2^n$ *when* $2^n \geqq c$.

Proof. Let $(u_n)_n$ and $(v_n)_n$ be sequences in $S(\mathfrak{A})$ as asserted in 22.1 for the functions $f^+ := f \vee 0$ and $f^- := (-f) \vee 0$. Define

$$f_n := u_n \chi_{[f \geqq 0]} - v_n \chi_{[f \leqq 0]},$$

so that the $f_n : X \to \mathbb{R}$ are measurable $\mathfrak{A}$ with finite value sets. One verifies all assertions.

The final theorem below looks similar but it of different type. We use the notations of chapter V.

22.4. THEOREM. *Let* $E \subset [0, \infty[^X$ *be a primitive lattice cone with* $1 \in E$, *and let* $\mathfrak{A} := \mathrm{A}\sigma\big(\mathfrak{T}(E)\big)$, *that is the smallest* σ *algebra in* X *such that all* $f \in E$ *are measurable* $\mathfrak{A}$. *Assume that* $H \subset [0, \infty[^X$ *contains* E *and is stable under monotone pointwise convergent sequences. Then* H *contains all functions* $f : X \to [0, \infty[$ *measurable* $\mathfrak{A}$.

We define E to consist of all functions sets $H \subset [0, \infty[^X$ as assumed in the theorem. This collection is nonvoid, since it contains the function set

$$\mathrm{M}(\mathfrak{A}) \cap [0, \infty[^X = \{ f \in [0, \infty[^X : f \text{ measurable } \mathfrak{A} \}.$$

Let $\hat{E}$ be the intersection of all $H \in$ E. It is obvious that $\hat{E}$ itself is a member of E. We shall prove that $\hat{E}$ contains all functions $f : X \to [0, \infty[$ measurable $\mathfrak{A}$, and hence coincides with the set $\mathrm{M}(\mathfrak{A}) \cap [0, \infty[^X$ of these functions. However, we do not claim that each $f \in \hat{E}$ is the pointwise limit of some sequence or even monotone sequence in E.

Proof. 1) $\hat{E}$ is a lattice cone. This is a familiar two-step conclusion. In the first step one fixes $f \in E$ and forms

$$H := \{u \in [0, \infty[^X : u + f, u \vee f, u \wedge f, cu \in \hat{E} \text{ for all } c \geqq 0\}.$$

One verifies that $H \in E$ and concludes that H contains $\hat{E}$. In the second step one does the same with $f \in \hat{E}$. Then the assertion follows. 2) We claim that

$$\chi_{[u<v]}, \chi_{[u\leqq v]} \in \hat{E} \quad \text{for all } u, v \in E.$$

In fact, first note that $(n(v - u \wedge v)) \wedge 1 \in E$ and $\uparrow \chi_{[u<v]}$, so that $\chi_{[u<v]} \in \hat{E}$. Then $\chi_{[u<v+1/n]} \downarrow \chi_{[u\leqq v]}$ for $n \to \infty$, so that $\chi_{[u\leqq v]} \in \hat{E}$. 3) We form $\mathfrak{M} := \{A \subset X : \chi_A \in \hat{E}\}$ and $\mathfrak{N} := \mathfrak{M} \cap (\mathfrak{M}\perp)$. From 2) we see that $[u \leqq v] \in \mathfrak{N}$ for all $u, v \in E$. In particular $[f \geqq t] \in \mathfrak{N}$ for all $f \in E$ and $t > 0$, that is $\mathfrak{T}(E) \subset \mathfrak{N}$. 4) From 1) and from the definition of $\hat{E}$ we conclude that $\mathfrak{M}$ fulfils $\cup\sigma$ and $\cap\sigma$. The same then follows for $\mathfrak{N}$. Combined with 3) this implies that $\mathfrak{N}$ is a σ algebra and $\mathfrak{A} = A\sigma(\mathfrak{T}(E)) \subset \mathfrak{N} \subset \mathfrak{M}$. 5) Thus we have $\chi_A \in \hat{E}$ for all $A \in \mathfrak{A}$. From 1) and 22.1 the assertion follows.

22.5. BIBLIOGRAPHICAL NOTE. The assertions 22.1 to 22.3 are mathematical folklore. Results of the type 22.4 are called monotone class theorems for functions. For related versions we refer to Revuz-Yor [1991] theorem 0.2.2 and to Hackenbroch-Thalmaier [1994] Satz 1.4.

The Fubini-Tonelli Theorems

In the last two sections on product formations we obtained four final results on sectional representations. These were

in section 20: 20.18 with 20.15;

in section 21: 21.12, and its specialization 21.20 to Radon measures;
   21.16 with 21.13, and its specialization 21.21 to Radon measures;
   21.19.

Between these results there are substantial differences. They are all important in their specific contexts, so that none of them should be put aside. Therefore we shall extend all four of them to theorems of the Fubini-Tonelli type. The proofs will follows the same pattern based on 22.1. We start with the traditional product situation of section 20.

22.6. THEOREM. *Assume that*

$$\alpha : \mathfrak{A} \to [0, \infty] \text{ is a cmeasure on a } \sigma \text{ algebra } \mathfrak{A} \text{ in } X,$$
$$\beta : \mathfrak{B} \to [0, \infty] \text{ is a cmeasure on a } \sigma \text{ algebra } \mathfrak{B} \text{ in } Y,$$

and that $\nu : \mathfrak{A} \otimes \mathfrak{B} \to [0, \infty]$ *is a narrow product measure of* α *and* β. *Let* $f : X \times Y \to [0, \infty]$ *be measurable* $\mathfrak{A} \otimes \mathfrak{B}$. *Then*

0) $f(x, \cdot) : Y \to [0, \infty]$ *is measurable* $\mathfrak{B}$ *for all* $x \in X$.

If $\Pr([f > 0]) \subset Y$ *is upward enclosable* $[\beta < \infty]^\sigma$ *then*

i) *the function* $f^0 : f^0(x) = \int f(x, \cdot)d\beta$ *for* $x \in X$ *is measurable* $\mathfrak{A}$.

If $[f > 0] \subset X \times Y$ *is upward enclosable* $[\alpha < \infty]^\sigma \times [\beta < \infty]^\sigma$ *then*

  ii) $\int f d\nu = \int f^0 d\alpha$.

The proof below and the subsequent ones will need two obvious formal remarks. 0) If $f : X \times Y \to [0, \infty]$ and $t > 0$ then $[f(x, \cdot) \geqq t] = [f \geqq t](x)$ for all $x \in X$. The same holds true for $>$ instead of $\geqq$. 1) If $f = \chi_E$ with $E \subset X \times Y$ then $f(x, \cdot) = \chi_{E(x)}$ for all $x \in X$.

Proof of 22.6. 0) follows from 20.15.0) combined with formal remark 0). i) Define H to consist of all functions $f : X \times Y \to [0, \infty]$ measurable $\mathfrak{A} \otimes \mathfrak{B}$ which fulfil i). From standard facts in section 13 it is clear that H is a cone and stable under isotone sequences. From 20.15.i) combined with formal remark 1) we see that H contains $f = \chi_E$ when $E \in \mathfrak{A} \otimes \mathfrak{B}$ has $\mathrm{Pr}(E) \subset Y$ upward enclosable $[\beta < \infty]^\sigma$. Therefore 22.1 implies that H contains all $f : X \times Y \to [0, \infty]$ with $\mathrm{Pr}([f > 0]) \subset Y$ upward enclosable $[\beta < \infty]^\sigma$. ii) Define H to consist of all functions $f : X \times Y \to [0, \infty]$ measurable $\mathfrak{A} \otimes \mathfrak{B}$ which fulfil i)ii). From 20.18 we conclude as before that H contains all $f : X \times Y \to [0, \infty]$ measurable $\mathfrak{A} \otimes \mathfrak{B}$ with $[f > 0] \subset X \times Y$ upward enclosable $[\alpha < \infty]^\sigma \times [\beta < \infty]^\sigma$.

The next theorems resume the situation of section 21. We assume as above that

$$\varphi : \mathfrak{S} \to [0, \infty[\text{ is an inner } \bullet \text{ premeasure with } \varphi(\varnothing) = 0,$$
$$\psi : \mathfrak{T} \to [0, \infty[\text{ is an inner } \bullet \text{ premeasure with } \psi(\varnothing) = 0,$$

and let $\Phi := \varphi_\bullet | \mathfrak{C}(\varphi_\bullet)$ and $\Psi := \psi_\bullet | \mathfrak{C}(\psi_\bullet)$, where $\bullet = \sigma\tau$. Furthermore let $\mathfrak{U} := (\mathfrak{S}\top\mathfrak{S}_\bullet)\bot$ and $\mathfrak{V} := (\mathfrak{T}\top\mathfrak{T}_\bullet)\bot$. We know from 21.9 that $\vartheta = \varphi \times \psi : \mathfrak{R} \to [0, \infty[$ is an inner $\bullet$ premeasure as well, and that $\theta := \vartheta_\bullet | \mathfrak{C}(\vartheta_\bullet)$ is an extension of $\Phi \times \Psi$.

22.7. THEOREM. *Assume that* $\mathfrak{P} \subset \mathfrak{U}$ *and* $\mathfrak{Q} \subset \mathfrak{V}$ *are lattices with* $\cup\bullet$. *Let* $f \in \mathrm{LM}((\mathfrak{P} \times \mathfrak{Q})^\bullet)$. *Then*

  0) $f(x, \cdot) \in \mathrm{LM}(\mathfrak{Q})$ *for all* $x \in X$;
  i) *the function* $f^0 : f^0(x) = \int f(x, \cdot) d\Psi$ *for* $x \in X$ *is in* $\mathrm{LM}(\mathfrak{P})$;
  ii) $\int f d\theta = \int f^0 d\Phi$.

Proof. 0) follows from 21.12.0) combined with formal remark 0). i) Define H to consist of all functions $f \in \mathrm{LM}((\mathfrak{P} \times \mathfrak{Q})^\bullet)$ which fulfil i)ii). We note that the classes $\mathrm{LM}(\cdot)$ which occur are cones by 11.1.3) and stable under isotone sequences by the definition. Then we conclude from 11.11 and 11.18, or from standard facts in section 13, that H is a cone and stable under isotone sequences. From 21.12.i)ii) combined with formal remark 1) we see that H contains $f = \chi_E$ when $E \in (\mathfrak{P} \times \mathfrak{Q})^\bullet$. Therefore 22.1 implies that H contains all $f \in \mathrm{LM}((\mathfrak{P} \times \mathfrak{Q})^\bullet)$.

22.8. THEOREM. *Assume that* $\mathfrak{P} \subset \mathfrak{U}$ *and* $\mathfrak{Q} \subset \mathfrak{V}$ *are lattices with* $\cup\bullet$. *Let* $f : X \times Y \to [0, \infty]$ *be measurable* $\mathrm{A}\sigma((\mathfrak{P} \times \mathfrak{Q})^\bullet)$. *Then*

  0) $f(x, \cdot) : Y \to [0, \infty]$ *is measurable* $\mathrm{A}\sigma(\mathfrak{Q})$ *for all* $x \in X$.

If $\mathrm{Pr}([f > 0]) \subset Y$ *is upward enclosable* $[\Psi < \infty]^\sigma$ *then*

i) *the function* $f^0 : f^0(x) = \int f(x, \cdot)d\Psi$ *for* $x \in X$ *is measurable* $A\sigma(\mathfrak{P})$;

ii) $\int f d\theta = \int f^0 d\Phi$.

This time the proof is so close to the former ones that we can save the details.

22.9. THEOREM. *Let* $f : X \times Y \to [0, \infty]$ *be measurable* $\mathfrak{C}(\vartheta_\bullet)$, *and let* $\mathrm{Pr}([f > 0]) \subset Y$ *be upward enclosable* $[\Psi < \infty]^\sigma$. *Then there exists a subset* $N \in \mathfrak{C}(\varphi_\bullet)$ *with* $\Phi(N) = 0$ *such that*

0) $f(x, \cdot) : Y \to [0, \infty]$ *is measurable* $\mathfrak{C}(\psi_\bullet)$ *for all* $x \in X \setminus N$;

i) *each function* $f^0 : X \to [0, \infty]$ *with* $f^0(x) = \int f(x, \cdot)d\Psi$ *for* $x \in X \setminus N$ *is measurable* $\mathfrak{C}(\varphi_\bullet)$;

ii) $\int f d\theta = \int f^0 d\Phi$.

Proof. Define H to consist of all functions $f : X \times Y \to [0, \infty]$ measurable $\mathfrak{C}(\vartheta_\bullet)$ for which there exists a subset $N \in \mathfrak{C}(\varphi_\bullet)$ with $\Phi(N) = 0$ and with 0)i)ii). We conclude from 10.16 that instead of N one can take any other null set which contains N. By standard facts in section 13 thus H is a cone and stable under isotone sequences. From 21.19.i)ii) combined with 10.16 and formal remark 1) we see that H contains $f = \chi_E$ when $E \in \mathfrak{C}(\vartheta_\bullet)$ has $\mathrm{Pr}(E) \subset Y$ upward enclosable $[\Psi < \infty]^\sigma$. Therefore 22.1 implies that H contains all $f : X \times Y \to [0, \infty]$ measurable $\mathfrak{C}(\vartheta_\bullet)$ such that $\mathrm{Pr}([f > 0]) \subset Y$ is upward enclosable $[\Psi < \infty]^\sigma$.

The above theorems have the usual consequences which assert that under the obvious assumptions the order of the two componental integration processes can be reversed. We shall not enter the details, and do the same in the next subsection on the Fubini theorems.

As in section 21 we conclude with the specialization of 22.7 and 22.8 to Radon measures. We assume that $\alpha : \mathrm{Bor}(X) \to [0, \infty]$ and $\beta : \mathrm{Bor}(Y) \to [0, \infty]$ are Borel-Radon measures on Hausdorff topological spaces X and Y, and that $\nu : \mathrm{Bor}(X \times Y) \to [0, \infty]$ is the unique Borel-Radon product measure of α and β obtained in 21.11.

22.10. THEOREM. *Let* $f \in \mathrm{LSC}(X \times Y, [0, \infty])$. *Then*

0) $f(x, \cdot) \in \mathrm{LSC}(Y, [0, \infty])$ *for all* $x \in X$;

i) *the function* $f^0 : f^0(x) = \int f(x, \cdot)d\beta$ *for* $x \in X$ *is in* $\mathrm{LSC}(X, [0, \infty])$;

ii) $\int f d\nu = \int f^0 d\alpha$.

22.11. THEOREM. *Assume that* $f : X \times Y \to [0, \infty]$ *is Borel measurable. Then*

0) $f(x, \cdot) : Y \to [0, \infty]$ *is Borel measurable for all* $x \in X$.

If $\mathrm{Pr}([f > 0]) \subset Y$ *is upward enclosable* $[\beta < \infty]^\sigma$ *then*

i) *the function* $f^0 : f^0(x) = \int f(x, \cdot)d\beta$ *for* $x \in X$ *is Borel measurable;*

ii) $\int f d\nu = \int f^0 d\alpha$.

The Fubini Theorems

The Fubini theorem for each of the present contexts arises from the respective Fubini-Tonelli theorem when one passes from measurable functions with values in $[0, \infty]$ to integrable functions with values in $\overline{\mathbb{R}}$. One knows that it is often less simple. In the present subsection we obtain the Fubini theorems

> for the context of section 20 from 22.6, and
> for the context of section 21 from 22.8 and 22.9,

while 22.7 has no Fubini counterpart. We even attempt a simultaneous proof.

22.12. THEOREM (Addendum to 22.6). *Let* $f : X \times Y \to \overline{\mathbb{R}}$ *be integrable* ν, *and let* $[f \neq 0] \subset X \times Y$ *be upward enclosable* $[\alpha < \infty]^\sigma \times [\beta < \infty]^\sigma$. *Then there exists a subset* $N \in \mathfrak{A}$ *with* $\alpha(N) = 0$ *such that*

0) $f(x, \cdot) : Y \to \overline{\mathbb{R}}$ *is integrable* β *for all* $x \in X \setminus N$;

i) *the function* $f^0 : X \to \mathbb{R}$, *defined to be*

$$f^0(x) = \left\{ \begin{array}{ll} \int f(x, \cdot)d\beta & \text{for } x \in X \setminus N \\ 0 & \text{for } x \in N \end{array} \right\}, \text{is integrable } \alpha;$$

ii) $\int f d\nu = \int f^0 d\alpha$.

In short words: There exists a null set $N \in \mathfrak{A}$ such that the iterated integral $\int\limits_{X \setminus N} \left(\int f(x, y)d\beta(y) \right) d\alpha(x)$ exists in the sense of integrable functions and is $= \int f d\nu$.

22.13. THEOREM (Addendum to 22.8 and 22.9). *We adopt the natural abbreviations*

$$\begin{array}{lllll} \text{in case 22.8:} & \alpha & := \Phi | A\sigma(\mathfrak{P}) & \text{on } \mathfrak{A} & := A\sigma(\mathfrak{P}), \\ & \beta & := \Psi | A\sigma(\mathfrak{Q}) & \text{on } \mathfrak{B} & := A\sigma(\mathfrak{Q}), \\ & \nu & := \theta | A\sigma((\mathfrak{P} \times \mathfrak{Q})^\bullet) & \text{on } \mathfrak{C} & := A\sigma((\mathfrak{P} \times \mathfrak{Q})^\bullet); \\ \text{in case 22.9:} & \alpha & := \Phi & \text{on } \mathfrak{A} & := \mathfrak{C}(\varphi_\bullet), \\ & \beta & := \Psi & \text{on } \mathfrak{B} & := \mathfrak{C}(\psi_\bullet), \\ & \nu & := \theta & \text{on } \mathfrak{C} & := \mathfrak{C}(\theta_\bullet). \end{array}$$

Let $f : X \times Y \to \overline{\mathbb{R}}$ *be integrable* ν, *and let* $\mathrm{Pr}([f \neq 0]) \subset Y$ *be upward enclosable* $[\Psi < \infty]^\sigma$. *Then there exists a subset* $N \in \mathfrak{A}$ *with* $\alpha(N) = 0$ *such that*

0) $f(x, \cdot) : Y \to \overline{\mathbb{R}}$ *is integrable* β *for all* $x \in X \setminus N$;

i) *the function* $f^0 : X \to \mathbb{R}$, *defined to be*

$$f^0(x) = \left\{ \begin{array}{ll} \int f(x, \cdot)d\beta & \text{for } x \in X \setminus N \\ 0 & \text{for } x \in N \end{array} \right\}, \text{is integrable } \alpha;$$

ii) $\int f d\nu = \int f^0 d\alpha$.

In short words: There exists a null set $N \in \mathfrak{A}$ such that the iterated integral $\int_{X \setminus N} \left(f(x,y)d\beta(y) \right) d\alpha(x)$ exists in the sense of integrable functions and is $= \int f d\nu$.

Proof of 22.12 and 22.13. 1) From the respective former theorems applied to $f^\pm : X \times Y \to [0, \infty]$ we obtain a subset $M \in \mathfrak{A}$ with $\alpha(M) = 0$ such that

0) $f^\pm(x, \cdot) : Y \to [0, \infty]$ are measurable $\mathfrak{B}$ for all $x \in X \setminus M$;

i) the functions $(f^\pm)^0 : X \to [0, \infty]$, defined to be

$$(f^\pm)^0(x) = \left\{ \begin{array}{ll} \int f^\pm(x, \cdot)d\beta & \text{for } x \in X \setminus M \\ 0 & \text{for } x \in M \end{array} \right\}, \text{ are measurable } \mathfrak{A};$$

ii) $\int f^\pm d\nu = \int (f^\pm)^0 d\alpha < \infty$.

One can take $M = \varnothing$ in cases 22.6 and 22.8. We note that $f = f^+ \dotplus (-f^-)$. 2) In view of ii) the functions $(f^\pm)^0$ are integrable α. Thus by 13.23.1) there exists a subset $N \in \mathfrak{A}$ with $\alpha(N) = 0$ such that $(f^\pm)^0(x) < \infty$ for all $x \in X \setminus N$. We can assume that $M \subset N$. It follows that for $x \in X \setminus N$ the functions $f^\pm(x, \cdot)$ and hence the function $f(x, \cdot)$ are integrable β. 3) Define $f^0 : X \to \mathbb{R}$ as in i) of the theorems. By 13.20.2) then $f^0(x) = (f^+)^0(x) - (f^-)^0(x)$ for $x \in X \setminus N$. Hence f^0 is integrable α, and once more from 13.20.2) we have

$$\int f^0 d\alpha = \int (f^+)^0 d\alpha - \int (f^-)^0 d\alpha = \int f^+ d\nu - \int f^- d\nu = \int f d\nu.$$

The proof is complete.

This time we shall not write out the specialization to Radon measures. We have said above that 22.7 has no Fubini counterpart. The part of 22.13 which comes from 22.8 has been formulated in such a manner that it coincides with its specialization when one fortifies the assumption upward enclosable $[\Psi < \infty]^\sigma$ to upward enclosable $[\beta < \infty]^\sigma$.

22.14. BIBLIOGRAPHICAL NOTE. The traditional Fubini-Tonelli and Fubini theorems 22.6 and 22.12 are in all textbooks.

For the specialization of the further Fubini-Tonelli and Fubini theorems to Borel-Radon measures on *locally compact* Hausdorff topological spaces we refer to Hewitt-Ross [1963] section 13, Bourbaki [1967] section 8, and Cohn [1980] section 7.6.

For Borel-Radon measures on arbitrary Hausdorff topological spaces there are versions of the present 22.10 and 22.9 in Bourbaki [1969] section 2.6. The full result 22.10 and the essence of 22.11 are in Berg-Christensen-Ressel [1984] theorem 2.1.12.

CHAPTER VIII

Applications of the New Contents and Measures

The present final chapter is independent of the extension theories of chapter II and of the subsequent chapters III and V to VII. It returns to chapter I and in particular to the new notions of contents and measures defined in section 2. The main purpose of these notions was to form the soil on which the outer and inner extension theories of chapter II became identical. The present chapter wants to demonstrate that the new notions can be useful in other contexts as well, in order to obtain more reasonable forms of the results, or to contribute to their proofs and to their mutual relations. We consider the domain of some famous decomposition theorems.

23. The Jordan and Hahn Decomposition Theorems

We start with the Jordan theorem which is for contents, like the Lebesgue decomposition theorem of the next section. The subsequent Hahn theorem will be for measures.

Introduction

Let $\mathfrak{A}$ be an algebra in a nonvoid set X. We start with the conventional versions of the Jordan decomposition problem.

The finite version assumes a modular set function $\varphi : \mathfrak{A} \to \mathbb{R}$ with $\varphi(\varnothing) = 0$. Wanted are representations $\varphi = \beta - \alpha$, where $\alpha, \beta : \mathfrak{A} \to [0, \infty[$ are finite ccontents. An obvious necessary condition for the existence of such a representation is that φ be bounded, because it implies that $-\alpha(X) \leqq \varphi \leqq \beta(X)$. The converse result will be that if φ is bounded then there exist representations, and a unique one which is optimal in an appropriate sense. We add a useful remark.

23.1. REMARK. 1) A modular set function $\varphi : \mathfrak{A} \to \mathbb{R}$ is bounded above iff it is bounded below. 2) There exist modular set functions $\varphi : \mathfrak{A} \to \mathbb{R}$ which are not bounded. For an example let $\mathfrak{A}$ consist of the finite and cofinite subsets of an infinite X, and define $\varphi : \mathfrak{A} \to \mathbb{R}$ to be

$$\varphi(A) = \left\{ \begin{array}{ll} \#(A) & \text{for } A \text{ finite} \\ -\#(A') & \text{for } A \text{ cofinite} \end{array} \right\}.$$

It is a simple verification that φ is modular.

The infinite version assumes, in order to avoid the difficulties with the addition in $\overline{\mathbb{R}}$ for which we have invented the operations $\dotplus$ and $\dotplus$, for instance a modular set function $\varphi : \mathfrak{A} \to]-\infty, \infty]$ with $\varphi(\varnothing) = 0$. Wanted are representations $\varphi = \beta - \alpha$, where this time of course $\alpha : \mathfrak{A} \to [0, \infty[$ is a finite ccontent and $\beta : \mathfrak{A} \to [0, \infty]$ is a ccontent. An obvious necessary condition for the existence of such a representation is that φ be bounded below, because it implies that $\varphi \geq -\alpha(X)$. The converse result will be that if φ is bounded below then there exist representations, and a unique one which is optimal in an appropriate sense.

We want to remove the asymmetry in the latter representations. We shall see that it is not due to the restriction from $\overline{\mathbb{R}}$ to $]-\infty, \infty]$, but that the condition $\varphi(\varnothing) = 0$ must be blamed for it. Thus we assume a modular set function $\varphi : \mathfrak{A} \to]-\infty, \infty]$ which is $\neq \infty$, so that in particular we do not require that $\varphi(\varnothing) < \infty$. Wanted are representations

$$\varphi(A) = \alpha(A') + \beta(A) \quad \text{for all } A \in \mathfrak{A},$$

where $\alpha, \beta : \mathfrak{A} \to]-\infty, \infty]$ are isotone modular set functions $\neq \infty$, that means contents in the new sense. Then α and β are bounded below with $\alpha(\varnothing), \beta(\varnothing) \in \mathbb{R}$, and hence $\varphi \geq \alpha(\varnothing) + \beta(\varnothing)$ is likewise bounded below. As before the converse result will be that if φ is bounded below then there exist representations, and (up to an additive real constant) a unique one which is optimal in an appropriate sense. The new representations fulfil

$$\varphi(\varnothing) = \alpha(X) + \beta(\varnothing) \quad \text{and} \quad \varphi(X) = \alpha(\varnothing) + \beta(X).$$

Thus $\varphi(\varnothing) < \infty$ is equivalent to $\alpha < \infty$, and $\varphi(X) < \infty$ is equivalent to $\beta < \infty$. We assert that in case $\varphi(\varnothing) = 0$ the new representations are equivalent to the former ones. In fact, a new representation furnishes the old one

$$\varphi(A) = \alpha(A') + \beta(A) = \big(\beta(A) - \beta(\varnothing)\big) - \big(\alpha(A) - \alpha(\varnothing)\big) \quad \text{for } A \in \mathfrak{A},$$

and an old representation furnishes the new one

$$\varphi(A) = \beta(A) - \alpha(A) = \big(\alpha(A') - \alpha(X)\big) + \beta(A) \quad \text{for } A \in \mathfrak{A}.$$

In the sequel we shall concentrate on the new concepts.

We conclude the introduction with a short discussion of what can happen when one passes from $]-\infty, \infty]$ to $\overline{\mathbb{R}}$.

23.2. REMARK. Assume that $\varphi : \mathfrak{A} \to]-\infty, \infty]$ is modular and not bounded below. Then it cannot be represented in the above sense. But φ can have representations

$$\varphi(A) = \alpha(A') \dotplus \beta(A) \quad \text{for all } A \in \mathfrak{A},$$

where $\alpha, \beta : \mathfrak{A} \to \overline{\mathbb{R}}$ are contents $\dotplus$ in the new sense. For an example let $\mathfrak{A}$ consist of the finite and cofinite subsets of an infinite X, and define $\varphi : \mathfrak{A} \to]-\infty, \infty]$ to be

$$\varphi(A) = \left\{ \begin{array}{ll} -\#(A) & \text{for } A \text{ finite} \\ \infty & \text{for } A \text{ cofinite} \end{array} \right\}.$$

One verifies that $\alpha : \mathfrak{A} \to [-\infty, 0]$ and $\beta : \mathfrak{A} \to [0, \infty]$, defined to be

$$\alpha(A) = \left\{ \begin{array}{ll} -\infty & \text{for } A \text{ finite} \\ -\#(A') & \text{for } A \text{ cofinite} \end{array} \right\} \text{ and}$$

$$\beta(A) = \left\{ \begin{array}{ll} 0 & \text{for } A \text{ finite} \\ \infty & \text{for } A \text{ cofinite} \end{array} \right\},$$

are as required. However, in the present case all such representations are pathological in a sense. This is the interpretation which will later be attributed to the final remark below.

We recall from 2.14 that each $\dot{+}$ content $\alpha : \mathfrak{A} \to \overline{\mathbb{R}}$ defines a certain ccontent $\alpha^\wedge : \mathfrak{A} \to [0, \infty]$. In particular if $\alpha(\varnothing) \in \mathbb{R}$ then $\alpha^\wedge = \alpha - \alpha(\varnothing)$.

23.3. REMARK. *Assume that* $\varphi : \mathfrak{A} \to]-\infty, \infty]$ *is modular and not bounded below. If* $\alpha, \beta : \mathfrak{A} \to \overline{\mathbb{R}}$ *are contents* $\dot{+}$ *with*

$$\varphi(A) = \alpha(A') \dot{+} \beta(A) \quad \text{for all } A \in \mathfrak{A},$$

then $\alpha^\wedge(A') + \beta^\wedge(A) = \infty$ *for all* $A \in \mathfrak{A}$.

Proof. Assume not, and fix $Q \in \mathfrak{A}$ with $\alpha^\wedge(Q') + \beta^\wedge(Q) < \infty$ and hence with $\alpha^\wedge(Q'), \beta^\wedge(Q) < \infty$. Also fix $P \in \mathfrak{A}$ with $\varphi(P) \in \mathbb{R}$ and hence with $\alpha(P'), \beta(P) \in \mathbb{R}$. We claim that $\alpha(\varnothing), \beta(\varnothing) \in \mathbb{R}$, which implies a contradiction. 1) By definition we have

$$\alpha^\wedge(Q') = \big(\alpha(P' \cup Q') - \alpha(P')\big) + \big(\alpha(P') - \alpha(P' \cap Q)\big),$$
$$\beta^\wedge(Q) = \big(\beta(P \cup Q) - \beta(P)\big) + \big(\beta(P) - \beta(P \cap Q')\big).$$

This implies that

1.i) $\alpha(P' \cup Q') \in \mathbb{R}$ and $\beta(P \cup Q) \in \mathbb{R}$,

1.ii) $\alpha(P' \cap Q) \in \mathbb{R}$ and $\beta(P \cap Q') \in \mathbb{R}$.

From 1.i) we see that $\alpha(Q') < \infty$ and $\beta(Q) < \infty$. On the other hand $\varphi(Q) = \alpha(Q') \dot{+} \beta(Q) > -\infty$. Therefore $\alpha(Q'), \beta(Q) \in \mathbb{R}$. 2) From

$$\alpha(P' \cup Q') \dot{+} \alpha(P' \cap Q') = \alpha(P') \dot{+} \alpha(Q'),$$
$$\beta(P \cup Q) \dot{+} \beta(P \cap Q) = \beta(P) \dot{+} \beta(Q),$$

and from the finiteness of the right sides we conclude that $\alpha(P' \cap Q') \in \mathbb{R}$ and $\beta(P \cap Q) \in \mathbb{R}$. 3) Now

$$\alpha(P' \cap Q') \dot{+} \alpha(P' \cap Q) = \alpha(P') \dot{+} \alpha(\varnothing),$$
$$\beta(P \cap Q) \dot{+} \beta(P \cap Q') = \beta(P) \dot{+} \beta(\varnothing).$$

On the left the first terms are finite by 2), and the second terms are finite by 1.ii). It follows that $\alpha(\varnothing), \beta(\varnothing) \in \mathbb{R}$ as claimed.

The Infimum Formation

Let as before $\mathfrak{A}$ be an algebra in a nonvoid set X.

23.4. THEOREM. *Assume that $\alpha, \beta : \mathfrak{A} \to]-\infty, \infty]$ are modular and bounded below. Define*

$$\lambda : \lambda(A) = \inf\{\alpha(A \cap P') + \beta(A \cap P) : P \in \mathfrak{A}\} \quad \text{for } A \in \mathfrak{A}.$$

Then $\lambda : \mathfrak{A} \to]-\infty, \infty]$ is modular and bounded below.

23.5. ADDENDUM. *Let $M \in \mathfrak{A}$ with $\lambda(M) < \infty$, and let $(P_l)_l$ be a sequence in $\mathfrak{A}$ with*

$$\alpha(M \cap P_l') + \beta(M \cap P_l) \to \lambda(M) \quad \text{for } l \to \infty.$$

Then $\alpha(A \cap P_l') + \beta(A \cap P_l) \to \lambda(A)$ for all $A \in \mathfrak{A}$ with $A \subset M$.

The proof requires a simple remark.

23.6. REMARK. *Let $\alpha : \mathfrak{A} \to \overline{\mathbb{R}}$ be modular $\dotplus$. If $P \subset A \subset Q$ in $\mathfrak{A}$ with $\alpha(P), \alpha(Q) \in \mathbb{R}$ then $\alpha(A) \in \mathbb{R}$.*

Proof of 23.6. For $B := Q|A|P = (Q \setminus A) \cup P \in \mathfrak{A}$ we have $A \cup B = Q$ and $A \cap B = P$. Therefore

$$\alpha(A) \dotplus \alpha(B) = \alpha(P) \dotplus \alpha(Q) \in \mathbb{R}.$$

The assertion follows.

Proof of 23.4 and 23.5. We can assume that α and β are $\not\equiv \infty$, since otherwise $\lambda \equiv \infty$. It is obvious that λ is bounded below. The proof that λ is modular will be combined with that of the addendum.

1) If $A, B \in \mathfrak{A}$ with $\lambda(A), \lambda(B) < \infty$ then $\lambda(A \cup B) < \infty$. In fact, by definition there exist

$$P \in \mathfrak{A} \text{ with } P \subset A \text{ and } \alpha(A \setminus P), \beta(P) \in \mathbb{R},$$
$$Q \in \mathfrak{A} \text{ with } Q \subset B \text{ and } \alpha(B \setminus Q), \beta(Q) \in \mathbb{R}.$$

Since α and β are modular we have $\beta(P \cup Q) \in \mathbb{R}$, and

$$\alpha\big((A \setminus P) \cap (B \setminus Q)\big) \in \mathbb{R} \quad \text{and} \quad \alpha\big((A \setminus P) \cup (B \setminus Q)\big) \in \mathbb{R}.$$

One verifies that

$$(A \setminus P) \cap (B \setminus Q) \subset (A \cup B) \setminus (P \cup Q) \subset (A \setminus P) \cup (B \setminus Q).$$

Thus 23.6 implies that $\alpha\big((A \cup B) \setminus (P \cup Q)\big) \in \mathbb{R}$. The results combine to furnish $\lambda(A \cup B) < \infty$. 2) We have to prove that

$$\lambda(A) + \lambda(B) = \lambda(A \cup B) + \lambda(A \cap B) \quad \text{for } A, B \in \mathfrak{A}.$$

In case $\lambda(A \cup B) = \infty$ this follows from 1), and in case $\lambda(A \cup B) < \infty$ it is a consequence of 23.5. Thus it remains to prove the addendum.

3) We need an intermediate computation. Let P, Q and $A \subset M$ be subsets of X, and put $Z := Q|A|P$. One verifies that

$$A \cap P'|A|M \cap Q' = A \cap Q' \quad \text{and} \quad A \cap P'|A'|M \cap Q' = M \cap Z',$$
$$A \cap P|A|M \cap Q = A \cap Q \quad \text{and} \quad A \cap P|A'|M \cap Q = M \cap Z.$$

Thus if $P, Q, A, M \in \mathfrak{A}$ and hence $Z \in \mathfrak{A}$ then

$$\left(\alpha(A \cap Q') + \beta(A \cap Q)\right) + \left(\alpha(M \cap Z') + \beta(M \cap Z)\right)$$
$$= \quad \alpha\left((A \cap P') \cup (M \cap Q')\right) + \alpha\left((A \cap P') \cap (M \cap Q')\right)$$
$$+ \quad \beta\left((A \cap P) \cup (M \cap Q)\right) + \beta\left((A \cap P) \cap (M \cap Q)\right)$$
$$= \quad \left(\alpha(A \cap P') + \beta(A \cap P)\right) + \left(\alpha(M \cap Q') + \beta(M \cap Q)\right).$$

4) Assume that $A \subset M$ in $\mathfrak{A}$. From 3) we have

$$\left(\alpha(A \cap Q') + \beta(A \cap Q)\right) + \lambda(M)$$
$$\leqq \quad \left(\alpha(A \cap P') + \beta(A \cap P)\right) + \left(\alpha(M \cap Q') + \beta(M \cap Q)\right) \quad \text{for } P, Q \in \mathfrak{A}.$$

The definition of $\lambda(A)$ furnishes

$$\left(\alpha(A \cap Q') + \beta(A \cap Q)\right) + \lambda(M)$$
$$\leqq \quad \lambda(A) + \left(\alpha(M \cap Q') + \beta(M \cap Q)\right) \quad \text{for } Q \in \mathfrak{A}.$$

Now assume that $\lambda(M) < \infty$, and let $(P_l)_l$ be as in 23.5. Then for $Q := P_l$ and $l \to \infty$ we obtain the assertion.

23.7. SPECIAL CASE. *Assume that* $\alpha, \beta : \mathfrak{A} \to [0, \infty]$ *are ccontents. Then* 23.4 *furnishes a ccontent* $\lambda : \mathfrak{A} \to [0, \infty]$. *It has the properties*

i) $\lambda \leqq \alpha, \beta$;

ii) *each ccontent* ϑ *on* $\mathfrak{A}$ *with* $\vartheta \leqq \alpha, \beta$ *fulfils* $\vartheta \leqq \lambda$.

Hence it is the unique ccontent on $\mathfrak{A}$ *with these properties.*

The assertions are all obvious. The above special case is the reason for the infimum type notation $\lambda =: \alpha \wedge \beta$ in all cases. Note that this formation is symmetric.

We are led to a basic definition. The ccontents $\alpha, \beta : \mathfrak{A} \to [0, \infty]$ are defined to be **singular** (to each other) iff $\alpha \wedge \beta = 0$, that is iff

$$\inf\{\alpha(P') + \beta(P) : P \in \mathfrak{A}\} =: (\alpha \wedge \beta)(X) = 0.$$

In this case we write $\alpha \perp \beta$. Moreover the $\dotplus$ contents $\alpha, \beta : \mathfrak{A} \to \overline{\mathbb{R}}$ are called **singular** iff $\alpha^\wedge, \beta^\wedge$ are singular. In this case we likewise write $\alpha \perp \beta$.

23.8. REMARK. *Let* $\alpha, \beta : \mathfrak{A} \to [0, \infty]$ *be ccontents. Then*

α, β *are singular* $\Longleftarrow$ *there exists* $P \in \mathfrak{A}$ *with* $\alpha(P') = \beta(P) = 0$.

Moreover we have $\Longrightarrow$ *when* $\mathfrak{A}$ *is a* σ *algebra and* α, β *are cmeasures.*

Proof. $\Leftarrow$) is obvious. $\Rightarrow$) In case $(\alpha \wedge \beta)(X) = 0$ there exists a sequence $(A_l)_l$ in $\mathfrak{A}$ with $\alpha(A_l'), \beta(A_l) \leqq 1/2^{l+1}$. For $P_n := \bigcup_{l=n}^{\infty} A_l \in \mathfrak{A}$ then on the one hand $\beta(P_n) \leqq 1/2^n$. On the other hand

for $l \geqq n : A_l \subset P_n$ or $P_n' \subset A_l'$ and hence $\alpha(P_n') \leqq 1/2^{l+1}$,

so that $\alpha(P_n') = 0$. Now $P_n \downarrow$ some $P \in \mathfrak{A}$. It follows that $\beta(P) = 0$ and $\alpha(P') = 0$.

23.9. BIBLIOGRAPHICAL NOTE. The above theorem 23.4 looks too narrow when compared with Rao-Rao [1983] proposition 2.5.2. There it is claimed that for each pair of modular set functions $\alpha, \beta : \mathfrak{A} \to] - \infty, \infty]$ with $\alpha(\varnothing) = \beta(\varnothing) = 0$, bounded below or not, the identical formation $\lambda : \mathfrak{A} \to \overline{\mathbb{R}}$ is a so-called *charge*, which in particular means that it cannot attain both of the values $\pm\infty$. But this assertion is false, as the counterexample below will show. Thus one-sided boundedness seems to play a key role, in accordance with the present exposition. The author did not attempt to overlook the possible impact of the false 2.5.2 on Rao-Rao [1983].

23.10. EXERCISE. Let X be the union of two disjoint infinite subsets S and T. Define $\mathfrak{S}$ to consist of the finite and cofinite subsets of S, and $\mathfrak{T}$ to consist of the finite and cofinite subsets of T, and let

$$\mathfrak{A} := \{A \subset X : A \cap S \in \mathfrak{S} \text{ and } A \cap T \in \mathfrak{T}\}.$$

As in earlier examples define $\varphi : \mathfrak{S} \to \mathbb{R}$ and $\psi : \mathfrak{T} \to [0, \infty]$ to be

$$\varphi(A) = \left\{ \begin{array}{ll} \#(A) & \text{for } A \subset S \text{ finite} \\ -\#(A') & \text{for } A \subset S \text{ cofinite} \end{array} \right\} \text{ and}$$

$$\psi(A) = \left\{ \begin{array}{ll} 0 & \text{for } A \subset T \text{ finite} \\ \infty & \text{for } A \subset T \text{ cofinite} \end{array} \right\}.$$

Then define $\alpha, \beta : \mathfrak{A} \to] - \infty, \infty]$ to be

$$\begin{aligned} \alpha(A) &= 2\varphi(A \cap S) + \psi(A \cap T), \\ \beta(A) &= \varphi(A \cap S) + \psi(A \cap T) \quad \text{for } A \in \mathfrak{A}. \end{aligned}$$

It is obvious that α and β are modular with $\alpha(\varnothing) = \beta(\varnothing) = 0$. To be verified is that the formation $\lambda : \mathfrak{A} \to \overline{\mathbb{R}}$ as in 23.4 satisfies $\lambda(S) = -\infty$ and $\lambda(T) = \infty$.

The Jordan Decomposition Theorem

We return to the context of the introduction. Assume that $\varphi : \mathfrak{A} \to] - \infty, \infty]$ is modular and bounded below $\neq \infty$. Then $I := \inf \varphi \in \mathbb{R}$. As above we understand a **representation** of φ to be a couple of contents $\alpha, \beta : \mathfrak{A} \to] - \infty, \infty]$ such that

$$\varphi(A) = \alpha(A') + \beta(A) \quad \text{for } A \in \mathfrak{A}.$$

This equation can be written

$$\varphi(A) = \alpha^{\wedge}(A') + \beta^{\wedge}(A) + \alpha(\varnothing) + \beta(\varnothing) \quad \text{for } A \in \mathfrak{A}.$$

Formation of the infimum over the $A \in \mathfrak{A}$ shows that $(\alpha^{\wedge}) \wedge (\beta^{\wedge})$ is a finite ccontent with

$$I = \big((\alpha^{\wedge}) \wedge (\beta^{\wedge}) \big)(X) + \alpha(\varnothing) + \beta(\varnothing).$$

Thus $I \geqq \alpha(\varnothing) + \beta(\varnothing)$, and $I = \alpha(\varnothing) + \beta(\varnothing)$ iff $\alpha \perp \beta$. We turn to the basic assertions on existence and uniqueness.

23.11. THEOREM. *Assume that* $\varphi : \mathfrak{A} \to]-\infty, \infty]$ *is modular and bounded below* $\not\equiv \infty$. *Define*

$$\sigma : \sigma(A) \;=\; \inf\{\varphi(A' \cap P) : P \in \mathfrak{A}\} - I \quad \text{and}$$
$$\tau : \tau(A) \;=\; \inf\{\varphi(A \cup P) : P \in \mathfrak{A}\} - I \quad \text{for } A \in \mathfrak{A}.$$

Then $\sigma, \tau : \mathfrak{A} \to [0, \infty]$ *are ccontents and fulfil*

$$\varphi(A) = \sigma(A') + \tau(A) + I \quad \text{for all } A \in \mathfrak{A}.$$

Thus $\sigma \perp \tau$. *Therefore* $(\sigma + a), (\tau + b)$ *with real constants* $a + b = I$ *form a representation of* φ *with* $(\sigma + a) \perp (\tau + b)$.

23.12. ADDENDUM. *Let* $(P_l)_l$ *be a sequence in* $\mathfrak{A}$ *with* $\varphi(P_l) \to I$. *Then*

$$\varphi(A' \cap P_l) \;\to\; \sigma(A) + I \quad \text{and}$$
$$\varphi(A \cup P_l) \;\to\; \tau(A) + I \quad \text{for all } A \in \mathfrak{A}.$$

Proof of 23.11 and 23.12. 0) For a set function $\vartheta : \mathfrak{A} \to \overline{\mathbb{R}}$ we define $\vartheta' : \mathfrak{A} \to \overline{\mathbb{R}}$ to be $\vartheta'(A) = \vartheta(A')$ for $A \in \mathfrak{A}$. It should not be confused with the upside-down transform $\vartheta \perp : \mathfrak{A} \perp = \mathfrak{A} \to \overline{\mathbb{R}}$ of ϑ defined before 2.7. There are obvious equivalences like those in 2.7, of which we shall make free use.

For the remainder of the proof we adapt the notations to those of 23.4 and 23.5. 1) We note that

$$\text{if } \alpha = 0 \text{ and } \beta = \varphi \;:\; \alpha(A' \cap P') + \beta(A' \cap P) = \varphi(A' \cap P) \text{ for } A, P \in \mathfrak{A},$$
$$(\alpha \wedge \beta)(A') = \sigma(A) + I \quad \text{for } A \in \mathfrak{A};$$
$$\text{if } \alpha = \varphi' \text{ and } \beta = 0 \;:\; \alpha(A' \cap P') + \beta(A' \cap P) = \varphi(A \cup P) \text{ for } A, P \in \mathfrak{A},$$
$$(\alpha \wedge \beta)(A') = \tau(A) + I \quad \text{for } A \in \mathfrak{A}.$$

Thus we see that $\sigma = (0 \wedge \varphi)' - I$ and $\tau = (\varphi' \wedge 0)' - I$. It follows from 23.4 that $\sigma, \tau : \mathfrak{A} \to]-\infty, \infty]$ are modular. The definitions themselves show that $\sigma, \tau \geqq 0$ and $\sigma(\varnothing) = \tau(\varnothing) = 0$. Thus 2.10 says that σ and τ are ccontents. 2) In both cases considered in 1) we obtain for $A = \varnothing$

$$\alpha(P') + \beta(P) \;=\; \varphi(P) \quad \text{for } P \in \mathfrak{A},$$
$$(\alpha \wedge \beta)(X) \;=\; I \in \mathbb{R}.$$

Thus 23.5 applied to $M := X$ asserts that for each sequence $(P_l)_l$ in $\mathfrak{A}$ with $\varphi(P_l) \to I$ we have

$$\alpha(A' \cap P_l') + \beta(A' \cap P_l) \to (\alpha \wedge \beta)(A') \quad \text{for all } A \in \mathfrak{A}.$$

This means that

$$\text{if } \alpha = 0 \text{ and } \beta = \varphi \;:\; \varphi(A' \cap P_l) \to \sigma(A) + I,$$
$$\text{if } \alpha = \varphi' \text{ and } \beta = 0 \;:\; \varphi(A \cup P_l) \to \tau(A) + I,$$

as claimed in 23.12. 3) For $A \in \mathfrak{A}$ we obtain from 2)

$$\varphi(A) + \varphi(P_l) = \varphi(A \cap P_l) + \varphi(A \cup P_l) \to \sigma(A') + \tau(A) + 2I,$$

since all terms are $> -\infty$. It follows that $\varphi(A) = \sigma(A') + \tau(A) + I$. This completes the proof.

23.13. THEOREM. *Assume that* $\varphi : \mathfrak{A} \to]-\infty, \infty]$ *is modular and bounded below* $\not\equiv \infty$. *Then the representations* $\alpha, \beta : \mathfrak{A} \to] - \infty, \infty]$ *of* φ *are in one-to-one correspondence with*

> *the finite ccontents* $\lambda : \mathfrak{A} \to [0, \infty[$ *and the couples* $a, b \in \mathbb{R}$
> *such that* $\lambda(X) + a + b = I$.

The correspondence is

$$\alpha := \sigma + a + \lambda \quad \text{and} \quad \beta := \tau + b + \lambda,$$

and its inverse is $\lambda := (\alpha^\wedge) \wedge (\beta^\wedge)$ *and* $a := \alpha(\varnothing), b := \beta(\varnothing)$. *In particular* $\alpha \perp \beta$ *iff* $\lambda = 0$, *that is iff* α, β *is one of the particular representations obtained in* 23.11.

Proof. 1) We start with a finite ccontent $\lambda : \mathfrak{A} \to [0, \infty[$ and $a, b \in \mathbb{R}$ with $\lambda(X) + a + b = I$, and form $\alpha := \sigma + a + \lambda$ and $\beta := \tau + b + \lambda$. Then $\alpha, \beta : \mathfrak{A} \to] - \infty, \infty]$ are contents. For $A \in \mathfrak{A}$ we have

$$\begin{aligned}
\alpha(A') + \beta(A) &= \big(\sigma(A') + \tau(A)\big) + (a + b) + \big(\lambda(A') + \lambda(A)\big) \\
&= \big(\varphi(A) - I\big) + (a + b) + \lambda(X) = \varphi(A),
\end{aligned}$$

so that α, β form a representation of φ. We claim that $(\alpha^\wedge) \wedge (\beta^\wedge)$ and $\alpha(\varnothing), \beta(\varnothing)$ lead back to λ and a, b. It is clear that $\alpha(\varnothing) = a$ and $\beta(\varnothing) = b$. Then for $A, P \in \mathfrak{A}$ we have

$$\begin{aligned}
\alpha^\wedge(A \cap P') + \beta^\wedge(A \cap P) &= \big(\alpha(A \cap P') - a\big) + \big(\beta(A \cap P) - b\big) \\
&= \sigma(A \cap P') + \tau(A \cap P) + \lambda(A).
\end{aligned}$$

Thus the infimum over $P \in \mathfrak{A}$ furnishes

$$\big((\alpha^\wedge) \wedge (\beta^\wedge)\big)(A) = (\sigma \wedge \tau)(A) + \lambda(A) = \lambda(A) \quad \text{for } A \in \mathfrak{A},$$

which is the assertion.

2) We start with a representation $\alpha, \beta : \mathfrak{A} \to] - \infty, \infty]$ of φ, and form $\lambda := (\alpha^\wedge) \wedge (\beta^\wedge)$ and $a := \alpha(\varnothing), b := \beta(\varnothing)$. We have seen above that $\lambda : \mathfrak{A} \to [0, \infty[$ is a finite ccontent with $\lambda(X) + a + b = I$. We claim that $\sigma + a + \lambda$ and $\tau + b + \lambda$ lead back to α and β. This will complete the proof.
i) We fix a sequence $(P_l)_l$ in $\mathfrak{A}$ with $\varphi(P_l) \to I$. This means that

$$\begin{aligned}
\alpha^\wedge(P_l') + \beta^\wedge(P_l) &= \alpha(P_l') + \beta(P_l) - (a + b) = \varphi(P_l) - (a + b) \\
&\to I - (a + b) = \lambda(X) = \big((\alpha^\wedge) \wedge (\beta^\wedge)\big)(X) \in \mathbb{R}.
\end{aligned}$$

Thus 23.5 applied to $M := X$ asserts that for all $A \in \mathfrak{A}$

$$\begin{aligned}
\alpha^\wedge(A \cap P_l') + \beta^\wedge(A \cap P_l) &\to \big((\alpha^\wedge) \wedge (\beta^\wedge)\big)(A) = \lambda(A), \\
\alpha(A \cap P_l') + \beta(A \cap P_l) &\to \lambda(A) + (a + b).
\end{aligned}$$

We combine this with 23.12 to obtain

$$\begin{aligned}
\varphi(A' \cap P_l) + \alpha(A \cap P_l') + \beta(A \cap P_l) &\to \sigma(A) + I + \lambda(A) + (a + b), \\
\varphi(A \cup P_l) + \alpha(A \cap P_l') + \beta(A \cap P_l) &\to \tau(A) + I + \lambda(A) + (a + b).
\end{aligned}$$

ii) We reformulate the left sides of the last two relations. For the first relation we obtain

$$\alpha(A \cup P_l') + \beta(A' \cap P_l) + \alpha(A \cap P_l') + \beta(A \cap P_l)$$
$$= \alpha(A) + \alpha(P_l') + \beta(P_l) + \beta(\varnothing) = \alpha(A) + \varphi(P_l) + b \to \alpha(A) + I + b.$$

It follows that $\alpha(A) = \sigma(A) + \lambda(A) + a$. For the second relation we obtain

$$\alpha(A' \cap P_l') + \beta(A \cup P_l) + \alpha(A \cap P_l') + \beta(A \cap P_l)$$
$$= \alpha(P_l') + \alpha(\varnothing) + \beta(A) + \beta(P_l) = \beta(A) + \varphi(P_l) + a \to \beta(A) + I + a.$$

It follows that $\beta(A) = \tau(A) + \lambda(A) + b$. The proof is complete.

We see that 23.11 furnishes the (up to an additive real constant) unique representation α, β of φ which fulfils $\alpha \perp \beta$. In the conventional special case $\varphi(\varnothing) = 0$ we have

$$0 = \varphi(\varnothing) = \sigma(X) + I \quad \text{and hence } \sigma < \infty,$$
$$\varphi(A) = \big(\sigma(X) - \sigma(A)\big) + \tau(A) + I = \tau(A) - \sigma(A) \quad \text{for } A \in \mathfrak{A},$$

that is $\varphi = \tau - \sigma$. Therefore we write $\tau =: \varphi^+$ and $\sigma =: \varphi^-$ in all cases.

The Existence of Minimal Sets

Let $\varphi : \mathfrak{S} \to \overline{\mathbb{R}}$ be a set function on a lattice $\mathfrak{S}$ in X. We recall that the different notions of continuous set functions, like most of the notions defined in section 2, assumed φ to be isotone. The present context requires an appropriate extension of the definition. Justified by success, we define $\varphi : \mathfrak{S} \to \overline{\mathbb{R}}$ to be **upward σ continuous** iff

$$\varphi(S) \leqq \liminf_{l \to \infty} \varphi(S_l) \text{ for all sequences } (S_l)_l \text{ in } \mathfrak{S} \text{ such that } S_l \uparrow \text{ or } \downarrow S \in \mathfrak{S}.$$

This is the previous notion when φ is isotone. However, we must admit that a more natural definition seems to be

$(\uparrow)$ $\varphi(S_l) \to \varphi(S)$ for all sequences $(S_l)_l$ in $\mathfrak{S}$ with $S_l \uparrow S \in \mathfrak{S}$.

Therefore we insert a short comparison of the two definitions. It is clear that the comparison requires that $\mathfrak{S}$ be at least an oval.

23.14. EXERCISE. Assume that $\varphi : \mathfrak{S} \to]-\infty, \infty]$ is a modular set function on an oval $\mathfrak{S}$. 1) The implication upward σ continuous $\Leftarrow (\uparrow)$ is true. 2) The implication upward σ continuous $\Rightarrow (\uparrow)$ need not be true, even when $\mathfrak{S}$ is a σ algebra and $\varphi \geqq 0$. Hint: Let $\alpha : \mathfrak{S} \to [0, \infty]$ be a cmeasure, and define $\varphi : \mathfrak{S} \to [0, \infty]$ to be $\varphi(S) = \alpha(S')$ for $S \in \mathfrak{S}$. Let $(A_l)_l$ be a sequence in $\mathfrak{S}$ such that $A_l \downarrow \varnothing$ and $\alpha(A_l) = \infty$ for all $l \in \mathbb{N}$, and consider the sequence $(S_l)_l$ of the complements $S_l := A_l'$. 2') The implication upward σ continuous $\Rightarrow (\uparrow)$ is true when for each $T \in \mathfrak{S}$ there exists $S \in \mathfrak{S}$ such that $S \subset T$ and $\varphi(S) < \infty$. In particular it suffices that $\mathfrak{S}$ is a ring and $\varphi(\varnothing) < \infty$.

The fundamental step which follows will lead from the Jordan theorem to the Hahn theorem.

23.15. THEOREM. *Assume that $\mathfrak{S}$ is a σ lattice and that $\varphi : \mathfrak{S} \to] -\infty, \infty]$ is submodular and upward σ continuous. Then there exists $P \in \mathfrak{S}$ such that $\varphi(P) = \inf \varphi$. In particular φ is bounded below.*

Proof. We can assume that $\varphi \not\equiv \infty$. Thus $I := \inf \varphi < \infty$. For $A \in \mathfrak{S}$ with $\varphi(A) < \infty$ we define

$$\delta(A) := \sup\{\varphi(A) - \varphi(S) : S \in \mathfrak{S} \text{ with } S \subset A\} \in [0, \infty].$$

1) We claim that

for each pair $A \in \mathfrak{S}$ with $\varphi(A) < \infty$ and $\varepsilon > 0$
there exists $B \in \mathfrak{S}$ with $B \subset A$ and $\varphi(B) \leq \varphi(A)$ and $\delta(B) \leq \varepsilon$.

Assume that this is false. Then there exists a pair $A \in \mathfrak{S}$ with $\varphi(A) < \infty$ and $\varepsilon > 0$ such that

each $B \in \mathfrak{S}$ with $B \subset A$ and $\varphi(B) \leq \varphi(A)$ has $\delta(B) > \varepsilon$,
that is it has $S \in \mathfrak{S}$ with $S \subset B$ and $\varphi(S) < \varphi(B) - \varepsilon$.

We iterate this procedure $B \mapsto S$, and start with $B := A$. Then we obtain a sequence of subsets $A = A_0 \supset A_1 \supset \cdots \supset A_{n-1} \supset A_n \supset \cdots$ in $\mathfrak{S}$ with $\varphi(A_n) < \varphi(A_{n-1}) - \varepsilon$ for $n \geq 1$. It follows that $A_n \downarrow$ some $S \in \mathfrak{S}$ and $\varphi(A_n) \downarrow -\infty$. But this cannot happen since φ is upward σ continuous. Thus the intermediate claim is proved.

2) Next we fix two sequences of real numbers

$$(c_n)_n \text{ with } c_n > I \text{ for } n \geq 1 \text{ and } c_n \downarrow I,$$

$$(\varepsilon_n)_n \text{ with } \varepsilon_n > 0 \text{ for } n \geq 1 \text{ and } \sum_{n=1}^{\infty} \varepsilon_n < \infty.$$

From the definition of I we obtain subsets $A_n \in \mathfrak{S}$ with $\varphi(A_n) < c_n$, and then from 1) subsets $B_n \in \mathfrak{S}$ with $\varphi(B_n) < c_n$ and $\delta(B_n) \leq \varepsilon$ for $n \geq 1$. We form

$$B_p^q := \bigcup_{l=p}^{q} B_l \in \mathfrak{S} \quad \text{for } 1 \leq p \leq q.$$

Then $B_p^{q+1} = B_p^q \cup B_{q+1}$ and hence

$$\varphi(B_p^q) + \varphi(B_{q+1}) \geq \varphi(B_p^{q+1}) + \varphi(B_p^q \cap B_{q+1}).$$

This implies first of all that $\varphi(B_p^q) < \infty$ for $1 \leq p \leq q$. From the definition of $\delta(\cdot)$ we obtain

$$\varphi(B_{q+1}) - \varphi(B_p^q \cap B_{q+1}) \;\leqq\; \delta(B_{q+1}) \leqq \varepsilon_{q+1},$$
$$\varphi(B_p^{q+1}) \;\leqq\; \varphi(B_p^q) + \varepsilon_{q+1}.$$

It follows that

$$\varphi(B_p^q) \leqq \varphi(B_p) + \sum_{l=p+1}^{q} \varepsilon_l < c_p + \sum_{l=p+1}^{q} \varepsilon_l \quad 1 \leq p \leq q.$$

3) For fixed $p \geq 1$ and $q \to \infty$ we have $B_p^q \uparrow P_p := \bigcup_{l=p}^{\infty} B_l \in \mathfrak{S}$. From 2) and since φ is upward σ continuous it follows that

$$\varphi(P_p) \leq c_p + \sum_{l=p+1}^{\infty} \varepsilon_l \quad \text{for } p \geq 1.$$

Now $P_p \downarrow$ some $P \in \mathfrak{S}$. Once more from the assumption that φ is upward σ continuous it follows that $\varphi(P) \leq I$. Therefore $\varphi(P) = I$. The proof is complete.

The Hahn Decomposition Theorem

The present subsection assumes that $\mathfrak{A}$ is a σ algebra in X. Let $\varphi : \mathfrak{A} \to]-\infty, \infty]$ be modular and upward σ continuous $\not\equiv \infty$. Then first of all 23.15 asserts that φ is bounded below and thus fulfils the assumptions of the Jordan type theorems 23.11 to 23.13. We retain the former notations $I := \inf \varphi \in \mathbb{R}$ and σ, τ. Above all 23.15 asserts that there exist subsets $P \in \mathfrak{A}$ such that $\varphi(P) = I$. The theorem below collects the principal consequences.

23.16. THEOREM. *Assume that* $\varphi : \mathfrak{A} \to]-\infty, \infty]$ *is modular and upward* σ *continuous* $\not\equiv \infty$. *Define* $\mathfrak{P}$ *to consist of the subsets* $P \in \mathfrak{A}$ *with* $\varphi(P) = I$. *0) The subset* $P \in \mathfrak{A}$ *is in* $\mathfrak{P}$ *iff* $\sigma(P') = \tau(P) = 0$. *1) If* $P \in \mathfrak{P}$ *then*

$$\sigma(A) = \varphi(A' \cap P) - I \quad and$$
$$\tau(A) = \varphi(A \cup P) - I \quad for\ all\ A \in \mathfrak{A}.$$

2) σ *and* τ *are cmeasures.*

Proof. 0) is obvious from 23.11, and 1) from 23.12. 2) It is evident from 1) that σ and τ are upward σ continuous in the new sense.

We add one more fact in the conventional special case $\varphi(\varnothing) = 0$. Then $\sigma < \infty$ and $\varphi = \tau - \sigma$. Therefore 23.16.0) implies for each $P \in \mathfrak{P}$ that

if $A \in \mathfrak{A}$ with $A \subset P$: $\tau(A) = 0$ and hence $\varphi(A) = -\sigma(A) \leq 0$,
if $A \in \mathfrak{A}$ with $A \subset P'$: $\sigma(A) = 0$ and hence $\varphi(A) = \tau(A) \geq 0$.

This is the essential point in the usual treatments of the Jordan and Hahn theorems.

23.17. EXERCISE. Under the present assumptions the paving $\mathfrak{P} \subset \mathfrak{A}$ is a σ oval.

We conclude with an equivalence result which justifies the present definition of upward σ continuous.

23.18. PROPOSITION. *Assume that* $\varphi : \mathfrak{A} \to]-\infty, \infty]$ *is modular and* $\not\equiv \infty$. *Then the following are equivalent.*

1) *There exist couples of measures* $\alpha, \beta : \mathfrak{A} \to]-\infty, \infty]$ *such that* $\varphi(A) = \alpha(A') + \beta(A)$ *for all* $A \in \mathfrak{A}$.

2) φ *is upward* σ *continuous.*

We see from 23.14.2) that the equivalence becomes false when one formulates 2) with ($\uparrow$). However, in virtue of 23.14.1)2') the equivalence remains true under the restriction $\varphi(\varnothing) < \infty$.

Proof of 23.18. The implication 2)$\Rightarrow$1) follows from 23.16.2). In order to prove 1)$\Rightarrow$2) consider a sequence $(S_l)_l$ in $\mathfrak{A}$ such that $S_l \uparrow$ or $\downarrow$ some $S \in \mathfrak{A}$. The assumption reads

$$\varphi(S_l) = \alpha(S_l') + \beta(S_l) \text{ for all } l \text{ and } \varphi(S) = \alpha(S') + \beta(S).$$

The assertion is $\varphi(S) \leqq \liminf_{l \to \infty} \varphi(S_l)$. The case $S_l \uparrow S$: On the one hand $\beta(S_l) \uparrow \beta(S)$. On the other hand we have $S_l' \downarrow S'$. If $\alpha(S_l') = \infty \; \forall l$ then $\varphi(S_l) = \infty \; \forall l$, which implies the assertion. Otherwise $\alpha(S_l') \downarrow \alpha(S')$. Since all values $\alpha(\cdot)$ and $\beta(\cdot)$ are $> -\infty$ it follows that $\varphi(S_l) \to \varphi(S)$. The case $S_l \downarrow S$ and hence $S_l' \uparrow S'$ has the same proof.

23.19. BIBLIOGRAPHICAL NOTE. The usual proofs of the conventional Jordan and Hahn theorems are for cmeasures and in the reverse order, and therefore do not make clear that the Jordan theorem is of much wider scope than the (crude) Hahn theorem. In the frame of ccontents a comprehensive older reference is Dunford-Schwartz [1958] chapter III, where the conventional Jordan theorem is in the finite version. The author cannot resist to remark that there is a counterpart for linear functionals which is a simple consequence of the Hahn-Banach theorem; see König [1972]. General conventional Jordan theorems are due to Schmidt [1982a][1982b] and to Rao-Rao [1983] theorem 2.5.3. The existence theorem 23.15 is an elaboration of the short paper of Doss [1980], which is based on an older idea of proof. At last we mention that there are more sophisticated versions of the Hahn theorem in the frame of ccontents. See for example Rao-Rao [1983] theorem 2.6.2 and Luxemburg [1991] theorem 8.1.

24. The Lebesgue Decomposition and Radon-Nikodým Theorems

The present final section serves to round off the last chapter and does not claim material innovation. We want to demonstrate that the main tool theorems of the last section, the infimum formation 23.4 and the minimum theorem 23.15, are also adequate means for the theorems of the present title. On the one side we want to be faithful, for example in that we establish the Lebesgue decomposition theorem for ccontents instead of cmeasures as most textbooks do. On the other side we want to avoid technical complications, as we did in other parts of this text, and therefore shall confine ourselves to basic versions. Also we shall not enter into the theories of bands and of Riesz spaces.

The Lebesgue Decomposition Theorem

We assume that $\mathfrak{A}$ is an algebra in X. Let $\alpha, \varphi : \mathfrak{A} \to [0, \infty]$ be ccontents. One defines α to be **absolutely continuous** with respect to φ iff

$$\sup\{\alpha(A) : A \in \mathfrak{A} \text{ with } \varphi(A) \leq \delta\} \to 0 \quad \text{for } \delta \downarrow 0.$$

In this case we write $\alpha \ll \varphi$.

24.1. REMARK. *Let* $\alpha, \varphi : \mathfrak{A} \to [0, \infty]$ *be ccontents. Then*

$$\alpha \ll \varphi \Longrightarrow \text{if } A \in \mathfrak{A} \text{ has } \varphi(A) = 0 \text{ then } \alpha(A) = 0.$$

Moreover we have $\Longleftarrow$ *when* $\mathfrak{A}$ *is a* σ *algebra and* α, φ *are cmeasures with* $\alpha < \infty$.

Proof. $\Rightarrow$) is obvious. $\Leftarrow$) Assume not. Then there exists a sequence $(A_l)_l$ in $\mathfrak{A}$ with $\varphi(A_l) \leq 1/2^{l+1}$ and $\alpha(A_l) \geq$ some $\varepsilon > 0$ for $l \geq 1$. For $P_n := \bigcup_{l=n}^{\infty} A_l \in \mathfrak{A}$ then $\varphi(P_n) \leq 1/2^n$ and $\alpha(P_n) \geq \varepsilon$. Now $P_n \downarrow$ some $P \in \mathfrak{A}$. Then $\varphi(P) = 0$, and $\alpha < \infty$ implies that $\alpha(P) \geq \varepsilon$. Thus we arrive at a contradiction.

24.2. THEOREM (Lebesgue Decomposition Theorem). *Let* $\varphi : \mathfrak{A} \to [0, \infty]$ *be a ccontent. Then each finite ccontent* $\lambda : \mathfrak{A} \to [0, \infty[$ *has a unique decomposition*

$$\lambda = \alpha + \beta \text{ into ccontents } \alpha, \beta : \mathfrak{A} \to [0, \infty[\text{ with } \alpha \ll \varphi \text{ and } \beta \perp \varphi.$$

We have $\alpha = \lim_{t \uparrow \infty} \lambda \wedge (t\varphi)$.

24.3. EXERCISE. In the above decomposition $\lambda = \alpha + \beta$ we have $\alpha \perp \beta$.

Proof of 24.2. We start with the existence assertion. 0) Define $\alpha_t := \lambda \wedge (t\varphi)$ for $t > 0$. We see from 23.7 that $\alpha_t : \mathfrak{A} \to [0, \infty[$ is a finite ccontent with $\alpha_t \leq \lambda$. Furthermore $\alpha_s \leq \alpha_t$ for $0 < s < t$. Thus the finite limit $\alpha(A) := \lim_{t \uparrow \infty} \alpha_t(A)$ exists for all $A \in \mathfrak{A}$ and defines a ccontent $\alpha : \mathfrak{A} \to [0, \infty[$ with $\alpha \leq \lambda$. Likewise $\beta := \lambda - \alpha$ is a finite ccontent $\leq \lambda$. 1) We claim that $\alpha \ll \lambda$. In fact, for $t > 0$ and $A \in \mathfrak{A}$ we have $\alpha(A) - \alpha_t(A) \leq \alpha(X) - \alpha_t(X)$ and hence

$$\alpha(A) \leq \big(\alpha(X) - \alpha_t(X)\big) + \alpha_t(A) \leq \big(\alpha(X) - \alpha_t(X)\big) + t\varphi(A).$$

For $t > 0$ and $\delta > 0$ therefore

$$\sup\{\alpha(A) : A \in \mathfrak{A} \text{ with } \varphi(A) \leq \delta\} \leq \big(\alpha(X) - \alpha_t(X)\big) + t\delta.$$

We perform $\delta \downarrow 0$ for fixed $t > 0$, and then $t \uparrow \infty$. The assertion follows. 2) We claim that $\beta \perp \varphi$. We fix $t > 0$ and obtain from $\alpha_t := \lambda \wedge (t\varphi)$ a sequence $(P_l)_l$ in $\mathfrak{A}$ such that $\lambda(P_l') + t\varphi(P_l) \to \alpha_t(X)$. Thus

$$\delta_l := \big(\lambda(P_l') - \alpha_t(P_l')\big) + \big(t\varphi(P_l) - \alpha_t(P_l)\big) \to 0,$$

where both brackets are $\geqq 0$. It follows that

$$
\begin{aligned}
& \beta(P_l') + \varphi(P_l) \\
= {}& \left(\lambda(P_l') - \alpha(P_l')\right) + (1/t)\left(t\varphi(P_l) - \alpha_t(P_l)\right) + (1/t)\alpha_t(P_l) \\
\leqq {}& \left(\lambda(P_l') - \alpha_t(P_l')\right) + (1/t)\left(t\varphi(P_l) - \alpha_t(P_l)\right) + (1/t)\alpha_t(P_l) \\
\leqq {}& (1 + 1/t)\delta_l + (1/t)\lambda(X).
\end{aligned}
$$

For $l \to \infty$ we obtain $(\beta \wedge \varphi)(X) \leqq (1/t)\lambda(X)$. Thus $t \uparrow \infty$ furnishes the assertion. This completes the existence proof.

We turn to the uniqueness assertion. Assume that $\lambda = \alpha_0 + \beta_0$ with ccontents $\alpha_0, \beta_0 : \mathfrak{A} \to [0, \infty[$ such that $\alpha_0 \ll \varphi$ and $\beta_0 \perp \varphi$. There exists a sequence $(P_l)_l$ in $\mathfrak{A}$ such that $\varphi(P_l) \to 0$ and both $\beta(P_l') \to 0$ and $\beta_0(P_l') \to 0$. Then both $\alpha(P_l) \to 0$ and $\alpha_0(P_l) \to 0$. It follows for $A \in \mathfrak{A}$ that $\alpha(A \cap P_l) \to 0$ and $\alpha_0(A \cap P_l) \to 0$, and hence

$$
\begin{aligned}
\lambda(A \cap P_l') &= \alpha(A \cap P_l') + \beta(A \cap P_l') \to \alpha(A), \\
\lambda(A \cap P_l') &= \alpha_0(A \cap P_l') + \beta(A \cap P_l') \to \alpha_0(A).
\end{aligned}
$$

Therefore $\alpha = \alpha_0$. The proof is complete.

24.4. EXERCISE. The assertion of 24.2 becomes false when one removes the assumption that $\lambda < \infty$, even in case that $\mathfrak{A}$ is a σ algebra and φ and λ are cmeasures with $\varphi < \infty$. Hint: Let $\mathfrak{A}$ consist of the countable and cocountable subsets of an uncountable set X. Define $\varphi : \mathfrak{A} \to [0, \infty[$ to be

$$
\varphi(A) = \left\{ \begin{array}{ll} 0 & \text{for } A \text{ countable} \\ 1 & \text{for } A \text{ cocountable} \end{array} \right\},
$$

and let $\lambda : \mathfrak{A} \to [0, \infty]$ be the counting measure.

24.5. BIBLIOGRAPHICAL NOTE. The unique basic textbook known to the author in which the Lebesgue decomposition theorem is for ccontents instead of cmeasures is Jacobs [1978] theorems VIII.4.2 and IX.3.8. For more details we refer to Rao-Rao [1983] chapter 6 with its notes and comments.

A fundamental extension of the Lebesgue decomposition theorem is its version for bands. For the notion of a band we refer to Jacobs [1978] chapter IX and to the textbooks on Riesz spaces, for example Meyer-Nieberg [1991]. The band version of the Lebesgue decomposition theorem appears in connection with its application to the abstract F. and M. Riesz theorem for complex function algebras. See Gamelin [1969] section II.7 and Barbey-König [1977] chapters II and III; more recent presentations are Rudin [1980] chapter 9 and Conway [1991] chapter V.

The Radon-Nikodým Theorem

The present subsection assumes that $\varphi : \mathfrak{A} \to [0, \infty]$ is a cmeasure on a σ algebra in a nonvoid set X. We recall 13.32: If $f : X \to [0, \infty]$ is measurable

$\mathfrak{A}$ then the set function

$$\alpha : \mathfrak{A} \to [0, \infty], \text{ defined to be } \alpha(A) = \int_A f d\varphi \text{ for } A \in \mathfrak{A},$$

is a cmeasure. Furthermore $\varphi(A) = 0 \Rightarrow \alpha(A) = 0$. Thus $\alpha \ll \varphi$ when $\alpha < \infty$, that is when f is integrable φ. We write $\alpha =: f\varphi$.

24.6. REMARK. We have $\int h d\alpha = \int (hf) d\varphi$ for all functions $h : X \to [0, \infty]$ measurable $\mathfrak{A}$. This is an immediate consequence of the approximation theorem 22.1.

We first note that the map $f \mapsto f\varphi$ is often, but not always injective in the appropriate sense.

24.7. REMARK. Assume that $f, g : X \to [0, \infty]$ are measurable $\mathfrak{A}$. i) If $f = g$ ae φ then $f\varphi = g\varphi$. ii) $f\varphi = g\varphi$ does not enforce that $f = g$ ae φ. iii) However, if $M \in \mathfrak{A}$ satisfies

$$\int_A f d\varphi = \int_A g d\varphi < \infty \quad \text{for all } A \in \mathfrak{A} \text{ with } A \subset M,$$

then $f = g$ ae φ on M.

Proof. i) is in 13.25.1). ii) will be seen in the next example. iii) Let $A := M \cap [f \geq g + \delta]$ with $\delta > 0$. Thus $f\chi_A \geq g\chi_A + \delta\chi_A$. The assumption implies that $\varphi(A) = 0$. Therefore $\varphi(M \cap [f > g]) = 0$; and likewise $\varphi(M \cap [f < g]) = 0$.

24.8. EXAMPLE. Let $\mathfrak{A}$ consist of the countable and cocountable subsets of an uncountable set X. Define $\varphi : \mathfrak{A} \to [0, \infty]$ to be

$$\varphi(A) = \left\{ \begin{array}{ll} 0 & \text{for } A \text{ countable} \\ \infty & \text{for } A \text{ cocountable} \end{array} \right\}.$$

For $f : X \to [0, \infty]$ measurable $\mathfrak{A}$ then

$$\int f d\varphi = \int f d\varphi = \left\{ \begin{array}{ll} 0 & \text{if } [f > 0] \text{ is countable} \\ \infty & \text{if } [f > 0] \text{ is cocountable} \end{array} \right\}.$$

Therefore all functions $f : X \to]0, \infty]$ measurable $\mathfrak{A}$ fulfil $f\varphi = \varphi$.

We come to the main point. The example below will show that a cmeasure $\alpha : \mathfrak{A} \to [0, \infty]$ such that $\alpha \ll \varphi$, even when it is finite, need not be of the form $\alpha = f\varphi$ for some $f : X \to [0, \infty]$ measurable $\mathfrak{A}$. After this comes the famous affirmative Radon-Nikodým theorem in form of several equivalence assertions.

24.9. EXAMPLE. We continue with example 24.8 and define $\alpha : \mathfrak{A} \to [0, \infty[$ as in 24.4 to be

$$\alpha(A) = \left\{ \begin{array}{ll} 0 & \text{for } A \text{ countable} \\ 1 & \text{for } A \text{ cocountable} \end{array} \right\}.$$

It is obvious that $\alpha \ll \varphi$. But 24.8 shows that α is not of the form $f\varphi$ for some $f : X \to [0, \infty]$ measurable $\mathfrak{A}$.

24.10. THEOREM (Radon-Nikodým Theorem). *For a finite cmeasure* $\alpha : \mathfrak{A} \to [0, \infty[$ *the following are equivalent.*

1) $\alpha \ll \varphi$, *that is* $\varphi(A) = 0 \Rightarrow \alpha(A) = 0$ *for* $A \in \mathfrak{A}$.

2) *There exists a function* $f : X \to [0, \infty]$ *measurable* $\mathfrak{A}$ *such that*

$$\alpha(A) = \int_A f d\varphi \quad \text{for all } A \in \mathfrak{A} \text{ with } \varphi(A) < \infty.$$

3) *There exists a function* $f : X \to [0, \infty[$ *measurable* $\mathfrak{A}$ *such that*

$$\alpha(A) \geqq \int_A f d\varphi \text{ for all } A \in \mathfrak{A}, \text{ and } \alpha(A) = \int_A f d\varphi \text{ whenever } \varphi(A) < \infty.$$

The implications 3)$\Rightarrow$2) and 2)$\Rightarrow$1) are obvious. The proof of 1)$\Rightarrow$3) will be based on the lemma which follows.

24.11. LEMMA. *Let* $\vartheta : \mathfrak{A} \to [0, \infty[$ *be a finite cmeasure such that*
i) $\vartheta \ll \varphi$ *and*
ii) *for* $P \in \mathfrak{A}$ *and* $\varepsilon > 0$: *If* $\varepsilon\varphi(A) \leqq \vartheta(A)$ *for all* $A \in \mathfrak{A}$ *with* $A \subset P$ *then* $\varphi(P) = 0$.
Then all $A \in \mathfrak{A}$ *with* $\varphi(A) < \infty$ *have* $\vartheta(A) = 0$.

Proof of 24.11. Assume not. Then there exists $E \in \mathfrak{A}$ with $\varphi(E) < \infty$ and $\vartheta(E) > 0$. We fix $\varepsilon > 0$ with $\varepsilon\varphi(E) < \vartheta(E)$ and put $\theta := \varepsilon\varphi - \vartheta$. Thus $\theta : \mathfrak{A} \to] - \infty, \infty]$ is modular with $\theta(\varnothing) = 0$, and upward σ continuous by 23.14.1). Thus 23.15 asserts that there exists $P \in \mathfrak{A}$ such that $\theta(P) \leqq \theta(A)$ for all $A \in \mathfrak{A}$. In particular $\theta(P) \leqq \theta(E) < 0$. For $A \subset P$ we have

$$\theta(A) + \theta(P) \leqq \theta(A) + \theta(P \setminus A) = \theta(P) \text{ and hence } \theta(A) \leqq 0.$$

Thus $\varphi(P) = 0$ from ii), and then $\vartheta(P) = 0$ from i). It follows that $\theta(P) = 0$ and hence a contradiction.

Proof of 24.10.1)$\Rightarrow$3). Define H to consist of the functions $h : X \to [0, \infty[$ measurable $\mathfrak{A}$ such that $\int_A h d\varphi \leqq \alpha(A)$ for all $A \in \mathfrak{A}$. H is nonvoid since $0 \in H$. Moreover all members of H are integrable φ. i) We claim that $u, v \in H \Rightarrow u \vee v \in H$. In fact, for $A \in \mathfrak{A}$ we have

$$\int_A (u \vee v) d\varphi = \int_{A \cap [u > v]} u d\varphi + \int_{A \cap [u \leqq v]} v d\varphi$$

$$\leqq \alpha(A \cap [u > v]) + \alpha(A \cap [u \leqq v]) = \alpha(A).$$

ii) Define $I := \sup\{\int h d\varphi : h \in H\}$, so that $0 \leqq I \leqq \alpha(X) < \infty$. We claim that $I = \int f d\varphi$ for some $f \in H$. To see this let $(u_l)_l$ be a sequence in H with $\int u_l d\varphi \to I$. By i) we can pass to the sequence of the functions $u_1 \vee \cdots \vee u_l \in H$ and hence assume that $u_l \uparrow$ some $u : X \to [0, \infty]$. We conclude from standard facts in section 13 that u is measurable $\mathfrak{A}$ and fulfils $\int_A u d\varphi \leqq \alpha(A)$ for $A \in \mathfrak{A}$ and $\int u d\varphi = I$. Thus the function $f := u\chi_{[u<\infty]}$ is as required.

iii) We fix $f \in H$ with $I = \int f d\varphi$ as obtained in ii). Then $f\varphi$ is a finite cmeasure $\leq \alpha$, and hence $\vartheta := \alpha - f\varphi$ a finite cmeasure as well. We claim that ϑ fulfils assumption ii) in 24.11. Then 24.11 will furnish the assertion 3). To see this fix $P \in \mathfrak{A}$ and $\varepsilon > 0$ such that $\varepsilon\varphi(A) \leq \vartheta(A)$ for all $A \in \mathfrak{A}$ with $A \subset P$. For all $A \in \mathfrak{A}$ then

$$\int_A (\varepsilon\chi_P + f) d\varphi = \varepsilon\varphi(A \cap P) + \int_{A \cap P} f d\varphi + \int_{A \cap P'} f d\varphi$$

$$\leq \vartheta(A \cap P) + \int_{A \cap P} f d\varphi + \alpha(A \cap P')$$

$$= \alpha(A \cap P) + \alpha(A \cap P') = \alpha(A).$$

Thus by definition $\varepsilon\chi_P + f \in H$. Therefore $\int (\varepsilon\chi_P + f) d\varphi \leq I = \int f d\varphi$ and hence $\varphi(P) = 0$. This is the present claim. The proof is complete.

24.12. CONSEQUENCE. *For a finite cmeasure* $\alpha : \mathfrak{A} \to [0, \infty[$ *the following are equivalent.*

1) $\alpha \ll \varphi$. *Moreover each* $A \in \mathfrak{A}$ *with* $\alpha(T) = 0$ *for all* $T \in \mathfrak{A}$ *with* $T \subset A$ *and* $\varphi(T) < \infty$ *has* $\alpha(A) = 0$.

2) *There exists a function* $f : X \to [0, \infty]$ *measurable* $\mathfrak{A}$ *such that* $\alpha = f\varphi$.

3) *There exists a function* $f : X \to [0, \infty[$ *integrable* φ *such that* $\alpha = f\varphi$.

Proof. 1)$\Rightarrow$3) Fix $f : X \to [0, \infty[$ as in 24.10.3), and form $P := [f > 0]$ and $N := [f = 0]$ in $\mathfrak{A}$. Let $A \in \mathfrak{A}$. i) For $t > 0$ we have $\varphi([f \geq t]) < \infty$ from 11.8.5) and hence

$$\alpha(A \cap [f \geq t]) = \int_{A \cap [f \geq t]} f d\varphi.$$

For $t \downarrow 0$ therefore $\alpha(A \cap P) = \int_{A \cap P} f d\varphi$. ii) For $T \in \mathfrak{A}$ with $T \subset A \cap N$ and $\varphi(T) < \infty$ we have $\alpha(T) = \int_T f d\varphi = 0$. Therefore $\alpha(A \cap N) = 0 = \int_{A \cap N} f d\varphi$. From i)ii) the assertion follows. 3)$\Rightarrow$2) is obvious. 2)$\Rightarrow$1) The function $f : X \to [0, \infty]$ as assumed in 2) is integrable φ. Thus $\varphi([f \geq t]) < \infty$ for $t > 0$ as above. Now

$$\alpha(A) \geq \alpha(A \cap [f \geq t]) = \int_{A \cap [f \geq t]} f d\varphi$$

$$= \int_{0\leftarrow}^{\to\infty} \varphi(A \cap [f \geq t] \cap [f \geq s]) ds \geq \int_t^{\to\infty} \varphi(A \cap [f \geq s]) ds.$$

Therefore

$$\alpha(A) = \sup\{\alpha(A \cap [f \geq t]) : t > 0\}.$$

In particular if $\alpha(A \cap [f \geq t]) = 0$ for all $t > 0$ then $\alpha(A) = 0$. The proof is complete.

We conclude with another equivalence version which is of particular interest.

24.13. THEOREM. *Assume that* $\varphi < \infty$. *Then for a cmeasure* $\alpha : \mathfrak{A} \to [0,\infty]$ *the following are equivalent.*

1) *If* $A \in \mathfrak{A}$ *has* $\varphi(A) = 0$ *then* $\alpha(A) = 0$.

2) *There exists a function* $f : X \to [0,\infty]$ *measurable* $\mathfrak{A}$ *such that* $\alpha = f\varphi$.

Proof. We have to prove 1)$\Rightarrow$2). We form

$$I := \sup\{\varphi(A) : A \in \mathfrak{A} \text{ with } \alpha(A) < \infty\} \leqq \varphi(X) < \infty,$$

and fix a sequence $(B_l)_l$ in $\mathfrak{A}$ with $\alpha(B_l) < \infty$ such that $\varphi(B_l) \to I$. We can assume that $B_l \uparrow$ some $B \in \mathfrak{A}$. i) For an $A \in \mathfrak{A}$ with $A \subset B'$ this enforces that $\alpha(A) < \infty \Rightarrow \varphi(A) = 0$. Thus we have

$$\text{either } \varphi(A) = 0 \text{ and hence } \alpha(A) = 0 \text{ from 1)},$$
$$\text{or } \varphi(A) > 0 \text{ and hence } \alpha(A) = \infty.$$

ii) We put $B_0 := \varnothing$. From 24.10 applied to the restrictions of φ and α to $B_l \setminus B_{l-1}$ we obtain for $l \geq 1$ functions $f_l : B_l \setminus B_{l-1} \to [0,\infty[$ integrable $\varphi|B_l \setminus B_{l-1}$ such that

$$\alpha(A) = \int_A f_l d\varphi \quad \text{for all } A \in \mathfrak{A} \cap (B_l \setminus B_{l-1}).$$

Then define $f : X \to [0,\infty]$ to be

$$f|B_l \setminus B_{l-1} := f_l \text{ for } l \geq 1 \text{ and } f|B' := \infty.$$

Thus f is measurable $\mathfrak{A}$, and in virtue of i) it is as required.

24.14. BIBLIOGRAPHICAL NOTE. The scheme of the present proof of 24.10 is from Cohn [1980] and Bauer [1992]. We note that the use of 23.15 helps to make it transparent. The well-known different proof of the Radon-Nikodým theorem due to von Neumann uses the elementary theory of Hilbert spaces; see for example Dudley [1989] section 5.5.

Bibliography

Adamski, W. [1982] Tight set functions and essential measure. Lect. Notes Math. *945*, 1-14. Springer.

Adamski, W. [1984a] Extension of tight set functions with applications in topological measure theory. Trans. Amer. Math. Soc. *283*, 353-368.

Adamski, W. [1984b] On the structure of the family of measurable sets. Math. Nachr. *115*, 159-165.

Adamski, W. [1987] On regular extensions of contents and measures. Journ. Math. Analysis Appl. *127*, 211-225.

Alexandroff, A.D. [1940-43] Additive set functions in abstract spaces. Mat. Sb. *50*, 307-348; *51*, 563-628; *55*, 169-238.

Anger, B. and C. Portenier [1992a] *Radon Integrals*. Progress in Math. *103*, Birkhäuser.

Anger, B. and C. Portenier [1992b] Radon integrals and Riesz representation. Measure Theory (Oberwolfach 1990). Rend. Circ. Mat. Palermo (2) *28* Suppl. 269-300.

Bachman, G. and A. Sultan [1980] On regular extensions of measures. Pacific Journ. Math. *86*, 389-395.

Barbey, K. and H. König [1977] *Abstract Analytic Function Theory and Hardy Algebras*. Lect. Notes Math. *593*, Springer.

Batt, J. [1973] Die Verallgemeinerungen des Darstellungssatzes von F. Riesz und ihre Anwendungen. Jahresber. Deutsche Math. Verein. *74*, 147-181.

Bauer, H. [1956] Über die Beziehungen einer abstrakten Theorie des Riemann-Integrals zur Theorie Radonscher Masse. Math. Z. *65*, 448-482.

Bauer, H. [1968] *Wahrscheinlichkeitstheorie und Grundzüge der Masstheorie*. 3. Aufl. de Gruyter.

Bauer, H. [1992] *Mass- und Integrationstheorie*. 2. Aufl. de Gruyter.

Behrends, E. [1987] *Mass- und Integrationstheorie*. Springer.

Berg, C., J.R.P. Christensen, and P. Ressel [1984] *Harmonic Analysis on Semigroups*. Grad. Texts Math. *100*, Springer.

Bledsoe, W.W. and A.P. Morse [1955] Product measures. Trans. Amer. Math. Soc. *79*, 173-215.

Bledsoe, W.W. and C.E. Wilks [1972] On Borel product measures. Pacific Journ. Math. *42*, 569-579.

Bourbaki, N. [1952] *Intégration.* Chap. I-IV. Hermann & Cie.

Bourbaki, N. [1956] *Intégration.* Chap. 5. Hermann & Cie.

Bourbaki, N. [1965] *Intégration.* Chap. 1-4, 2ième Ed. Hermann.

Bourbaki, N. [1967] *Intégration.* Chap. 5, 2ième Ed. Hermann.

Bourbaki, N. [1969] *Intégration.* Chap. IX. Hermann.

Carathéodory, C. [1914] Über das lineare Mass von Punktmengen - eine Verallgemeinerung des Längenbegriffs. Nachr. K. Ges. Wiss. Göttingen, Math.-Nat. Kl. 404-426.

Choksi, J.R. [1958] On compact contents. Journ. London Math. Soc. *33*, 387-398.

Choquet, G. [1953-54] Theory of capacities. Ann. Inst. Fourier (Grenoble) *5*, 131-295.

Choquet, G. [1959] Forme abstraite du théorème de capacitabilité. Ann. Inst. Fourier (Grenoble) *9*, 83-89.

Cohn, D.L. [1980] *Measure Theory.* Birkhäuser.

Conway, J.B. [1991] *The Theory of Subnormal Operators.* Math. Surveys and Monogr. *36*, Amer. Math. Soc.

Daniell, P.J. [1917-18] A general form of integral. Ann. of Math. *19*, 279-294.

Dellacherie, C. and P.A. Meyer [1978] *Probabilities and Potential* A. Math. Studies *29*, North-Holland.

Dellacherie, C. and P.A. Meyer [1988] *Probabilities and Potential* C. Math. Studies *151*, North-Holland.

Denneberg, D. [1994] *Non-Additive Measure and Integral.* Kluwer.

Doss, R. [1980] The Hahn decomposition theorem. Proc. Amer. Math. Soc. *80*, 377.

Dudley, R.M. [1989] *Real Analysis and Probability.* Wadsworth & Brooks/ Cole.

Dunford, N. [1935] Integration in general analysis. Trans. Amer. Math. Soc. *37*, 441-453.

Dunford, N. and J.T. Schwartz [1958] *Linear Operators* I: General Theory. Interscience.

Engelking, R. [1989] *General Topology.* Rev. Ed. Heldermann.

Floret, K. [1981] *Mass- und Integrationstheorie.* Teubner.

Fox, G. and P. Morales [1983] Extension of group-valued set functions defined on lattices. Fund. Math. *116*, 143-155.

Fremlin, D.H. [1975] Topological measure theory: Two counter-examples. Math. Proc. Camb. Phil. Soc. *78*, 95-106.

Găină, St. [1986] Dualité dans la théorie de la mesure. Rev. Roumaine Math. Pures Appl. *31*, 677-694.

Gamelin, Th.W. [1969] *Uniform Algebras.* Prentice Hall.

Gardner, R.J. and W.F. Pfeffer [1984] *Borel Measures*. In: Handbook of Set-Theoretic Topology (ed. K. Kunen and J.E. Vaughan) 961-1043. Elsevier.

Glazkov, V.N. [1988] Inner and outer measures. Sib. Math. Zhurn. *29*, 197-201. Engl. Transl. Plenum [1988] 152-155.

Glicksberg, I. [1952] The representation of functionals by integrals. Duke Math. Journ. *19*, 253-261.

Godfrey, M.C. and M. Sion [1969] On products of Radon measures. Canad. Math. Bull. *12*, 427-444.

Greco, G. [1982] Sulla rappresentazione di funzionali mediante integrali. Rend. Sem. Mat. Padova *66*, 21-42.

Hackenbroch, W. and A. Thalmaier [1994] *Stochastische Analysis*. Teubner.

Halmos, P.R. [1950] *Measure Theory*. van Nostrand.

Henry, J.-P. [1969] Prolongements de mesures de Radon. Ann. Inst. Fourier (Grenoble) *19*, 237-247.

Hewitt, E. and K.A. Ross [1963] *Abstract Harmonic Analysis* I. Grundl. Math. Wiss. *115*, Springer.

Hewitt, E. and K. Yosida [1952] Finitely additive measures. Trans. Amer. Math. Soc. *72*, 46-66.

Hoffmann-Jørgensen, J. [1994] *Probability with a view toward Statistics* I. Chapman & Hall.

Jacobs, K. [1978] *Measure and Integral*. Academic Press.

Johnson, R.A. [1966] On product measures and Fubini's theorem in locally compact spaces. Trans. Amer. Math. Soc. *123*, 112-129.

Kelley, J.L. and T.P. Srinivasan [1971] Pre-measures on lattices of sets. Math. Ann. *190*, 233-241.

Kelley, J.L. and T.P. Srinivasan [1988] *Measure and Integral* 1. Grad. Texts Math. *116*, Springer.

Kelley, J.L., M.K. Nayak, and T.P. Srinivasan [1973] Pre-measures on lattices of sets II. Vector and Operator Valued Measures and Applications (Utah 1972) 155-164. Academic Press.

Kindler, J. [1986] A Mazur-Orlicz theorem for submodular set functions. Journ. Math. Analysis Appl. *120*, 533-564.

Kindler, J. [1987] Supermodular and tight set functions. Math. Nachr. *134*, 131-147.

Kisyński, J. [1968] On the generation of tight measures. Studia Math. *30*, 141-151.

Knowles, J.D. [1967] Measures on topological spaces. Proc. London Math. Soc. *17*, 139-156.

Kölzow, D. [1965] Charakterisierung der Masse, welche zu einem Integral im Sinne von Stone oder von Bourbaki gehören. Arch. Math. *16*, 200-207.

Kölzow, D. [1966] Adaptions- und Zerlegungseigenschaften von Massen. Math. Z. *94*, 309-321.

Kölzow, D. [1967] Charakterisierung der Masse, welche zu einem Integral im Sinne von Bourbaki gehören. II. Normalität. Arch. Math. *18*, 45-60.

Kölzow, D. [1968] *Differentiation von Massen*. Lect. Notes Math. *65*, Springer.

König, H. [1969/70] Analysis III. Lecture Notes, Universität des Saarlandes.

König, H. [1972] Sublineare Funktionale. Arch. Math. *23*, 500-508.

König, H. [1985] On the basic extension theorem in measure theory. Math. Z. *190*, 83-94.

König, H. [1987] On the abstract Hahn-Banach theorem due to Rodé. Aequat. Math. *34*, 89-95.

König, H. [1991] The transporter theorem - a new version of the Dynkin class theorem. Arch. Math. *57*, 588-596.

König, H. [1992a] Daniell-Stone integration without the lattice condition and its application to uniform algebras. Ann. Univ. Saraviensis Ser. Math. *4*, 1-91.

König, H. [1992b] On inner/outer regular extensions of contents. Measure Theory (Oberwolfach 1990). Rend. Circ. Mat. Palermo (2) *28* Suppl. 59-85.

König, H. [1992c] New constructions related to inner and outer regularity of set functions. Topology, Measures, and Fractals (Warnemünde 1991) 137-146. Math. Res. *66*, Akademie Verlag.

König, H. [1993] A simple example on non-sequentialness in topological spaces. Amer. Math. Monthly *100*, 674-675.

König, H. [1995] The Daniell-Stone-Riesz representation theorem. Operator Theory in Function Spaces and Banach Lattices (the A.C. Zaanen Anniversary Volume) 191-222. Operator Theory: Adv. Appl. *75*, Birkhäuser.

Leinert, M. [1995] *Integration und Mass*. Vieweg.

Lembcke, J. [1970] Konservative Abbildungen und Fortsetzung regulärer Masse. Z. Wahrsch. Th. verw. Geb. *15*, 57-96.

Lipecki, Z. [1974] Extensions of tight set functions with values in a topological group. Bull. Acad. Pol. Sci. Sér. Sci. Math. Astr. Phys. *22*, 105-113.

Lipecki, Z. [1983] On unique extensions of positive additive set functions. Arch. Math. *41*, 71-79.

Lipecki, Z. [1987] Tight extensions of group-valued quasi-measures. Coll. Math. *51*, 213-219.

Lipecki, Z. [1988] On unique extensions of positive additive set functions. II. Arch. Math. *50*, 175-182.

Lipecki, Z. [1990] Maximal and tight extensions of positive additive set functions. Math. Nachr. *146*, 167-173.

Łoś, J. and E. Marczewski [1949] Extensions of measure. Fund. Math. *36*, 267-276.

Luxemburg, W.A.J. [1991] Integration with respect to finitely additive measures. Positive Operators, Riesz Spaces, and Economics (Pasadena 1990) 109-150. Springer.

Mařík, J. [1957] The Baire and Borel measure. Czech. Math. Journ. *7(82)*, 248-253.

Meyer, P.A. [1966] *Probability and Potential.* Blaisdell.

Meyer-Nieberg, P. [1991] *Banach Lattices.* Springer.

Pettis, B.J. [1951] On the extension of measures. Ann. of Math. *54*, 186-197.

Pollard, D. and F. Topsøe [1975] A unified approach to Riesz type representation theorems. Studia Math. *54*, 173-190.

Rao, K.P.S. Bhaskara and M. Bhaskara Rao [1983] *Theory of Charges.* Academic Press.

Revuz, D. and M. Yor [1991] *Continuous Martingales and Brownian Motion.* Springer.

Ridder, J. [1971] Dualität in den Methoden zur Erweiterung von beschränkt-wie von total additiven Massen II. Indag. Math. *33*, 399-410.

Ridder, J. [1973] Dualität in den Methoden zur Erweiterung von beschränkt-wie von total-additiven Massen III. Indag. Math. *35*, 393-396.

Rodé, G. [1978] Eine abstrakte Version des Satzes von Hahn-Banach. Arch. Math. *31*, 474-481.

Royden, H.L. [1963] *Real Analysis.* Macmillan.

Rudin, W. [1980] *Function Theory in the Unit Ball of* $\mathbb{C}^n$. Grundl. Math. Wiss. *241*, Springer.

Sapounakis, A. and M. Sion [1983] On generation of Radon like measures. Measure Theory and its Applications (Sherbrooke 1982) 283-294. Lect. Notes Math. *1033*, Springer.

Sapounakis, A. and M. Sion [1987] Construction of regular measures. Ann. Univ. Saraviensis Ser. Math. *1*, 157-196.

Sion, M. [1963] On capacitability and measurability. Ann. Inst. Fourier (Grenoble) *13*, 83-98.

Sion, M. [1969] Outer measures with values in a topological group. Proc. London Math. Soc. *19*, 89-106.

Srinivasan, T.P. [1955] On extension of measures. J. Indian Math. Soc. *19*, 31-60.

Schmidt, K.D. [1982a] A general Jordan decomposition. Arch. Math. *38*, 556-564.

Schmidt, K.D. [1982b] On a result of Cobzas on the Hahn decomposition. Arch. Math. *39*, 564-567.

Schwartz, L. [1973] *Radon Measures on Arbitrary Topological Spaces and Cylindrical Measures.* Oxford Univ. Press.

Stone, M.H. [1948-49] Notes on integration I-IV. Proc. Nat. Acad. Sci. USA *34*, 336-342; *34*, 447-455; *34*, 483-490; *35*, 50-58.

Stroock, D.W. [1994] *A Concise Introduction to the Theory of Integration.* Sec. Ed. Birkhäuser.

Tarashchanskii, M.T. [1989] Uniqueness of extension of a regular measure. Ukr. Mat. Zhurn. *41*, 372-377. Engl. Transl. Plenum [1989] 329-333.

Topsøe, F. [1970a] Compactness in spaces of measures. Studia Math. *36*, 195-212.

Topsøe, F. [1970b] *Topology and Measure.* Lect. Notes Math. *133*, Springer.

Topsøe, F. [1976] Further results on integral representations. Studia Math. *55*, 239-245.

Topsøe, F. [1978] On construction of measures. Topology and Measure I (Zinnowitz 1974) Part 2, 343-381. Ernst-Moritz-Arndt Univ. Greifswald.

Topsøe, F. [1983] Radon measures, some basic constructions. Measure Theory and its Applications (Sherbrooke 1982) 303-311. Lect. Notes Math. *1033*, Springer.

Varadarajan, V.S. [1965] Measures on topological spaces. Amer. Math. Soc. Transl. *48*, 161-228.

Weir, A.J. [1974] *General Integration and Measure.* Cambridge Univ. Press.

Wheeler, R.J. [1983] A survey of Baire measures and strict topologies. Expos. Math. *2*, 97-190.

Zaanen, A.C. [1967] *Integration.* North-Holland.

List of Symbols

Index

Subsequent Articles of the Author

(1) Image measures and the so-called image measure catastrophe.
Positivity 1(1997), 255-270.

(2) The product theory for inner premeasures.
Note Mat. 17(1997), 235-249.

(3) Measure and Integration: Mutual generation of outer and inner premeasures.
Annales Univ. Saraviensis Ser. Math. 9(1998) No. 2, 99-122.

(4) Measure and Integration: Integral representations of isotone functionals.
Annales Univ. Saraviensis Ser. Math. 9(1998) No. 2, 123-153.

(5) Measure and Integration: Comparison of old and new procedures.
Arch. Math. 72(1999), 192-205.

(6) What are signed contents and measures?
Math. Nachr. 204(1999), 101-124.

(7) Upper envelopes of inner premeasures.
Annales Inst. Fourier 50(2000), 401-422.

(8) On the inner Daniell-Stone and Riesz representation theorems.
Documenta Math. 5(2000), 301-315.

(9) Sublinear functionals and conical measures.
Arch. Math. 77(2001), 56-64.

(10) Measure and Integration: An attempt at unified systematization.
Workshop on Measure Theory and Real Analysis, Grado 2001.
Rend. Istit. Mat. Univ. Trieste 34(2002), 155-214.

(11) New facts around the Choquet integral.
Sém. d'Analyse Fonctionelle, U Paris VI 2002.

(12) The (sub/super)additivity assertion of Choquet.
Studia Math. 157(2003), 171-197.

(13) Projective limits via inner premeasures and the true Wiener measure.
Mediterr. J. Math. 1(2004), 3-42.

(14) Stochastic processes in terms of inner premeasures.
Note Mat. 25(2005/06) n. 2, 1-30.

(15) New versions of the Radon-Nikodým theorem.
Arch. Math. 86(2006), 251-260.

(16) The Lebesgue decomposition theorem for arbitrary contents.
Positivity 10(2006), 779-793.

(17) The new maximal measures for stochastic processes.
J. Analysis Appl. 26(2007), 111-132.

(18) Stochastic processes on the basis of new measure theory.
Sém. d'Initiation à l'Analyse, U Paris VI 2004.
Conf. Positivity IV - Theory and Applications, TU Dresden 2005,
Proc. 2006 pp. 79-92.

(19) New versions of the Daniell-Stone-Riesz representation theorem.
Positivity 12(2008), 105-118.

(20) Measure and Integral: New foundations after one hundred years.
Functional Analysis and Evolution Equations (The Günter Lumer
Volume). Birkhäuser 2007, pp. 405-422.

(21) Fubini-Tonelli theorems on the basis of inner and outer
premeasures.

(22) Measure and Integration: Characterization of the new maximal
contents and measures.

(23) Measure and Integration: The basic extension theorems.